IMPERIAL UNITS

ac	acre
bbl	barrel
cu ft	cubic foot
cu in.	cubic inch
cu yd	cubic yard
cwt	hundred weight
fbm	foot board measure
ft	foot or feet
gal	gallon(s)
in.	inch(es)
lb	pound
lf	linear foot (feet)
mi	mile(s)
mph	miles per hour
psi	pounds per square inch
sq ft	square foot (feet)
sq in.	square inch(es)
sq yd	square yard(s)
mf bm	thousand foot board measure
m gal	thousand gallons
yd	yard(s)

METRIC UNITS

C	Celsius
cm	centimeter
ha	hectare
kg	kilogram(s)
km	kilometer(s)
kN	kilonewton(s)
kPa	kilopascal(s)
L	liter(s)
m	meter(s)
m^2	square meter
m^3	cubic meter
mm	millimeter(s)
t	tonne

Surveying
Principles and Applications

NINTH EDITION

Barry F. Kavanagh
Seneca College, Emeritus

Tom B. Mastin
California Polytechnic State University, San Luis Obispo

Boston   Columbus   Indianapolis   New York   San Francisco   Upper Saddle River
Amsterdam   Cape Town   Dubai   London   Madrid   Milan   Munich   Paris   Montréal   Toronto
Delhi   Mexico City   São Paulo   Sydney   Hong Kong   Seoul   Singapore   Taipei   Tokyo

Editorial Director: Vernon R. Anthony
Editorial Assistant: Nancy Kesterson
Director of Marketing: David Gesell
Senior Marketing Manager: Harper Coles
Senior Marketing Coordinator: Alicia Wozniak
Marketing Assistant: Les Roberts
Senior Managing Editor: JoEllen Gohr
Associate Managing Editor: Alexandrina
 Benedicto Wolf
Production Project Manager: Maren L. Miller
Production Manager: Susan Hannahs
Art Director: Jayne Conte

Image Permission Coordinator: Mike Lackey
Photo Researcher: Kerri Wilson, PreMedia
 Global USA, Inc.
Text Researcher: Jen Roach, PreMedia
 Global USA, Inc.
Cover Designer: Bruce Kenselaar
Full-Service Project Management: Abinaya
 Rajendran, Integra Software Services, Inc.
Composition: Integra Software Services, Ltd.
Cover and Text Printer/Bindery: Courier/Westford
Text Font: 10/12, Minion Pro

Credits and acknowledgments borrowed from other sources and reproduced, with permission, in this textbook appear on the appropriate page within the text.

Many of the designations by manufacturers and sellers to distinguish their products are claimed as trademarks. Where those designations appear in this book, and the publisher was aware of a trademark claim, the designations have been printed in initial caps or all caps.

Library of Congress Cataloging-in-Publication Data
Kavanagh, Barry F.
  Surveying : principles and applications.—Nineth edition / Barry F. Kavanagh, Seneca College,
Emeritus, Tom B. Mastin, California Polytechnic State University, San Luis Obispo.
    pages cm
  ISBN 978-0-13-700940-4 (alk. paper)—ISBN 0-13-700940-2 (alk. paper)
  1. Surveying. I. Mastin, Tom B. II. Title.
TA545.K37 2014
526.9—dc23

                                                          2012027019

10 9 8 7 6 5 4 3 2 1

ISBN 10:   0-13-700940-2
ISBN 13: 978-0-13-700940-4

WHAT'S NEW IN THIS EDITION

Responding to the advice of reviewers and faculty who have adopted this text for use in their programs, we have shortened and streamlined the text from seventeen to fourteen chapters.

- Optical theodolites, total stations, and total station applications have now been brought together in Chapter 5.
- Chapter 7, "Satellite Positioning Systems"; Chapter 9, "Geographic Information Systems"; and Chapter 11, "Remote Sensing," have been substantially revised.
- The treatment of hydrographic surveying has been shortened and included at the end of Chapter 8, "Topographic Surveying."
- Chapter 13, "Engineering Surveying," has been revised to combine all the engineering work into one chapter.
- Chapter 14, "Land Surveys," has been rewritten to reflect coauthor Tom Mastin's knowledge and experience.
- In addition, all chapters were carefully reviewed and updated to ensure that the latest in technological advances were included. New end-of-chapter questions have been added and end-of-chapter problems have been expanded and refreshed. The websites given in selected chapters and in Appendix E have been updated and expanded.

The text is divided into four parts:

- Part I, Surveying Principles, includes chapters on the basics of surveying, leveling, distance measurement (taping and electronic distance measurement), angles and directions, theodolites, total stations, traverse surveys, satellite positioning, topographic surveying and mapping, geographic information systems, and control surveys.
- Part II, Remote Sensing, includes chapters on satellite imagery and airborne imagery.

- Part III, Surveying Applications, includes chapters on engineering surveys and land surveys.
- Part IV, Appendices, includes the following information: random errors, trigonometric definitions and identities, glossary, answers to selected chapter problems, Internet websites, a color photo gallery (located at the end of the book), typical field projects, and early surveying.

Finally, this edition introduces coauthor Tom B. Mastin of the California Polytechnical State University. Tom's background includes many years of experience in academics, land surveying, and engineering surveying, and he is a great addition to the team producing this text.

INSTRUCTOR SUPPLEMENTS

The following online supplements are available for instructors:

- Online PowerPoints
- Online Instructor's Manual

To access supplementary materials online, instructors need to request an instructor access code. Go to http://www.pearsonhighered.com/irc to register for an instructor access code. Within 48 hours of registering, you will receive a confirming e-mail including an instructor access code. Once you have received your code, locate your text in the online catalog and click on the Instructor Resources button on the left side of the catalog product page. Select a supplement, and a login page will appear. Once you have logged in, you can access instructor material for all Prentice Hall textbooks. If you have any difficulties accessing the site or downloading a supplement, please contact Customer Service at http://247pearsoned.custhelp.com/.

ACKNOWLEDGMENTS

We are grateful for the comments and suggestions received from those who adopted previous editions of this text.

In addition, particular thanks are due to Kamal Ahmed, University of Washington; R. H. Birkett, PS, Macomb Community College, South Campus; Arvin Farid, Ph.D., P.E., Boise State University; Selvaraj S. Immanuel, University of Evansville; Mohamad Mustafa, Savannah State University; P. Warren Plugge, Ph.D., Central Washington University; Paul Pope, Ph.D., Los Alamos National Laboratory; and Brian Smith, University of Virginia, for their assistance with the ninth edition text review.

The following surveying, engineering, and equipment manufacturers have provided generous assistance:

- American Society for Photogrammetry and Remote Sensing
- Applanix, Richmond Hill, Ontario
- Bird and Hale, Ltd., Toronto, Ontario
- Canadian Institute of Geomatics, Ottawa, Ontario
- Carl Zeiss, Inc., Thornwood, New York
- CST/Berger, Watseka, Illinois
- Environmental Systems Research Institute, Inc. (ESRI), Redlands, California
- Geomagnetic Laboratory, Geological Survey of Canada, Ottawa—Larry Newitt
- International Systemap Corp., Vancouver, British Columbia
- Laser Atlanta, Norcross, Georgia

- Leica Geosystems, Inc., Norcross, Georgia
- MicroSurvey International, Kelowna, British Columbia
- National Geodetic Survey (NGS), Silver Spring, Maryland
- National Society of Professional Surveyors
- OPTECH, Toronto, Ontario
- Pacific Crest Corporation, Santa Clara, California
- Position, Inc., Calgary, Alberta
- Sokkia Corporation, Olathe, Kansas
- Topcon Positioning Systems, Pleasanton, California
- Trimble, Sunnyvale, California
- Tripod Data Systems, Corvallis, Oregon
- U.S. Geological Survey, Denver, Colorado— John M. Quinn
- U.S. Geological Survey, Sioux Falls, South Dakota—Ron Beck
- Wahl, L. Jerry, Sun/Polaris Ephemeris Tables, http://www.cadastral.com

Comments and suggestions about this text are welcome. Please contact us at:

Barry F. Kavanagh
barry.kavanagh@cogeco.ca

Tom B. Mastin
tmastin@calpoly.edu

Contents

Field Note Index

SURVEYING PRINCIPLES

CHAPTER ONE

BASICS OF SURVEYING

1.1 OVERVIEW

The concept of surveying has been around ever since some members of the human race stopped hunting and gathering, and began to stay for extended periods in one geographic area, where they could support themselves with various agricultural endeavors. The early practices of land ownership required some mechanism to mark (and re-mark when necessary) the boundaries of individual landowners and thus reduce conflicts over competing land claims. Many early settlements occurred near bodies of water where shorelines shifted over time due to flooding and other natural occurrences, thus requiring continual surveys to re-mark boundaries; see Appendix H for early Egyptian surveying techniques.

What do surveyors do? Surveyors take and analyze measurements.

What do surveyors measure? Surveyors measure distances, angles, and positions.

What distances do surveyors measure? Surveyors measure horizontal distances, slope distances, and vertical distances.

What angles do surveyors measure? Surveyors measure angles in the horizontal and vertical planes.

What positions do surveyors measure? Surveyors measure the two-dimensional positions of points on or near the surface of the earth referenced to a defined Cartesian grid or to a geographic grid (latitude and longitude), and they measure elevation dimensions referenced to mean sea level (MSL); as

well, they measure three-dimensional positions of points on or near the earth's surface referenced to a defined ellipsoidal model of the earth called the Geodetic Reference System (GRS80).

Who can perform surveys? The two largest fields of surveying are land surveying (boundary or property surveying) and engineering surveying. North American surveyors engaged in establishing, or reestablishing, legal boundaries must be licensed by the state or province in which they are working (see Chapter 14). They must pass state/provincial exams and have years of field training before being licensed. On the other hand, engineering surveys are performed by surveyors who have civil engineering (or civil engineering technology) education, together with suitable field experience, and are thus prepared to perform a wide variety of preengineering and construction layout surveys. Some university programs and state/provincial professional organizations have established programs (including *Geomatics*) and objectives designed to produce "professional" surveyors capable of working in both fields. Another, more recent, indication of the changing scene is that some "layout surveys" on large projects, once performed by both engineering and land surveyors, are now being accomplished by construction equipment operators who are guided by in-cab interactive monitors or other *line and grade* signaling devices. See Chapter 13.

1.2 SURVEYING DEFINED

Surveying is the art and science of measuring distances, angles, and positions, on or near the surface of the earth. It is an art in that only a surveyor who possesses a thorough understanding of surveying techniques will be able to determine the most efficient methods needed to obtain optimal results over a wide variety of surveying problems. Surveying is scientific to the degree that rigorous mathematical techniques are used to analyze and adjust the field survey data. The accuracy, and

thus reliability, of the survey depends not only on the field expertise of the surveyor, but also on the surveyor's understanding of the scientific principles underlying and affecting all forms of survey measurement.

Figure 1-1 is an aerial photo of undeveloped property. Figure 1-2 is an aerial photo of the same property after development. All the straight and curved lines that have been added to the post-development photo, showing modifications and/or additions to roads, buildings, highways,

FIGURE 1-1 Aerial photograph of undeveloped property

FIGURE 1-2 Aerial photograph of same property after development

residential areas, commercial areas, property boundaries, and so on, are all the direct or indirect result of surveying.

1.3 TYPES OF SURVEYS

Plane surveying is that type of surveying in which the surface of the earth is considered to be a plane for all X and Y dimensions. All Z dimensions (height) are referenced to the mean surface of the earth (MSL) or to the surface of the earth's

reference ellipsoid (GRS80). Most engineering and property surveys are classed as plane surveys, although some route surveys that cover long distances (e.g., highways and railroads) will have corrections applied at regular intervals (e.g., 1 mile) to correct for the earth's curvature.

Geodetic surveying is that type of surveying in which the surface of the earth is considered to be an ellipsoid of revolution for X and Y dimensions. As in plane surveying, the Z dimensions (height) can be referenced to the surface of the earth's reference ellipsoid (GRS80) or can be converted to refer to the mean surface of the earth (MSL). Traditional geodetic surveys were very precise surveys of great magnitude (e.g., national boundaries and control networks). Modern surveys such as data gathering, control, and layout which utilize satellite positioning [e.g., the global positioning system (GPS)] are also based on the earth's reference ellipsoid (GRS80) and, as such, could be classed as being geodetic surveys. Such geodetic measurements must be mathematically converted to local coordinate grids and to MSL elevations to be of use in leveling and other local surveying projects.

1.4 CLASSES OF SURVEYS

Control surveys are used to reference both preliminary and layout surveys. Horizontal control can be arbitrarily placed, but it is usually tied directly to property lines, roadway centerlines, or coordinated control stations. Vertical control is often a series of benchmarks, permanent points whose elevations above a datum (e.g., MSL) have been carefully determined. It is accepted practice to take more care in control surveys with respect to precision and accuracy; great care is also taken to ensure that the control used for a preliminary survey can be readily reestablished at a later date, whether it be needed for further preliminary work or for a related layout survey.

Preliminary surveys (data gathering) gather geospatial data (distances, positions, and angles) to locate physical features (e.g., water boundaries, trees, roads, structures, or property markers) so that the data can be plotted to scale on a map or plan. Preliminary surveys also include the determination of differences in elevation (vertical distances) so that elevations and contours may also be plotted. [Also see digital terrain models (DTMs) in Chapter 8.]

Layout surveys involve marking on the ground (using wood stakes, iron bars, aluminum and concrete monuments, nails, spikes, etc.) the features shown on a design plan. The layout can be for boundary lines, as in land division surveying,

or it can be for a wide variety of engineering works (e.g., roads, pipelines, structures, bridges); the latter group is known as construction surveying. In addition to marking the proposed horizontal location (*X* and *Y* coordinates) of the designed feature, data will also be given for the proposed (design) elevations which are referenced to MSL.

1.5 DEFINITIONS

1. *Topographic surveys:* preliminary surveys used to locate and map the natural and man-made surface features of an area. The features are located relative to one another by tying them all into the same control lines or control grid. See Chapter 8.

2. *Hydrographic surveys:* preliminary surveys that are used to tie in underwater features to surface control points. Usually shorelines, marine features, and water depths are shown on the hydrographic map or electronic chart. See Chapter 8.

3. *Route surveys:* preliminary, layout, and control surveys that range over a narrow but long strip of land. Typical projects that require route surveys are highways, railroads, electricity transmission lines, and channels. See Chapter 13.

4. *Property surveys:* preliminary, layout, and control surveys that are involved in determining boundary locations or in laying out new property boundaries (also known as *cadastral* or *land surveys*). See Chapter 14.

5. *Final ("as-built") surveys:* similar to preliminary surveys. Final surveys tie in features that have just been constructed to provide a final record of the construction and to check that the construction has proceeded according to the design plans.

6. *Aerial surveys:* preliminary and final surveys using both traditional aerial photography and aerial imagery. Aerial imagery includes the use of digital cameras, multispectral scanners, LiDAR, and radar. See Chapter 12.

7. *Construction surveys:* layout surveys for engineering works. See Chapter 13.

1.6 SURVEYING INSTRUMENTATION

The instruments most commonly used in field surveying are (1) *satellite positioning receiver*; (2) *total station*; (3) *level* and *rod*; (4) *theodolite*; and (5) *steel tape*.

1. *Global Navigation Satellite System* (GNSS) is a term used world-wide to describe the various satellite positioning systems now in use, or in various stages of implementation and planning. *Global positioning system* (GPS) is the term used to describe the U.S. NAVSTAR positioning system, which was the original fully-operational GNSS. *GLONASS* describes the Russian satellite positioning system, which is also now fully operational; *Galileo* describes the European Union satellite positioning system, which is soon to be implemented; and *Beidou*, or *Compass*, describes the Republic of China's regional satellite system now being rapidly expanded to a global positioning system.

A satellite positioning receiver (see Figure 1-3) captures signals transmitted by four or more positioning satellites in order to determine position coordinates (e.g., northing, easting, and elevation) of a survey station. Satellite positioning is discussed in Chapter 7. Some satellite positioning receivers are already programmed to capture signals from three systems: GPS, GLONASS, and the soon-to-be-implemented Galileo.

2. In the 1980s, the *total station* first appeared. This instrument combines electronic distance measurement (EDM), which was developed in the 1950s, with an electronic theodolite. In addition to electronic distance and angle measuring capabilities, this instrument is equipped with a central processor, which enables the computation of horizontal and vertical distances. The central processor also monitors instrument status and executes software programs that enable the surveyor to perform a wide variety of surveying applications. Total stations measure horizontal and vertical angles as well as horizontal and vertical distances. All data can be captured into attached (cable or wireless) *electronic field books* or into onboard storage as the

FIGURE 1-3 Zepher™ geodetic GPS antenna, with a 5700 GPS receiver and radio communications equipment. Zepher antennas are said to have accuracy potentials similar to those of choke ring antennas—at lower costs
(Courtesy of Trimble)

data are received. See Figure 1-4. Total stations are described in detail in Chapter 5.

3. Theodolites (sometimes called transits, short for transiting theodolites) are instruments designed for use in measuring horizontal and vertical angles and for establishing linear and curved alignments in the field. The theodolite has evolved through three distinct phases.

1. An open-faced, vernier-equipped (for angle determination) theodolite is commonly called *a transit*; a plumb bob is used to center the transit over the station mark. See Figure H-8.

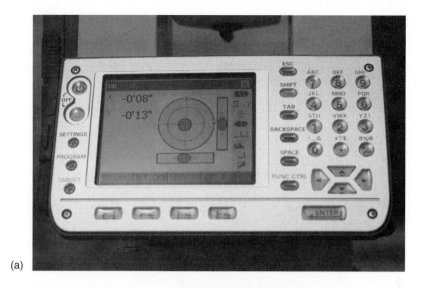

(a)

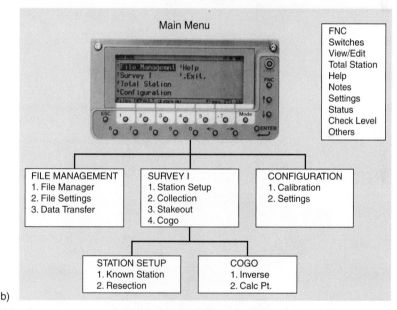

(b)

FIGURE 1-4 (a) Total station screen showing electronic level bubbles; these highly sensitive bubbles along with laser plummets allow the total stations to be precisely set over a point. (b) Menu schematic for the Nikon DTM 750

(Courtesy of Cansell Survey Equipment Co., Toronto)

2. In the 1950s, the vernier transit gave way to the *optical theodolite*. This instrument came equipped with optical scales, permitting direct digital readouts or micrometer-assisted readouts;

an optical plummet was used to center the instrument over the station mark. See Figure H-13.

3. In the 1970s, *electronic theodolites* first appeared. These instruments used photoelectric sensors

capable of sensing vertical and horizontal angles and then displaying these angles in degrees, minutes, and seconds. Optical plummets and later laser plummets are used to center the instrument over the station mark. See Figure 1-5. Electronic theodolites are discussed in detail in Chapter 5. Optical and vernier transits are discussed in detail in Section H.3.

FIGURE 1-5 Early model of an electronic theodolite with interfaced Distomat EDM (mounted on the telescope).

(Courtesy of imagebroker.net/SuperStock)

Level Rod (Foot)

FIGURE 1-6 Level and rod

(Courtesy of Sokkia Co. Ltd.)

4. The level and rod are used to determine elevations in a wide variety of surveying, mapping, and engineering applications. See Figure 1-6. Leveling is discussed in Chapter 2.

5. Steel tapes are relatively precise measuring instruments, and are used mostly for short measurements in both preliminary and layout surveys. Steel tapes, and their use, are discussed, in detail, in Chapter 3.

Other instruments are used in remote-sensing techniques to acquire geospatial images; panchromatic, multispectral scanning, radar, and LiDAR imaging can be based on both airborne and satellite platforms. See Chapters 12 and 13.

1.7 OVERVIEW OF A MODERN SURVEYING DATA SYSTEM— THE SCIENCE OF GEOMATICS

Advances in computer science have had a tremendous impact on all aspects of modern technology. The effects on the collection and processing of data in both field surveying and remotely sensed imagery have been significant. Survey data once laboriously collected with tapes, transits, and levels (recorded manually in field books) can now be quickly and efficiently collected using total stations and precise satellite positioning receivers (see Chapter 7). These latter techniques can provide the high-accuracy results usually required in control surveys (see Chapter 10), engineering surveys (see Chapter 13), and land surveys (see Chapter 14). When high accuracy is not a prime requirement, as in some geographic information systems (GISs) surveys and many mapping surveys, data can be efficiently collected from less precise (1 m) satellite positioning receivers and, as well, from satellite and airborne imaging platforms (see Chapters 11 and 12). The broad picture encompassing all aspects of data collection, processing, analysis, and presentation is now referred to as the field of *geomatics*.

Geomatics is a term used to describe the science and technology dealing with geospatial data, including *collection, analysis, sorting, management, planning and design, storage,* and *presentation.* It has applications in all disciplines and professions that use geospatial data, for example, planning, geography, geodesy, infrastructure engineering, agriculture, natural resources, environment, land division and registration, project engineering, and mapping.

This computerized technology has changed the way field data are collected and processed. To appreciate the full impact of this new technology, one has to view the overall operation, that is, from field to computer, computer processing, and data portrayal in the form of maps and plans. Figure 1-7 gives a schematic overview of an integrated survey data system, a geomatics data model. This model shows that all the branches and specializations are tied together by their common interest in earth measurement data and in their common dependence on computer science and information technology (IT).

1.7.1 Data-Gathering Components

The upper portion of the Figure 1-7 schematic shows the various ways that data can be collected and transferred to the computer. In addition to total station techniques, field surveys can be performed using conventional surveying instruments (theodolites, EDMs, and levels), with the field data entered into a data collector instead of conventional field books. This manual entry of field data lacks the speed associated with interfaced equipment, but after the field data (including attribute data) have been entered, all the advantages of electronic techniques are available to the surveyor. The raw field data, collected and stored in the total station, as well as the attribute data entered by the surveyor, are transferred to the computer; the raw data download program is supplied by the instrument manufacturer, but acquiring and using the program [(cordinate geometry (COGO), GIS, etc.] required to translate the raw data into properly formatted field data is the responsibility of the surveyor. As well, topographic point positioning is now often being captured using GPS techniques; newer controllers

(data collectors) can be used with both total stations and GPS receivers. When the collected terrain data have been downloaded into a computer, COGO programs and/or image analysis programs can be used to determine the positions (northing, easting, and elevation) of all points. Also at this stage, additional data points (e.g., inaccessible ground points) can be computed and added to the data file.

Existing maps and plans have a wealth of lower-precision data that may be relevant for an area survey database. If such maps and plans are available, the data can be digitized on a digitizing table or by digital scanners and added to the northing, easting, and elevation (N, E, and Z) coordinates files (keeping a record of the precision levels of such scanned data). In addition to distances and elevations, the digitizer can provide codes, identifications, and other attribute data for each digitized point. One of the more important features of the digitizer is its ability to digitize maps and plans drawn at various scales and store the distances and elevations in the computer at their ground (or grid) values.

The stereo analysis of aerial photos (see Chapter 12) is a very effective method of collecting topographic ground data, particularly in high-density areas, where the costs for conventional field surveys would be high. Many municipalities routinely fly all major roads and develop plans and profiles that can be used for design and construction. With the advent of computerized surveying systems, the stereo analyzers can coordinate all horizontal and vertical features and transfer these Y (north), X (east), and Z (height) coordinates to database files.

Satellite imagery is received from the U.S. (e.g., EOS and Landsat), French, European, Japanese, Canadian, Chinese, and South American satellites, and can be processed by a digital image analysis system that classifies terrain into categories of soil and rock types and vegetative cover; these and other data can be digitized and added to the database files. See Chapters 11 and 12.

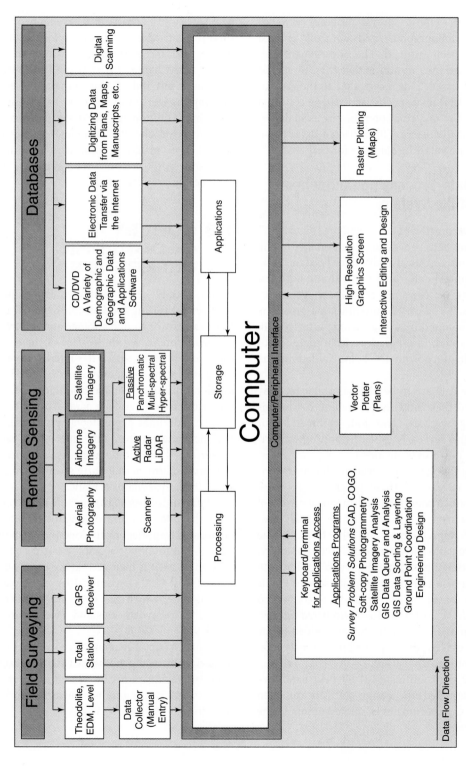

FIGURE 1-7 Geomatics data model, showing the collection, processing, analysis, design, and plotting of geodata

In this section, several different ways of collecting topographic (and control) ground data have been outlined. The one element that they all have in common is that they all can be computer-based. This means that all the ground data (geodata) for a specific area can be collected and stored in one computer, but be available to many potential users. The collected data for an area are known as the database for that area.

1.7.2 Data-Processing Components of the System

The central portion of the schematic (Figure 1-7) depicts the data-processing components of the system. Initially, as already described, the total station control data can be closed and adjusted by means of various coordinate geometry programs. Additionally, missing data positions can be computed by using various intersection, resection, and interpolation techniques, with the resultant coordinates being added to the database.

If the data are to be plotted, a plot file may be created that contains point plot commands (including symbols) and join commands for straight and curved lines; labels and other attribute data may also be included.

Design programs are available for most construction endeavors. These programs can work with the stored coordinates to provide a variety of possible designs, which can then be quickly analyzed with respect to costs and other factors. Some design programs incorporate interactive graphics, which permit a plot of the survey to be shown to scale on a high-resolution graphics screen. Points and lines can be moved, created, edited, and so forth, with the final positions coordinated right on the screen and the new coordinates added to the coordinates files. See Chapter 8 for information on digital plotting.

Once the coordinates of all field data have been determined, the design software can then compute the coordinates of all key points in the new facility (e.g., roads, site developments). The surveyor can take this design information back to the field and perform a construction layout survey (see Chapter 13) showing the contractor how much cut and fill (for example) is required to bring the land to the proposed elevations and locations. Additionally, the design data can now be transferred directly (sometimes wirelessly) to construction machine controllers for machine guidance and control functions in heavy engineering projects.

1.8 SURVEY GEOGRAPHIC REFERENCE

It has already been mentioned that surveying involves measuring the location of physical land features relative to one another and relative to a defined reference on the surface of the earth. In the broadest sense, the earth's reference system is composed of the surface divisions denoted by geographic lines of latitude and longitude. Latitude lines run east/west and are parallel to the equator. The latitude lines are formed by projecting the latitude angle out from the center of the earth to its surface. The latitude angle itself is measured (90° maximum) at the earth's center, north or south from the equatorial plane.

Longitude lines all run north/south, converging at the poles. The lines of longitude (meridians) are formed by projecting the longitude angle at the equator out to the surface of the earth. The longitude angle itself is measured at the earth's center, east or west (180° maximum) from the plane of 0° longitude, which has been arbitrarily placed through Greenwich, England (see Figures 1-8 and 1-9).

This system of geographic coordinates is used in navigation and geodesy, but those engaged in plane surveying normally use either coordinate grid systems or the original township fabric as a basis for referencing.

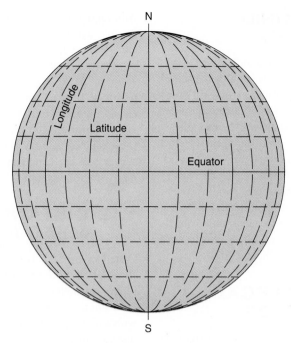

FIGURE 1-8 Sketch of earth showing lines of latitude and longitude

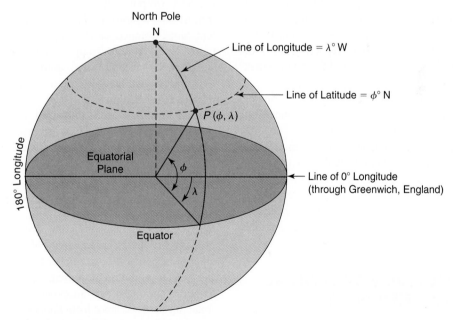

FIGURE 1-9 Sketch showing location of point P, referenced by geographical coordinates

1.9 SURVEY GRID REFERENCE

All states and provinces have adopted a grid system best suited to their needs. The grid itself is limited in size so that no serious errors will accumulate when the curvature of the earth is ignored. Advantages of the grid systems are the ease of calculation (plane geometry and trigonometry) and the availability of one common datum for *X* and *Y* dimensions in a large (thousands of square miles) area. The coordinates in most grid systems can be referenced to a zone's central meridian and to the equator so that translation to geographic coordinates is always readily accomplished. This topic is discussed in more detail in Chapter 10.

1.10 SURVEY LEGAL REFERENCE

Public lands in North America were originally laid out for agricultural use by the settlers. In the United States and parts of Canada, the townships were laid out in 6-mile squares; however, in the first established eastern areas, including the original colonies of the United States, a wide variety of township patterns exist—reflecting both the French and English heritage and the relative inexperience of planners.

The townships themselves were subdivided into sections and ranges (lots and concessions in Canada), each uniquely numbered. The basic township sections or lots were either 1 mile square or some fraction thereof. Eventually, the townships were (and still are) further subdivided in real-estate developments. All developments are referenced to the original township fabric, which has been reasonably well preserved through ongoing resurveys. This topic is discussed in detail in Chapter 14.

1.11 SURVEY VERTICAL REFERENCE

The previous sections described how the *X* and *Y* dimensions (horizontal) of any feature could be referenced for plane surveying purposes.

Although vertical dimensions can be referenced to any datum (including the earth's reference ellipsoid; see Section 7.12), the reference datum most used is that of MSL. MSL is assigned an elevation of 0.000 feet (ft) (or meters), and all other points on the earth are usually described as being elevations above or below zero. Permanent points whose elevations have been precisely determined (*benchmarks*) are available in most areas for survey use. See Chapters 2, 5, and 10 for further discussion of this topic.

1.12 DISTANCE MEASUREMENT

Distances between two points can be *horizontal*, *slope*, or *vertical* and are recorded in feet (foot units) or meters (SI units) (see Figure 1-10). *Horizontal* and *slope distances* can be measured with a fiberglass or steel tape or with an electronic distance measuring device. When surveying, the horizontal distance is always required (for plan plotting and design purposes); if a slope distance between two points has been taken and recorded, it must then be converted to its horizontal equivalent. Slope distances can be trigonometrically converted to horizontal distances by using either the slope angle (accomplished automatically when total stations are used) or the difference in elevation (vertical distance) between the two points. *Vertical distances* can be measured with a tape, as in construction work, with a surveyor's level and leveling rod (see Figure 1-6) or with a total station.

1.13 UNITS OF MEASUREMENT

Historically, the many different measuring units used on earth have caused no end of confusion. An attempt to standardize weights and measures led to the creation of the metric system in the 1790s. The length of the meter was supposed to be one ten-millionth of the distance from the North Pole to the equator. In 1866, the U.S. Congress made the use of metric weights and measures legal. The meter was then equal to 39.37 in so 1 foot (U.S. survey foot)

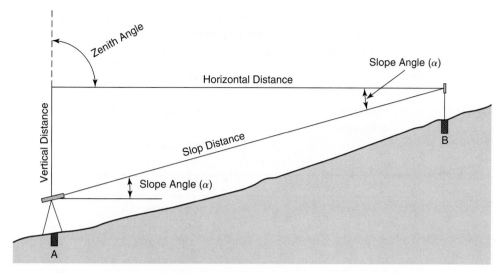

FIGURE 1-10 Distance measurement

equaled 0.3048006 m. In 1959, the United States officially adopted the International foot, whereby 1 foot equals 0.3048 m exactly. One U.S. survey foot equals 1.000002 International feet. In 1960, the metric system was modernized and called the Système International d'Unités (SI). In the United States, the complete changeover to the metric system will take many years, perhaps several generations. The impact is that, from now on, most surveyors will have to be proficient in both the foot and metric systems. Additional equipment costs in this dual system are limited mostly to measuring tapes and leveling rods.

SI units were a modernization (1960) of the long-used metric units. This modernization included a redefinition of the meter and the addition of some new units. One example is the newton (see Table 2-1). With the United States committed, to some degree, to switching to metric units, all industrialized nations are now using the metric system.

Table 1-1 describes and contrasts metric and foot units. Degrees, minutes, and seconds are used almost exclusively in both metric and foot systems for angular measurement. In some European countries, however, the circle has also been graduated into 400 gon (also called *grad*). Angles, in that system, are expressed to four decimals (e.g., a right angle = 100.0000 gon).

1.14 LOCATION METHODS

A great deal of surveying effort is spent in measuring points of interest relative to some reference line so that these points may be shown later on a scaled plan. The illustrations in Figure 1-11 show some common location techniques. Point *P* in Figure 1-11(a) is located relative to known line *AB* by determining *CB* or *CA*, the right angle at *C*, and distance *CP*. This is known as the *right-angle offset tie* (also known as the *rectangular tie*). Point *P* in Figure 1-11(b) is located relative to known line *AB* by determining the angle (θ) at *A* and the distance *AP*. This is known as the *angle-distance tie* (also known as the *polar tie*). Point *P* in Figure 1-11(c) can also be located relative to known line *AB* by determining *either* the angles at *A* and *B* to *P* or by determining the distances *AP* and *BP*. Both methods are called *intersection* techniques.

Alternately, a point can be tied in using positioning techniques. For example, point *P* could be located by simply holding a pole-mounted GPS receiver/antenna directly on the point and then waiting until a sufficient number of measurements indicate that the point has been located (coordinates determined) to the required level

Table 1-1   Measurement Definitions and Equivalencies

Linear Measurements	Foot Units
1 mile = 5,280 feet	1 foot = 12 inches
= 1,760 yards	1 yard = 3 feet
= 320 rods	1 rod = 16½ feet
= 80 chains	1 chain = 66 feet
	1 chain = 100 links
1 acre = 43.560 ft^2 = 10 square chains	

Linear Measurement		Metric (SI) Units
1 kilometer	=	1,000 meter
1 meter	=	100 centimeter
1 centimeter	=	10 millimeter
1 decameter	=	10 centimeter
1 hectare (ha)	=	10,000 m^2
1 square kilometer	=	1,000,000 m^2
		100 hectares

Foot-to-Metric Conversion*

1 ft = 0.3048 m (exactly)	1 inch = 25.4 mm (exactly)*
1 km = 0.62137 miles (approx.)	
1 hectare (ha) = 2.471 acres (approx.)	
1 km^2 = 247.1 acres (approx.)	

Angular Measurement

1 revolution = 360°

1° (degree) = 60′

1′ (minute) = 60″ (seconds)

*Prior to 1959, the United States used the relationship 1 m = 39.37 in. This resulted in a U.S. survey foot of approximately 0.3048006 m, which was used in earlier public land surveys.

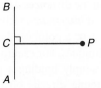

(a) Right-angle offset tie

(b) Angle-distance tie

(c) Intersection tie

FIGURE 1-11 Location ties

of precision. In addition, points can be located using aerial imagery, with the assistance of GPS and/or with inertial measuring units (IMUs)—see Chapter 12.

1.15 ACCURACY AND PRECISION

Accuracy is the relationship between the value of a measurement and the "true" value (see Section 1.17 for "true" values) of the dimension being measured. *Precision* describes the refinement of the measuring process and the ability to repeat the same measurement with consistently small variations in the measurements (i.e., no large discrepancies). Figure 1-12 depicts targets with hit marks for both a rifle and a shotgun, which illustrates the concepts of precision and accuracy.

The concepts of accuracy and precision are also illustrated in the following example: A building wall that is known to be 157.22 ft long is measured by two methods. In the first case the wall is measured very carefully using a fiberglass tape graduated to the closest 0.1 ft. The result of this operation is a measurement of 157.3 ft. In the second case the wall is measured with the same care, but with a more precise steel tape graduated to the closest 0.01 ft. The result of this operation is a measurement of 157.23 ft. In this example, the more precise method (steel tape) resulted in the more accurate measurement.

	"True" distance (ft)	Measured distance (ft)	Error (ft)
Fiberglass tape	157.22	157.3	0.08
Steel tape	157.22	157.23	0.01

It is conceivable, however, that more precise methods can result in less accurate answers. In the preceding example, if the steel tape had previously been broken and then incorrectly repaired (say that an even foot had been dropped), the results would still be relatively precise but very inaccurate.

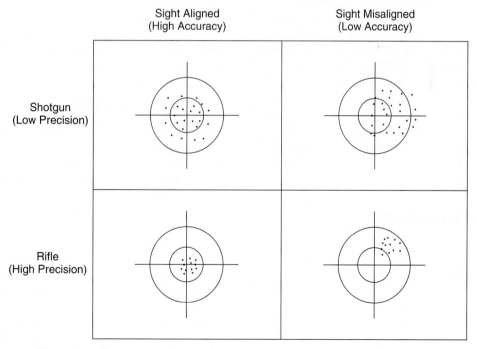

FIGURE 1-12 Illustration showing precision and accuracy

1.16 ACCURACY RATIO

The *accuracy ratio* of a measurement or series of measurements is the ratio of error of closure to the distance measured. The *error of closure* is the difference between the measured location and the theoretically correct location. The theoretically correct location can be determined from repeated measurements or mathematical analysis. Since relevant systematic errors and mistakes can and should be eliminated from all survey measurements, the error of closure will be composed of random errors.

To illustrate, a distance was measured and found to be 250.56 ft. The distance was previously known to be 250.50 ft. The error is 0.06 ft in a distance of 250.50 ft.

$$\text{Accuracy ratio} = 0.06/250.50$$
$$= 1/4{,}175 \approx 1/4{,}200$$

The accuracy ratio is expressed as a fraction whose numerator is unity and whose denominator is rounded to the closest 100 units.

Survey specifications are discussed in Chapter 10. Many land and engineering surveys have in the past been performed at 1/5,000 or 1/3,000 levels of accuracy. With the trend to polar layouts from coordinated control, accuracy ratios on the order of 1/10,000 and 1/20,000 were often specified. It should be emphasized that *for each of these specified orders of accuracy, the techniques and instrumentation used must also be specified.* Sometimes the accuracy or error ratio is expressed in parts per million (ppm). One ppm is simply the ratio of 1/1,000,000 and 50 ppm is 50/1,000,000 or 1/20,000.

1.17 ERRORS

It can be said that no measurement (except for counting) can be free of error. For every measuring technique used, a more precise and potentially more accurate method can be found. For purposes of calculating errors, the "true" value is determined statistically after repeated measurements. In the simplest case, the true value for a distance is taken as the mean value for a series of repeated measurements. This topic is discussed further in Appendix A.

Systematic errors are defined as being those errors whose magnitude and algebraic sign can be determined. The fact that these errors can be determined allows the surveyor to eliminate them from the measurements and thus improve the accuracy. An error due to the effects of temperature on a steel tape is an example of a systematic error. If the temperature is known, the shortening or lengthening effects on a steel tape can be precisely determined.

Random errors are associated with the skill and vigilance of the surveyor. Random (also known as *accidental*) errors are introduced into each measurement mainly because no human being can perform perfectly. Some random errors, by their very nature, tend to cancel themselves; when surveyors are skilled and careful in measuring, random errors will be of little significance except for high-precision surveys. However, random errors resulting from unskilled or careless work do cause problems. As noted earlier, some random errors, even large random errors, tend to cancel themselves mathematically—this does not result in accurate work, only in work that appears to be accurate. Even if the random errors canceled exactly, the final averaged measurement will be imprecise.

1.18 MISTAKES

Mistakes are blunders made by survey personnel. Examples of mistakes include transposing figures (recording a tape value of 68 as 86), miscounting the number of full tape lengths in a long measurement, measuring to or from the wrong point, and the like. Students should be aware that mistakes *will* occur. Mistakes must be discovered and eliminated, preferably by the people who make them. All survey measurements are suspect until they have been verified. Verification may be as simple as repeating the measurement, or verification can result from geometric or trigonometric analysis of related measurements. As a rule, *every* measurement is immediately checked or repeated. This immediate repetition enables the surveyor to eliminate most mistakes and, at the same time, improve the precision of the measurement.

1.19 STATIONING

When performing route surveys, measurements can be taken along a baseline and at right angles to that baseline. Distances along a survey baseline are referred to as stations or chainages, and distances at right angles to that baseline (offset distances) are simple dimensions. The beginning of the survey baseline is the zero end, and is denoted by $0 + 00$; a point 100 ft (m) from the zero end is denoted as $1 + 00$; a point 131.26 ft (m) from the zero end is $1 + 31.26$; and so on. If the stationing is extended back of the $0 + 00$ mark (rarely), the stations would be $0 - 50$, $-1 + 00$, and so on.

In the preceding discussion, the *full stations* are at 100 ft (m) multiples and the *half-stations* would be at even 50-ft intervals. In the metric system, 20-m intervals are often used as partial stations. In those countries using metric units, most municipalities have kept the 100-unit station (i.e., $1 + 00 = 100$ m), whereas many highway agencies have adopted the 1,000-unit station (i.e., $1 + 000 = 1,000$ m).

Figure 1-13 shows a building tied in to the centerline (℄) of Elm Street and shows the baseline (℄) distances as stations and the offset distances as simple dimensions. The sketch also shows that $0 + 00$ is the intersection of the center-lines of the two streets.

1.20 FIELD NOTES

Modern field surveying equipment, employing *electronic data collection*, stores point positioning angles, distances, and descriptive attributes, all of which are later electronically transferred to the computer. Surveyors have discovered that, even when using electronic data collection, selected hand-written field notes can be very useful; such field notes, which can be used to describe key points and/or a sampling of point positioning data, are invaluable as a check on critical aspects of the fieldwork and as an aid in the data editing that always occurs in the data-processing stage of an electronic survey. See also Sections 5.4, 5.5, and 5.6. When surveys are performed not using electronic data transfer, one of the most important aspects of surveying is the taking of neat, legible, and complete hand-written field notes. The notes will be used to plot scale drawings of the area surveyed and also will be used to provide a permanent record of the survey proceedings. An experienced surveyor's notes will be complete without redundancies, well arranged to aid in comprehension, and neat and legible to ensure that the correct information is conveyed. Sketches will be used to illustrate the survey and thus resolve any possible ambiguities. Hand-written field notes can be placed in bound field books or in loose-leaf binders. Loose-leaf notes are preferred for small projects, as they are easily filed alphanumerically by project name or number. Bound books are used to advantage on large projects, such as

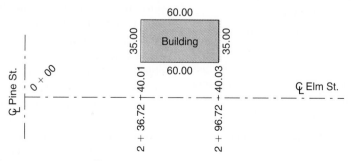

FIGURE 1-13 Baseline stations and offset distances

highway construction or other heavy construction operations, where the data can readily fill one or more field books.

Comments: Bound Books

1. Name, address, e-mail address, and phone number should be in ink on the outside cover.

2. Pages should be numbered consecutively throughout.

3. Space is reserved at the front of the field book for title, index, and diary.

4. Each project must show the date, title, surveyors' names, and instrument identification numbers.

Loose-Leaf Books

1. Name, address, e-mail address, and phone number should be in ink on the binder.

2. Each page must be titled and dated, with identification by project number, surveyors' names, and instrument numbers.

All Field Notes

1. Entries are to be in pencil in the range 2H to 4H. The harder pencil (4H) is more difficult to use but will not smear. The softer pencil (2H) is easy to use for most people, but will smear somewhat if care is not exercised. Most surveyors use 2H or 3H lead. Pencils softer than 2H are not used in field notes.

2. All entries should be neatly printed. Uppercase lettering can be reserved for emphasis, or it is sometimes used throughout.

3. All arithmetic computations are to be checked and signed.

4. Sketches are used to clarify the field notes. Although the sketches are not scale drawings, they are usually drawn roughly to scale to help order the inclusion of details.

5. Sketches are not freehand. Straightedges and curve templates are used for all line work.

6. Sketches should be properly oriented by the inclusion of a north arrow (preferably pointing up the page or to the left).

7. Avoid crowding the information onto the page. This practice is one of the chief causes of poor-looking notes.

8. Mistakes in the entry of **measured data** are to be carefully lined out, **not erased.**

9. Mistakes in entries other than measured data (e.g., descriptions, sums, or products of measured data) may be erased and reentered neatly.

10. Show the word "COPY" at the top of copied pages.

11. Lettering on sketches is to be read from the bottom of the page or from the right side.

12. Measured data are to be entered in the field notes at the time the measurements are taken.

13. The notekeeper verifies all data by repeating them aloud as he or she is entering them in the notes. The surveyor who originally gave the data to the notekeeper will listen and respond to the verification call-out.

14. If the data on an entire page are to be voided, the word "VOID," together with a diagonal line, is placed on the page. A reference page number is shown for the location of the new data.

1.21 FIELD MANAGEMENT

Survey crews (parties) often comprise a *party chief*, an *instrument operator*, and a *survey assistant*; two-person crews are more common when modern electronic equipment is being used. The party chief controls the survey to ensure that objectives (including specifications) are being met; in some cases, the party chief can control the survey best by operating the instrument, and in other cases, especially robotic surveys where the prism holder can control the instrument at the remote station, the party chief can ensure not only that the instrument is operated properly, but that the proper point positioning locations are occupied and measured. GPS surveys (see Chapter 7) can be effectively carried out using two surveyors, and in some applications, just one roving surveyor. All layout surveys require an assistant to physically mark the appropriate points on the ground.

Questions

1.1 How do plane surveys and geodetic surveys differ?

1.2 Are preliminary (or data-gathering) surveys plane surveys or geodetic surveys? Explain your response.

1.3 What kinds of data are collected during preliminary surveys?

1.4 How is a total station different from an electronic theodolite?

1.5 Describe an integrated survey system (geomatics data model).

1.6 Why is it that surveyors must measure or determine the horizontal distances, rather than just the slope distances, when showing the relative locations of two points?

1.7 Describe how a very precise measurement can be inaccurate.

1.8 Describe the term *error*; how does this term differ from *mistake*?

1.9 Describe several different ways of locating a physical feature in the field so that it later can be plotted in its correct position on a scaled plan.

CHAPTER TWO

LEVELING

2.1 GENERAL BACKGROUND

Leveling is the procedure for determining differences in elevation between points that are some distance from each other. An **elevation** is the vertical distance above or below a reference datum. Elevations can be determined using the leveling techniques described in this chapter, the total station techniques described in Chapter 5, the GPS vertical positioning techniques described in Section 7.11, or the remote-sensing techniques described in Chapter 11. The traditional vertical reference datum is *mean sea level (MSL)*. In North America, 19 years of observations at tidal stations in 26 locations on the Atlantic, Pacific, and Gulf of Mexico shorelines were reduced and adjusted to provide the National Geodetic Vertical Datum (NGVD) of 1929. This datum was further refined to reflect gravimetric and other anomalies in the 1988 general control readjustment (North American Vertical Datum [NAVD88]). Because of the inconsistencies found in widespread determinations of

MSL, the new datum has been tied to one representative tidal gage benchmark (BM) known as *Father Point* located in the St. Lawrence River Valley in the province of Quebec. Although the NAVD datum does not agree precisely with MSL at some specific points on the earth's surface, the term *mean sea level* is generally used to describe the datum. MSL is assigned a vertical value (elevation) of 0.000 ft or 0.000 m. See Figure 2-1. Typical specifications for vertical control in the United States and Canada are shown in Tables 2-1 and 2-2.

A **vertical line** is a line from the surface of the earth to the earth's center. It is also referred to as a *plumb line* or a *line of gravity*.

A **level line** is a line in a level surface. A level surface is a curved surface parallel to the mean surface of the earth. A level surface is best visualized as being the surface of a large body of water at rest.

A **horizontal line** is a straight line perpendicular to a vertical line.

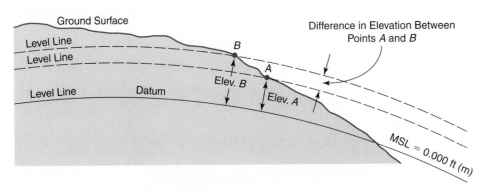

FIGURE 2-1 Leveling concepts

Table 2-1 National Ocean Survey, U.S. Coast and Geodetic Surveys: Classification, Standards of Accuracy, and General Specifications for Vertical Control

Classification	First Order	Second Order		Third Order
	Class I, Class II	Class I	Class II	
Principal uses:				
Minimum standards; higher accuracies may be used for special purposes	Basic framework of the National Network and of metropolitan area control Extensive engineering projects Regional crustal movement investigations Determining geopotential values	Secondary control of the National Network and of metropolitan area control Large engineering projects Local crustal movement and subsidence investigations Support for lower-order control	Control densification, usually adjusted to the National Network Local engineering projects Topographic mapping Studies of rapid subsidence Support for local surveys	Miscellaneous local control; may not be adjusted to the National Network Small engineering projects Small-scale topographic mapping Drainage studied and gradient establishment in mountainous areas
Recommended spacing of lines:				
National Network	Net A: 100–300 km, *Class I* Net B: 50–100 km, *Class II*	Secondary network: 20–50 km	Area control: 10–25 km	As needed
Metropolitan control	2–8 km	0.5–1 km	As needed	As needed
Other purposes	As needed	As needed	As needed	As needed
Spacing of marks along lines	1–3 km	1–3 km	Not more than 3 km	Not more than 3 km
Gravity requirement*	0.20 × 10⁻³ gpu	—	—	—
Instrument standards	Automatic or tilting levels with parallel plate micrometers; invar scale rods	Automatic or tilting levels with optical micrometers or three-wire levels; invar scale rods	Geodetic levels and invar scale rods	Geodetic levels and rods
Field procedures	Double-run; forward and backward, each section	Double-run; forward and backward, each section	Double- or single-run	Double- or single-run
Section length	1–2 km	1–2 km	1–3 km for double-run	1–3 km for double-run
Maximum length of sight	50 m *Class I;* 60 m *Class II*	60 m	70 m	90 m

(Continued)

23

Table 2-1 (Continued)

Classification	First Order — Class I, Class II	Second Order — Class I	Second Order — Class II	Third Order
Field procedures†				
Maximum difference in lengths of forward and backward sights				
Per setup	2 m *Class I;* 5 m *Class II*	5 m	10 m	10 m
Per section (cumulative)	4 m *Class I;* 10 m *Class II*	10 m	10 m	10 m
Maximum length of line between connections	Net A: 300 km Net B: 100 km	50 km	50 km double-run; 25 km single-run	25 km double-run; 10 km single-run
Maximum closures‡				
Section; forward and backward	3 mm $\sqrt{K}$ *Class I;* 4 mm $\sqrt{K}$ *Class II;*	6 mm $\sqrt{K}$	8 mm $\sqrt{K}$	12 mm $\sqrt{K}$
Loop or line	4 mm $\sqrt{K}$ *Class I;* 5 mm $\sqrt{K}$ *Class II**	6 mm $\sqrt{K}$	8 mm $\sqrt{K}$	12 mm $\sqrt{K}$

*See text for discussion of instruments.

†The maximum length of line between connections may be increased to 100 km double-run for second order, class II, and to 50 km for double-run for third order in those areas where the first-order control has not been fully established.

‡Check between forward and backward runnings, where K is the distance in kilometers.

Notes:
1. K = kilometers, m = miles = the one-way distance between BMs measured along the leveling route.
2. To maintain the specified accuracy, long narrow loops should be avoided. The distance between any two BMs measured along the actual route should not exceed four times the straight-line distance between them.
3. Branch, spur, or open-ended lines should be avoided because of the possibility of undetected gross errors.
4. For precise work, the sections should be leveled once forward and once backward independently using different instruments and, if possible, under different weather conditions and at different times of the day.
5. A starting BM must be checked against another independent BM by two-way leveling before the leveling survey can commence. If the check is greater than the allowable discrepancy, both BMs must be further checked until the matter is resolved.
6. When a parallel-plate micrometer is used for special- or first-order leveling, double-scale invar rods must be used; the spacing of the smallest graduations must be equivalent to the displacement of the parallel-plate micrometer. When the three-wire method is used for first- or second-order leveling, rods with the checkerboard design must be used.
7. Line of sight should be not less than 0.5 m above the ground (special and first order).
8. Alternate reading of backsight and foresight at successive setups should be taken.
9. Third- and lower-order surveys should use the two-rod system; read only one wire and try to balance backsight and foresight distances.
10. Results equivalent to fourth-order spirit leveling can sometimes be obtained by measurement of vertical angles in conjunction with traverses, trilateration, or triangulation. Best results are obtained on short (<16 km) lines with simultaneous (within the same minute) measurement of the vertical angles at both ends of the line, using a 1-second theodolite.

Source: Adapted from the 1974 publication *Classification, Standards of Accuracy, and General Specifications of Geodetic Control Surveys.* Detailed specifications related to these tables are available in the publication *Specifications to Support Classification, Standards of Accuracy, and General Specifications of Geodetic Control Surveys* by the Federal Geodetic Control Committee. Both publications may be obtained from U.S. Department of Commerce, National Oceanic and Atmospheric Administration, National Ocean Survey, Rockville, Maryland.

Table 2-2   Classification Standards of Accuracy and General Specifications for Vertical Control

Classification	Special Order	First Order	Second Order (First-Order Procedures Recommended)	Third Order	Fourth Order
Allowable discrepancy between forward and backward levelings	± 3 mm $\sqrt{K}$ ± 0.012 ft $\sqrt{m}$	± 4 mm $\sqrt{K}$ ± 0.017 ft $\sqrt{m}$	± 8 mm $\sqrt{K}$ ± 0.035 ft $\sqrt{m}$	± 24 mm $\sqrt{K}$ ± 0.10 ft $\sqrt{m}$	± 120 mm $\sqrt{K}$ ± 0.5 ft $\sqrt{m}$
Instruments					
Self-leveling high-speed compensator	Equivalent to 10″/2 mm level vial 10″/2 mm	Equivalent to 10″/2 mm 20″/mm 10″/2 mm	Equivalent to level vial below 20″/2 mm	Equivalent to sensitivity below 40″ to 50″/2 mm	Equivalent to sensitivity below 40″ to 50″/2 mm
Level vial telescopic magnification	40×	40×	40×		
Parallel plate micrometer					
Sun shade and instrument cover				Graduations on wood, metal, alloy, or fiberglass are satisfactory	
Rods					
Invar and double scale	×	×	×		
Invar-checkerboard footplates or steel pins for turning points		×	×		
Circular bubble attached to rod	×	×	×		
Rod supports	×	×	×		
Difference between backsight and foresight distances and their total for the section not to exceed:	5 m	10 m	10 m	Balanced	Balanced
Maximum length of sight					
1-mm-wide rod mark	50 m	×	×		
1.6-mm-wide rod mark	60 m				
Parallel-plate method		80 m	80 m	N/A*	N/A
Three-wire method		110 m	110 m	N/A	N/A

*N/A, not applicable.

Source: Adapted from "Specifications and Recommendations for Control Surveys and Survey Markers." Surveys and Mapping Branch, Department of Energy, Mines, and Resources, Ottawa, Ontario, Canada, 1973.

2.2 THEORY OF DIFFERENTIAL LEVELING

Differential leveling is used to determine differences in elevation between points that are some distance from each other by using a surveyor's level together with a graduated measuring rod. The surveyor's level consists of a crosshair-equipped telescope and an attached circular bubble, or spirit level tube, all of which are mounted on a sturdy tripod. The surveyor can sight through the telescope to a rod graduated in feet or meters and determine a measurement reading at the point where the crosshair intersects the rod. See Figure 2-2. If the rod reading at $A = 6.27$ ft and the rod reading at $B = 4.69$ ft, the difference in elevation between A and B is

$6.27 - 4.69 = 1.58$ ft. Also, if the elevation of A is 61.27 ft (above MSL), then the elevation of B is $61.27 + 1.58 = 62.85$ ft. That is, 61.27 (elevation A) $+ 6.27$ (rod reading at A) $- 4.69$ (rod reading at B) $= 62.85$ (elevation B).

Figure 2-3 shows a potential problem. Whereas elevations are referenced to level lines (surfaces), the line of sight through the telescope of a surveyors' level is theoretically a horizontal line. All rod readings taken with a surveyors' level contain an error c over a distance K. The curvature of the level lines shown in Figures 2-1 through 2-3 is greatly exaggerated for illustrative purposes. In fact, the divergence between a level line and a horizontal line is quite small. For example, over a distance of 1,000 ft, the divergence is only 0.024 ft, and for a distance of 300 ft, the divergence is only 0.002 ft (0.0007 m in 100 m).

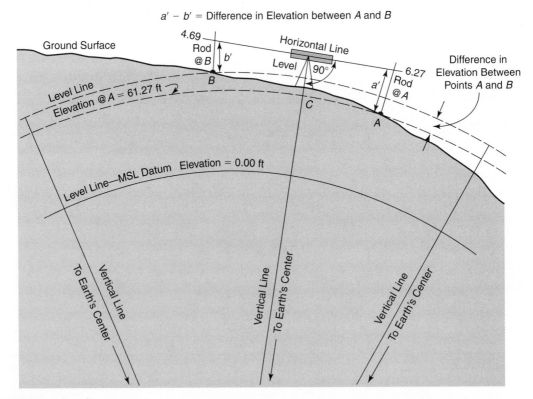

FIGURE 2-2 Leveling process

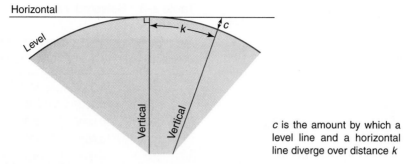

c is the amount by which a level line and a horizontal line diverge over distance k

FIGURE 2-3 Relationship between a horizontal line and a level line

2.3 CURVATURE AND REFRACTION

The previous section introduced the concept of curvature error, that is, the divergence between a level line and a horizontal line over a specified distance. When considering the divergence between level and horizontal lines, one must also account for the fact that all sight lines are refracted downward by the earth's atmosphere. Although the magnitude of the refraction error depends on atmospheric conditions, it is generally considered to be about one-seventh of the curvature error. You can see in Figure 2-4 that the negative

refraction error of $r = AB$ compensates for part of the curvature error of $c = AE$, resulting in a net error due to curvature and refraction $c + r$ BE; r = refraction error (negative) and is the distance that a line of sight is bent downward over distance CA; c = curvature error which is the divergence between a horizontal line CA, drawn through C, and a level line K, also drawn through C.

From Figure 2-4, the curvature error can be computed as follows:

$$(R + c)^2 = R^2 + K^2$$
$$R^2 + 2Rc + c^2 = R^2 + K^2$$
$$c(2R + c) = K^2$$

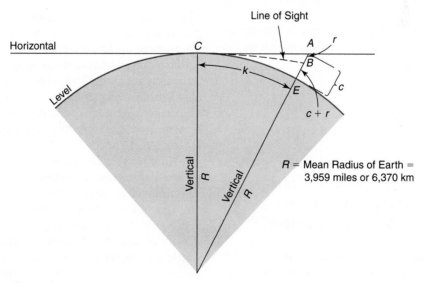

R = Mean Radius of Earth = 3,959 miles or 6,370 km

FIGURE 2-4 Effects of curvature and refraction

$$c = \frac{K^2}{2R + c} \approx \frac{K^2}{2R} \qquad (2\text{-}1)$$

In the term $2R + c$, c is so small when compared to R that it can be safely ignored. Consider $R = 6,370$ km:

$$c = \frac{K^2}{2 \times 6,370} = 0.0000785K^2 \text{ km}$$

$$= 0.0785K^2 \text{ m}$$

Refraction (r) is affected by atmospheric pressure, temperature, and geographic location but, as noted earlier, it is usually considered to be about one-seventh of the curvature error (c). If $r = -0.14c$, $c + r = 0.0675K^2$, where $K \approx CA$ (Figure 2-4) is the length of sight in kilometers. The combined effects of curvature and refraction $c + r$ can be determined from the following formulas:

$$(c + r)_m = 0.0675K^2 \quad (c + r)_m \text{ in meters}$$
$$K \text{ in kilometers} \quad (2\text{-}2)$$

$$(c + r)_{ft} = 0.574K^2 \quad (c + r)_{ft} \text{ in feet}$$
$$K \text{ in miles} \quad (2\text{-}3)$$

$$(c + r)_{ft} = 0.0206M^2 \quad (c + r)_{ft} \text{ in feet}$$
$$M \text{ in thousands of feet} \quad (2\text{-}4)$$

Example 2-1: Calculate the error due to curvature and refraction for the following distances:

   a. 2,500 ft

   b. 400 ft

   c. 2.7 miles

   d. 1.8 km

Solution:

   a. $c + r = 0.0206 \times 2.5^2 = 0.13$ ft

   b. $c + r = 0.0206 \times 0.4^2 = 0.003$ ft

   c. $c + r = 0.574 \times 2.7^2 = 4.18$ ft

   d. $c + r = 0.0675 \times 1.8^2 = 0.219$ m

You can see from the values in Table 2-3 that $c + r$ errors are relatively insignificant for differential leveling. Even for precise leveling,

Table 2-3   Selected Values for $c + r$ and Distance

Distance (m)	$(c + r)_m$	Distance (ft)	$(c + r)_{ft}$
30	0.0001	100	0.000
60	0.0002	200	0.001
100	0.0007	300	0.002
120	0.001	400	0.003
150	0.002	500	0.005
300	0.006	1,000	0.021
1 km	0.068	1 mi	0.574

where distances of rod readings are seldom in excess of 200 ft (60 m), it would seem that this error is of only marginal importance. We will see in Section 2.11 that the field technique of balancing the distances of rod readings (from the instrument) effectively cancels out this type of error.

2.4 TYPES OF SURVEYING LEVELS

2.4.1 Automatic Level

The automatic level (see Figure 2-5) employs a gravity-referenced prism or mirror compensator to orient the line of sight (line of collimation) automatically. The instrument is quickly leveled when a circular spirit level is used. When the bubble has been centered (or nearly so), the compensator takes over and maintains a horizontal line of sight, even if the telescope is slightly tilted (see Figure 2-5).

Automatic levels are extremely popular in present-day surveying operations and are available from most survey instrument manufacturers. They are easy to set up and use, and can be obtained for use at almost any required precision. A word of caution: All automatic levels employ a fragile compensator. This component normally entails freely moving prisms or mirrors, some of which are hung by fine wires. If a wire or fulcrum

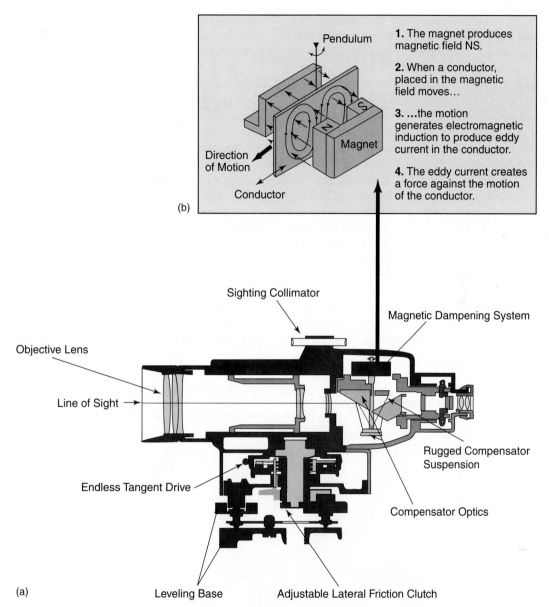

FIGURE 2-5 (a) Schematic of an engineer's automatic level. (b) Magnetic dampening system
(Courtesy of Sokkia Corp., Overland Park, Kansas)

breaks, the compensator becomes inoperative, and all subsequent rod readings will be incorrect.

The operating status of the compensator can be verified by tapping the end of the telescope or by slightly turning one of the leveling screws (one manufacturer provides a push button), causing the telescopic line of sight to temporarily veer from horizontal. If the compensator is operative, the crosshair appears to deflect momentarily before returning to its original rod reading. Constant checking of the compensator avoids costly mistakes caused by broken components.

Most current surveying instruments now come equipped with a three-screw leveling base. Whereas the support for a four-screw leveling base is the center bearing, the three-screw instruments are supported entirely by the foot screws themselves. Adjustment of the foot screws of a three-screw instrument effectively raises or lowers the height of the instrument line of sight. Adjustment of the foot screws of a four-screw instrument (see Section H-3) does not affect the height of the instrument line of sight because the instrument is supported by the center bearing. The surveyor should be aware that adjustments made to a three-screw level in the midst of a setup operation effectively changes the elevation of the line of sight and can cause significant errors on very precise surveys (e.g., BM leveling or industrial surveying).

The bubble in the circular spirit level is centered by adjusting one or more of the three independent screws. Figure 2-6 shows the positions for a telescope equipped with a tube level when using three leveling foot screws. If you think of this configuration when you are leveling the circular spirit level, you can easily predict the movement of the bubble. Some manufacturers have provided levels and the theodolites with only two leveling screws. Figure 2-6 shows that the telescopic positions are identical for both two- and three-screw instruments.

Levels used to establish or densify vertical control are designed and manufactured to give precise results. The magnifying power, setting accuracy of the tubular level or compensator, quality of optics, and so on are all improved to provide for precise rod readings. The least count on leveling rods is 0.01 ft or 0.001 m. Precise levels are usually equipped with optical micrometers so that readings can be determined one or two places beyond the rod's least count.

Many automatic levels utilize a concave base that, when attached to its dome-head

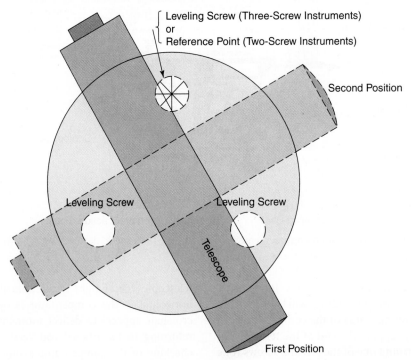

FIGURE 2-6 Telescope positions when leveling a two- or three-screw instrument

tripod top, can be roughly leveled by sliding the instrument on the tripod top. This rough leveling can be accomplished in a few seconds. If the bull's-eye bubble is nearly centered by this maneuver and the compensator is activated, the leveling screws may not be needed at all to level the instrument.

2.4.2 Digital Level

Figure 2-7 shows a digital level and barcode rod. This level features a digital electronic image-processor that uses a charge-coupled device (CCD) for determining heights and distances, with the automatic recording of data for later

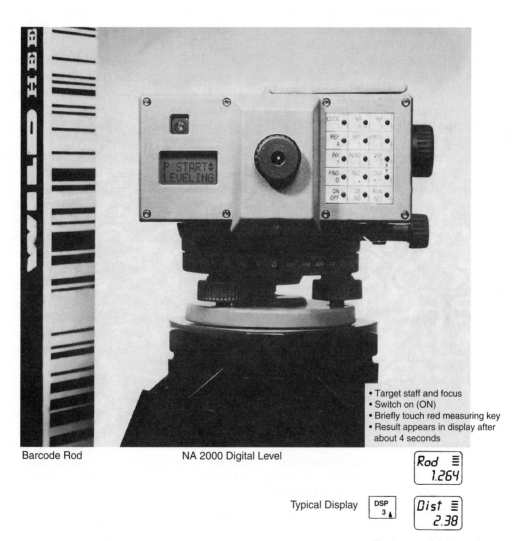

Barcode Rod   NA 2000 Digital Level

- Target staff and focus
- Switch on (ON)
- Briefly touch red measuring key
- Result appears in display after about 4 seconds

Typical Display

Rod ≡
1.264

Dist ≡
2.38

- Touch measuring key again to take next reading

FIGURE 2-7 Wild NA 2000 digital level and barcode rod
(Courtesy of Leica Geosystems Inc.)

transfer to a computer. The digital level is an automatic level (pendulum compensator) capable of normal optical leveling with a rod graduated in feet or meters. When used in electronic mode, with the rod face graduated in barcode, this level can, with the press of a button, capture and process the image of the barcode rod for distances in the range of about 0.5 m to about 100 m. This simple one-button operation initiates storage of the rod reading and distance measurement and the computation of the point elevation. The instrument's central processing unit (CPU) compares the processed image of the rod reading with the image of the whole rod, which is stored permanently in the level's memory module, thus determining height and distance values. Data can be stored in internal onboard memory or on easily transferable PC cards and then transferred to a computer via an RS232 connection or by transferring the PC card from the digital level to the computer. Most levels can operate effectively with only 30 cm of the rod visible. Work can proceed in the dark by illuminating the rod face with a small flashlight. The rod shown in Figure 2-7 is 4.05 m long (others are 5 m long); it is graduated in barcode on one side and in either feet or meters on the other side.

After the instrument has been leveled, the image of the barcode must be focused properly by the operator. Next, the operator presses the measure button to begin the image processing, which takes about 3 s (see Figure 2-7 for typical displays). Although the heights and distances are automatically determined and recorded (if desired), the horizontal angles are read and recorded manually.

Preprogrammed functions include level loop survey, two-peg test, self test, set time, and set units. Coding can be the same as that used with total stations (see Chapter 5), which means that the processed leveling data can be transferred directly to the computer database. The barcode can be read in the range of 1.8–100 m away from the instrument; the rod can be read optically as close as 0.5 m. If the rod is not plumb or is held upside down, an error message flashes on the screen. Other error messages include "Instrument

not level," "Low battery," and "Memory almost full." Rechargeable batteries are said to last for more than 2,000 measurements. Distance accuracy is in the range of 1:1,000, whereas leveling accuracy (for the more precise digital levels) is stated as having a standard deviation for a 1-km double run of 0.3–1.0 mm for electronic measurement and 2.0 mm for optical measurement. Manufacturers report that use of the digital level increases productivity by about 50 percent, with the added bonus of the elimination of field-book entry mistakes. The more precise digital levels can be used in first- and second-order leveling, whereas the less precise digital levels can be used in third-order leveling and construction surveys.

2.4.3 Tilting Level

The (mostly obsolete) tilting level is roughly leveled by observing the bubble in the circular spirit level. Just before each rod reading is to be taken, and while the telescope is pointing at the rod, the telescope is precisely leveled by manipulating a tilting screw, which effectively raises or lowers the eyepiece end of the telescope. The level is equipped with a tube level that is precisely leveled by operating the tilting screw. The bubble is viewed through a separate eyepiece or, as is the case in Figure 2-8, through the telescope. The image of the bubble is longitudinally split in two and viewed with the aid of prisms. One half of each end of the bubble can be seen, and after adjustment, the two half ends are brought to coincidence and appear as a continuous curve. When coincidence has been achieved, the telescope has been precisely leveled.

It has been estimated by Leica Geosystems that the accuracy of centering a level bubble with reference to the open tubular scale graduated at intervals of 2 mm is about one-fifth of a division, or 0.4 mm. With coincidence-type (split bubble) levels, however, this accuracy increases to about one-fortieth of a division, or 0.05 mm. As you can see, these levels are useful where a relatively high degree of precision is required; however, if tilting levels are used on ordinary work (e.g., earthwork),

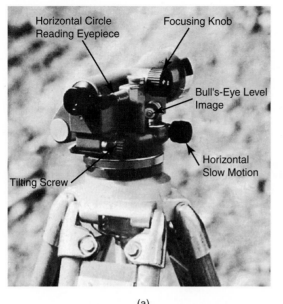

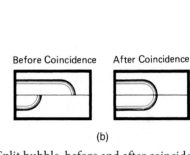

Before Coincidence   After Coincidence

(a)   (b)

FIGURE 2-8 (a) Kern engineering tilting level, GK 23 C. (b) Split bubble, before and after coincidence
(Courtesy of Kern Instruments—Leica Geosystems)

the time and expense involved in setting the split bubble to coincidence for each rod reading can scarcely be justified. These levels have by now been replaced mostly by automatic levels.

The level tube is used in many levels; it is a sealed glass tube filled mostly with alcohol or a similar substance with a low freezing point. The upper (and sometimes lower) surface has been ground to form a circular arc. The degree of precision possessed by a surveyor's level is partly a function of the sensitivity of the level tube; the sensitivity of the level tube is related directly to the radius of curvature of the upper surface of the level tube. The larger the radius of curvature, the more sensitive is the level tube.

Sensitivity is usually expressed as the central angle subtending one division (usually 2 mm) marked on the surface of the level tube. The sensitivity of many engineer's levels is 30″: that is, for a 2-mm arc, the central angle is 30″($R = 13.75$ m or 45 ft) (see Figure 2-9). The sensitivity of levels used for precise work is usually 10″, that is, $R = 41.25$ m or 135 ft.

In addition to possessing more sensitive level tubes, precise levels have improved optics, including a greater magnification power. The relationship between the quality of the optical system and the sensitivity of the level tube can be simply stated: For any observable movement of the bubble in the level tube, there should be an observable movement of the crosshair on the leveling rod.

2.5 LEVELING RODS

Leveling rods are manufactured from wood, metal, or fiberglass and are graduated in feet or meters. The foot rod can be read directly to 0.01 ft, whereas the metric rod can usually be read directly only to 0.01 m, with millimeters being estimated. Metric rod readings are normally booked to the closest 1/2 or 1/3 cm (i.e., 0.000, 0.003, 0.005, 0.007, and 0.010); more precise values can be obtained by using an optical micrometer. One-piece rods are used for more precise work. The most precise work requires the face of the rod to be an invar strip held in place under temperature-compensating spring

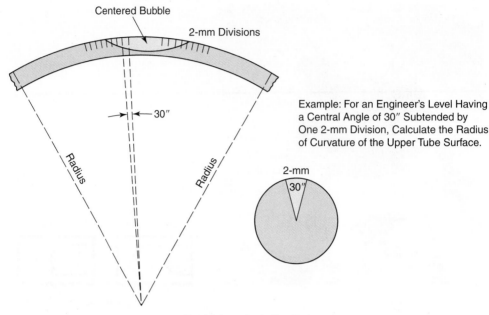

Establish an Angle/Arc Ratio

$$\frac{30''}{360°} = \frac{0.002}{2\pi R}, \ R = \frac{360}{0.00833} \times \frac{0.002}{2\pi}, \ R = 13.75 \text{ m (or 45 ft)}$$

FIGURE 2-9 Level tube showing the relationship between the central angle per division and the radius of curvature of the level tube upper surface
(Courtesy of Leica Geosystems Inc.)

tension (invar is a metal that has a very low rate of thermal expansion).

Most leveling surveys utilize two- or three-piece rods graduated in either feet or meters. The sole of the rod is a metal plate that can withstand the constant wear and tear of leveling. The zero mark is at the bottom of the metal plate. The rods are graduated in a wide variety of patterns, all of which readily respond to logical analysis. The surveyor is well advised to study an unfamiliar rod at close quarters prior to leveling to ensure that the graduations are thoroughly understood. See Figure 2-10 for examples of different graduation markings.

The rectangular sectioned rods are of either the folding (hinged) or the sliding variety. Newer fiberglass rods have oval or circular cross sections and fit telescopically together for heights of 3, 5, and 7 m, from a stored height of 1.5 m (equivalent foot

rods are also available). BM leveling utilizes folding (one-piece) rods or invar rods, both of which have built-in handles and rod levels. When the bubble is centered, the rod is plumb. All other rods can be plumbed by using a rod level (see Figure 2-11).

2.6 DEFINITIONS FOR DIFFERENTIAL LEVELING

• A **benchmark (BM)** is a permanent point of known elevation. BMs are established by using precise leveling techniques and instrumentation. BMs are bronze disks or plugs usually set into vertical wall faces. It is important that the BM be placed in a structure that has substantial footings (at least below minimum frost depth penetration) that will resist vertical

Level Rod Faces

Rod faces pictured are approximately one-half actual size.

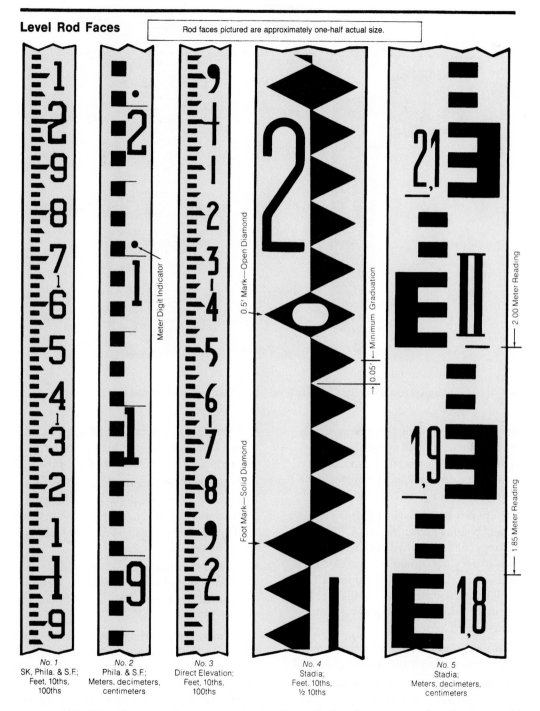

No. 1	No. 2	No. 3	No. 4	No. 5
SK, Phila. & S.F.; Feet, 10ths, 100ths	Phila. & S.F.; Meters, decimeters, centimeters	Direct Elevation; Feet, 10ths, 100ths	Stadia; Feet, 10ths, ½ 10ths	Stadia; Meters, decimeters, centimeters

FIGURE 2-10 Traditional rectangular cross-section leveling rods showing a variety of graduation markings (Courtesy of Sokkia Co. Ltd.)

(a)   (b)

FIGURE 2-11 (a) Circular rod level. (b) Circular level, shown with a leveling rod

movement due to settling or upheaval. BM elevations and locations are published by federal, state or provincial, and municipal agencies and are available to surveyors for a nominal fee. See Figure 2-15.

- A **temporary benchmark (TBM)** is a semipermanent point of known elevation. TBMs can be flange bolts on fire hydrants, nails in the roots of trees, top corners of concrete culvert headwalls, and so on. The elevations of TBMs are not normally published, but they are available in the field notes of various surveying agencies. See Figure 2-12.

- A **turning point (TP)** is a point temporarily used to transfer an elevation (see Figure 2-14).

- A **backsight (BS)** is a rod reading taken on a point of known elevation to establish the elevation of the instrument line of sight. See Figure 2-12.

- The **height of instrument (HI)** is the elevation of the line of sight through the level (i.e., elevation of BM + BS = HI). See Figure 2-12.

- A **foresight (FS)** is a rod reading taken on a turning point, benchmark, or temporary benchmark to determine its elevation; that is, HI − FS = elevation of TP (or BM or TBM). See Figure 2-12.

- An **intermediate sight (IS)** is a rod reading taken at any other point where the elevation is required (see Figures 2-17 and 2-18); that is, HI − IS = elevation.

Most engineering leveling projects are initiated to determine the elevations of intermediate points (as in profiles, cross sections, etc.). The addition of backsights to elevations to obtain heights of instrument and the subtraction of foresights from heights of instrument to obtain new elevations are known as note reductions.

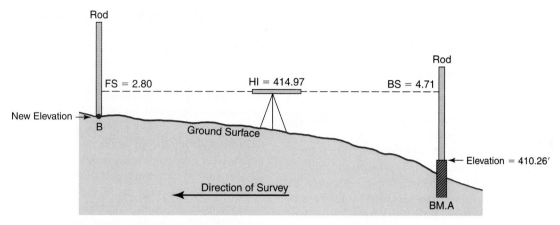

FIGURE 2-12 Leveling procedure: one setup

2.7 TECHNIQUES OF LEVELING

When in leveling, as opposed to theodolite work, the instrument can usually be set up in a relatively convenient location. If the level has to be set up on a hard surface, such as asphalt or concrete, the tripod legs will be spread out more to provide a stable setup. When the level is to be set up on a soft surface (e.g., turf), the tripod is first set up so that the tripod top is nearly horizontal, and then the tripod legs are pushed firmly into the earth. The tripod legs are snugly tightened to the tripod top so that a leg, when raised, will fall back under the force of its own weight. Undertightening can cause an unsteady setup, just as overtightening can cause torque strain. On hills, it is customary to place one leg uphill and two legs downhill; the instrument operator stands facing uphill while setting up the instrument. The tripod legs can be adjustable or straight leg. The straight-leg tripod is recommended for leveling because it contributes to a more stable setup. After the tripod has been set roughly level, with the feet firmly pushed into the ground, the instrument can be leveled.

Three-screw instruments are attached to the tripod via a threaded bolt that projects up from the tripod top into the leveling base of the instrument. The three-screw instrument, which usually has a circular bull's-eye bubble level, is leveled as described in Section 2.4.1 and Figure 2-6. Unlike the four-screw instrument, the three-screw instrument can be manipulated by moving the screws one at a time, although experienced surveyors usually manipulate at least two at a time. After the bubble has been centered, the instrument can be revolved to check that the circular bubble remains centered. For techniques used with four-screw instruments, see Appendix H.

Once the level has been set up, preparation for the rod readings can take place. The eyepiece lenses are focused by turning the eyepiece focusing ring until the crosshairs are as black and as sharp as possible. (It helps to have the telescope pointing to a light-colored background for this operation.) Next, the rod is brought into focus by turning the telescope focusing screw until the rod graduations are as clear as possible. If both of these focusing operations have been carried out correctly, the crosshairs appear to be superimposed on the leveling rod. If either focusing operation (eyepiece or rod focus) has not been carried out properly, the crosshairs appear to move slightly up and down as the observer's head moves slightly up and down. The apparent movement of the crosshairs can cause incorrect readings and is known as **parallax.**

Figure 2-12 shows one complete leveling cycle; actual leveling operations are no more complicated than what you see in the figure. Leveling operations typically involve numerous repetitions of this leveling cycle, with some operations requiring that additional (intermediate) rod readings be taken at each instrument setup.

$$\text{Existing elevation} + BS = HI \qquad (2\text{-}5)$$

$$HI - FS = \text{new elevation} \qquad (2\text{-}6)$$

These two equations completely describe the differential leveling process.

When you are leveling between BMs or turning points, the level is set approximately midway between the BS and FS locations to eliminate (or minimize) errors due to curvature and refraction (Section 2.2) and errors due to a faulty line of sight (Section 2.11). To ensure that the rod is plumb, either the surveyor uses a rod level (Figure 2-11) or he gently "waves the rod" toward and away from the instrument. The correct rod reading will be the lowest reading observed. The surveyor must ensure that the rod does not sit up on the back edge of the base and effectively raise the zero mark on the rod off the BM (or TP). The instrument operator is sure that the rod has been properly waved if the readings decrease to a minimum value and then increase in value (see Figure 2-13).

Refer again to Figure 2-12. To determine the elevation of B:

Elevation BM *A*	410.26
Backsight rod reading at BM *A*	+ 4.71 BS
Height (elevation) of instrument line of sight	414.97 HI
Foresight rod reading at TBM *B*	−2.80 FS
Elevation TBM *B*	412.17

After the rod reading of 4.71 is taken at BM *A*, the elevation of the line of sight of the instrument is known to be 414.97 (410.26 + 4.71). The elevation of TBM *B* can be determined by holding the rod at *B*, sighting the rod with the instrument, and reading the rod (2.80 ft). The elevation of TBM *B* is therefore 414.97 − 2.80 = 412.17 ft. In addition to determining the elevation of TBM *B*, the elevations of any other points lower than the line of sight and visible from the level can be determined in a similar manner.

The situation depicted in Figure 2-14 shows the technique used when the point whose elevation is to be determined (BM 461) is too far from the point of known elevation (BM 460) for a one-setup solution. The elevation of an intermediate point (TP 1) is determined, allowing the surveyor to move the level to a location where BM 461 can be "seen." Real-life situations may require numerous setups and the determination of the elevation of many turning points before getting close enough to determine the elevation of the desired point. When the elevation of the desired point has been determined, the surveyor must then either continue the survey to a point (BM) of known elevation or return (loop) the survey to the point of commencement. The survey must be closed onto a point of known elevation so that the accuracy and acceptability of the survey can be determined. If the closure is not within allowable limits, the survey must be repeated.

The arithmetic can be verified by performing the **arithmetic check** (page check). All BSs are added and all FSs are subtracted. When the sum of BS is added to the original elevation and then the sum of FS is subtracted from that total, the remainder should be the same as the final elevation calculated (see Figure 2-15).

$$\text{Starting elevation} + \Sigma BS - \Sigma FS$$
$$= \text{ending elevation}$$

In the 1800s and early 1900s, leveling procedures like the one described here were used to survey locations for railroads that traversed North America between the Atlantic Ocean and the Pacific Ocean.

2.8 BENCHMARK LEVELING (VERTICAL CONTROL SURVEYS)

BM leveling is the type of leveling employed when a system of BMs is to be established or when an existing system of BMs is to be extended or densified. For example, perhaps a BM is required in a new location, or perhaps an existing BM has been destroyed and a suitable replacement is required.

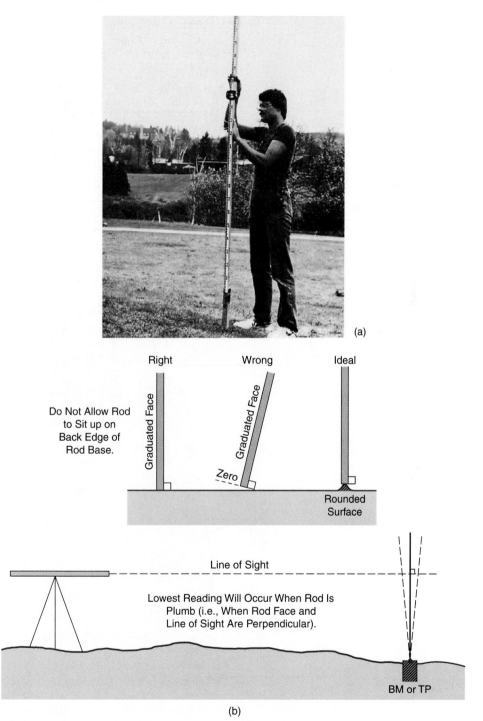

FIGURE 2-13 (a) Waving the rod. (b) Waving the rod slightly to and from the instrument allows the instrument operator to take the most precise (lowest) reading

40

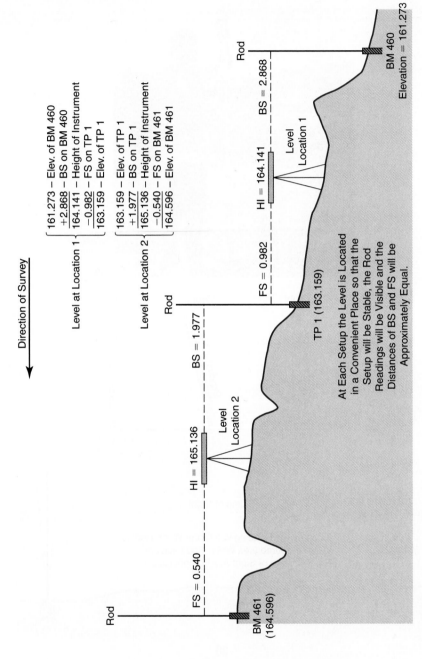

FIGURE 2-14 Leveling procedure: more than one setup

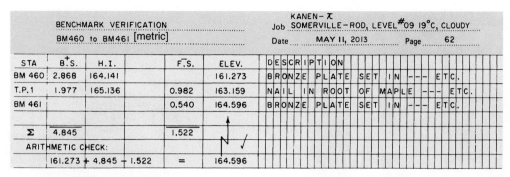

FIGURE 2-15 Leveling field notes and arithmetic check. (Data from Figure 2-14)

BM leveling is typified by the relatively high level of precision specified, both for the instrumentation and for the technique itself.

The specifications shown in Tables 2-1 and 2-2 cover the techniques of precise leveling. Precise levels with coincidence tubular bubbles of a sensitivity of 10″ per 2-mm division (or equivalent for automatic levels) and with parallel-plate micrometers are used almost exclusively for this type of work. Invar-faced rods, together with a base plate [see Figure 2-16(a)], rod level, and supports, are used in pairs to minimize the time required for successive readings.

Tripods for this type of work are longer than usual, enabling the surveyor to keep the line of sight farther above the ground and thus minimizing interference and errors due to refraction. Ideally the work is performed on a cloudy, windless day, although work can proceed on a sunny day if the instrument is protected from the sun and its possible differential thermal effects.

At the national level, BMs are established by federal agencies utilizing first-order methods and first-order instruments. The same high requirements are also specified for state and provincial grids. But as work proceeds from the whole to the part (i.e., down to municipal or regional grids), the rigid specifications are relaxed somewhat. For most engineering works, BMs are established (from the municipal or regional grid) at third-order specifications. BMs established to control isolated construction projects may be at even lower orders of accuracy.

It is customary in BM leveling at all orders of accuracy to verify first that the starting BM's elevation is correct. This can be done by two-way leveling to the closest adjacent BM, or through the use of precise GNSS measurements. This check is particularly important when the survey loops back to close on the starting BM and no other verification is planned. When BM surveys are planned in areas of possible heavy construction (e.g., roads, highways, railways, industrial and residential developments), authorities are now arranging for satellite positioning surveys to determine northing and easting (and elevation) coordinates. These combined horizontal and vertical control monuments are placed clear of walls so that construction equipment used in future machine guidance and control can position over the BM in order to calibrate the vehicles' GNSS receivers (see Chapter 13). Such control monuments aid the authorities to more closely determine local geoid undulations (see Chapter 7, Geoid Modeling). This improves the accuracy of converting from geodetic heights, given by GNSS measurements, to orthometric heights (distances above MSL), determined by BM leveling.

2.9 PROFILE AND CROSS-SECTION LEVELING

In engineering surveying, we often consider a route (road, sewer, pipeline, channel, etc.) from three distinct perspectives. The **plan view** of

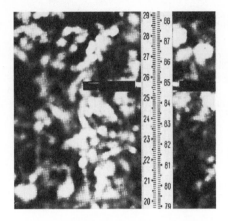

(a)

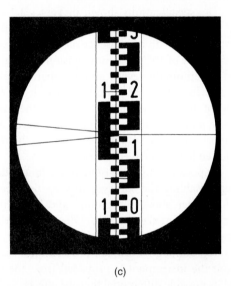

(b)   (c)

FIGURE 2-16 (a) Invar rod, also showing the footplate, which ensures a clearly defined rod position. (b) Philadelphia rod, which can be read directly by the instrument operator or the rod holder after the target has been set; Frisco rod, with two or three sliding sections having replaceable metal scales. (c) Metric rod; horizontal crosshair reading on 1.143 m

(Courtesy of Kern Instruments–Leica Geosystems)
(Courtesy of Keuffel & Esser Company, Morristown, NJ, USA)
(Courtesy of Leica Geosystems Inc., Norcross, Georgia)

route location is the same as if we were in an aircraft looking straight down. The **profile** of the route is a side view or elevation (see Figures 2-17 and 2-18) in which the longitudinal surfaces are highlighted (e.g., road, top and bottom of pipelines). The **cross section** shows the end view of a section at a station (0 + 60 in Figure 2-19 and 0 + 60 in Figures 2-20–2-22) and is at right angles to the centerline (review Section 1.19). Together, these three views (plan, profile, and

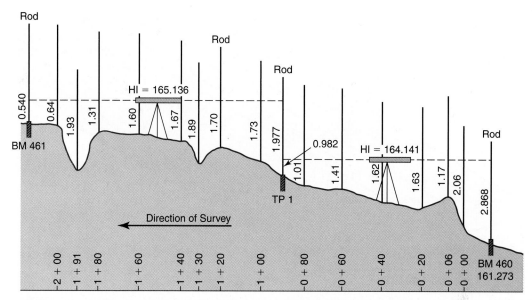

FIGURE 2-17 Example of profile leveling; see Figure 2-18 for the survey notes

PROFILE OF PROPOSED ROAD 0 + 00 to 2 + 00 [metric]

SMITH—NOTES
BROWN—π
JONES—ROD

Job 21 °C – SUNNY   LEVEL   L–14
Date   AUG 3, 2013   Page   72

STA.	B.S.	H.I.	I.S	F.S.	ELEV.	DESCRIPTION
BM 460	2.868	164.141			161.273	BRONZE PLATE SET IN --- ETC.
0 + 00			2.06		162.08	₵
0 + 06			1.17		162.97	₵ -TOP OF BERM
0 + 20			1.63		162.51	₵
0 + 40			1.62		162.52	₵
0 + 60			1.41		162.73	₵
0 + 80			1.01		163.13	₵
T.P. 1	1.977	165.136		0.982	163.159	NAIL IN ROOT OF MAPLE --- ETC.
1 + 00			1.73		163.41	₵
1 + 20			1.70		163.44	₵
1 + 30			1.89		163.25	₵ BOTTOM OF GULLY
1 + 40			1.67		163.47	₵
1 + 60			1.60		163.54	₵
1 + 80			1.31		163.83	₵
1 + 91			1.93		163.21	₵ BOTTOM OF GULLY
2 + 00			0.64		164.50	₵
BM 461				0.540	164.596	BRONZE PLATE SET IN --- ETC.

164.591– PUBLISHED ELEV.
Σ=4.845   Σ=1.522   E = 164.596
ARITHMETIC CHECK: 161.273 + 4.845 −1.522   164.591
= 164.596   0.005
ALLOWABLE ERROR (3RD ORDER)
= 12 mm $\sqrt{k}$, = .012 $\sqrt{.2}$ = .0054 m
ABOVE ERROR (.005) SATISFIES 3RD ORDER.

FIGURE 2-18 Profile field notes

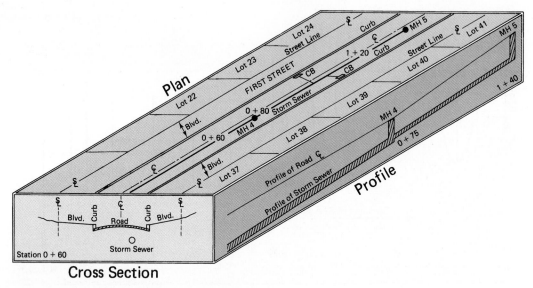

FIGURE 2-19 Relationship of plan, profile, and cross-section views

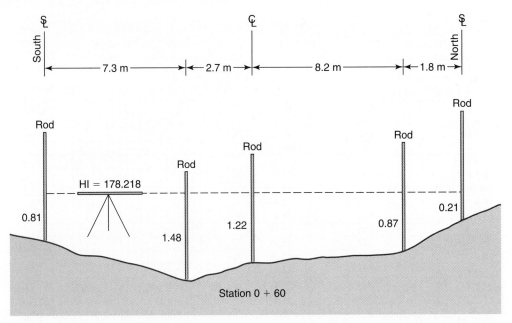

FIGURE 2-20 Cross-section surveying

cross section) completely define the route in x, y, and z coordinates.

Profile levels are taken along a path that holds interest for the designer. In roadwork, preliminary surveys often profile the proposed location of

the centerline (℄); see Figure 2-17. The proposed ℄ is staked out at an even interval (50–100 ft or 20–30 m). The level is set up in a convenient location so that the BM, and as many intermediate points as possible, can be sighted. Rod readings

					SMITH—NOTES BROWN—兀	JONES—ROD TYLER—TAPE

CROSS—SECTIONS FOR PROPOSED — Job 14 °C   CLOUDY   LEVEL #6

LOCATION OF DUNCAN ROAD — Date   NOV 24, 2013   Page   23

STA.	B.S.	H.I.	I.S.	F.S.	ELEV.	DESCRIPTION
BM 28	2.011	178.218			176.207	BRONZE PLATE SET IN SOUTH WALL 0.50m.
		(178.22)				ABOVE GROUND, CIVIC #2242, 23RD AVE.
0 + 60						
10m LT			0.81		177.41	S . ₵
2.7m LT			1.48		176.74	BOTTOM OF SWALE
₵			1.22		177.00	₵
8.2m RT			0.87		177.35	CHANGE IN SLOPE
10 m RT			0.21		178.01	N . ₵
0 + 80						
10m LT			1.02		177.20	S . ₵
3.8m LT			1.64		176.58	BOTTOM OF SWALE
₵			1.51		176.71	₵
7.8m RT			1.10		177.12	CHANGE IN SLOPE
10 m RT			0.43		177.79	N . ₵

FIGURE 2-21 Cross-section notes (municipal format)

are taken at the even station locations and at any other point where the ground surface has a significant change in slope. When the rod is moved to a new location and it cannot be seen from the instrument, a turning point is necessary so that the instrument can be moved ahead and the remaining stations leveled.

The turning point can be taken on a wood stake, the corner of a concrete step or concrete headwall, a lug on the flange of a hydrant, and so on. The turning point should be a solid, well-defined point that can be described precisely and, it is hoped, found again in the future. In the case of leveling across fields, it usually is not possible to find turning point features of any permanence. In that case, stakes are driven in and then abandoned when the survey is finished. In the example shown in Figure 2-18, the survey was closed acceptably to

BM 461. Had there been no BM at the end of the profile survey, the surveyor would have looped back and closed into the initial BM. The note on Figure 2-18 at the BM 461 elevation shows that the correct elevation of BM 461 is 164.591 m. This means that there was a survey error of 0.005 m over a distance of 200 m. Using the standards shown in Tables 2-1 and 2-2, this result qualifies for consideration at the third level of accuracy in both the United States and Canada.

The intermediate sights (ISs) in Figure 2-18 are shown in a separate column, and the elevations at the intermediate sights show the same number of decimals as are shown in the rod readings. Rod readings on turf, ground, and the like are usually taken to the closest 0.1 ft or 0.01 m. Rod readings taken on concrete, steel, asphalt, and so on are usually taken to the closest 0.01 ft or 0.003 m.

SMITH—NOTES JONES—ROD
BROWN—π TYLER—TAPE

CROSS-SECTIONS FOR PROPOSED
LOCATION OF DUNCAN HIGHWAY

Job 14 °C CLOUDY LEVEL 6
Date NOV 24, 2013 Page 23

STA.	B.S.	H.I.	I.S	F.S.	ELEV.	
	2.011	178.218	[metric]		176.207	BRONZE PLATE SET IN SOUTH WALL
		(178.22)				0.50 ABOVE GROUND, CIVIC #2242, 23RD AVE.

Cross-section data:

	LEFT		℄	RIGHT	
0 + 60	10.0	2.7		8.2	10.0
	0.81	1.48	1.22	0.87	0.21
	177.41	176.74	177.00	177.35	178.01
0 + 80	10.0	3.8		7.8	10.0
	1.02	1.64	1.51	1.10	0.43
	177.20	176.58	176.71	177.12	177.79

FIGURE 2-22 Cross-section notes (highway format)

It is a waste of time and money to read the rod more precisely than conditions warrant. Refer to Chapter 8 for details on plotting the profile.

When the final route of the facility has been selected, further surveying is required. Once again the ℄ is staked (if necessary), and cross sections are taken at all even stations. In roadwork, rod readings are taken along a line perpendicular to ℄ at each even station. The rod is held at each significant change in surface slope and at the limits of the job. In uniformly sloping land areas, rod readings required at each cross-sectioned station are often at ℄ and the two street lines (for roadwork). Chapter 8 shows how the cross sections are plotted and then utilized to compute volumes of cut and fill.

Figure 2-20 illustrates the rod positions required to define the ground surface suitably at 2 + 60, at right angles to ℄. Figure 2-21 shows the field notes for this typical cross section, in a format favored by municipal surveyors. Figure 2-22 shows the same field data entered in a cross-section note form favored by many highway agencies. Note that the HI (178.218) has been rounded to two decimals (178.22) in Figures 2-21 and 2-22 to facilitate reduction of the two-decimal rod readings. The rounded value is placed in brackets to distinguish it from the correct HI, from which the next FS will be subtracted, or from which any three-decimal intermediate rod readings are subtracted.

Road and highway construction often requires the location of granular (sand, gravel) deposits for use in the highway roadbed. These **borrow pits** (gravel pits) are surveyed to determine the volume of material "borrowed" and transported to the site. Before any excavation takes place, one or more reference baselines are established, and two BMs (at minimum) are located in convenient locations. The reference lines are located in secure locations where neither the stripping and stockpiling of topsoil nor the actual excavation of the granular material will endanger the stake (see Figure 2-23). Cross sections are taken over (and beyond) the area of proposed excavation. These *original* cross sections are used as data against which interim and final survey measurements are compared to determine total excavation. The volumes calculated from the cross sections (see Chapter 8) are often converted to tons (tonnes) for payment purposes. In many locations, weigh scales are located at the pit to aid in converting the volumes and as a check on the calculated quantities.

The original cross sections are taken over a grid at 50-ft (20-m) intervals. As the excavation proceeds, additional rod readings (in addition to 50-ft grid readings) for the top and bottom of the excavation are required. Permanent targets can be established to assist in the visual alignment of the cross-section lines running perpendicular to the baseline at each 50-ft station. If permanent targets have not been erected, a surveyor on the baseline can keep the rod on line by using a prism or estimated right angles.

2.10 RECIPROCAL LEVELING

Section 2.7 advises the surveyor to keep BS and FS distances roughly equal so that instrumental and natural errors cancel out. In some situations, such as river or valley crossings, it is not always possible to balance BS and FS distances. The reciprocal leveling technique is illustrated in Figure 2-24.

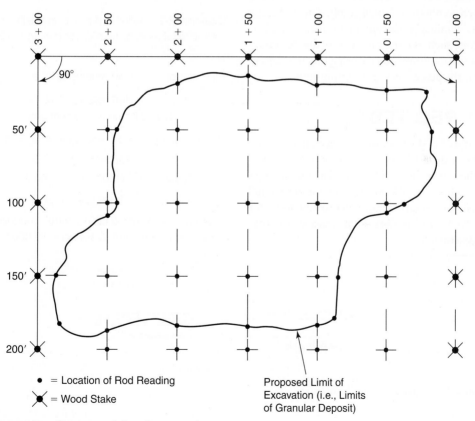

● = Location of Rod Reading

✖ = Wood Stake

Proposed Limit of Excavation (i.e., Limits of Granular Deposit)

FIGURE 2-23 Baseline control for a borrow pit survey

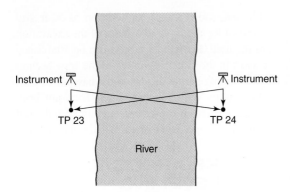

FIGURE 2-24 Reciprocal leveling

The level is set up, and readings are taken on TP 23 and TP 24. Precision can be improved by taking several readings on the far point (TP 24) and then averaging the results. The level is then moved to the far side of the river, and the process is repeated. The differences in elevation thus obtained are averaged to obtain the final result. The averaging process eliminates instrumental errors and natural errors, such as curvature. Errors due to refraction can be minimized by ensuring that the elapsed time for the process is kept to a minimum.

2.11 PEG TEST

The purpose of the peg test is to check that the line of sight through the level is horizontal (i.e., parallel to the axis of the bubble tube). The line-of-sight axis is defined by the location of the horizontal crosshair (see Figure 2-25). Refer to Chapter 5 for a description of the horizontal crosshair orientation adjustment.

To perform the peg test, the surveyor first places two stakes at a distance of 200–300 ft (60–90 m) apart. The level is set up midway (paced) between the two stakes, and rod readings are taken at both locations [see Figure 2-26(a), first setup]. If the line of sight through the level is not horizontal, the errors in rod readings (Δe_1) at both points A and B will be identical because the level is halfway between the points. Because the errors are identical, the calculated difference in elevation between points A and B (difference in rod readings) will be the *true* difference in elevation. The level is then moved to within 5 or 6 ft, or 2 m (minimum focusing distance of the level) from one of the points (A) and set up with a rod reading (a_2) determined. Any line-of-sight error generated over that very short distance will be relatively insignificant compared to the next rod reading at B. The rod is then held at B and a rod reading (b_2) obtained. See Example 2-2.

Example 2-2: What is the error in the line of sight for the level used to take the following readings?

Solution:

First setup: Rod reading at A, a_1 = 1.075

Rod reading at B, b_1 = 1.247

True difference in elevations = 0.172

Second setup: Rod reading at A, a_2 = 1.783

Rod reading at B, b_2 = 1.946*

Apparent difference in elevation = 0.163

Error (Δe_2) in 60 m = 0.009

This is an error of −0.00015 m/m. Therefore, the *collimation correction* (C factor) = +0.00015 m/m.

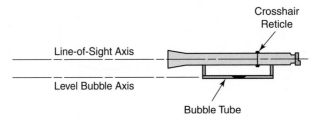

FIGURE 2-25 Optical axis and level tube axis

---

*Had there been no error in the instrument line of sight, the rod reading at b_2 would have been 1.955; that is, 1.783 + 0.172.

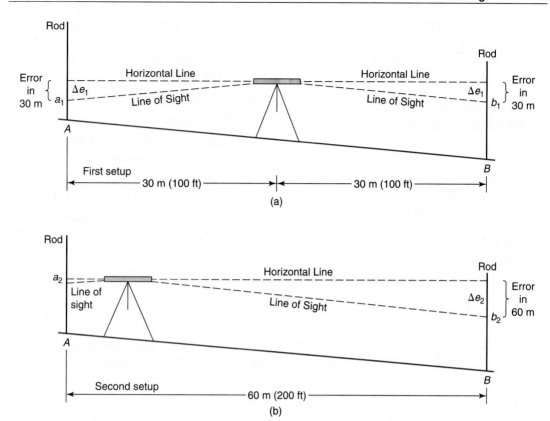

FIGURE 2-26 Peg test

In Section 2.7, you were told to try to keep the BS and FS distances equal. The peg test illustrates clearly the benefits to be gained by use of this technique. If the BS and FS distances are kept roughly equal, errors due to a faulty line of sight simply do not have the opportunity to develop. For example, if the level used in the peg test of Example 2-2 is used in the field with a BS distance of 80m and an FS distance of 70m, the net error in the rod readings will be $10 \times 0.00015 = 0.0015$ (0.002, rounded); that is, for a relatively large differential between BS and FS distances and a large line-of-sight error (0.009 for 60 m), the effect on the survey is negligible for ordinary work. The peg test can also be accomplished by using the techniques of reciprocal leveling (Section 2.10).

2.12 THREE-WIRE LEVELING

Leveling can be performed by utilizing the stadia crosshairs found on most levels and theodolites (see Figure 2-27). Each backsight (BS) and foresight (FS) is recorded by reading the rod at the stadia hairs in addition to at the horizontal crosshair. The three readings thus obtained are averaged to obtain the desired value. The stadia hairs (lines) are positioned an equal distance above and below the main crosshair and are spaced to give 1.00 ft (m) of interval for each 100 ft (m) of horizontal distance that the rod is away from the level. The recording of three readings at each sighting enables the surveyor to perform a relatively precise survey while utilizing ordinary levels.

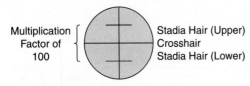

Multiplication Factor of 100
Stadia Hair (Upper)
Crosshair
Stadia Hair (Lower)
100 × Stadia Hair Interval = Distance

FIGURE 2-27 Reticle crosshairs

Readings to the closest thousandth of a foot (mm) are estimated and recorded. The leveling rod used for this type of work should be calibrated to ensure its integrity.

Figure 2-28 shows typical notes for BM leveling. A realistic survey would include a completed loop or a check into another BM of known elevation. If the collimation correction as calculated in Section 2.11 (+0.00015 m/m) is applied to the survey shown in Figure 2-28, the correction to the elevation is as follows:

$$C = +0.00015 \times (62.9 - 61.5) = +0.0002$$

Sum of FS corrected to 5.7196 + 0.0002 = 5.7198

To calculate the elevation of BM 201 (from Figure 2-31), see the following:

$$\begin{aligned} \text{Elevation of BM 17} &= 186.2830 \\ +\Sigma \text{BS} &= \underline{+2.4143} \\ & 188.6973 \\ -\Sigma \text{FS (corrected)} &= \underline{-5.7198} \\ \text{Elevation of BM 201} &= 182.9775 \end{aligned}$$

(corrected for collimation)

B.M. LEVELING–3 WIRE
B.M. 17 to B.M. 201
(RETURN RUN ON P.48)

JONES–NOTES
SMITH–X
BROWN–ROD
GREEN–ROD

Job ROD 19, INST. L.33 8 °C, CLOUDY
Date MAR 3, 2013 Page 47

STA.	B.S.	DIST.	F.S.	DIST.	ELEV.	DESCRIPTION
BM 17					186.2830	BRONZE PLATE SET IN WALL --- ETC.
	0.825		1.775			
	0.725	10.0	1.673	10.2	+0.7253	
	0.626	9.9	1.572	10.1	187.0083	
	2.176	19.9	5.020	20.3	−1.6733	
	+0.7253		−1.6733			
TP 1					185.3350	N. LUG TOP FLANGE FIRE HYD. N/S
	0.698		1.750			MAIN ST. OPP. CIVIC #181.
	0.571	12.7	1.620	13.0	+0.5710	
	0.444	12.7	1.490	13.0	185.9060	
	1.713	25.4	4.860	26.0	−1.6200	
	+0.5710		−1.6200			
TP 2					184.2860	N. LUG TOP FLANGE FIRE HYD. N/S
	1.199		2.509			MAIN ST. OPP. CIVIC #163.
	1.118	8.1	2.427	8.2	+1.1180	
	1.037	8.1	2.343	8.4	185.4040	
	3.354	16.2	7.279	16.6	−2.4263	
	+1.1180		−2.4263			
BM 201					182.9777	BRONZE PLATE SET IN ESTLY FACE
						OF RETAINING WALL --- ETC.
Σ	+2.4143	61.5m	−5.7196	62.9m		
ARITHMETIC CHECK: 186.283 + 2.4143 − 5.7196 =					182.9777 ✓	

FIGURE 2-28 Survey notes for three-wire leveling

When levels are used for precise purposes, it is customary to determine the collimation correction at least once each day.

2.13 TRIGONOMETRIC LEVELING

The difference in elevation between A and B (Figure 2-29) can be determined if the vertical angle (α) and the slope distance (S) are measured.

$$V = S \sin \alpha \qquad (2\text{-}7)$$

$$\text{Elevation at } \overline{\wedge} + \text{hi} \pm V - RR$$
$$= \text{elevation at rod} \qquad (2\text{-}8)$$

The hi in this case is not the elevation of the line of sight (HI), as it is in differential leveling. Instead, hi here refers to the distance from point A up to the optical center of the theodolite, measured with a steel tape or rod.

Trigonometric leveling can be used where it is not feasible to use a level, or as an alternative to leveling. For example, a survey crew running a ℄ profile for a route survey comes to a point where the ℄ runs off a cliff. In that case, a theodolite can be set up on the ℄ with the vertical angle and distance (slope and vertical) measured to a ℄ station at the lower elevation. Modern practice involving the use of total stations (see Chapter 6) routinely gives the elevations of sighted points by processing the differences in elevation between the total station point and the sighted points, along with the horizontal distances to those points. Total stations having dual-axis compensation can produce very accurate results. Section 5.4 discusses total station techniques of trigonometric leveling.

Example 2-3: Use Figure 2-30 and Equations 2-7 and 2-8 to determine the elevation of the instrument.

Solution:

$$V = S \sin \alpha$$
$$= 82.18 \sin 30°22'$$
$$= 41.54 \text{ ft}$$

$$\text{Elevation at } \overline{\wedge} + \text{hi} \pm V - RR$$
$$= \text{elevation at rod}$$

$$361.29 + 4.72 - 41.54 - 4.00 = 320.47$$

Note that the RR in Example 2-3 could have been 4.72, the value of the hi. If that reading had been visible, the surveyor would have sighted on it to eliminate +hi and −RR from the calculation; that is,

$$\text{Elevation at } \overline{\wedge} - V = \text{elevation at rod} \qquad (2\text{-}8a)$$

In Example 2-3, the station (chainage) of the rod station can also be determined.

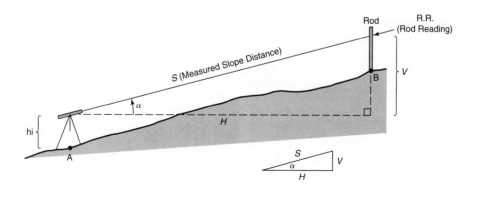

FIGURE 2-29 Trigonometric leveling

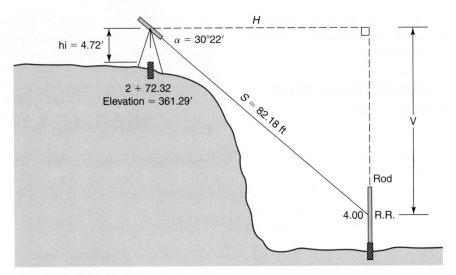

FIGURE 2-30 Example of trigonometric leveling (see Section 2.13)

2.14 LEVEL LOOP ADJUSTMENTS

We noted in Section 2.7 that level surveys had to be closed within acceptable tolerances or the survey would have to be repeated. The tolerances for various orders of surveys were shown in Tables 2-1 and 2-2. If a level survey is performed to establish new BMs, it would be desirable to proportion any acceptable error suitably throughout the length of the survey. Since the error tolerances shown in Tables 2-1 and 2-2 are based on the distances surveyed, adjustments to the level loop will be based on the relevant distances, or on the number of instrument setups, which is a factor directly related to the distance surveyed.

Example 2-4: A level circuit is shown in Figure 2-31. A survey is needed for local engineering projects. It starts at BM 20; the elevations of new BMs 201, 202, and 203 were determined; and then the level survey was looped back to BM 20, the point of commencement (the survey could have terminated at any established BM). See Table 2-4 for the starting elevation at BM 20 and for the field elevations and adjusted elevations at BMs 201, 202, and 203. What is the permissible error?

Solution: According to Table 2-1, the allowable error for a second-order, class II (local engineering projects) survey is $0.008\sqrt{K}$. Thus, $0.008\sqrt{4.7} = 0.017$ m is the permissible error. The error in the survey was found to be −0.015 m

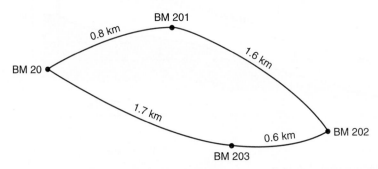

FIGURE 2-31 Level loop: Total distance around loop is 4.7 km

Table 2-4   Level Loop Adjustments

BM	Loop Distance: Cumulative (km)*	Field Elevation	Correction: $\frac{\text{Cumulative Distance}}{\text{Total Distance}} \times E**$	Adjusted Elevation
20		186.273 (fixed)		186.273
201	0.8	184.242	+0.8/4.7 × 0.015 = +0.003 =	184.245
202	2.4	182.297	−2.4/4.7 × 0.015 = +0.008 =	182.305
203	3.0	184.227	+3.0/4.7 × 0.015 = +0.010 =	184.237
20	4.7	186.258	+4.7/4.7 × 0.015 = +0.015 =	186.273

*This is the one-way distance referred to in Table 2-1, note 1.

**E = 186.273 − 186.258 = −0.015 m.

over a total distance of 4.7 km, in this case, an acceptable error. This acceptable error must be distributed suitably over the length of the survey. The error is proportioned according to the fraction of cumulative distance over total distance, as in Table 2-4. More complex adjustments are normally performed by computer, using the adjustment method of least squares.

2.15 SUGGESTIONS FOR ROD WORK

The following list of suggestions will help to ensure accurate rod work:

1. The rod should be properly extended and clamped. Take care to ensure that the bottom of the sole plate does not become encrusted with mud, dirt, and so on, which can result in mistaken readings. If a rod target is being used, ensure that it is properly positioned and that it cannot slip.

2. The rod should be held plumb for all rod readings. Either a rod level is used or the rod is waved gently to and from the instrument so that the lowest (indicating a plumb rod) reading can be determined. This practice is particularly important for all backsights and foresights.

3. Ensure that all points used as turning points are suitable, in other words, describable, identifiable, and capable of having the elevation determined to the closest 0.01 ft or 0.001 m. The TP should be nearly equidistant from the two proposed instrument locations.

4. Make sure that the rod is held in precisely the same position for the backsight as it was for the foresight for all turning points.

5. If the rod is held temporarily near, but not on, a required location, the face of the rod should be turned away from the instrument so that the instrument operator cannot take a mistaken reading. This type of mistaken reading usually occurs when the distance between the two surveyors is too far to allow for voice communication and sometimes even for good visual contact.

2.16 SUGGESTIONS FOR INSTRUMENT WORK

The following list offers suggestions to ensure that your instrument work is accurate:

1. Use a straight-leg (nonadjustable) tripod, if possible.

2. Tripod legs should be tightened (if possible) so that when one leg is extended horizontally, it falls slowly back to the ground under its own weight.

3. The instrument can be comfortably carried resting on one shoulder. If tree branches or other obstructions (e.g., door frames) threaten the safety of the instrument, it should be cradled under one arm with the instrument forward, where it can be seen. Heavier instruments, such as total stations, must be carried separately, off the tripod.

4. When setting up the instrument, gently force the legs into the ground by applying weight on the tripod shoe spurs. On rigid surfaces (e.g., concrete), the tripod legs should be spread farther apart to increase stability.

5. When the tripod is to be set up on a side hill, two legs should be placed downhill and the third leg placed uphill. The instrument can be set up roughly level by careful manipulation of the third, uphill leg.

6. The location of the level setup should be chosen so that you can "see" the maximum number of rod locations, particularly BS and FS locations.

7. Prior to taking rod readings, the crosshair should be focused sharply. It helps to point the instrument toward a light-colored background (e.g., the sky).

8. If apparent movement of the crosshairs on the rod (parallax) is observed, carefully check the crosshair focus adjustment and the objective focus adjustment for consistent results.

9. Read the rod consistently, at either the top or the bottom of the crosshair.

10. Never move the level before a foresight is taken; otherwise, all work done from that HI will have to be repeated.

11. Check that the level bubble remains centered or that the compensating device (in automatic levels) is operating.

12. Rod readings (and the line of sight) should be kept at least 18 in. (0.5 m) above the ground surface to help minimize refraction errors when you perform a precise level survey.

2.17 MISTAKES IN LEVELING

Mistakes in level loops can be detected by performing arithmetic checks, and also by closing in on the starting BM or on any other BM whose elevation is known. Mistakes in rod readings that do not form part of a level loop, such as in intermediate sights taken in profiles, cross sections, or construction grades, are a much more irksome problem. It is bad enough to discover that a level loop contains mistakes and must be repeated. It is a far more serious problem, however, to have to redesign a highway profile because a key elevation contains a mistake, or to have to break out a concrete bridge abutment (the day after the concrete was poured) because the grade stake elevation contained a mistake. Because intermediate rod readings cannot be inherently checked, it is essential that the opportunities for mistakes be minimized.

Common mistakes in leveling include the following: misreading the foot (meter) value, transposing figures, not holding the rod in the correct location, entering the rod readings incorrectly (i.e., switching BS and FS), entering the station of the rod reading incorrectly, and making mistakes in the note reduction arithmetic. Mistakes in arithmetic can be eliminated almost completely by having the other crew members check the reductions and initial each page of notes checked. Mistakes in the leveling operation cannot be eliminated totally, but they can be minimized if the crew members are aware that mistakes can (and probably will) occur. All crew members should be constantly alert to the possible occurrence of mistakes, and all crew members should try to develop rigid routines for doing their work so that mistakes, when they do eventually occur, will be all the more noticeable.

Problems

2.1 Compute the error due to curvature and refraction for the following distances:

  (a) 500 ft
  (b) 4,000 ft
  (c) 300 m
  (d) 2.2 mi
  (e) 2,800 m
  (f) 3 km

2.2 Determine the rod readings indicated on the foot and metric rod illustrations in Figure 2-32. The foot readings are to the closest 0.01 ft, and the metric readings are to the closest one-half or one-third cm.

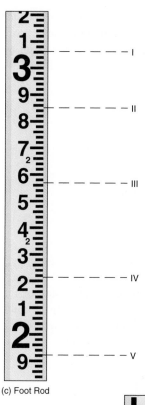

(c) Foot Rod

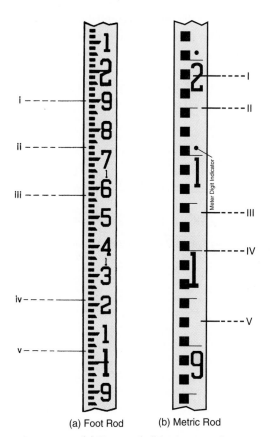

(a) Foot Rod   (b) Metric Rod

FIGURE 2-32 (a) Foot rod. (b) Metric rod

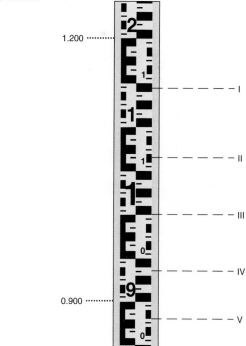

(d) Metric Rod

2.3 An offshore drilling rig is being towed out to sea. What is the maximum distance away that the navigation lights can still be seen by an observer standing at the shoreline? The observer's eye height is 5′0″ and the uppermost navigation light is 147 ft above the water.

2.4 Prepare a set of level notes for the survey in Figure 2-33. Show the arithmetic check.

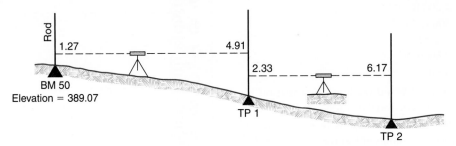

FIGURE 2-33

2.5 Prepare a set of profile leveling notes for the survey in Figure 2-34. In addition to computing all elevations, show the arithmetic check and the resulting error in closure.

2.6 Complete the set of differential leveling notes in Table 2-5, and perform the arithmetic check.

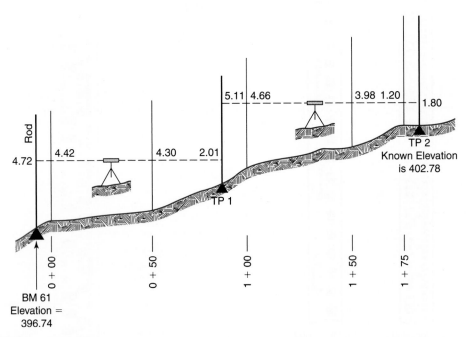

FIGURE 2-34

Table 2-5

Station	BS	HI	FS	Elevation
BM 100	2.71			314.88
TP 1	3.62		4.88	
TP 2	3.51		3.97	
TP 3	3.17		2.81	
TP 4	1.47		1.62	
BM 100			1.21	

2.7 If the loop distance in Problem 2.6 is 1,000 ft, at what order of survey do the results qualify? Use Table 2-1 or Table 2-2.

2.8 Reduce the set of differential leveling notes in Table 2-6, and perform the arithmetic check.

Table 2-6

Station	BS	HI	IS	FS	Elevation
BM 20	8.27				177.77
TP 1	9.21			2.60	
0 + 00			11.3		
0 + 50			9.6		
0 + 61.48			8.71		
1 + 00			6.1		
TP 2	7.33			4.66	
1 + 50			5.8		
2 + 00			4.97		
BM 21				3.88	

2.9 If the distance leveled in Problem 2.8 is 1,000 ft, for what order of survey do the results qualify if the elevation of BM 21 is known to be 191.40? See Tables 2-1 and 2-2.

2.10 Reduce the set of profile notes in Table 2-7, and perform the arithmetic check.

Table 2-7

Station	BS	HI	IS	FS	Elevation
BM 22	1.203				181.222
0 + 00					
℄			1.211		
10 m left, ℄			1.430		
10 m right, ℄			1.006		
0 + 20					
10 m left, ℄			2.93		
7.3 m left			2.53		
4 m left			2.301		
℄			2.381		
4 m right			2.307		
7.8 m right			2.41		
10 m right, ℄			2.78		
0 + 40					
10 m left, ℄			3.98		
6.2 m left			3.50		
4 m left			3.103		
℄			3.187		
4 m right			3.100		
6.8 m right			3.37		
10 m right			3.87		
TP 1				2.773	

2.11 Reduce the set of municipal cross-section notes in Table 2-8.

Table 2-8

Station	BS	HI	IS	FS	Elevation
BM 41	4.11				302.99
TP 13	4.10			0.89	
12 + 00					
50 ft left			3.9		
18.3 ft left			4.6		

(Continued)

Table 2-8 (Continued)

Station	BS	HI	IS	FS	Elevation
℄			6.33		
20.1 ft right			7.9		
50 ft right			8.2		
13 + 00					
50 ft left			5.0		
19.6 ft left			5.7		
℄			7.54		
20.7 ft right			7.9		
50 ft right			8.4		
TP 14	7.39			1.12	
BM S.22				2.41	

2.12 Complete the set of highway cross-section notes in Table 2-9.

2.13 Complete the set of highway cross-section notes in Table 2-10.

2.14 A level is set up midway between two wood stakes that are about 300 ft apart. The rod reading on stake A is 8.72 ft, and it is 5.61 ft on stake B. The level is then moved to point B and set up about 6 ft or 2 m away. A reading of 5.42 ft is taken on the rod at B. The level is then sighted on the rod held on stake A, where a reading of 8.57 ft is noted.

(a) What is the correct difference in elevation between the tops of stakes A and B?

Table 2-9

Station	BS	HI	FS	Elevation	Left	℄	Right
BM 37	7.20			377.97			
5 + 50					50' 26.7'		28.4' 50'
					4.6 3.8	3.7	3.0 2.7
6 + 00					50' 24.1'		25.0' 50'
					4.0 4.2	3.1	2.7 2.9
6 + 50					50' 26.4'		23.8' 50'
					3.8 3.7	2.6	1.7 1.1
TP 1			6.71				

Table 2-10

Station	BS	HI	FS	Elevation	Left	℄	Right
BM 107	7.71			399.16			
80 + 50					60' 28'		32' 60'
					9.7 8.0	5.7	4.3 4.0
81 + 00					60' 25'		30' 60'
					10.1 9.7	6.8	6.0 5.3
81 + 50					60' 27'		33' 60'
					11.7 11.0	9.2	8.3 8.0
TP 1			10.17				

(b) If the level had been in perfect adjustment, what reading would have been observed at *A* from the second setup?

(c) What is the line-of-sight error in 300 ft?

(d) Describe how you would eliminate the line-of-sight error from the telescope.

2.15 A preengineering baseline was run down a very steep hill (see Figure 2-35). Rather than measure horizontally downhill with the steel tape, the surveyor measures the vertical angle with a theodolite and the slope distance with a 200-ft steel tape. The vertical angle is −21°26′ turned to a point on a plumbed range pole that is 4.88 ft above the ground. The slope distance from the theodolite to the point on the range pole is 148.61 ft. The theodolite's optical center is 4.66 ft above the upper baseline station at 110 + 71.25.

(a) If the elevation of the upper station is 318.71, what is the elevation of the lower station?

(b) What is the stationing chainage of the lower station?

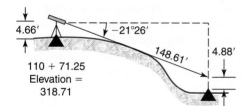

FIGURE 2-35

2.16 You must establish the elevation of point *B* from point *A* (elevation 216.612 m). *A* and *B* are on opposite sides of a 12-lane highway. Reciprocal leveling is used, with the following results:

Table 2-11

Station	BS	HI	FS	Elevation
BM 130	0.702			188.567
TP 1	0.970		1.111	
TP 2	0.559		0.679	
TP 3	1.744		2.780	
BM K110	1.973		1.668	
TP 4	1.927		1.788	
BM 132			0.888	

Setup at *A* side of highway:

Rod reading on *A* = 0.673 m

Rod readings on *B* = 2.416 and 2.418 m

Setup at *B* side of highway:

Rod reading on *B* = 2.992 m

Rod readings on *A* = 1.254 and 1.250 m

(a) What is the elevation of point *B*?

(b) What is the leveling error?

2.17 Reduce the set of differential leveling notes in Table 2-11, and perform the arithmetic check.

(a) Determine the order of accuracy (see Table 2-1 or Table 2-2).

(b) Adjust the elevation of BM K110. The length of the level run was 780 m, with setups that are equally spaced. The elevation of BM 132 is 187.536 m.

CHAPTER THREE

DISTANCE MEASUREMENT

3.1 METHODS OF DISTANCE DETERMINATION

There are two ways to locate the position of a topographic feature in relation to other topographic features. First, the position of each feature can be captured directly using GPS techniques and/or by remote-sensing data capture (see Chapters 7 and 12). Second, topographic features can be related to each other by measuring the distance between them, or by measuring the distance to any number of topographic features from some common baseline or control net. Most surveying field distance measurements are accomplished using either electronic distance measurement (EDM) or taping (steel tapes or fiber glass tapes). This chapter and chapter 8 describes these techniques in some detail.

The early Egyptians used ropes, knotted at specific intervals, to directly measure out property lines after each flooding of the Nile River (see Appendix H). That technique illustrates the direct application of a measuring standard against the distance to be measured; modern surveyors now use fiberglass and steel tapes to perform the same direct measurement function. In addition to these direct methods of distance determination, there are several indirect methods of determining distances; these techniques (including EDM) use related measurements to deduce the required distance. Distances between points can also be determined using geometric or trigonometric computations working with related distance and angle measurements. Ground distances can be roughly determined by simply measuring distances on a scaled map or plan and then converting the scaled distances to their ground distance equivalents.

3.2 DISTANCE MEASURING TECHNIQUES

3.2.1 Pacing

Pacing is a very useful (although imprecise) technique of distance measurement. Surveyors can determine the length of pace that, for them, can be comfortably repeated. An individual's length of pace can be determined by repeatedly pacing between two marks a set distance apart (say, 100 ft or 30 m). Pacing is particularly useful when looking for previously set survey markers; the plan distance from a found marker to another marker can be paced off so that the marker can be located visually using a magnetic or electronic bar locator and/or by digging with a shovel. Pacing is also useful when checking the positions of property and construction layout markers. Carefully done on a horizontal surface, pacing can result in accuracy ratios of about 1:100.

3.2.2 Odometer

When beginning a survey, the surveyor often has to distinguish between fence lines abutting a road; the automobile odometer can be quite useful in measuring distance from a known corner to the fence-marked property lines that define the area to be surveyed. Also, a measuring wheel (12–24-in. diameter) equipped with an odometer can be used in traffic investigations to measure distances at an accident scene. These measuring wheels can also be used by assessors and other real-estate personnel to record property frontages and areas.

3.2.3 Distances Obtained from Positioning Techniques

As noted earlier, ground position can be determined using satellite positioning and by using various remote-sensing techniques. Once the position coordinates are known, it is simple enough to compute the distances between those positions. See Section 7.5.

3.2.4 Electronic Distance Measurement (EDM)

EDM can be classed as an indirect technique of measurement because the distance measurement is deduced either after noting the phase delays between transmitted and received light signals or by measuring the time of travel for pulsed light transmissions. Many EDM instruments function by sending a light wave or microwave along the path to be measured and then measuring the phase differences between the transmitted and received signals. In the case of microwaves, identical instruments—one transmitting and one receiving—are positioned at each end of the line to be measured. In the case of light waves, only one instrument is required, with a reflecting prism located at the other end of the line. Over shorter distances, light waves can be reflected back to the instrument right from the measured object itself (reflectorless EDM); many of those instruments employ pulsed laser emissions. The distance here depends on the measurement of the elapsed time between transmission and reception of the laser signals. Figure 3-1 shows a handheld laser instrument (Leica's Disto), which utilizes a visible laser beam to measure distances indoors or outdoors up to a range of 200 m (650 ft), with accuracies from 1.5 to 3 mm.

FIGURE 3-1 Handheld laser
(Courtesy of Leica Geosystems Inc.)

3.3 OTHER INDIRECT MEASURING TECHNIQUES

3.3.1 Stadia

Stadia is now an obsolete form of indirect measurement that uses a telescopic crosshair configuration to assist in determining distances (see Appendix H). Additional crosshairs (stadia hairs) are positioned in the telescope an equal distance above and below the main crosshair such that when the interval (as measured on a leveling rod) between the upper and lower stadia hairs is multiplied by a constant (usually 100), the ground distance is determined. Additional treatment is required for inclined sightings.

3.3.2 Subtense Bar

A subtense bar is a tripod-mounted horizontal bar having sighting targets set precisely 2.000 m apart. The subtense bar is set over a point, leveled, and then turned so that it is precisely perpendicular to the sight from a precise (capable of measuring to at least 1″) theodolite. The bar has a sighting device that permits the theodolite operator to determine

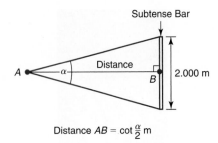

Distance $AB = \cot \frac{\alpha}{2}$ m

FIGURE 3-2 Subtense bar

perpendicularity. The device is reasonably accurate over short ($\leq$500 ft) distances. Figure 3-2 shows the geometry involved in this indirect distance determination. It is apparent that the horizontal distance in meters $= 1/\tan \frac{1}{2} \alpha = \cot \frac{1}{2} \alpha$. Because the angle α is the measurement between the vertical planes containing the subtense bar targets, the distance obtained is always the horizontal distance, regardless of the angle of inclination to the targets. EDM instruments have now replaced the subtense bar for field measurements, but the device continues to be useful as a calibration tool in electronic coordination, a technique using electronic theodolites interfaced to a computer, to position precisely manufacturing components (e.g., welding robots, on assembly lines).

3.4 GUNTER'S CHAIN

When North America was first surveyed (in the eighteenth and nineteenth centuries), the distance measuring device in use was the Gunter's chain. It was 66 ft long and was composed of 100 links. The length of 66 ft was apparently chosen because of its relationships to other units in the Imperial System:

$$80 \text{ chains} = 1 \text{ mile}$$
$$10 \text{ square chains} = 1 \text{ acre} (10 \times 66^2 = 43,560 \text{ ft}^2)$$
$$4 \text{ rods} = 1 \text{ chain}$$

Many of North America's old plans and deeds contain measurements in chains and links, so surveyors occasionally have to convert these distances to feet or meters for current projects.

Example 3-1: An old plan shows a dimension of 3 chains, 83 links. Convert this value to (a) feet and (b) meters.

Solution:

a. $3.83 \times 66 = 252.78$ ft

b. $3.83 \times 66 \times 0.3048 = 77.047$ m

3.5 TAPING

3.5.1 General

Although most distances are now measured using total stations or other EDM devices, there are still many applications for the use of fiberglass and steel tapes. Taping is used for short distances and in many construction applications. Fiberglass tapes can give accuracies in the half-centimeter range and steel tapes can measure to the nearest hundredth of a foot or the nearest millimeter. Typical engineering surveying accuracy ratios in the range of 1:3,000 to 1:5,000 can be readily attained when measuring with a steel tape if the proper techniques are employed (see Section 3.16). For precise tape measuring, steel tapes are always used. However, for many applications where a lower precision is acceptable, various types of fiberglass tapes are used. Typical uses of fiberglass tapes involve topographic tie-ins, fencing measurements, and other low-cost construction quantities. See Figure 3-3(a) and (b).

3.5.2 Steel Tapes

Steel tapes [see Figure 3-3(d)] are manufactured in both foot and metric units and come in various lengths, markings, and weights. In foot units, tapes come in many lengths, but the 100-ft length is the most commonly used. Tapes are also used in 200-ft and 300-ft lengths for special applications. In metric units, the most commonly used tape length is 30 m, although lengths of 20, 50, and 100 m can also be obtained.

Steel tapes come in two common cross sections: heavy duty is 8 mm $\times$ 0.45 mm, or $^{5}/_{16}$ in. $\times$ 0.18 in.; normal usage is 6 mm $\times$ 0.30 mm, or $^{1}/_{4}$ in. $\times$ 0.012 in. Metric tapes are used occasionally in the United States; in Canada, they

(a)

(b)

Graduated in 10ths and Metric

Printed on two sides—one side in 10ths and 100ths of a foot; the second side in metric with increments in meters, cm, and 2 mm.

Graduated in 8ths and Metric

Printed on two sides—one side in feet, inches, and 8ths; the second side in metric with increments in meters, cm, and 2 mm.

(c)

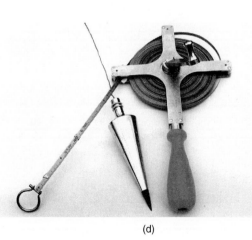

(d)

(e)

FIGURE 3-3 Fiberglass and steel tapes. (a) Closed case. (b) Open reel. (c) Tape graduations. (d) Steel tape and plumb bob (Courtesy of CST/Berger, Illinois.) (e) Measuring to a water main access frame using a steel tape and plumb bob

have been in use for several decades. The 100-ft tape and the 30-m tapes have similar handling characteristics (see Section 3.8).

Generally, the heavy-duty tapes (drag tapes) were used in route surveys (e.g., highways, railroads) and are designed for use off the reel. Leather thongs are tied through the eyelets at both ends of the tape to aid in measuring. The lighter-weight tapes can be used on (usually) or off the reel and are usually found in municipal and structural surveys. If they are used on the reel, the wind-up handle can be flipped over to contact the handle frame and serve as a brake, thus assisting in the measuring process [see Figure 3-3(b), 3-3(d) and 3-3(e)].

Steel tapes are usually graduated in one of three ways (see Figure 3-4):

1. Preferred by many, this tape is graduated throughout in feet and hundredths (0.01) of a foot or in meters and millimeters; see Figures 3-4(a) and 3-3(d) and (e).

2. The cut tape is marked throughout in feet, with the first and last feet graduated in tenths and hundredths of a foot; see Figure 3-4(b).

The metric tape is marked throughout in meters and decimeters, with the first and last decimeters further graduated in millimeters. A measurement is made between two points with the cut tape by one surveyor holding that even foot (decimeter) mark on one point, which allows the other surveyor to read the graduations on the first foot (or decimeter) on the other point. For example, in Figure 3-4(b), the distance from A to B is determined by holding 39 ft at B and reading 0.18 at A. Distance $AB = 38.82$ ft (i.e., 39 ft "cut" 0.18 = 38.82). Each measurement involves this "cut" subtraction from the even foot or meter mark being held at the far end of the measurement. Extra care is needed here to avoid subtraction mistakes.

3. The add tape is similarly marked, except that an additional foot or decimeter is placed before the zero mark at the beginning of the tape. In Figure 3-4(c), the distance from A to B is determined by holding 38 ft at B and reading 0.82 at A. Thus, distance AB is 38.82 ft (i.e., 38 "add" 0.82).

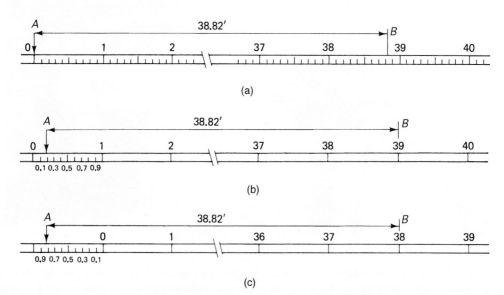

FIGURE 3-4 Various tape markings (hundredth marks not shown). (a) Fully graduated tape. (b) Cut tape. (c) Add tape

3.6 TAPING ACCESSORIES

3.6.1 Plumb Bob

Plumb bobs are normally made of brass and weigh from 8 to 18 oz., with the 10 oz. and 12 oz. plumb bobs most widely used [see Figure 3-3(d) and (e)]. Plumb bobs come with about 6 ft of string and a sharp, replaceable screw-on point. Plumb bobs are used in taping to permit the surveyor to hold the tape horizontal when the ground is sloping. A graduation mark on the horizontal tape can be transferred down to a point on the ground using the plumb bob string. A graduation can be read on the tape as the plumb bob string is moved until the plumb bob is directly over the ground mark being measured. [See Figures 3-3(e) and 3-5.] Also, a plumb bob (with or without a string-mounted target) can be used to provide precise theodolite and total station sightings.

3.6.2 Hand Level

Hand levels [see Figure 3-6(a)] are small rectangular or cylindrical sighting tubes equipped with tubular bubbles and horizontal crosshairs that permit the surveyor to make low-precision horizontal sightings. The bubble location and the crosshair can be viewed together via a 45° mounted mirror. Hand levels can be used to assist the surveyor in keeping the tape horizontal while the tape is held

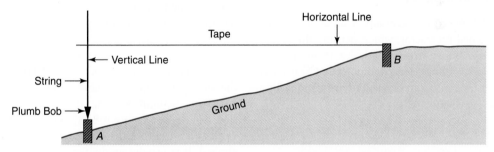

FIGURE 3-5 Use of a plumb bob

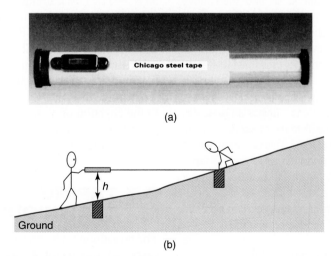

(a)

(b)

FIGURE 3-6 (a) Hand level. (b) Hand-level application
(Courtesy of CST/Berger, Illinois.)

off the ground. The hand level is held by the surveyor at the lower elevation and a sight is taken on the uphill surveyor, who is used as a measuring standard. For example, if the lower-elevation surveyor is sighting horizontally at the other surveyor's waist, and if both are roughly the same height, then the surveyor with the hand level is lower than his or her partner by the distance from his eye to his waist. See Figure 3-6(b). The low end of the tape is held up in the air that distance (say, 1.8 ft or 0.55 m) and, using a plumb bob, the distance is measured and recorded with the high end of the tape held right on the mark; see Figure 3-5. The hand level, along with a graduated rod, can also be used for rough leveling surveys (see Chapter 2).

3.6.3 Clinometer

The clinometer is essentially a hand level with an attached protractor, which permits the determination of low-precision (closest 10 minutes of arc) angle readings. The clinometer can be used as a hand level or it can be used to determine slope angles [see Figure 3-7(a)].

Example 3-2: How can you determine the height of a building using a clinometer and a tape?

Solution: To determine the height of a building, set 45° on the protractor scale and then move backward or forward until the top of the structure is sighted on the crosshair. At the same time, the bubble (mounted on the protractor) will appear to be superimposed on the crosshair. Refer to Figure 3-7(b), where it can be seen that the height of the building is $h_1 + h_2$. Also, from trigonometry, we know that the tan of 45° is 1 (this is an isosceles triangle), so the distance from the observer's eye to the building wall is equal to h_1. This distance can be measured with a tape and then added to the tape-measured distance (h_2) from the eye-height mark on the wall down to the ground to arrive at the overall building height. Similar techniques can be used to determine electrical conductor heights.

The clinometer was also useful when working on route surveys where long tapes (e.g., 300 ft or 100 m) were being used. The long tape can be held mostly supported on the ground and the slope angle can be taken for each tape length. By having the long tape mostly supported, the tension requirements can be reduced to a comfortable level. See Figure 3-7(c).

3.6.4 Additional Accessories

Range poles are 6-ft or 8-ft wood or steel poles with steel points. These poles are usually painted alternately red and white in 1-ft sections [see Figure 3-8(a)]. The range pole can be used to provide theodolite and total station sightings for angle and line work. These poles were also used in taping to help with alignment for distances longer than one tape length. The pole was set behind the measurement terminal point and the rear tape person could keep the forward tape person on line by simply sighting on the pole and then waving the forward tape person left or right until they are on line.

The clamp handle [see Figure 3-8(b)] helps the surveyor to grip the tape at any intermediate point without kinking the tape. The tension handle [see Figure 3-8(c)] is used in precise work to ensure that the correct tension is being applied. It is usually graduated to 30 lbs in ½-lb increments. With metric tapes, the relationship 50 N = 11.24 lbs is useful.

Chaining pins (marking arrows) come in a set of 11. They are painted alternately red and white and are 14–18 in. long. Chaining pins are used to mark intermediate points on the ground while making long measurements. The chaining pin is set sloping 45° to the ground and at right angles to the direction of measurement. Sloping the pin permits precise measurement with a plumb bob to the point where the pin enters the ground.

3.7 TAPING TECHNIQUES

The measurement begins with the head surveyor carrying the zero end of the tape forward toward the final point, until the tape has been unwound. At this point, the rear surveyor calls "tape" to

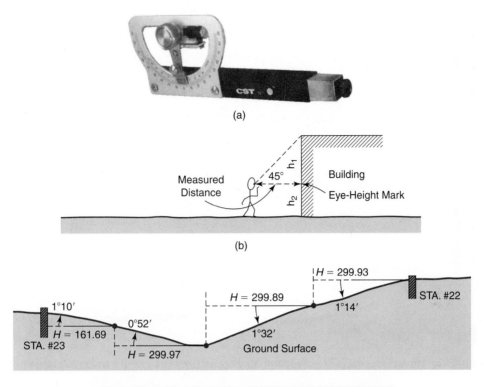

(a)

(b)

(c)

FIGURE 3-7 (a) Abney hand level; scale graduated in degrees with a vernier reading to 10 minutes. (Courtesy of CST/Berger, Illinois.) (b) Abney hand-level application in height determination. (c) Abney hand-level typical application in taping

Station	Course	Slope Angle Line 22–23	Horizontal Distance
From #22	300	−1°14′	299.93
	300	−1°32′	299.89
	300	+0°52′	299.97
To #23	161.72	+1°10′	161.69
		Line 22–23 =	1061.48

alert the head surveyor to stop walking and to prepare for measuring. If a drag tape is used, the tape is removed from the reel and a leather thong is attached to the reel end to facilitate measuring. If the tape is not designed to come off the reel, the winding handle is folded to the lock position and the reel is used to help hold the tape. The head surveyor is put on line by the rear surveyor, who is sighting forward to a range pole or other target that has been erected at the final mark. In very precise work, the intermediate marks can be aligned by theodolite. The rear surveyor holds the appropriate graduation (e.g., 100.00 ft 30.000 m), or other appropriate even-unit graduation, against the mark from which the measurement is being taken. The head surveyor, after ensuring that the tape is straight, slowly increases tension to the proper amount

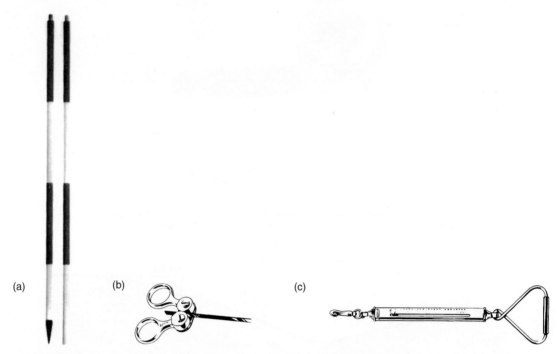

FIGURE 3-8 Taping accessories. (a) Two-section range pole. (b) Tape clamp handle. (c) Tension handle

and then marks the ground with a chaining pin or other marker. Once the mark has been made, both surveyors repeat the measuring procedure to check the measurement. If necessary, adjustments are made and the check procedure is repeated. See Figure 3-9.

If the ground is not level (determined by estimation or by the use of a hand level), one or both surveyors must use a plumb bob (see Figure 3-5). Normally, the only occasion when both surveyors have to use plumb bobs is when the ground rises or obstacles exist between the two surveyors

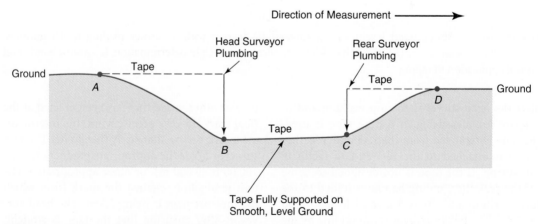

FIGURE 3-9 Horizontal taping; plumb bob used at one end

Both Surveyors Plumbing

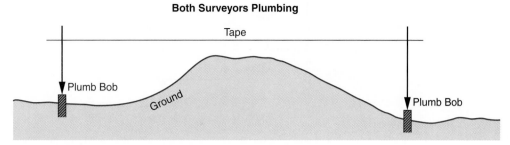

FIGURE 3-10 Horizontal taping; plumb bob used at both ends

(see Figure 3-10). Because of the additional random errors introduced when plumbing, plumb bobs are not used unless it is necessary. Figure 3-11 shows the relationship of the slope and the horizontal and vertical distances between any two points.

When plumbing, the tape is usually held at waist height, although any height between the shoulders and the ground is common. Holding the tape above shoulder height creates more chance for error because the surveyor must move his or her eyes up and down to check the ground mark and the tape graduation. The plumb bob string is usually held on the tape with the left thumb (in the case of right-handed people). Care must be taken not to cover the graduation mark completely. As the tension is increased, it is common for the surveyor to sometimes take up some of the tension with the left thumb, causing it to slide along the

tape. If the graduations have been covered with the left thumb, the surveyor is often not aware that the thumb (and string) has moved, resulting in an erroneous measurement. When plumbing, it is advisable to hold the tape close to the body and thus provide good leverage for applying or holding tension and to transfer the graduations accurately from tape to ground, and vice versa.

If the rear surveyor is using a plumb bob, he or she shouts out "tape," "mark," or some other signal that, at that instant, the plumb bob is steady and over the mark. If the head surveyor is also using a plumb bob, he or she must wait to take a reading until both plumb bobs are simultaneously over their respective marks.

Plumbing is a challenging aspect of taping. Students may encounter difficulty in simultaneously holding the plumb bob steady over the point and applying appropriate tension while also ensuring that the plumb bob string is precisely set at the correct tape graduation and then reading the tape graduation at the plumb bob string. To help steady the plumb bob, it is held only a short distance above the ground point and is repeatedly touched down. This momentary touching down will dampen the plumb bob oscillations and generally steady the plumb bob. The plumb bob should not be allowed to rest on the point because this will result in an erroneous measurement.

In practice, most measurements are taken with the tape held horizontally. If the slope is too great to allow an entire tape length to be employed, shorter increments are measured until all the required distance has been measured.

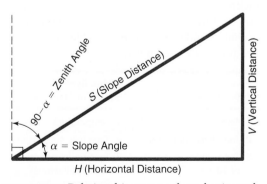

FIGURE 3-11 Relationship among slope, horizontal, and vertical distances

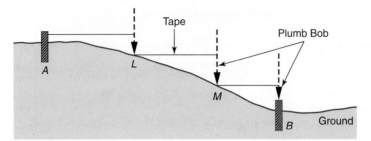

FIGURE 3-12 Breaking tape

This operation is known as *breaking tape* (see Figure 3-12). The sketch shows distance *AB*, composed of increments *AL*, *LM*, and *MB*. The exception to the foregoing occurs when preliminary route surveys (e.g., electricity transmission lines) are performed using a 300-ft (100-m) steel tape. It is customary to measure slope distances, which allows the surveyors to keep this relatively heavy tape more or less fully supported on the ground. To allow for reduction to the horizontal, each tape length is accompanied by its slope angle, usually determined by using a clinometer—also known as an Abney hand level; see Figure 3-7(a).

In summary, the rear surveyor handles the following tasks:

1. Visually aligns the head surveyor by sighting to a range pole or other target placed at the forward station.

2. Holds the tape on the mark, either directly or with the aid of a plumb bob. If a plumb bob is being used, the rear surveyor repeatedly calls out "tape," "mark," or a similar word to signify to the head surveyor that, for that instant in time, the plumb bob (tape mark) is precisely over the station.

3. Calls out the station and tape reading for each measurement and listens for verification from the head surveyor as the information is being entered in the field book or data collector.

4. Keeps a count of all full tape lengths included in each overall measurement.

5. Maintains the equipment (e.g., wipes the tape clean at the conclusion of the day's work or as conditions warrant).

The head surveyor is responsible for these tasks:

1. Carries the tape forward and ensures that the tape is free of loops, which could lead to tape breakage.

2. Prepares the ground surface for the mark (e.g., clears away grass, leaves, etc.).

3. Applies proper tension after first ensuring that the tape is straight.

4. Places marks (chaining pins, wood stakes, iron bars, nails, rivets, cut crosses, etc.).

5. Takes and records measurements of distances, temperature, and other factors.

3.8 STANDARD CONDITIONS FOR THE USE OF STEEL TAPES

Because steel tapes can give different measurements when used under various tension, support, and temperature conditions, it is necessary to provide standards for their use. Standard taping conditions are shown below:

Foot System, 100-ft Steel Tape	Metric System, 30.000-m Steel Tape
1. Temperature = 68°F	1. Temperature = 20°C
2. Tape fully supported	2. Tape fully supported
3. Tape under a tension of 10 lbs	3. Tape under a tension of 50 N (newtons) Because a 1 lb force = 4.448 N, 50 N = 11.24 lbs.

In the real world of field surveying, the above-noted standard conditions seldom occur at the same time. The temperature is usually something other than standard, and in many instances the tape cannot be fully supported (one end of the tape is often held off the ground to keep it horizontal). If the tape is not fully supported, the standard tension of 10 lbs does not apply. When standard conditions are not present, systematic errors will be introduced into the tape measurements. The following sections illustrate how these systematic errors and random errors are treated.

3.9 TAPING CORRECTIONS: GENERAL BACKGROUND

As noted in Chapter 1, no measurements can be performed perfectly, so all measurements (except for counting) must contain some errors. Surveyors must use measuring techniques that minimize random errors to acceptable levels, and they must make corrections to systematic errors that can affect the accuracy of the survey. Typical taping errors are summarized below:

Systematic Taping Errors*	Random Taping Errors†
1. Slope	1. Slope
2. Erroneous length	2. Temperature
3. Temperature	3. Tension and sag
4. Tension and sag	4. Alignment
	5. Marking and plumbing

*See Sections 3.10–3.14.

†See Sections 3.15 and 3.16.

3.10 SYSTEMATIC SLOPE CORRECTIONS

As noted in the previous section, taping is usually performed by keeping the tape horizontal. In some situations, however, distances are deliberately measured on a slope and then converted to their horizontal equivalents. To convert slope distances to horizontal distances, either the slope angle (α) or the vertical distance (difference in elevation) must also be known (see Figure 3-11).

$$\frac{H\,(\text{horizontal})}{S\,(\text{slope})} = \cos\alpha \quad \text{or} \quad H = S\cos\alpha \quad (3\text{-}1)$$

$$\frac{H}{S} = \sin(90 - \alpha) \quad \text{or} \quad H = S\sin(90 - \alpha) \quad (3\text{-}1a)$$

[where $(90 - \alpha)$ is the zenith angle]

When the vertical distance or difference in elevation is known, the expression becomes:

$$H^2 = S^2 - V^2 \quad \text{or} \quad H = \sqrt{S^2 - V^2} \quad (3\text{-}2)$$

Slope can also be expressed as a **gradient**, or rate of grade. The gradient is expressed as a ratio of the vertical distance over the horizontal distance, the same ratio (opposite/adjacent) which defines the tan of the vertical angle θ, shown in the unnumbered figures accompanying Example 3-5. When this ratio is multiplied by 100, it is called a percentage gradient. For example, if the ground rises 1 ft (m) in 100 ft (m), it is said to have a +1% slope (i.e., 1/100 × 100 = 1). If the ground falls 2.6 ft (m) in 195.00 ft (m), it is said to have a slope of −1.33%. If the elevation of a point on a gradient is known, the elevation of any other point on the gradient can be calculated. See Example 3-3.

Example 3-3: A road center line ℄ gradient falls from station 0 + 00, elevation = 564.22 ft, to station 1 + 50 at a rate of −2.5%. What is the ℄ elevation at station 1 + 50?

Solution:

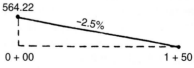

Difference in elevation = 150 (2.5/100) = −3.75

Elevation at 1 + 50 = 564.22 − 3.75 = 560.47 ft

If the elevations of at least two points on a grade line are known, as well as the distances between them, the slope gradient can be determined as shown in Example 3-4.

Example 3-4: A road runs from station 1 + 00, elevation = 471.37 ft, to station 4+37.25, elevation = 476.77 ft. What is the slope of the ℄ grade line?

Solution:

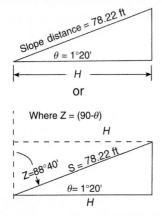

Elevation difference = +5.40

Distance = 337.25

Gradient = (+5.40/337.25)100 = +1.60%

Example 3-5: *Slope Corrections*

a. The slope distance (S) and slope angle (α), or the zenith angle (90°−α), are given. A slope distance between two points is 78.22 ft and the slope angle is + 1° 20′ (the equivalent zenith angle is 88° 40′). What is the corresponding horizontal distance?

or

Where Z = (90-θ)

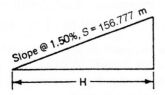

b. The slope distance (S) and the gradient (slope percentage) are given. A line has a slope distance of 156.777 m from one point to another point at a slope rate of +1.5%. What is the corresponding horizontal distance (H) between the points?

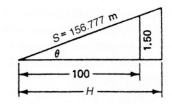

c. The slope distance between two points is measured to be 199.908 m and the vertical distance between the points (i.e., the difference in elevation) is +2.435 m. What is the horizontal distance (H) between the points?

d. The slope distance between two points is found to be 83.52 ft, and the vertical distance is found to be 3.1 ft. What is the horizontal distance between the two points?

Solutions:

a. $H = S\cos\alpha$ $H/S = \sin Z$

$H = 78.22\cos 1°20'$ $H = S\sin Z$

$H = 78.20\,\text{ft}$ $H = 78.22\sin 88°40'$

$H/S = \cos\alpha$ $H = 78.20\,\text{ft}$

b. $1.50/100 = \tan\alpha$

$\alpha = 0.85937°$

$H/156.777 = \cos 0.85937°$

$H = 156.777\cos 0.85937°$

$H = 156.759\,\text{m}$

$H^2 = S^2 = V^2$

$H = \sqrt{199.908^2 - 2.435^2}$

$H = 199.893\,\text{m}$

$H = \sqrt{83.52^2 - 3.1^2}$

$H = 81.46$

$H = 81.5\,\text{ft}$

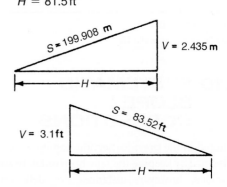

Usually, the solution is rounded to the same number of decimals shown in the least precise measurement.

3.11 ERRONEOUS TAPE LENGTH CORRECTIONS

For all but precise work, tapes supplied by the manufacturer are considered to be correct under standard conditions. Through extensive use, tapes do become kinked and stretched, and in need of repair. The length can become something other than that specified. When this occurs, the tape must be corrected, or the measurements taken with the erroneous tape must be corrected.

Example 3-6: A measurement was recorded as 171.278 m with a 30-m tape that was only 29.996 m under standard conditions. What is the corrected measurement?

Solution:

Correction per tape length $= -0.004$

Number of times the tape was used $= 171.278/30$

Total correction $= -0.004 \times 171.278/30 = -0.023$ m

Corrected distance $= 171.278 - 0.023 = 171.255$ m

or

$= 29.996/30 \times 171.278 = 171.255$ m

Example 3-7: You must lay out the side of a building, a distance of 210.08 ft. The tape to be used is known to be 100.02 ft under standard conditions.

Solution:

Correction per tape length $= 0.02$ ft

Number of times that the tape is to be used $= 2.1008$

Total correction $= 0.02 \times 2.1008 = +0.04$ ft

When the problem involves laying out a distance, the sign of the correction must be reversed before being applied to the layout measurement. We must find the distance that, when corrected by $+0.04$, will give 210.08 ft, that is, $210.08 - 0.04 = 210.04$ ft. This is the distance to be laid out with that tape (100.02 ft) so that the corner points will be exactly 210.08 ft apart.

Four variations of this problem are possible: correcting a measured distance while using (1) a long tape or (2) a short tape, or precorrecting a layout distance using (3) a long tape or (4) a short tape. To minimize confusion about the algebraic sign of the correction, it helps to consider the problem with the distance reduced to only one tape length (100 ft or 30 m).

In Example 3-6, a recorded distance of 171.278 m was measured with a tape only 29.996 m long. The total correction was found to be 0.023 m. If doubt exists about the sign of 0.023, ask yourself what the procedure would be for correcting only one tape length. In Example 3-6, after one tape length had been measured, it would have been recorded that 30 m had been measured. If the tape were only 29.996 m long, then the field book entry of 30 m must be corrected by -0.004 m. The magnitude of the tape error is determined by comparing the tape with a tape that has been certified (by the National Bureau of Standards, Gaithersburg, Maryland, or the National Research Council, Ottawa, Ontario, Canada). In practice, tapes that require corrections for ordinary work are either repaired or discarded.

3.12 TEMPERATURE CORRECTIONS

Section 3.6 notes the conditions under which tape manufacturers specify the accuracy of their tapes. One of these standard conditions is that of temperature. In the United States and Canada, tapes are standardized at 68°F, or 20°C. Temperatures other than standard result in an erroneous tape length.

The thermal coefficient of the expansion of steel (k) is 0.00000645 per unit length per degree

Fahrenheit (°F), or 0.0000116 per unit length per degree Celsius (°C). The general formula is:

$$C_t = k(T - T)L$$

For foot units, the formula is:

$$C_t = 0.00000645(T - 68)L \qquad (3\text{-}3)$$

where C_t = correction due to temperature, in feet
T = temperature of tape (°F) during measurement
L = distance measured, in feet

For metric units, the formula is:

$$C_t = 0.0000116(T - 20)L \qquad (3\text{-}4)$$

where C_t = correction due to temperature, in meters
T = temperature of tape (°C) during measurement
L = distance measured, in meters

Example 3-8:   A distance was recorded as being 471.37 ft at a temperature of 38°F. What is the distance when corrected for temperature?

Solution:

$C_t = 0.00000645(38 - 68)471.37 = -0.09$
Corrected distance = 471.37 − 0.09 = 471.28 ft

Example 3-9:   You must lay out two points in the field that will be exactly 100.000 m apart. Field conditions indicate that the temperature of the tape is 27°C. What distance will be laid out?

Solution:

$C_t = 0.0000116(27 - 20)100.000 = +0.008$ m

Because this is a layout (precorrection) problem, the correction sign must be reversed. In other words, we are looking for the distance that, when corrected by +0.008, will give us 100.000 m: Layout distance is 100.000 − 0.008 = 99.992 m.

For most survey work, accuracy requirements do not demand high precision in determining the actual temperature of the tape. Usually, it is sufficient to estimate air temperature. However, for more precise work (say, 1:15,000 and higher), care is required in determining the actual temperature of the tape, which can be significantly different than the temperature of the air.

3.13 TENSION AND SAG CORRECTIONS

The three conditions under which tapes are normally standardized are given in Section 3.8. If a tension other than standard is applied, a tension (pull) error exists. The tension correction formula is:

$$C_P = \frac{(P - P_s)L}{AE} \qquad (3\text{-}5)$$

where the variables are defined in Table 3-1.

If a tape has been standardized while fully supported and is being used without full support, an error called **sag** occurs. The force of gravity pulls the center of the unsupported section downward in the shape of a catenary, thus creating an error $B'B$.

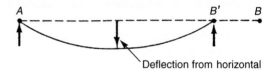

Deflection from horizontal

The sag correction formula is:

$$C_s = \frac{-w^2 L^3}{24P^2} = \frac{-W^2 L}{24P^2} \qquad (3\text{-}6)$$

where the variables are defined in Table 3-1.

3.13.1 Examples of Tension and Sag Corrections

Example 3-10:   A 100-ft tape is used with a 20-lb force pull, instead of the standard tension of 10 lb. If the cross-sectional area of the tape is 0.003 in., what is the tension error for each tape length used?

Solution:

$$C_p = \frac{(20 - 10)100}{29,000,000 \times 0.003} = +0.011 \text{ ft}$$

If a distance of 421.22 ft had been recorded, the total correction would be:

$$4.2122 \times 0.011 = +0.05 \text{ ft}$$

The corrected distance would be 421.27 ft.

Table 3-1 Correction Formula Terms Defined [Foot, Metric, and Metric (SI) Units]

Unit	Description	Foot	Metric (old)	Metric (SI)
C_P	Correction due to tension per tape length	ft	m	m
C_s	Sag correction per tape length	ft	m	m
L	Length of tape under consideration	ft	m	m
P_s	Standard tension	lb (force)	kg (force)	N (newtons)
	Typical standard tension	10 lb (f)	4.5–5 kg (f)	50 N
P	Applied tension	lb (f)	kg (f)	N
A	Cross-sectional area	in.^{2}	cm^2	m^2
E	Average modulus of elasticity of steel tapes	29×10^6 lb (f)/in.^{2}	21×10^5 kg (f)/cm^2	20×10^{10} N/m^2
	Average modulus of elasticity of invar tapes	21×10^6 lb (f)/in.^{2}	14.8×10^5 kg (f)/cm^2	14.5×10^{10} N/m^2
w	Weight of tape per unit length	lb (f)/ft	kg (f)/m	N/m
W	Weight of tape	lb (f)	kg (f)	N

Example 3-11: A 30-m tape is used with a 100-N force, instead of the standard tension of 50 N. If the cross-sectional area of the tape is 0.02 cm^2, what is the tension error per tape length?

Solution:

$$C_p = \frac{(100 - 50)\,30}{0.02 \times 21 \times 10^5 \times 9.807} = +0.0036 \text{ m}$$

If a distance of 182.716 m had been measured under these conditions, the total correction would be:

$$\text{Total } C_p = \frac{182.716}{30} \times 0.0036 = +0.022 \text{ m}$$

The corrected distance would be 182.738 m.

Example 3-12: (See formula 3-6)
A 100-ft steel tape weighs 1.6 lb and is supported only at the ends with a tension of 10 lb. What is the sag correction?

Solution:

$$C_s = \frac{-1.6^2 \times 100}{24 \times 10^2} = -0.11 \text{ ft}$$

If the tension were increased to 20 lb, the sag is reduced to:

$$C_s = \frac{-1.6^2 \times 100}{24 \times 20^2} = -0.03 \text{ ft}$$

Example 3-13: Calculate the length between two supports if the recorded length is 50.000 m, the mass of the tape is 1.63 kg, and the applied tension is 100 N.

Solution:

$$C_s = \frac{-(1.63 \times 9.807)^2 \times 50.000}{24 \times 100^2} = -0.053 \text{ m}$$

Therefore, the length between supports = 50.000 − 0.053 = 49.947 m.

3.13.2 Normal Tension

The error in a measurement due to sag can be eliminated by increasing the tension. Tension that eliminates sag errors is known as **normal tension** (P_n). It ranges from 19 lb (light 100-ft tapes) to 31 lb (heavy 100-ft tapes).

Normal tension can be determined experimentally for individual tapes. The most popular steel tapes (100 ft) require a normal tension of about 24 lb. For most 30-m tapes (lightweight), a normal tension of 90 N (20 lb or 9.1 kg) is appropriate. To determine the normal tension for a 100-ft steel tape:

1. Lay out the tape on a flat, horizontal surface; the floor of an indoor corridor is ideal.

2. Select (or mark) a well-defined point on the surface at which the 100-ft mark is held.

3. Attach a tension handle at the zero end of the tape; apply standard tension, say 10 lb, and mark the surface at 0.00 ft.

4. Repeat the process, switching personnel duties to ensure that the two marks are in fact exactly 100.00 ft apart.

5. Raise the tape off the surface to a comfortable height (waist). The surveyor at the 100-ft end holds a plumb bob over the point. At the same time, the surveyor(s) at the zero end slowly increases tension (a third surveyor could perform this function) until the plumb bob is over the zero mark on the surface. The normal tension is then read off the tension handle.

This process is repeated several times until a set of consistent results is obtained.

3.14 RANDOM ERRORS ASSOCIATED WITH SYSTEMATIC TAPING ERRORS

As mentioned in Section 3.9, random errors can coexist with systematic errors. For example, when dealing with the systematic error caused by variations in temperature, the surveyor can determine the prevailing temperature in several ways:

1. The air temperature can be estimated.

2. The air temperature can be taken from a pocket thermometer.

3. The actual temperature of the tape can be determined by a tape thermometer held in contact with the tape.

For an error of 15°F in temperature, the error in a 100-ft tape would be:

$$C_t = 0.00000645(15)100 = 0.01 \text{ ft}$$

Because $0.01/100 = 1/10,000$, even an error of 15°F would be significant only for higher-order

surveys. If metric equipment were being used, a comparable error would be 8.5°C; that is:

$$C_t = 0.0000116(8.5)30 = 0.003 \text{ m}$$

For precise work, however, random errors in determining temperature are significant. Tape thermometers are recommended for precise work because of difficulties in estimating and because of the large differentials possible between air temperature and the actual temperature of the tape on the ground.

Consider the treatment of systematic errors dealing with slope versus horizontal dimensions. For a slope angle of 2°40′ read with an Abney hand level (clinometer) to the closest 10 minutes, we can say that an uncertainty of 5 minutes exists:

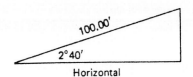

Horizontal distance $= 100 \cos 2°40′ = 99.89 \text{ ft}$

In this case, an uncertainty of 5 minutes in the slope angle introduces an uncertainty of only 0.01 ft in the answer.

$$\begin{aligned} \text{Horizontal distance} &= 100 \cos 2°45′ = 99.88 \text{ ft} \\ &= 100 \cos 2°35′ = 99.90 \text{ ft} \end{aligned}$$

Once again, this error (1/10,000) would be significant only for higher-order surveys.

3.15 RANDOM TAPING ERRORS

In addition to the systematic and random errors already discussed, random errors associated directly with the skill and care of the surveyors sometimes occur. These errors result from the inability of the surveyor to work to perfection in the areas of alignment, plumbing, and marking, and when estimating horizontal.

Alignment errors occur when the tape is inadvertently aligned off the true path (see Figure 3-13).

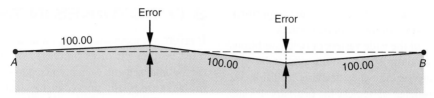

FIGURE 3-13 Alignment errors

Under ordinary surveying conditions, the rear surveyor can keep the head surveyor on line by sighting a range pole marking the terminal point. It would take an alignment error of about 1.5 ft to produce an error of 0.01 ft in 100 ft. It is not difficult to keep the tape aligned by eye to within a few tenths of a foot (0.2–0.3 ft), so alignment is not usually a major concern. It should be noted that although most random errors are compensating, alignment errors are cumulative. Misalignment can occur randomly on the left or on the right, but in both cases the result of the misalignment is to make the measured course too long. Alignment errors can be nearly eliminated on precise surveys by using a theodolite to align all intermediate points.

Marking and plumbing errors are the most significant of all random taping errors. Even experienced surveyors must exercise great care to place a plumbed mark accurately to within 0.02 ft of true value over a distance of 100 ft. Horizontal measurements taken with the tape fully supported on the ground can be determined more accurately than measurements taken on a slope requiring the use of plumb bobs. Rugged terrain conditions that require many breaks in the taping process will cause these errors to multiply significantly.

Errors are also introduced when surveyors *estimate the tape height* for a plumbed measurement. The effect of this error is identical to that of the alignment error previously discussed, although the magnitude of these errors is often larger than that of alignment errors. Skilled surveyors can usually estimate a height to within 1 ft (0.3 m) over a distance of 100 ft (30 m). However, even experienced surveyors can be seriously in error when measuring across side hills, where one's perspective with respect to the horizon can be seriously distorted. These errors can be nearly eliminated by using a hand level, or by moving the tape up/down along plumb line to check for the minimum distance reading.

3.16 TECHNIQUES FOR ORDINARY TAPING PRECISION

Ordinary taping precision is referred to as precision that can result in 1:5,000 accuracy. The techniques used for ordinary taping, once mastered, can easily be maintained. It is possible to achieve an accuracy level of 1:5,000 with little more effort than is required to attain the 1:3,000 level. Because the bulk of all engineering surveying need only be at either the 1:3,000 or 1:5,000 level, experienced surveyors will often use 1:5,000 techniques even for the 1:3,000 level. This practice permits good measuring work habits to be reinforced continually without appreciably increasing survey costs.

Modern electronic surveying equipment can routinely attain levels of precision and accuracy well beyond the 1:3,000 and 1:5,000 levels. It is advisable, however, to keep in mind the project needs in small-area surveys, and to consider how much precision and accuracy are really needed for such assignments as mapping projects, road construction, resource surveys, and structural surveys.

Greater precision and accuracy can be achieved with high-precision equipment, but if high-precision equipment (e.g., precise GPS equipment) is rented at high cost just to perform drainage surveys, money will be wasted.

Because of the wide variety of field conditions that exist, absolute specifications cannot be prescribed. However, the specifications in Table 3-2 can be considered as typical for ordinary 1:5,000 taping. To determine the total random error (Σe) in one tape length, we take the square root of the sum of the squares of the individual maximum errors (see Appendix A):

FOOT	METRIC
0.005^2	0.0014^2
0.006^2	0.0018^2
0.005^2	0.0015^2
0.001^2	0.0004^2
0.015^2	0.0046^2
$\underline{0.005^2}$	$\underline{0\ .0015^2}$
0.000337	0.000031

$$\Sigma e = \sqrt{0.000337} = 0.018\,\text{ft}$$
$$\text{or}$$
$$\sqrt{0.000031} = 0.0056\,\text{m}$$

$$\text{Accuracy} = \frac{0.018}{100} = \frac{1}{5,400} \quad \text{or} \quad \frac{0.0056}{30} = \frac{1}{5,400}$$

Typical errors shown in Table 3.2

In this calculation, it is understood that corrections due to systematic errors have already been applied.

3.17 MISTAKES IN TAPING

If errors are associated with inexactness, mistakes must be thought of as blunders. Whereas errors can be analyzed and even predicted to some degree, mistakes are totally unpredictable. Just one undetected mistake can nullify the results of an entire survey, so it is essential to perform the work so that the opportunity for mistakes to develop can be minimized and verification of the results can occur.

The opportunities for the occurrence of mistakes are minimized by setting up and then rigorously following a standard method of performing the measurement. The more standardized and routine the measurements, the more likely it is that the surveyor will immediately spot a mistake. The immediate double-checking of all measurement manipulations reduces the opportunities for mistakes to go undetected and at the same time increases the precision of the measurement. In addition to the immediate checking of all measurements, the surveyor looks constantly for independent methods of verifying the survey results. Gross mistakes can often be detected by comparing the survey results with distances scaled from existing plans. The simple check technique of pacing can be an invaluable tool for rough verification of measured distances, especially construction layout distances. The possibilities for verification are limited only by the surveyor's diligence and resourcefulness.

Table 3-2   Specification for 1:5,000 Accuracy (Systematic Errors Removed)

Source of Error	Maximum Effect on One Tape Length	
	100 ft	30 m
Temperature estimated to closest 7°F(4°C)	±0.005 ft	±0.0014 m
Care is taken to apply at least normal tension (lightweight tapes) and tension is known within 5 lb (20 N)	±0.006 ft	±0.0018 m
Slope errors are no larger than 1 ft/100 ft or 0.30 m/30 m	±0.005 ft	±0.0015 m
Alignment errors are no larger than 0.5 ft/100 ft or 0.15 m/30 m	±0.001 ft	±0.0004 m
Plumbing and marking errors are at a maximum of 0.015 ft/100 ft or 0.0046 m/30 m	±0.015 ft	±0.0046 m
Length of tape is known within ±0.005 ft (0.0015 m)	±0.005 ft	±0.0015 m

Common mistakes in taping include:

1. Measuring to or from the wrong marker. All members of the survey crew must be vigilant in ensuring that measurements begin or end at the appropriate permanent or temporary marker. Markers include legal steel bars, construction stakes or steel bars, nails, and the like.

2. Reading the tape incorrectly. Mistakes are sometimes made by reading. Transposing figures is a common mistake (e.g., reading or writing 56 instead of 65).

3. Losing proper count of the full tape lengths involved in a measurement. The counting of full tape lengths is primarily the responsibility of the rear surveyor and can be as simple as counting the chaining pins that have been collected as the work progresses. If the head surveyor is also keeping track of full tape lengths, mistakes such as failing to pick up all chaining pins can be spotted and corrected easily.

4. Recording the values in the notes incorrectly. The notekeeper sometimes hears the rear surveyor's call-out correctly but then transposes the figures as they are being entered in the notes. This mistake can be eliminated if the notekeeper calls out the value as it is recorded. The rear surveyor listens for this call-out to ensure that the values called out are the same as the data originally given.

5. Calling out values ambiguously. The rear surveyor can call out 20.27 as twenty (pause) two, seven. This might be interpreted as 22.7. To avoid mistakes, this value should be called out as twenty, decimal (point), two, seven.

6. Not identifying the zero point of the tape correctly when using fiberglass or steel tapes. This mistake can be avoided if the surveyor checks unfamiliar tapes before use. The tape itself can be used to verify the zero mark.

7. Making arithmetic mistakes in sums of dimensions and in error corrections (e.g., for temperature and slope). These mistakes

can be identified and corrected if each crew member is responsible for checking (and signing) all survey notes.

3.18 ELECTRONIC DISTANCE MEASUREMENT

EDM, first introduced in the 1950s, has undergone continual refinement. The early instruments, which were capable of very precise measurements over long distances, were large, heavy, complicated, and expensive. Rapid advances in related technologies have provided lighter, simpler, and less expensive instruments. Early EDM instruments were manufactured for use alone in tribrachs [the lower part of survey instruments which contain the leveling screws and the optical plummet; see Figure 3-14(e)], or mounted on theodolites [see Figure 3-14(a)], EDMs are incorporated into total stations coaxially with the instruments' optical system during the manufacturing process (see Figure 3-15 and Chapter 5).

Technological advances in electronics continue at a rapid rate, as evidenced by market surveys indicating that many new electronic instruments have been on the market for only a few years. Current EDM instruments use infrared light or laser light. Microwave systems (now largely outdated) use a receiver/transmitter at both ends of the measured line, whereas infrared and laser systems utilize a transmitter at one end of the measured line and a reflecting prism at the other end. Some laser EDM instruments measure short distances (100–350 m) without a reflecting prism (note that some newer models can measure to 2,000 m without using reflecting prisms). They reflect the light directly off the feature (e.g., building wall) being measured. Microwave instruments are sometimes still used in hydrographic surveys and usually have an upper measuring range of 50 km. Hydrographic EDM and positioning techniques have largely been replaced, in a few short years, by global positioning system (GPS) techniques (see Chapter 7).

(a)

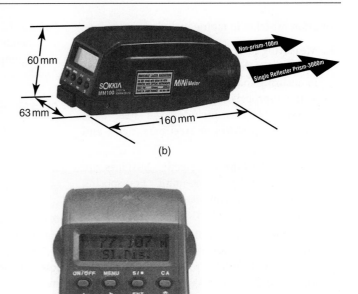

60 mm

63 mm

160 mm

Non-prism-100m

Single Reflector Prism-3000m

(b)

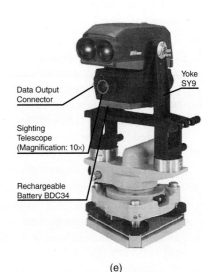

(c)

Mounting on the DT4F is by convenient hotshoe

Connection via which power supply and data communication are provided.

(d)

Yoke SY9

Data Output Connector

Sighting Telescope (Magnification: 10×)

Rechargeable Battery BDC34

(e)

FIGURE 3-14 Sokkia MiNi Meter MM100 laser add-on EDM. (a) EDM shown mounted on the telescope of Sokkia DT4F electronic 50″ theodolite. (b) MiNi Meter dimensions. (c) Display screen showing menu button, which provides access to programs permitting input for atmospheric corrections, height measurements, horizontal and vertical distances, self-diagnostics, etc. (d) Illustration showing "hot shoe" electronics connection and counterweight. (e) MiNi Meter shown mounted directly into a tribrach with attached sighting telescope, data output connector, and battery

(Courtesy of Sokkia Corporation, Overland Park, KS.)

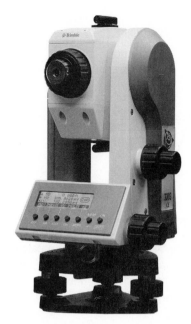

FIGURE 3-15 Trimble total stations. The Trimble 3303 and 3305 total stations incorporate DR (direct reflex) technology (no prism required) and include the choice of the integrated Zeiss Elta control unit, detachable Geodimeter control unit or handheld TSCe data collector, and a wide range of software options

(Courtesy of Trimble Geomatics & Engineering Division, Dayton, OH.)

3.19 ELECTRONIC ANGLE MEASUREMENT

The electronic digital theodolite, first introduced in the late 1960s, helped set the stage for modern field data collection and processing. (See Figure 5-1 for a typical electronic theodolite.) When the electronic theodolite is used with a built-in EDM (see the Zeiss Elta in Figure 3-15) or an add-on and interfaced EDM (the Wild T-1000 in Figure 3-16), the surveyor has a very powerful instrument. The addition of an onboard microprocessor that automatically monitors the instrument's operating status and manages built-in surveying programs, and a data collector (built-in or interfaced) that stores and processes measurements and attribute data, provides what is known as a **total station**. Total stations are discussed next in Chapter 5.

FIGURE 3-16 Wild T-1000 electronic theodolite, shown with D1 1000 Distomat EDM and the GRE 3 data collector

(Courtesy of Leica Geosystems Inc.)

Infrared and laser EDM instruments come in long range (10–20 km), medium range (3–10 km), and short range (0.5–3 km). EDM instruments can be mounted on the standards or the telescope of most theodolites. They can also be mounted directly in a tribrach (see Figure 3-14). When used with an electronic theodolite, the combined instruments can provide both the horizontal and the vertical position of one point relative to another. The slope distance provided by an add-on EDM instrument can be reduced to its horizontal and vertical equivalents by utilizing the slope angle provided by the theodolite. In total station instruments, this reduction is accomplished automatically.

3.20 PRINCIPLES OF ELECTRONIC DISTANCE MEASUREMENT (EDM)

Figure 3-17 shows a wave of wavelength λ. The wave is traveling along the x-axis with a velocity of $299,792.5 \pm 0.4\,\text{km/s}$(in vacuum). The frequency of the wave is the time taken for one complete wavelength:

$$\lambda = \frac{c}{f} \qquad (3\text{-}7)$$

where λ = wavelength, in meters
c = velocity, in m/s
f = frequency, in hertz (one cycle per second)

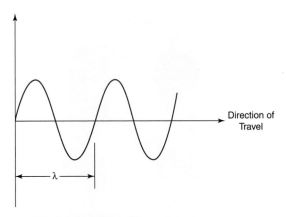

FIGURE 3-17 Light wave

Figure 3-18 shows the modulated electromagnetic wave leaving the EDM instrument and being reflected (light waves) or retransmitted (microwaves) back to the EDM instrument.

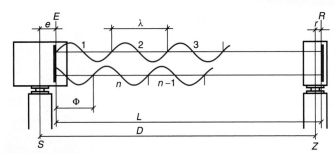

S   Station
Z   Target
E   Reference plane within the distance meter for phase comparison between transmitted and received wave
R   Reference plane for the reflection of the wave transmitted by the distance meter
a   Addition constant
e   Distance meter component of addition constant
r   Reflector component of addition constant
λ Modulation wavelength
Φ Fraction to be measured of a whole wavelength of modulation ($\Delta \lambda$)
L   The distance between the reference planes of the instrument and reflecting prism
D   The distance between the instrument station and the prism station, i.e., the distance L corrected for the off-center constants at the instrument and at the prism

The addition constant *a* applies to a measuring equipment consisting of a distance meter and reflector. The components *e* and *r* are only auxiliary quantities.

FIGURE 3-18 Principles of EDM measurement

(Courtesy of Kern Instruments—Leica Geosystems.)

The double distance (2L) is equal to a whole number of wavelengths (nλ) plus the partial wavelength (φ) occurring at the EDM instrument:

$$L = \frac{n\lambda + \varphi}{2} \qquad (3\text{-}8)$$

The partial wavelength (φ) is determined in the instrument by noting the phase delay required to match precisely the transmitted and the reflected or retransmitted waves. The instrument can send out three or four waves at different frequencies/wavelengths (modulation), and then by substituting the resulting values of (λ) and (φ) into Equation (3-8) for the three or four different wavelengths, the value of n can be found. The instruments are designed to carry out this procedure in a matter of seconds and then to display the value of L in digital form. Other EDM instruments use pulsed laser emissions, which require those instruments to determine the distance by measuring the time it takes for the transmitted pulsed signal to be returned to the instrument (see Section 3.24).

The velocity of light through the atmosphere can be affected by (1) temperature, (2) atmospheric pressure, and (3) water vapor content. In practice, the corrections for temperature and pressure can be determined manually by consulting nomographs similar to that shown in Figure 3-19, or the corrections can be performed automatically on some EDM instruments by their onboard processor/calculator after the values for temperature and pressure have been internally sensed or entered by the operator.

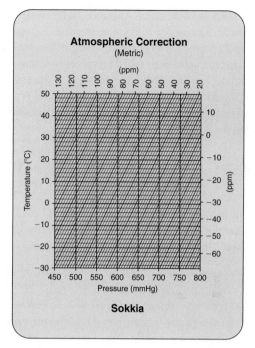

FIGURE 3-19 Atmospheric correction graph (Courtesy of Sokkia Co. Ltd.)

For short distances using light-wave EDM instruments, atmospheric corrections are relatively insignificant. For long distances using light-wave instruments and especially microwave instruments, atmospheric corrections can become quite important. Table 3-3 shows the comparative effects of atmospheric factors on both light waves and microwaves.

At this point, it is also worth noting that several studies of general EDM use show that more

Table 3-3   Atmospheric Effects on Light Waves and Microwaves

		Error (Parts per Million)	
Parameter	Error	Light Wave	Microwave
T, temperature	+1°C	−1.0	−1.25
P, pressure	+1 mm Hg	+0.4	+0.4
E, partial water, vapor pressure	1 mm Hg	−0.05	+7 at 20°C
			+17 at 45°C

than 90 percent of all distance determinations involve distances of 1,000 m or less and that more than 95 percent of all layout measurements involve distances of 400 m or less. The values in Table 3-3 seem to indicate that, for the type of measurements normally encountered, instrumental errors and centering errors hold much more significance than do the atmosphere-related errors.

3.21 EDM INSTRUMENT CHARACTERISTICS

Following are the characteristics of recent models. Generally the more expensive instruments have longer distance ranges and higher precision.

Distance range: 1,000 m to 3,000 km (single prism with average atmospheric conditions).

Accuracy range: $\pm(15\,mm + 5\,ppm \times D)$ m.s.e. (mean square error) for short-range inexpensive EDM instruments, up to $\pm(2\,mm + 2\,ppm \times D)$ m.s.e. for precise instruments.

Working temperature range: −20°C (−4°F) to +50°C (+122°F).

(Note: North of latitude 44°, winter temperatures often fall below −20°C.)

Non-prism measurements: available on some models; distances from 100 to 350 m (see Section 3.24).

3.22 PRISMS

Prisms are used with EDM instruments to reflect the transmitted signal (see Figure 3-20). A single reflector is a cube corner prism that has the characteristic of reflecting light rays precisely back to the emitting EDM instrument. This retrodirect capability means that the prism can be somewhat misaligned with respect to the EDM instrument and still be effective. A cube corner prism is formed by cutting the corners off a solid glass cube. The quality of the prism is determined by the flatness of the surfaces and the perpendicularity of the 90° surfaces.

FIGURE 3-20 Various target and reflector systems in tribrach mounts

(Courtesy of Topcon Positioning Systems, Inc., Pleasanton, CA.)

Prisms can be tribrach-mounted on a tripod or bipod-mounted, centered by optical or laser plummet, or attached to a prism pole held vertical on a point with the aid of a bull's-eye level; however, prisms must be tribrach-mounted or bipod-mounted if a higher level of accuracy is required. In older control surveys, tribrach-mounted prisms can be detached from their tribrachs and then interchanged with a theodolite (and EDM instrument) similarly mounted at the other end of the line being measured. This interchangeability of prism and theodolite (also targets) speeds up the work because the tribrach mounted on the tripod is centered and leveled only once. Equipment that can be interchanged and mounted on tribrachs already set up is known as **forced centering equipment**.

Prisms mounted on adjustable-length prism poles are very portable and, as such, are particularly well suited for stakeout surveys and topographic surveys. Figure 3-21 shows a prism pole (supported by a bipod) being centered and set vertical over a survey point. It is important that prisms mounted on poles or tribrachs be permitted to tilt up and down so that they can be made

FIGURE 3-21 Prism pole (supported by a bipod) being set vertical over the survey point

(Courtesy of SECO Manufacturing Co. Inc., Redding, CA. Photo by Eli Williem.)

perpendicular to EDM signals that are being sent from much higher or lower positions; this characteristic is particularly important for short sights.

3.23 EDM INSTRUMENT ACCURACIES

EDM instrument accuracies shown in Section 3.21 are stated in terms of a constant instrumental error and a measuring error proportional to the distance being measured. Typically, accuracy is claimed as $\pm[5$ mm $+ 5$ parts per million (ppm)] or $\pm(0.02$ ft $+ 5$ ppm). The ± 5 mm (0.02 ft) error is the instrument error that is independent of the length of the measurement, whereas the 5 ppm (5 mm/km) error denotes the distance-related error.

The proportional part error (ppm) is insignificant for most work, and the constant part of the error assumes less significance as the distance being measured lengthens. At 100 m, an error of ± 5 mm represents 1/20,000 accuracy, whereas for 1,000 m, the same instrumental error represents 1/200,000 accuracy.

When dealing with accuracy, both the EDM instrument and the prism reflectors must be corrected for off-center characteristics. The measurement being recorded goes from the electrical center of the EDM instrument to the back of the prism (allowing for refraction through glass) and then back to the electrical center of the EDM instrument. The difference between the electrical center of the EDM instrument and the plumb line through the tribrach center is compensated for by the manufacturer at the factory. The prism constant (30–40 mm) is eliminated either by the EDM instrument manufacturer at the factory or in the field.

The instrument/prism constant value can be field checked in the following manner. A long line (>1 km) is laid out with end stations and an intermediate station (see Figure 3-22). The overall distance AC is measured, along with partial lengths AB and BC. The constant value will be present in all measurements; therefore:

$$AC - AB - BC = \text{instrument/} \\ \text{prism constant} \quad (3\text{-}9)$$

The constant can also be determined by measuring a known baseline, if one can be conveniently accessed.

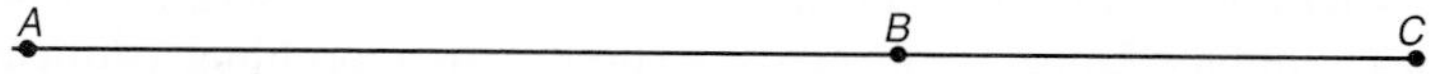

A ————————— B ——— C

FIGURE 3-22 Method of determining the instrument-reflector constant

3.24 EDM INSTRUMENTS WITHOUT REFLECTING PRISMS

Some EDM instruments (see Figure 3-23) can measure distances without using reflecting prisms—the measuring surface itself is used as a reflector. Both phase-shift technology (infrared) and time-of-flight technology (pulsed lasers) are used for reflectorless measurement. When the reflecting surface is uneven or is not at right angles to the measuring beam, varying amounts of the light pulses are not returned to the instrument. As many as 20,000 pulses per second are employed to ensure that sufficient data is received. Another consideration is that various surfaces have different reflective properties; for example, a bright white surface at right angles to the measuring beam may reflect almost 100 percent of the light, but most natural surfaces reflect light at a rate of only 18 percent.

EDM instruments can be used conventionally with reflecting prisms for distances up to 4 km.

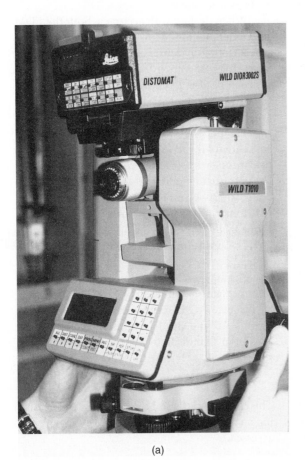

(a)

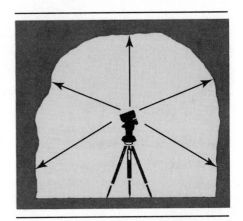

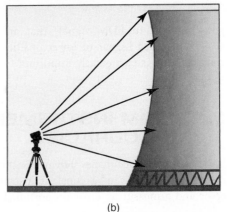

(b)

FIGURE 3-23 Distance measurement without reflectors. (a) Wild T1010 Electronic theodolite, together with an interfaced DIOR 3002S prismless EDM (angle accuracy is 3 seconds). (b) Illustrations of two possible uses for this technique. Upper: tunnel cross sections. Lower: profiling a difficult-access feature (Courtesy of Leica Geosystems Inc.)

When used without prisms, the range drops to 100–1,200 m, depending on the equipment, the light conditions (cloudy days and night darkness provide better measuring distances), the angle of reflection from the surface, and the reflective properties of the measuring surface. (Instrument manufacturer Topcon's website shows reflectorless distance measurement for some robotic total stations at 2,000 m.) With prisms, the available accuracy is about $\pm(3 \text{ mm} + 1 \text{ ppm})$; without prisms, the available accuracy ranges from $\pm(3 \text{ mm} + 3 \text{ ppm})$ to about $\pm 10 \text{ mm}$. Targets with light-colored and flat surfaces perpendicular to the measuring beam (e.g., building walls) provide the best ranges and accuracies.

These instruments also provide quick results (0.8 s in rapid mode and 0.3 s in tracking mode), which means that applications for moving targets are possible. Applications are expected for near-shore hydrographic surveying and in many areas of heavy construction. Already this technique is being used, with an interfaced data collector, to measure cross sections in mining applications automatically; plotted cross sections and excavated volumes are automatically generated by digital plotter and computer. Other applications include cross-sectioning above-ground excavated works and material stockpiles; measuring to dangerous or difficult access points—for example, bridge components, cooling towers, and dam faces; and automatically measuring liquid surfaces—for example, municipal water reservoirs and catchment ponds. It is conceivable that these new techniques may have some potential in industrial surveying, where production line rates require this type of monitoring. These instruments are used with an attached visible laser, which helps to identify positively the feature being measured; that is, the visible laser beam is set on the desired feature so the surveyor can be sure that the correct surface is being measured and not some feature just beside it or just behind it. Because the measurement is so fast, care must be taken not to measure mistakenly to some object that may temporarily intersect the measuring signal, for example, trucks or other traffic. See Section 5.17 for additional information on reflectorless distance measurement with total stations.

Questions

3.1 How do random and systematic errors differ?

3.2 Describe three surveying applications where measurements can be made with a cloth or fiberglass tape.

3.3 Give two examples of the suitable use of each of the following measuring techniques or instruments:

   a. Pacing
   b. Steel tape
   c. EDM
   d. Remote sensing
   e. Subtense
   f. Scaling
   g. Fiberglass tape
   h. Odometer

3.4 What measuring or positioning techniques would you use in the following examples? Explain your choice.

   a. Topographic survey of a large tract of land (>100 acres)
   b. Topographic survey for a proposed factory or shopping mall
   c. Construction layout survey for a row of houses
   d. Measurements for payment for new concrete curb construction
   e. Measurements for payment for the laying of sod or the clearing of trees

3.5 Steel taping and EDM can both be used for precise distance measurements. List the types of situations where one or the other is the more effective technique.

Problems

3.1 A Gunter's chain was used to take the following measurements. Convert these distances to feet and meters.

    a. 65 chains, 15 links;

    b. 105 chains, 11 links;

    c. 129.33 chains;

    d. 7 chains, 41 links.

3.2 A 100-ft "cut" steel tape was used to measure between two property markers. The rear surveyor held 52 ft, while the head surveyor cut 0.27 ft. What is the distance between the markers?

3.3 The slope measurement between two points is 36.255 m and the slope angle is 1°50′. Compute the horizontal distance.

3.4 A distance of 210.23 ft was measured along a 3 percent slope. Compute the horizontal distance.

3.5 The slope distance between two points is 21.835 m and the difference in elevation between the points is 3.658 m. Compute the horizontal distance.

3.6 The ground clearance of an overhead electrical cable must be determined. Surveyor B is positioned directly under the cable (surveyor B can make a position check by sighting past the string of a plumb bob, held in his or her outstretched hand, to the cable); surveyor A sets the clinometer to 45° and then backs away from surveyor B until the overhead electrical cable is on the crosshair of the leveled clinometer. At this point, surveyors A and B determine the distance between them to be 17.4 ft (see figure). Surveyor A then sets the clinometer to 0° and sights surveyor B; this horizontal line of sight cuts surveyor B at the knees, a distance of 2.0 ft above the ground. Determine the ground clearance of the electrical cable.

3.7 A 100-ft steel tape known to be only 99.98 ft long (under standard conditions) was used to record a measurement of 276.22 ft. What is the distance corrected for the erroneous tape?

3.8 A 30-m steel tape, known to be 30.004 m (under standard conditions), was used to record a measurement of 129.085 m. What is the distance corrected for the erroneous tape length?

3.9 It is required to lay out a rectangular commercial building 200.00 ft wide and 350.00 ft long. If the steel tape being is 100.02 ft long (under standard conditions), what distances should be laid out?

3.10 A survey distance of 287.13 ft was recorded when the field temperature was 96°F. What is the distance, corrected for temperature?

3.11 Station 2 + 33.33 must be marked in the field. If the steel tape to be used is only 99.98 ft (under standard conditions) and if the temperature is 87°F at the time of the measurement, how far from the existing station mark at 0 + 79.23 will the surveyor have to measure to locate the new station?

3.12 The point of intersection of the center line of Elm Rd. with the center line of First St. was originally recorded (using a 30-m steel tape) as being at 6 + 11.23. How far from existing station mark 5 + 00 on First St. would a surveyor have to measure along the center line to reestablish the intersection point under the following conditions?

Temperature to be −6°C, with a tape that is 29.995 m under standard conditions.

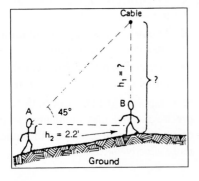

In Problems 3.13 through 3.17, compute the corrected horizontal distance (assume normal tension was used in all cases).

	Temperature	Tape Length	Slope Data	Slope Measurement
3.13	−14°F	99.98 ft	Difference in elevation = 6.35 ft	198.61 ft
3.14	46°F	100.00 ft	Slope angle 3°18′	219.51 ft
3.15	26°C	29.993 m	Slope angle −3°42′	177.032 m
3.16	0°C	30.002 m	Slope at 1.50%	225.000 m
3.17	100°F	100.03 ft	Slope at −0.80%	349.65 ft

For Problems 3.18 through 3.22, compute the required layout distance.

	Temperature	Tape Length	Design Distance	Required Layout Distance
3.18	29°F	99.98 ft	366.45 ft	?
3.19	26°C	30.012 m	132.203 m	?
3.20	13°C	29.990 m	400.000 m	?
3.21	30°F	100.02 ft	500.00 ft	?
3.22	100°F	100.04 ft	88.92 ft	?

3.23 A 50-m tape is used to measure between two points. The average weight of the tape per meter is 0.320 N. If the measured distance is 48.888 m, with the tape supported at the ends only and with a tension of 100 N, find the corrected distance.

3.24 A 30-m tape has a mass of 544 g and is supported only at the ends with a force of 80 N. What is the sag correction?

3.25 A 100-ft steel tape weighing 1.8 lb and supported only at the ends with a tension of 24 lb is used to measure a distance of 471.16 ft. What is the distance corrected for sag?

3.26 A distance of 72.55 ft is recorded using a steel tape supported only at the ends with a tension of 15 lb and weighing 0.016 lb per foot. Find the distance corrected for sag.

3.27 In order to verify the constant of a particular prism, a straight line EFG is laid out. The EDM instrument is first set up at E, with the following measurements recorded: EG = 426.224 m, EF = 277.301 m. The EDM instrument is then set up at F, where distance FG is recorded as FG = 148.953 m. Determine the prism constant.

3.28 The EDM slope distance between two points is 2556.28 ft, and the vertical angle is +2°45′30″ (the vertical angles were read at both ends of the line and then averaged using a coaxial theodolite/EDM combination). If the elevation of the instrument station is 322.87 ft and the heights of the theodolite/EDM and the target/reflector are all equal to 5.17 ft, compute the elevation of the target station and the horizontal distance to that station.

3.29 A line AB is measured at both ends as follows:

Instrument at A, slope distance = 879.209 m; vertical angle = +1°26′50″

Instrument at B, slope distance = 879.230 m; vertical angle = −1°26′38″

The heights of the instrument, reflector, and target are equal for each observation.

a. Compute the horizontal distance AB.
b. If the elevation at A is 163.772 m, what is the elevation at B?

3.30 A coaxial EDM instrument at station K (elevation $= 198.92$ ft) is used to sight stations L, M, and N, with the heights of the instrument, target, and reflector being equal for each sighting. The results are as follows:

Compute the elevations of L, M, and N (correct for curvature and refraction).

Instrument at station L, vertical angle $= +3°30'$,	EDM distance $= 2200.00$ ft
Instrument at station M, vertical angle $= -1°30'$,	EDM distance $= 2200.00$ ft
Instrument at station N, vertical angle $= 0°00'$,	EDM distance $= 2800.00$ ft

CHAPTER FOUR

ANGLES AND DIRECTIONS

4.1 GENERAL BACKGROUND

We noted in Section 1.13 that the units of angular measurement employed in North American practice are degrees, minutes, and seconds. For the most part, angles in surveying are measured with a transit/theodolite or total station, although angles can also be measured less precisely with clinometers, sextants (hydrographic surveys), or compasses.

4.2 REFERENCE DIRECTIONS FOR VERTICAL ANGLES

Vertical angles, which are used in slope distance corrections (Section 3.10) or in height determination (Section 2.13), are referenced to the (1) horizon by plus (up) or minus (down) angles, (2) zenith, or (3) nadir (see Figure 4-1). *Zenith* and *nadir* are terms describing points on a celestial sphere (i.e., a sphere or infinitely large radius with its center at the center of the earth). The zenith is directly above the observer and the nadir is directly below the observer; the zenith, nadir, and observer are all on the same vertical line (see also Figure 10-30).

4.3 MERIDIANS

A line on the mean surface of the earth joining the north and south poles is called a **meridian.** In Section 1.8, we noted that surveys could be referenced to lines of latitude and longitude. All lines of longitude are meridians. The term *meridian*

can be more precisely defined by noting that it is the line formed by the intersection with the earth's surface of a plane that includes the earth's axis of rotation. The meridian, as described, is known as the **geographic meridian. Magnetic meridians** are parallel to the directions taken by freely moving magnetized needles, as in a compass. Whereas geographic meridians are fixed, magnetic meridians vary with time and location. **Grid meridians** are lines that are parallel to a grid reference meridian (central meridian). The concept of a grid for survey reference was introduced in Section 1.9 and is described in detail in Chapter 10.

Figure 4-2 shows geographic meridians (also known as meridians of longitude), which all converge to meet at the pole, and grid meridians, which are all parallel to the central (geographic) meridian. In the case of a small-scale survey of only limited importance, meridian directions are sometimes assumed, and the survey is referenced to that assumed direction. We saw in Section 4.2 that vertical angles are referenced to a horizontal line (plus or minus) or to a vertical line (from either the zenith or nadir direction). By contrast, we now see that all horizontal directions are referenced to meridians.

4.4 HORIZONTAL ANGLES

Horizontal angles are usually measured with a theodolite or total station whose precision usually ranges from 1 to 20 seconds of arc. These instruments are described in detail in Chapter 5. Angles

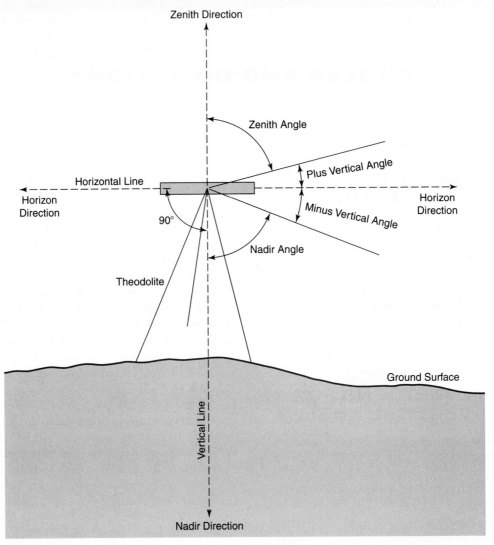

FIGURE 4-1 The three reference directions for vertical angles: horizontal, zenith, and nadir

can be measured between lines forming a closed traverse, between lines forming an open traverse, or between a line and a point so that the point's location may be determined.

For all closed polygons of n sides, the sum of the interior angles will be $(n - 2)180°$. In Figure 4-3 the interior angles of a five-sided closed polygon have been measured. For a five-sided polygon, the sum of the interior angles must be $(5 - 2)180° = 540°$; the angles shown in Figure 4-3 do, in fact, total

$(540)°$. In practical field problems, however, the total is usually marginally more or less than $(n - 2)180°$, and it is then up to the surveyor to determine if the error of angular closure is within tolerances as specified for that survey. The adjustment of angular errors is described in Chapter 6.

Note that the exterior angles at each station in Figure 4-3 could have been measured instead of the interior angles, as shown. (The exterior angle at A of $272°55'$ is shown.) Generally, exterior angles are

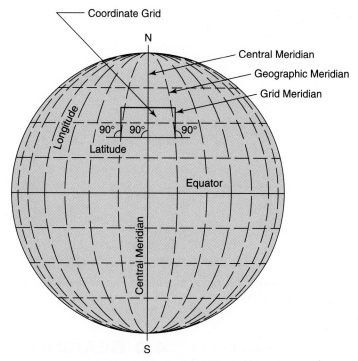

FIGURE 4-2 The relationship between geographic and grid meridians

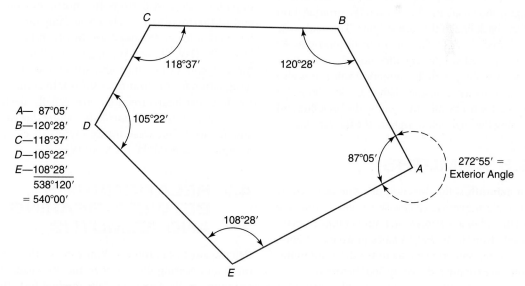

A— 87°05′
B—120°28′
C—118°37′
D—105°22′
E—108°28′
538°120′
= 540°00′

272°55′ =
Exterior Angle

FIGURE 4-3 Closed traverse showing the interior angles

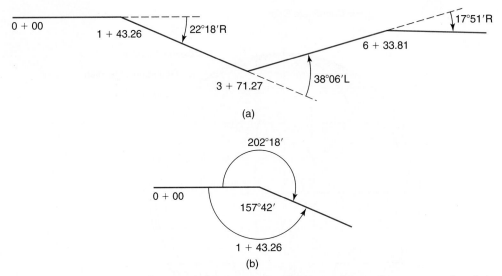

FIGURE 4-4 (a) Open traverse showing deflection angles. (b) Part of same first traverse showing angle right (202°18′) and angle left (157°42′)

measured to serve only occasionally as a check on the interior angle. Refer to Chapter 5 for the actual field techniques used in the measurement of angles.

An open traverse (usually found in route surveys) is illustrated in Figure 4-4(a). The **deflection angles** shown are measured from the prolongation of the back line to the forward line. The angles are measured either to the left (L) or to the right (R) of the projected line. The direction (L or R) must be shown along with the numerical value. It is also possible to measure the change in direction [see Figure 4-4(b)] by directly sighting the back line and turning the angle left or right to the forward line.

4.5 AZIMUTHS

An **azimuth** is the direction of a line as given by an angle measured clockwise (usually) from the north end of a meridian. Azimuths range in magnitude from 0° to 360°. Values in excess of 360°, which are sometimes encountered in computations, are simply reduced by 360° before final listing. Figure 4-5 illustrates the concept of azimuths by showing four line directions in addition to the four cardinal directions (N, S, E, and W).

4.6 BEARINGS

A **bearing** is the direction of a line as given by the acute angle between the line and a meridian. The bearing angle, which can be measured clockwise or counterclockwise from the north or south end of the meridian, is always accompanied by letters that locate the quadrant in which the line falls (NE, NW, SE, or SW). Figure 4-6 illustrates the concepts of bearings and shows the proper designation for the four lines shown. In addition, the four cardinal directions are usually designated by the terms *due north, due south, due east,* and *due west.* Due west, for example, can also be designated as N 90° W (or S 90° W).

4.7 RELATIONSHIPS BETWEEN BEARINGS AND AZIMUTHS

Refer again to Figure 4-5. You can see that azimuth and bearing directions in the NE quadrant are numerically equal. That is, for line 0–1, the azimuth is 52° and the bearing is N 52° E. In the SE quadrant, you can see that 0–2, which has an

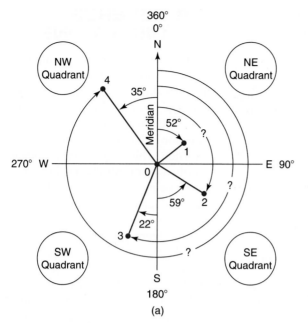

FIGURE 4-5 (a) Given azimuth data. (b) Azimuths calculated from given data

Line	Azimuth
0−1	52°
0−2	121°
0−3	202°
0−4	325°

(b)

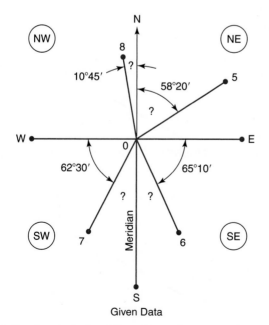

Line	Bearing
0−5	N 58°20′ E
0−6	S 24°50′ E
0−7	S 27°30′ W
0−8	N 10°45′ W

Bearings

Given Data

FIGURE 4-6 Bearings calculated from given data

azimuth of 121°, has a bearing (acute angle from meridian) of S 59° E. In the SW quadrant, the azimuth of 0–3 is 202° and the bearing is S 22° W. In the NW quadrant, the azimuth of 0–4 is 325° and the bearing is N 35° W.

To convert from azimuths to bearings, first, determine the proper quadrant letters (see Figure 4-5):

1. For 0° to 90°, use NE (quadrant 1 in most software programs).
2. For 90° to 180°, use SE (quadrant 2 in most software programs).
3. For 180° to 270°, use SW (quadrant 3 in most software programs).
4. For 270° to 360°, use NW (quadrant 4 in most software programs).

Then the numerical value is determined by using the following relationships:

1. NE quadrant: bearing = azimuth
2. SE quadrant: bearing = 180° − azimuth
3. SW quadrant: bearing = azimuth − 180°
4. NW quadrant: bearing = 360° − azimuth

Bearing can be converted to azimuth by using these relationships:

1. NE quadrant: azimuth = bearing
2. SE quadrant: azimuth = 180° − bearing
3. SW quadrant: azimuth = 180° + bearing
4. NW quadrant: azimuth = 360° − bearing

4.8 REVERSE DIRECTIONS

It can be said that every line has two directions. The line shown in Figure 4-7 has direction *AB* or it has direction *BA*. In surveying, a direction is called *forward* if it is oriented in the direction of fieldwork or computation staging. If the direction is the reverse of that, it is called a *back* direction. The designations of forward and back are often arbitrarily chosen, but if more than one line is being considered, the forward and backward designations must be consistent for all adjoining lines.

In Figure 4-8, the line *AB* has a bearing of N 62°30′ E, whereas the line *BA* has a bearing of S 62°30′ W; that is, *to reverse a bearing, simply reverse the direction letters.* In this case, N and E become S and W, and the numerical value (60°30′) remains unchanged. In Figure 4-9 the line *CD* has an azimuth of 128°20′. Analysis of the sketch leads quickly to the conclusion that

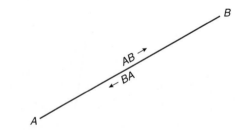

FIGURE 4-7 Reverse directions

Line	Bearing
AB	N 62°30′ E
BA	S 62°30′ E

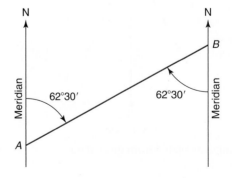

FIGURE 4-8 Reverse bearings

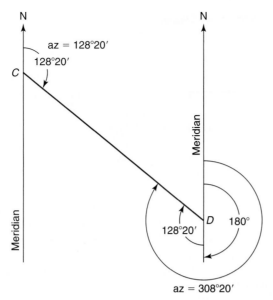

FIGURE 4-9 Reverse azimuths

Line	Azimuth
CD	128°20′
DC	308°20′

the azimuth of *DC* is 308°20′; that is, *to reverse an azimuth, simply add 180° to the original direction.* If the original azimuth is greater than 180°, 180° can be subtracted from it to reverse its direction. The key factor to remember is that a forward and back azimuth must differ by 180°.

4.9 AZIMUTH COMPUTATIONS

The data in Figure 4-10 will be used to illustrate the computation of azimuths. Before azimuths or bearings are computed, it is usual to check that the figure is geometrically closed: that the sum of the interior angles = $(n − 2)180°$. In Figure 4-10, the five angles do add to 540°00′, and *AB* has a given azimuth of 333°00′. At this point, a decision must be made about how the computation will proceed. Using the given azimuth and the angle at *B*, the azimuth of *BC* can be computed (counterclockwise direction); or using the given azimuth and the angle at *A*, the azimuth of *AE* can be computed (clockwise direction). Once a direction for solving the problem has been established, all the computed

directions must be consistent with that general direction. A neat, well-labeled sketch should accompany each step of the computation.

Analysis of the preceding azimuth computations gives the following observations:

1. If the computation is proceeding in a *counterclockwise direction, add the interior angle to the back azimuth of the previous course.* See Figure 4-11.

2. If the computation is proceeding in a *clockwise direction, subtract the interior angle from the back azimuth of the previous course.* See Figure 4-12.

If the bearings of the sides are also required, they can now be derived from the computed azimuths. See Table 4-1.

Note the following points about azimuth calculations:

1. For reversal of direction (i.e., clockwise versus counterclockwise), the azimuths for the same side differ by 180° whereas bearings for the same side remain numerically the same and have the letters (N/S, E/W) reversed.

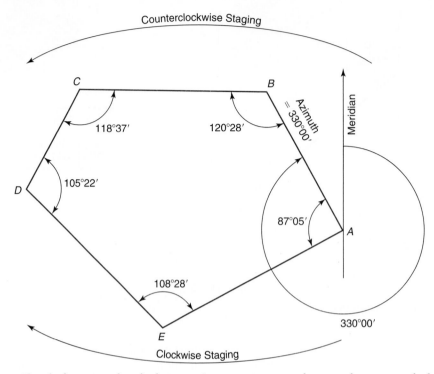

FIGURE 4-10 Sketch for azimuth calculations. Computations can be staged to proceed clockwise or counterclockwise

2. The bearings calculated directly from computed azimuths have no built-in check. The only way that these bearing calculations can be verified is to double-check the computation. *Constant reference to a good problem diagram helps to reduce the incidence of mistakes.*

4.10 BEARING COMPUTATIONS

As with azimuth computations, the solution can proceed in a clockwise or counterclockwise manner. In Figure 4-13, side *AB* has a given bearing of N 30°00′ W, and the bearing of either *BC* or *AE* may be computed first. Because there is no systematic method of directly computing bearings, each bearing computation will be regarded as a separate problem. *It is essential that a neat, well-labeled diagram accompany*

each computation. The sketch of each individual bearing computation should show the appropriate interior angle together with one bearing angle. The required bearing angle should also be shown clearly.

In Figure 4-14(a), the interior angle (*B*) and the bearing angle (30°) for side *BA* are shown; the required bearing angle for side *BC* is shown as a question mark. Analysis of the sketch shows that the required bearing angle (?) = 180° − (120°28′ − 30°00′) = 89°32′ and that the quadrant is NW. The bearing of *BC* = N 89°32′ W.

In Figure 4-14(b), the bearing for *CB* is shown as S 89°32′ E, which is the reverse of the bearing that was just calculated for side *BC*. When the meridian line is moved from *B* to *C* for this computation, it necessitates the reversal of direction. Analysis of the completely labeled sketch shows that the required bearing angle

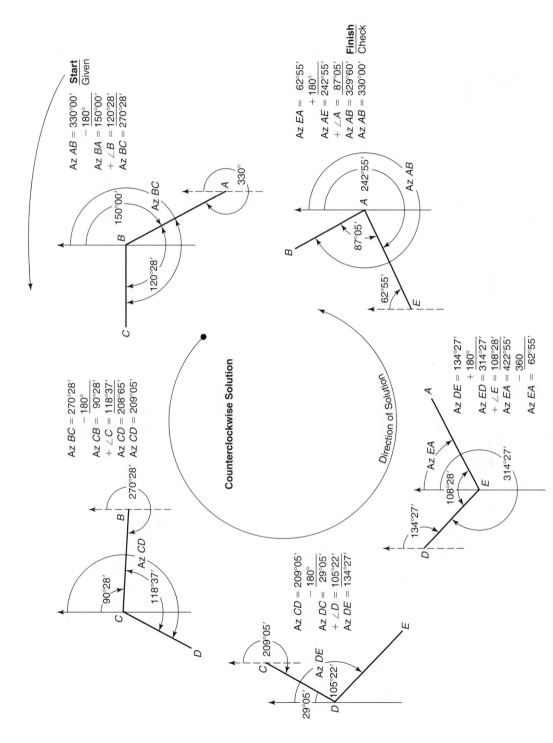

FIGURE 4-11 Azimuth calculations: counterclockwise solution

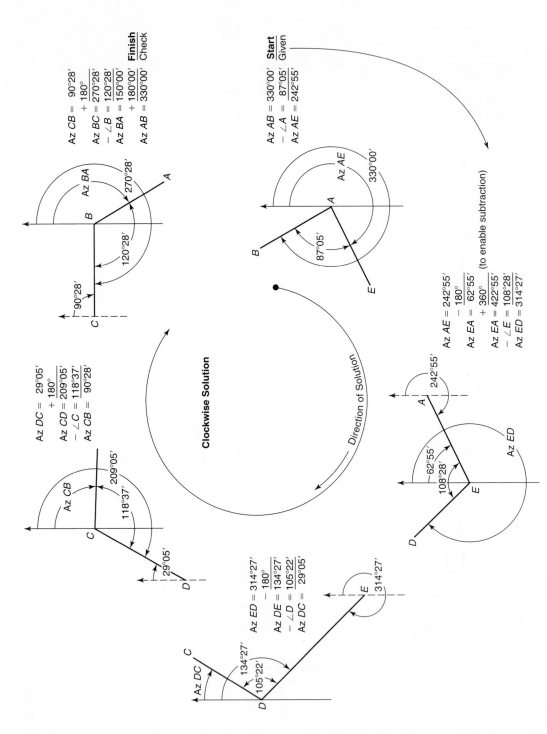

FIGURE 4-12 Azimuth calculations: clockwise solution

Table 4-1

Counterclockwise Solution		
Course	Azimuth	Bearing
BC	270°28′	N 89°32′ W
CD	209°05′	S 29°05′ W
DE	134°27′	S 45°33′ E
EA	62°55′	N 62°55′ E
AB	330°00′	N 30°00′ W
Clockwise Solution		
Course	Azimuth	Bearing
AE	242°55′	S 62°55′ W
ED	314°27′	N 45°33′ W
DC	29°05′	N 29°05′ E
CB	90°28′	S 89°32′ E
BA	150°00′	S 30°00′ E

for $CD(?) = (118°37′ − 89°32′) = 29°05′$ and that the direction is SW. The bearing of CD is S 29°05′ W.

Analysis of Figure 4-14(c) shows that the bearing angle of line $DE(?) = 180° − (105°22′ + 29°05′) = 45°33′$ and that the direction of DE is SE. The bearing of DE is S 45°33′ E. Analysis of Figure 4-14(d) shows that the bearing angle of line EA (?) is $(108°28′ − 45°33′) = 62°55′$ and that the direction is NE. The bearing of EA is N 62°55′ E.

The problem's original data included the bearing of AB as N 30°00′ W. The bearing of AB will now be computed using the interior angle at A and the bearing just computed for the previous course (EA). The bearing angle of $AB = 180° − (62°55′ + 87°05′) = 30°00′$ and the direction is NW. The bearing of AB is N 30°00′ W [see Figure 4-14(e)]. This last computation serves as a check on all our computations.

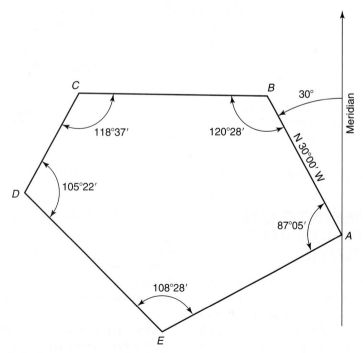

FIGURE 4-13 Sketch for bearing computations

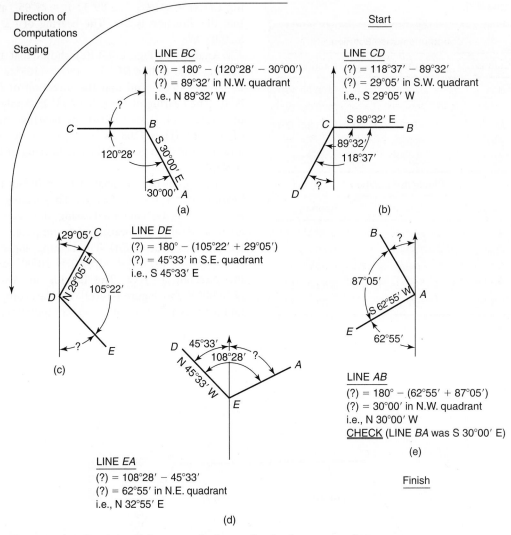

Direction of
Computations
Staging

Start

LINE BC
(?) = 180° − (120°28′ − 30°00′)
(?) = 89°32′ in N.W. quadrant
i.e., N 89°32′ W

LINE CD
(?) = 118°37′ − 89°32′
(?) = 29°05′ in S.W. quadrant
i.e., S 29°05′ W

(a)

(b)

LINE DE
(?) = 180° − (105°22′ + 29°05′)
(?) = 45°33′ in S.E. quadrant
i.e., S 45°33′ E

(c)

LINE AB
(?) = 180° − (62°55′ + 87°05′)
(?) = 30°00′ in N.W. quadrant
i.e., N 30°00′ W
CHECK (LINE BA was S 30°00′ E)

(e)

LINE EA
(?) = 108°28′ − 45°33′
(?) = 62°55′ in N.E. quadrant
i.e., N 32°55′ E

Finish

(d)

FIGURE 4-14 Sketches for each bearing calculation for the discussion of Section 4.10

4.11 COMMENTS ON BEARINGS AND AZIMUTHS

Both bearings and azimuths may be used to give the direction of a line. North American tradition favors the use of bearings over azimuths; most legal plans (plats) show directions in bearings. In the previous sections, bearings were derived from computed azimuths (Section 4.9), or bearings were computed directly from the given data (Section 4.10). The *advantage* of computing bearings directly from the given data in a closed traverse is that the final computation (of the given bearing) provides a check on all the problem computations, ensuring (normally) the correctness of all the computed bearings. By contrast, if bearings are derived from computed azimuths, there is no intrinsic check on the correctness of the derived bearings.

The *disadvantage* associated with computing bearings directly from the data in a closed traverse is that there is no uncomplicated systematic approach to the overall solution. Each bearing computation is unique, requiring individual analysis. It is sometimes difficult to persuade people to prepare neat, well-labeled sketches for computations involving only intermediate steps in a problem. Without neat, well-labeled sketches for each bearing computation, the potential for mistakes is quite large. And when mistakes do occur in the computation, the lack of a systematic approach to the solution often means that much valuable time is lost before the mistake is found and corrected. By contrast, the computation of azimuths involves a highly systematic routine: *add (subtract) the interior angle from the back azimuth of the previous course.* If the computations are arranged as shown in Section 4.9, mistakes that

may be made in the computation will be found quickly. See Figure 4-15 for a summary of results.

With the widespread use of computers and sophisticated handheld calculators, azimuths will likely be used more to give the direction of a line. It is easier to deal with straight numeric values rather than the alphanumeric values associated with bearings, and it is more efficient to have the algebraic sign generated by the calculator or computer rather than trying to remember if the direction was north, south, east, or west.

We will see in Chapter 6 that bearings or azimuths are used in calculating the geometric closure of a closed survey. The absolute necessity of eliminating mistakes from the computation of line directions will become more apparent in that chapter. It will be apparent that the computation of direction (bearings or azimuths) is only the first step in what could be a very involved computation.

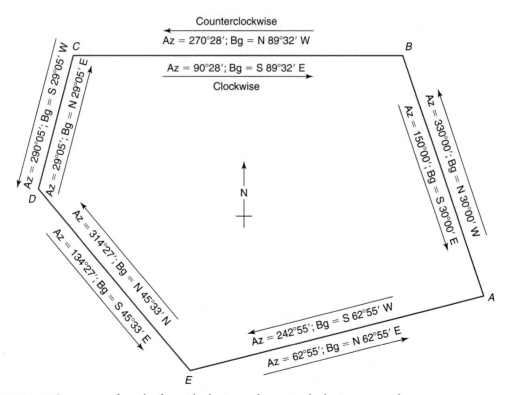

FIGURE 4-15 Summary of results from clockwise and counterclockwise approaches

4.12 MAGNETIC DIRECTION

Most original township surveys in North America were carried out using a Gunter's chain and a magnetic compass. As a result, most original plans show directions referenced to magnetic north. Modern surveyors must be able to convert from magnetic directions to geographic (or geodetic) directions.

Because the magnetic north pole is some distance from the geographic north pole, the compass's magnetized needle does not typically point to geographic north. The horizontal angle between the direction taken by the compass needle and geographic north is the *magnetic declination*. In North America, magnetic declination ranges from about 15° east on the west coast of the continental United States to about 15° west on the east coast, and from about 26° east on the west coast of Canada to about 26° west on the east coast (see Figure 4-16).

The location of the north magnetic pole does not stay in one place; it is constantly moving. The Canadian Geological Survey last measured the location of the magnetic north pole in 2005 and found it to be at 82.7° north latitude and 114° west longitude. The pole appears now to be moving northwesterly at a rate of 40 km per year and will eventually move from Canadian territory to Russian territory. Magnetic declination thus varies with both location and time.

Careful records are kept over the years so that, although magnetic variations are not well understood, it is possible to predict magnetic declination changes over a short span of time. Accordingly, many countries issue *isogonic charts*, usually every 5 or 10 years, on which lines are drawn (isogonic lines) that join points of the earth's surface having equal magnetic declination. See Figure 4-16. Additional lines are shown on the chart that join points on the earth's surface that are experiencing equal annual changes in magnetic declination. Due to the uncertainties of determining magnetic declination and the effects of local attraction (e.g., ore bodies) on the compass needle, magnetic directions are not employed for any but the lowest order of surveys. Also, due to the difficulty in predicting

magnetic influences, users should be cautious about applying annual change corrections beyond 5 years from the epoch of the current isogonic chart.

More recently, it has become possible to determine magnetic declination for a specific location in the United States by using the National Geophysical Data Center (NGDC) website at http://www.ngdc.noaa.gov/seg/geomag/declination.shtml and in Canada at http://www.geolab.nrcan.gc.ca/. These websites permit a user to obtain the magnetic declination for a specific location by entering the latitude and longitude coordinates of a location or by simply entering its zip code. In addition, one can obtain the magnetic declination value for the current date or for any date in the past (only back to 1960 in Canada). The computations shown in Figure 4-17(a) and (b) are based on an initial zip code entry of 10020 (a section of New York City); the geographic coordinates of that zip code are then shown as Latitude: 40.759729° N and Longitude: 73.982347° W. Figure 4-17(a) shows the magnetic declination, calculated for the date May 29, 2007, to be 13°11′ W, changing by 1′ E per year. Figure 4-17(b) shows magnetic declination values calculated every 5 years from 1850 for that same location using the website's Historical Declination function. The magnetic declinations shown for each year are valid only for January 1 of each year. Figure 4-17(c) shows a similar computation for a location in Canada. Here the initial entry can be the selection of a town or city, selected from a pull-down menu, Niagara-on-the-Lake, Ontario, in this case. After selecting the town, the geographic coordinates appear, Latitude: 43°15′ N and Longitude: 79°04′ W. The calculated magnetic declination is then displayed, 10°40′ W for the specified date May 29, 2007.

Isogonic charts and website calculators can be valuable aids when original surveys (magnetic) are to be retraced. Most original township surveys in North America were magnetically referenced. When magnetically referenced surveys are to be retraced, it is necessary to determine the magnetic declination for that area at the time of the original survey and at the time of the retracement survey.

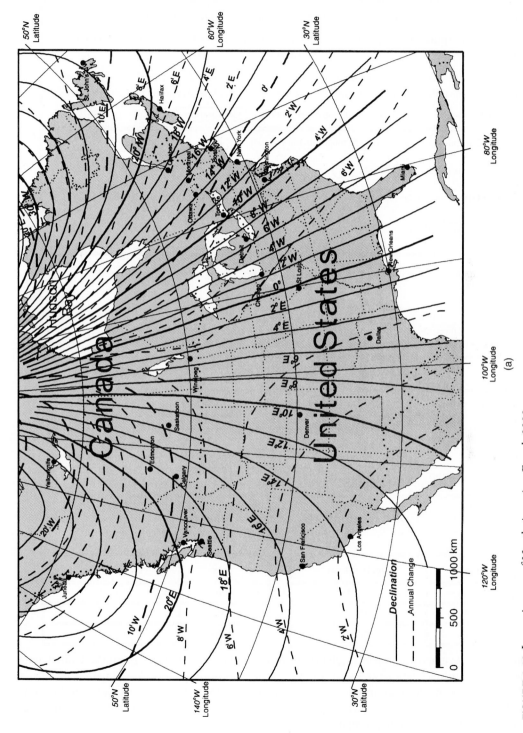

FIGURE 4-16 Isogonic map of North America, Epoch 2000
(Compilation by Natural Resources Canada, Geomagnetic Laboratory, Ottawa.)

Estimated Value of Magnetic Declination

To compute the magnetic declination, you must enter the location and date of interest.

If you are unsure about your city's latitude and longitude, look it up online! In the United States try entering your zip code in the box below or visit the U.S. Gazetteer. Outside the United States try the Getty Thesaurus.

Search for a place in the United States by zip code: **10020**   (**Get Location**)

Enter Location: (latitude 90 S to 90 N, longitude 180 W to 180 E). See Instructions for details.

Latitude: 40.759729   ⦿ N   ◯ S   **Longitude:** 73.982347   ◯ E   ⦿ W

Enter Date (1900–2010): Year: 2007   Month (1–12): **5**   Day (1–31): **29**

(**Compute Declination**)

Declination = 13°11′ W changing by 0°1′ E/year

For more information, visit: www.ngdc.noaa.gov/seg/geomag/declination.shtml
Answers to some **frequently asked questions | Instructions** for use | **Today's Space Weather**

(a)

Location		Degree	Minute		Date
Zip Code. 10020	Northern Latitude:	40	46	Start Date:	1850
Get Location	Western Longitude:	73	57	End Date:	2007

Compute Historical Declination

Results

Year	Declination	Year	Declination
1850	5°38′ W	1880	7°35′ W
1860	6°17′ W	1890	8°5′ W
1870	6°56′ W	1900	8°40′ W

FIGURE 4-17 (a) Magnetic declination calculated for a specific location and date in the United States. (b) List of magnetic declinations at 5-year intervals at same location as in (a). (c) Magnetic declination calculated for a specific location in Canada. Magnetic declination calculators are found in the United States in the places shown at http://www.ngdc.noaa.gov/seg/geomag/declination.shtml and in Canada in the places shown at http://www.geolab.nrcan.gc.ca/

Year	Declination
1905	9°4′ W
1910	9°31′ W
1915	9°57′ W
1920	10°15′ W
1925	10°39′ W
1930	10°59′ W
1935	11°13′ W
1940	11°14′ W
1945	11°16′ W
1950	11°13′ W
1955	11°15′ W
1960	11°22′ W

Year	Declination
1965	11°30′ W
1970	11°40′ W
1975	12°57′ W
1980	12°21′ W
1985	12°38′ W
1990	12°55′ W
1995	13°15′ W
2000	13°21′ W
2005	13°16′ W
2007	13°12′ W

(b)

Geomagnetism
Magnetic Declination Calculator

Enter latitude and longitude or choose a city: Niagara-on-the-Lake

Year: 2007   Month 05   Day 29

Latitude: 43   15   minutes   (◉ North   ○ South)
              degrees

Longitude: 79   degrees   4   minutes   (◉ West   ○ East)

(**Calculate Magnetic Declination**)

Year: 2007 05 29   Latitude: 43°15′ North   Longitude: 79°4′ West

Calculated magnetic declination: 10°40′ West

(c)

FIGURE 4-17 (*Continued*)

Example 4-1: *Use of an Isogonic Map*
A magnetic bearing was originally recorded for a specific lot line (*AB*) in Seattle as N 10°30′ E. The magnetic declination at that time was 15°30′ E. You must retrace the survey from the original notes during the first week of September 2001.

a. What compass bearing will be used for the same lot line during the retracement survey?

b. What will be the geographic bearing for the same survey line in the retracement survey?

Solution: Using the isogonic chart in Figure 4-16, the following data are interpolated for the Seattle area:

$$\text{Declination}(2000.0) = 18°50′\,E$$
$$\text{Annual change} = 08′\,W$$

Declination September 2001
$$= 18°50′ - (08′ \times 1.75)$$
$$= 18°36′\,E$$

Declination during
$$\text{original survey} = \underline{15°30′\,E}$$
$$\text{Difference in declination} = \underline{3°06′}$$

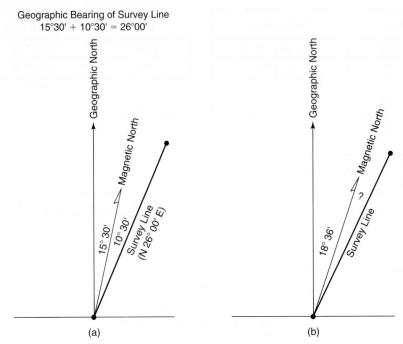

Geographic Bearing of Survey Line
15°30' + 10°30' = 26°00'

(a)

(b)

FIGURE 4-18 Magnetic declination problem. (a) Original survey. (b) Retracement survey

Geographic
    bearing = 10°30' + 15°30' = N 26°00' E
Magnetic bearing, September 2001
        = 10°30' − 3°06' = N 7°24' E

or

Magnetic bearing, September 2001
        = 26°00' − 18°36' = N 7°24' E.

See Figure 4-18.

Other situations when modern surveyors need a compass are determining the meridian by observation of the North Star, Polaris (see Chapter 10) and preparing station visibility diagrams for proposed GPS surveys (see Figure 7-10). Various compasses are available, from the simple to the more complex. Figure 4-19 shows a Brunton compass, which is popular with many surveyors and geologists. It can be handheld or mounted on a tripod. It can also be used as a clinometer, with vertical angles read to the closest 5 minutes.

FIGURE 4-19 Brunton compass, combining the features of a sighting compass, prismatic compass, hand level, and clinometer. Can be staff-mounted for more precise work

Problems

4.1 A closed five-sided field traverse has the following interior angles: $A = 121°13'00''$; $B = 136°44'30''$; $C = 77°05'30''$; $D = 94°20'30''$. Find the angle at E.

4.2 Convert the following azimuths to bearings.
  a. 210°30′
  b. 128°22′
  c. 157°55′
  d. 300°30′
  e. 0°08′
  f. 355°10′
  g. 182°00′

4.3 Convert the following bearings to azimuths.
  a. N 20°20′ E
  b. N 1°33′ W
  c. S 89°28′ E
  d. S 82°36′ W
  e. N 89°29′ E
  f. S 11°38′ W
  g. S 41°14′ E

4.4 Convert the azimuths given in Problem 4.2 to reverse (back) azimuths.

4.5 Convert the bearings given in Problem 4.3 to reverse (back) bearings.

4.6 An open traverse that runs from A through H has the following deflection angles: $B = 6°25'$ R; $C = 3°54'$ R; $D = 11°47'$ R; $E = 20°02'$ L; $F = 7°18'$ L; $G = 1°56'$ R. If the bearing of AB is N 19°09′ E compute the bearings of the remaining sides.

4.7 Closed traverse $ABCD$ has the following bearings: $45AB =$ N 45°20′ E; $BC =$ S 47°32′ E; $CD =$ S 14°55′ W; $DA =$ N 68°00′ W.

Compute the interior angles and provide a geometric check for your work.

Use the following sketch (Figure 4-20) and interior angles for Problems 4.8 through 4.11.

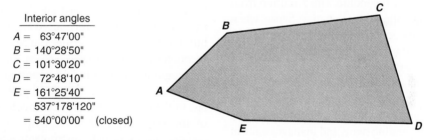

Interior angles
$A =$ 63°47'00"
$B =$ 140°28'50"
$C =$ 101°30'20"
$D =$ 72°48'10"
$E =$ 161°25'40"
537°178'120"
$= 540°00'00''$ (closed)

FIGURE 4-20 Sketch of interior angles for Problems 4.8 through 4.11

4.8 If the bearing of AB is N 42°11′10″ E, compute the bearings of the remaining sides. Provide two solutions: one solution proceeding clockwise and the other solution proceeding counterclockwise.

4.9 If the azimuth of AB is 42°11′10″, compute the azimuths of the remaining sides. Provide two solutions: one solution proceeding clockwise and the other proceeding counterclockwise.

4.10 If the bearing of AB is N 45°48′56″ E, compute the bearings of the remaining sides proceeding in a clockwise direction.

4.11 If the azimuth of AB is 45°48′56″, compute the azimuths of the remaining sides proceeding in a counterclockwise direction.

4.12 On January 2, 2000, the compass reading on a survey line in the Seattle, Washington, area was N 39°30′ E. Use Figure 4-16(a) to determine the following:

  a. The compass reading for the same survey line on July 2, 2004.

  b. The geodetic bearing of the survey line.

CHAPTER FIVE

TOTAL STATIONS
AND THEODOLITES

5.1 INTRODUCTION

As noted in Section 1.5, the term *theodolite* or *transit* (transiting theodolite) can be used to describe those survey instruments designed to precisely measure horizontal and vertical angles. In addition to measuring horizontal and vertical angles, theodolites can also be used to mark out straight and curved lines in the field. The electronic digital theodolite, first introduced in the late 1960s, helped set the stage for modern field data collection and processing. See Figure 5-1 for a typical electronic theodolite. The evolution from electronic theodolite to total station began when the electronic theodolite was first fitted with built-in electronic distance measurement (EDM), or an add-on and interfaced EDM, so the surveyor had an extremely capable instrument.

A *total station* also contains an onboard microprocessor that automatically monitors the instrument's operating status and manages built-in surveying programs, and a data collector (built-in or interfaced) that stores and processes measurements and attribute data. (See Figures 5-2 and 5-6.) Surveyors can utilize the onboard computation programs to determine horizontal and vertical distances, sighted-point and instrument station coordinates, and a wide variety of other surveying and engineering applications programmed into the instrument's central processor. See later in this chapter for more on total station applications.

During the twentieth century, theodolites went through three distinct evolutionary stages:

- The open-face, vernier-equipped engineers' transit (American transit); see Figure H-4.

- The enclosed, optical-readout theodolites with direct digital readouts or micrometer-equipped readouts (for more precise readings); see Figure H-13.

- The enclosed electronic theodolite with direct readouts; see Figure 5-1.

Most recently manufactured theodolites are electronic, but many of the earlier optical instruments and even a few vernier instruments still survive in the field (and in the classroom), a tribute to the excellent craftsmanship of the instrument makers. In past editions of this text, the instruments were introduced chronologically, but in this edition the vernier transits and optical theodolites are introduced last (in Appendix H), in recognition of their fading importance.

The electronic theodolite will probably be the last in the line of transits and theodolites. Because of the versatility and lower costs of electronic equipment, future field instruments will be more along the line of the **total station** (see Figures 5-2 and 5-3), which combine all the features of a theodolite with the additional capabilities of electronically measuring distances and angles and storing all measurements along with relevant attribute data for future transfer to the computer. Advanced total stations also include interfaced global positioning receiver (see Figure 5-4) to permit precise horizontal and vertical positioning (see Chapter 7). The sections in this chapter deal with total station and theodolite setup and general usage, such as prolonging a straight line, bucking-in, and prolonging a line past an obstacle. Total station operations are described more fully later in this chapter.

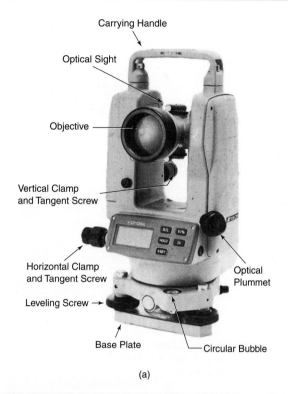

Carrying Handle

Optical Sight

Objective

Vertical Clamp
and Tangent Screw

Horizontal Clamp
and Tangent Screw

Optical
Plummet

Leveling Screw

Base Plate

Circular Bubble

(a)

ROTARY ENCODER SYSTEM FOR ELECTRONIC THEODOLITES AND TRANSITS

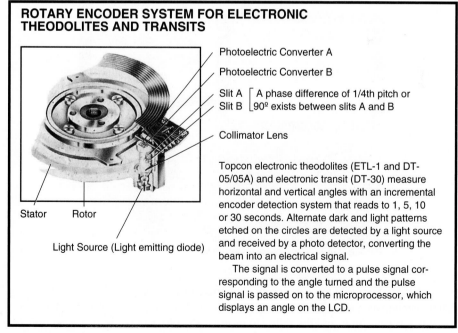

Stator   Rotor

Light Source (Light emitting diode)

Photoelectric Converter A

Photoelectric Converter B

Slit A ⎡ A phase difference of 1/4th pitch or
Slit B ⎣ 90º exists between slits A and B

Collimator Lens

Topcon electronic theodolites (ETL-1 and DT-05/05A) and electronic transit (DT-30) measure horizontal and vertical angles with an incremental encoder detection system that reads to 1, 5, 10 or 30 seconds. Alternate dark and light patterns etched on the circles are detected by a light source and received by a photo detector, converting the beam into an electrical signal.

The signal is converted to a pulse signal corresponding to the angle turned and the pulse signal is passed on to the microprocessor, which displays an angle on the LCD.

(b)

FIGURE 5-1 Topcon DT-05 electronic digital theodolite. (a) Theodolite. (b) Encoder system for angle readouts
(Courtesy of Topcon Instrument Corp., Paramus, NJ)

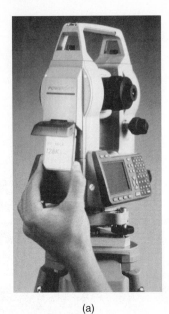

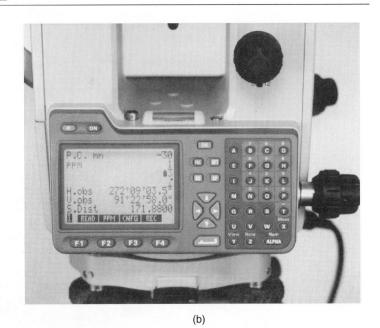

(a)

(b)

■ Graphic "Bulls-Eye" Level

A graphically displayed "bulls-eye" lets the user quickly and efficiently level the instrument.

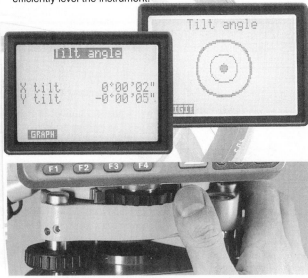

(c)

FIGURE 5-2 (a) Sokkia SET1000 total station having angle display of 0.5″ (1″ accuracy) and a distance range to 2,400 m using one prism. The instrument comes with a complete complement of surveying programs, dual-axis compensation, and the ability to measure (to 120 m) to reflective sheet targets, a feature suited to industrial surveying measurements. (b) Keyboard and liquid crystal display (LCD). (c) Graphics bull's-eye level allows the user to level the instrument while observing the graphics display

(Courtesy of Sokkia Corp., Overland Park, KS)

Radio

FIGURE 5-3 Total station with cable connected electronic field book. Also shown is a two way radio (2-mile range) with push-to-talk headset

FIGURE 5-4 Leica Smart Station total station with integrated GPS

(Courtesy of Leica Geosystems Inc.)

5.2 ELECTRONIC THEODOLITES

Unlike the optical theodolites described in Appendix H, electronic theodolites usually have only one horizontal motion. That is, when the theodolite is rotated the alidade turns on the circle assembly to show a continuous change in angle values. Electronic theodolites have the capability of permitting the surveyor to set the horizontal scale set to zero (or any other value) *after* the theodolite has been sighted and clamped on a reference point. Angle readouts can be to 1″, with precision from 1″ to 20″. The surveyor should check the specifications of new instruments to determine their precision rather than simply accept the minimum readout graduation (some instruments with 1″ readouts may be capable of only 5″ precision).*

Digital readouts eliminate the uncertainty associated with the reading and interpolation of scale and micrometer settings found in optical theodolites. Horizontal angles can be turned left or right, and repeat-angle averaging is available on some models. Figures 5-1 and 1-5 show typical electronic theodolites. The display windows for horizontal and vertical angles can be located at both the front and rear of some instruments for easy access. Figure 5-1 also shows how angles are sensed electronically using the incremental encoder detection technique. After turning on some instruments, the operator must activate the vertical circle by turning the telescope slowly across the horizon; newer instruments do not require this referencing procedure. The vertical circle can be set with zero at the zenith or

---

*Accuracies are now specified by most surveying instrument manufacturers by reference to DIN 18723. DIN (Deutsches Institut für Normung) is known in English-speaking countries as the German Institute for Standards. These accuracies are tied directly to surveying practice. For example, to achieve a claimed accuracy (stated in terms of one standard deviation) of ±5 seconds, the surveyor must turn the angle four times (two each on face 1 and 2 with the telescope inverted on face 2). This practice assumes that collimation errors, centering errors, and the like have been eliminated before measuring the angles. DIN specifications can be purchased at http://www2.dim.de/index.php?lang=en.

at the horizon. The factory setting of zenith can be changed in the menu display, or by setting the appropriate dip switch as described in the instrument's manual. The status of the battery charge can be monitored on the display panel, giving the operator ample warning of the need to replace or recharge the battery. The horizontal and vertical circles can be clamped and fine adjusted using the clamps and tangent screws shown in Figure 5-1.

The procedures for measuring an angle with an electronic theodolite follow: First, after the theodolite is set properly over the instrument station and leveled (see Section 5.4), sight on the left-hand (usually) station, clamp the horizontal movement, and then set the vertical crosshair precisely on the station by using the fine adjustment screw; second, set the scales to zero (usually) or some other required value by turning the appropriate screws; third, loosen the clamp and turn the instrument until it is pointing at the second (right-hand) station, set the clamp, and then use the fine adjusting screw to precisely sight the vertical crosshair on that station; fourth, read and book (store) the angle. To help eliminate mistakes and to improve precision, angles are usually measured twice: once with the telescope normal (right-side up) and once with the telescope reversed (upside down).

To turn an angle twice (called "doubling"), after booking the first angle, press the hold button to keep the booked angle displayed on the instrument's scale, then transit the telescope, loosen the clamp, and re-sight the backsight (on the original station) as described above. Finally, to turn the double angle, simply press (release) the hold button and re-sight the foresight station; after fine adjusting the vertical crosshair on the foresight station, read and book the double angle. The mean angle is determined by halving the doubled angle. See Figure 5-5 for typical field notes.

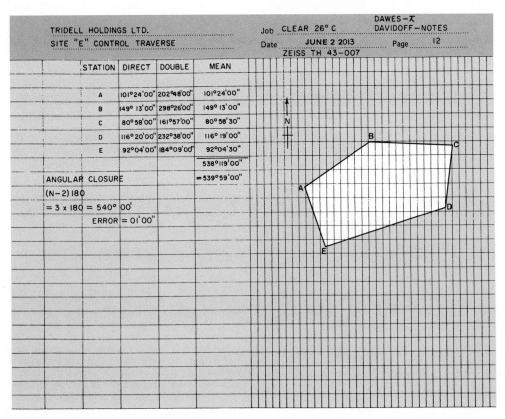

FIGURE 5-5 Field notes for repeated angles

These simple electronic theodolites have replaced optical transits and theodolites (which replaced the vernier transit). They are simpler to use and less expensive to purchase and repair, and their use of electronic components indicates a continuing drop in both purchase and repair costs. Some of these instruments have various built-in functions that enable the operator to perform other theodolite operations, such as determining "remote object elevation" and "distance between remote points."

5.3 TOTAL STATION

Total stations (see Figures 5-2 and 5-3) can measure and record horizontal and vertical angles together with slope distances. The microprocessors in the total stations can perform different mathematical measurements, for example, averaging multiple angle measurements; averaging multiple distance measurements; determining horizontal and vertical distances; determining X, Y, and Z coordinates, remote object elevations (i.e., heights of sighted features), and distances between remote points; and performing atmospheric and instrument corrections.

Figure 5-2 shows a typical total station, one of a series of instruments that have angle accuracies from 0.5 to 5 seconds, and distance ranges (to one prism) from 1,600 to 3,000 m. We noted in Chapter 2, that modern levels employ compensation devices to correct the line of sight automatically to its horizontal position; total stations come equipped with single-axis or dual-axis compensation (see Figure 5-6). With total stations, dual-axis compensation measures and corrects for left/right (lateral) tilt and forward/back (longitudinal) tilt. Tilt errors affect the accuracies of vertical angle measurements. Vertical angle errors can be minimized by averaging the readings of two-face measurements; with dual-axis compensation, however, the instrument's processor can determine tilt errors and remove their effects from the measurements, thus much improving the surveyor's efficiency. The presence of dual-axis compensation also permits the inclusion of electronic

instrument leveling [see Figures 5-2(c) and 5-36(b)], a technique that is much faster than the repetitive leveling of plate bubbles in the two positions required by instruments without dual-axis compensation. Some total stations are equipped to store collected data in onboard data cards (see Figures 1-4 and 5-2), whereas other total stations come equipped with data collectors that

■ **Simultaneous Detection of Inclination in Two Directions and Automatic Compensation**

The built-in dual-axis tilt sensor constantly monitors the inclination of the vertical axis in two directions. It calculates the compensation value and automatically corrects the horizontal and vertical angles. (The compensation range is ±3'.)

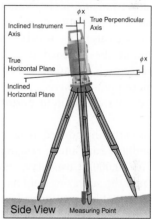

Side View   Measuring Point

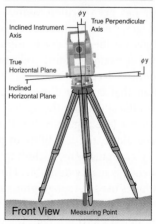

Front View   Measuring Point

FIGURE 5-6 Dual-axis compensation illustration (Courtesy of Sokkia Corp., Overland Park, KS)

are connected by cable (or wirelessly) to the instrument (see Figures 5-3 and 5-39).

Modern instruments have a wide variety of built-in instrument-monitoring, data collection and layout programs, and rapid battery charging that can charge the battery in 70 min. Data are stored onboard in internal memory and/or on memory cards (data storage typically ranges from 10,000 points to 80,000 points, depending on the make and model). Data can be transferred from the total station to the computer either directly or via a USB link; or data can be transferred from data storage cards, first to a card reader/writer, and from there to the computer.

All the data collectors (built-in and hand-held) described here are capable of doing much more than just collecting data. The capabilities vary a great deal from one model to another. The computational characteristics of the main total stations themselves also vary widely. Some total stations (without the attached data collector) can compute remote elevations and distances between remote points, whereas others require the interfaced data collector to perform those functions.

Many recently introduced data collectors are really handheld computers, ranging in price from $600 to $2,000. Costs have moderated recently because the surveying field has seen the influx of a variety of less expensive handheld or "pocket" PCs for which total station data collection software has been written. Some data collectors have graphics touch screens (stylus or finger) that permit the surveyor to input or edit data rapidly. For example, some collectors allow the surveyor to select (by touching) the line-work function that enables displayed points to be joined into lines, without the need for special line-work coding (see Figure 5-29). If the total station is used alone, the capability of performing all survey computations, including closures and adjustments, is highly desirable. However, if the total station is used in topographic surveys, as a part of a system (field data collection, data processing, and digital plotting), then the extended computational capacity of some data collectors

(e.g., traverse closure computations) becomes less important. If the total station is used as part of a topographic surveying system (field data collection–data processing–digital plotting), the data collector need collect only the basic information: that is, slope distance, horizontal angle, vertical angle–determining coordinates, and the sighted points' attribute data, such as point number, point description, and any required secondary tagged data. Computations and adjustments to the field data are then performed on the office computer by using one of the many coordinate geometry programs now available for surveyors and engineers. More sophisticated data collectors are needed in many applications, for example, machine guidance, construction layouts, GIS (geographic information system) surveys, and real-time GPS (global positioning system) surveys.

Most early total stations (and some current models) use the absolute method for reading angles. These instruments are essentially optical coincidence instruments with photo-electronic sensors that scan and read the circles, which are divided into preassigned values, from 0 to 360° (or 0–400 grad or gon). Other models employ a rotary encoder technique of angle measurement [see Figure 5-1(b)]. These instruments have two (one stationary and one rotating) glass circles divided into many graduations (5,000–20,000). Light is projected through the circles, and converted to electronic signals which the instrument's processor converts to the angle measurement between the fixed and rotating circles. Accuracies can be improved, if needed, by averaging the results of the measurements on opposite sides of the circles. Some manufacturers determine the errors in the finished instrument (dual-axis compensated) and create a processor program to remove such errors during field use, thus giving one-face measurements the accuracy of two-face measurements. Both systems enable the surveyor to assign zero degrees (or any other value) conveniently to an instrument setting after the instrument has been sighted-in on the target backsight.

Total stations have onboard microprocessors that monitor the instrument status (e.g., level and plumb orientation, battery status, return signal strength) and make corrections to measured data for instrument orientation, when warranted. In addition, the microprocessor controls the acquisition of angles and distances and then computes horizontal distances, vertical distances, coordinates, and the like. Some total stations are designed so that the data stored in the data collector can be downloaded quickly to a computer via a USB connection. The manufacturer usually supplies the download program; a second program is required to translate the raw data into a format that is compatible with the surveyor's coordinate geometry (i.e., processing) programs.

Most total stations also enable the topographic surveyor to capture the slope distance and the horizontal and vertical angles to a point by simply pressing one button. The point number (most total stations have automatic point number incrementation) and point description for that point can then be entered and recorded. In addition, the wise surveyor prepares field notes showing the overall detail and the individual point locations. These field notes help the surveyor to keep track of the completeness of the work and are invaluable during data editing and preparation of the plot file for a digital plotter.

5.4 INSTRUMENT SETUP

The steps in a typical setup procedure for theodolites and total stations are listed below:

1. Place the instrument over the point with the tripod plate as level as possible and with two tripod legs on the downhill side, if applicable.

2. Stand back a pace or two and see if the instrument appears to be over the station; if it does not, adjust the location and check again from a pace or two away.

3. Move to a position 90° around from the original inspection location and repeat step 2. (This simple act of "eyeing-in" the instrument from two directions 90° apart takes only seconds but can save a great deal of time eventually.)

4. Check that the station point can now be seen through the optical plummet (or that the laser plummet spot is reasonably close to the setup mark). Then push in the tripod legs firmly by pressing down on the tripod shoe spurs.

5. While looking through the optical plumb (or at the laser spot), manipulate the leveling screws until the crosshair (bull's-eye) of the optical plummet or the laser spot is directly on the station mark.

6. Level the theodolite circular bubble by adjusting the tripod legs up or down. This is accomplished by noting which leg, when slid up or down, moves the circular bubble toward the bull's-eye. Upon adjusting that leg, either the bubble will move into the circle (the instrument is almost level) or it will slide around until it is exactly opposite another tripod leg. That leg should then be adjusted up or down until the bubble moves into the circle. If the bubble does not move into the circle, repeat the process. If this manipulation has been done correctly, the bubble will be roughly centered after the second leg has been adjusted; it is seldom necessary to adjust the legs more than three times. (Comfort can be taken from the fact that these manipulations take less time to perform than they do to read about!)

7. Perform a check through the optical plummet or note the location of the laser spot to confirm that it is still close to being over the station mark.

8. Turn one (or more) leveling screws to be sure that the circular bubble is now exactly centered (if necessary).

9. Loosen the tripod clamp bolt slightly and slide the instrument on the flat tripod top (if necessary) until the optical plummet or laser spot is exactly centered on the station mark.

Retighten the tripod clamp bolt and reset the circular bubble, if necessary. When sliding the instrument on the tripod top, do not twist the instrument, but move it in a rectangular fashion. This ensures that the instrument will not go seriously off level if the tripod top itself is not close to being level.

10. The instrument can now be leveled precisely by centering the tubular bubble. Set the tubular bubble so that it is aligned in the same direction as two of the foot screws. Turn these two screws (together or independently) until the bubble is centered. Then turn the instrument 90°, at which point the tubular bubble will be aligned with the third leveling screw. Finally turn that third screw to center the bubble. The instrument now should be level, although it is always checked by turning the instrument through 180°. On instruments with dual-axis compensation (see previous section), final leveling can be achieved by viewing the electronic display [see Figures 5-2(b) and 5-36(b)] and then turning the appropriate leveling screws. This latter technique is faster because the instrument does not have to be rotated repeatedly.

5.5 GEOMETRY OF THE THEODOLITE AND THE TOTAL STATION

The vertical axis of these instruments goes up through the center of the spindles and is oriented over a specific point on the earth's surface. The circle assembly and alidade rotate about this axis. The horizontal axis of the telescope is perpendicular to the vertical axis, and the telescope and vertical circle tilt on it. The line of sight (line of collimation) is a line joining the intersection of the reticle crosshairs and the center of the objective lens. The line of sight is perpendicular to the horizontal axis and should be exactly horizontal when the telescope level bubble is centered and when the vertical circle is set at 90°/270°, or 0° for vernier transits.

5.6 ADJUSTMENT OF THE THEODOLITE AND THE TOTAL STATION

Figure 5-7 shows the geometric features of both the theodolite and the total station. The most important relationships are as follows:

1. The axis of the plate bubble should be in a plane perpendicular to the vertical axis.

2. The vertical crosshair should be perpendicular to the horizontal axis (tilting axis).

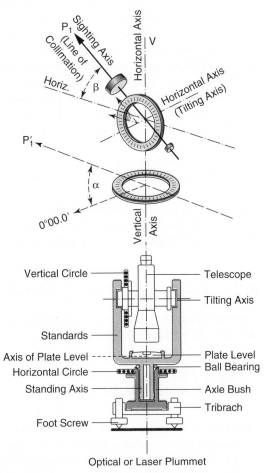

FIGURE 5-7 Geometry of the theodolite and total station

(Courtesy of Leica Geosystems Inc.)

3. The line of sight should be perpendicular to the horizontal axis.

4. The horizontal axis should be perpendicular to the vertical axis (standards adjustment).

Note: Modern surveying instruments have now become so complex that all adjustments should be left to factory-trained personnel.

5.6.1 Plate Bubbles

We noted that after a bubble has been centered, its adjustment is checked by rotating the instrument through 180°; if the bubble does not remain centered, it can be set properly by bringing the bubble halfway back using the foot screws. For example, if the check of a tubular bubble accuracy position shows that it is out by four division marks, the bubble can now be set properly by turning the foot screws until the bubble is only two division marks off center. The bubble should remain in this off-center position as the telescope is rotated, indicating that the instrument is in fact level. Although the instrument can now be used safely, it is customary to remove the error by adjusting the bubble tube.

The bubble tube can be adjusted by turning the capstan screws at one end of the bubble tube until the bubble becomes centered precisely. The entire leveling and adjusting procedure is repeated until the bubble remains centered as the instrument is rotated and checked in all positions. All capstan screw adjustments are best done in small increments; that is, if the end of the bubble tube is to be lowered, first loosen the lower capstan screw a slight turn (say, one-eighth). Then tighten (snug) the top capstan screw to close the gap. This incremental adjustment is continued until the bubble is centered precisely.

5.7 LAYING OFF ANGLES

5.7.1 Case 1

The angle is to be laid out no more precisely than the least count of the theodolite or total station. Consider a route survey where a deflection angle (31°12′ R) between two successive ℄ (centerline) alignments has been determined from aerial photos in order to position the roadway clear of natural obstructions [see Figure 5-8(a)]. The surveyor sets the instrument at the PI (point of intersection of the two straight-line tangents) and sights the back line with the telescope reversed and the horizontal circle set to zero. Next the surveyor transits (plunges) the telescope and turns off the required deflection angle and sets a point on line. The deflection angle is 31°12′ R and a point is set at a'. The surveyor then loosens the clamp (keeping the recorded angle on the scale), sights again at the back line, transits the telescope, and turns off the required value (31°12′ × 2 = 62°24′). It is very likely that this line of sight will not precisely match the first sighting at a'; if the line does not cross a' a new mark a'' is made and the instrument operator given a sighting on the correct point a midway between a' and a''.

5.7.2 Case 2

The angle is to be laid out more precisely than the least count of the instrument will permit directly. Assume that an angle of 27°30′45″ is required in a heavy construction layout, and that a 1-minute instrument is being used. In Figure 5-8(b), the instrument is set up at A zeroed on B, with an angle of 27°31′ turned to set point C'. The angle to point C' is measured a suitable number of times so that an accurate value of that angle can be determined.

Assume that the scale reading after four repetitions is 110°05′, giving a mean angle value of 27°31′15″ for angle BAC', an excess of 0°00′30″. If the layout distance of AC is 250.00 ft, point C can be located precisely by measuring from C' a distance $C'C$:

$$C'C = 250.00 \tan 0°00′30″$$
$$= 0.036 \text{ ft}$$

After point C has been located, its position can be verified by repeating angles to it from point B.

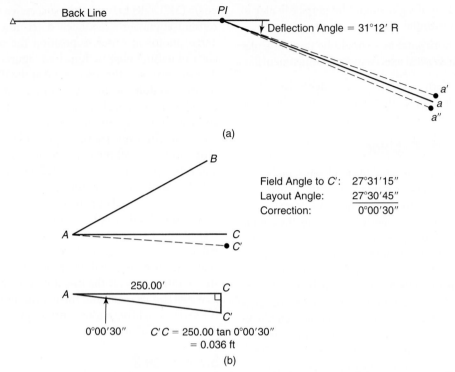

(a)

Field Angle to C':   27°31′15″
Layout Angle: 27°30′45″
Correction: 0°00′30″

250.00′

$C'C = 250.00 \tan 0°00′30″$
$= 0.036$ ft

(b)

FIGURE 5-8 Laying off an angle. (a) Case I, angle laid out to the least count of the instrument. (b) Case 2, angle to be laid out more precisely than the least count that the instrument will permit

5.8 PROLONGING A STRAIGHT LINE

Prolonging a straight line (also known as **double centering**) is a common survey procedure used every time a straight line must be extended. The best example of this requirement is in route surveying, where straight lines are routinely prolonged over long distances and often over very difficult terrain. The technique of reversion (the same technique used in repeating angles) is used to ensure that the straight line is accurately prolonged.

In Figure 5-9, the straight line BA is to be prolonged to C. With the instrument at A, a sight is made carefully on station B. The telescope is transited, and a temporary point is set at C_1. The instrument is rotated (telescope still inverted) back to station B and a new sighting is made on that station. The telescope is transited again (it is now in its original erect position) and a temporary mark is made at C_2, adjacent to C_1. The correct location of station C is established midway between C_1 and C_2 by measuring with a steel tape.

Over short distances, well-adjusted instruments show no appreciable displacement between points C_1 and C_2. Over the longer distances normally encountered in this type of work, however, all theodolites and total stations display a displacement

FIGURE 5-9 Double centering to prolong a straight line

between direct and reversed sightings; the longer the forward sighting, the greater the displacement.

Note: When using modern instruments with dual-axis compensation, the above techniques for prolonging a straight line can be modified. With dual-axis compensation, the central processing unit of the instrument can measure the effects of the standing axis tilt and correct all subsequent horizontal and vertical angles. Also, the surveyor can prolong the straight line by turning 180° from the backsight to the forward station (C). When this is done in both face 1 and face 2 (telescope transited), points C_1 and C_2 should be quite close together, thus confirming that the instrument has been calibrated correctly.

5.9 BUCKING-IN

It is sometimes necessary to establish a straight line between two points that themselves are not intervisible (i.e., a theodolite or total station set up at one point cannot, because of an intervening hill, be sighted at the other required point). It is usually possible to find an intermediate position from which both points can be seen. In Figure 5-10, points A and B are not intervisible, but point C is in an area from which both A and B can be seen. The **bucking-in** (also known as balancing-in or interlining) procedure is as follows. The instrument is set up in the area of C (at C_1) and as close to line AB as is possible to estimate. The instrument is roughly leveled and a

sight is taken on point A. Then the telescope is transited and a sight is taken toward B. The line of sight will, of course, not be on B but on point B_1, some distance away. Noting roughly the distance B_1B and the fractional position of the instrument along AB, an estimate is made about how far proportionately the instrument is to be moved to be on the line AB. The instrument is once again roughly leveled (position C_2), and the sighting procedure is repeated.

This trial-and-error technique is repeated until, after sighting A, the transited line of sight falls on point B or close enough to point B so that it can be set precisely by shifting the instrument on the tripod top. When the line has been established, a point is set at or near point C so that the position can be saved for future use. The entire procedure of bucking-in can be accomplished in a surprisingly short period of time. All but the final instrument setups are only roughly leveled, and at no time does the instrument have to be set up over a point.

5.10 INTERSECTION OF TWO STRAIGHT LINES

The intersection of two straight lines is also a very common survey technique. In municipal surveying, street surveys usually begin (0 + 00) at the intersection of the centerlines of two streets, and the station and angle of the intersections of all subsequent street centerlines are routinely determined.

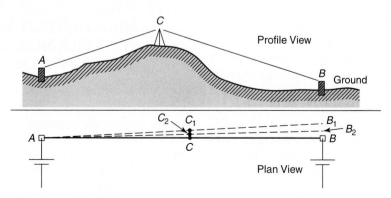

FIGURE 5-10 Bucking-in, or interlining

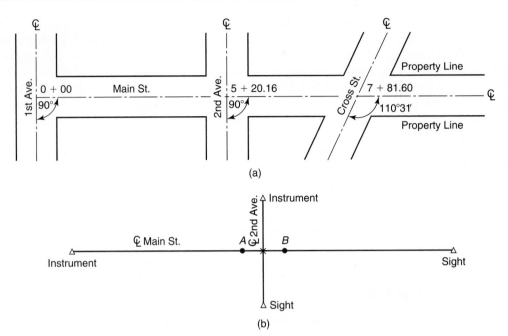

FIGURE 5-11 Intersection of two straight lines. (a) Example of the intersection of the centerlines of streets. (b) Intersecting technique

Figure 5-11(a) illustrates the need for intersecting points on a municipal survey, and Figure 5-13(b) illustrates just how the intersection point is located. In Figure 5-11(b), with the instrument set on a Main Street station and a sight taken also on the Main Street ℄ (the longer the sight, the more precise the sighting), two points (2–4 ft apart) are established on the Main Street ℄ on each side of where the surveyor estimates that the 2nd Avenue ℄ will intersect. The instrument is then moved to a 2nd Avenue station, and a sight is taken some distance away, on the far side of the Main Street ℄. The surveyor can stretch a plumb bob string over the two points (*A* and *B*) established on the Main Street ℄, and the instrument operator can note where on the string the 2nd Avenue ℄ intersects. If the two points (*A* and *B*) are reasonably close together (2–3 ft), the surveyor can use the plumb bob itself to take a line from the instrument operator on the plumb bob string; otherwise, the instrument operator can take a line with a pencil or any other suitable sighting target.

The intersection point is then suitably marked (e.g., nail and flagging on asphalt, wood stake with tack on ground), and then the angle of intersection and the station (chainage) of the point can be determined. After marking and referencing the intersection point, the surveyors remove temporary markers *A* and *B*.

5.11 PROLONGING A MEASURED LINE BY TRIANGULATION OVER AN OBSTACLE

In route surveying, obstacles such as rivers or chasms must be traversed. Whereas the alignment can be conveniently prolonged by double centering, the station (chainage) may be deduced from the construction of a geometric figure. In Figure 5-12, the distance from 1 + 121.271 to the station established on the far side of the river can be determined by solving the constructed triangle (triangulation).

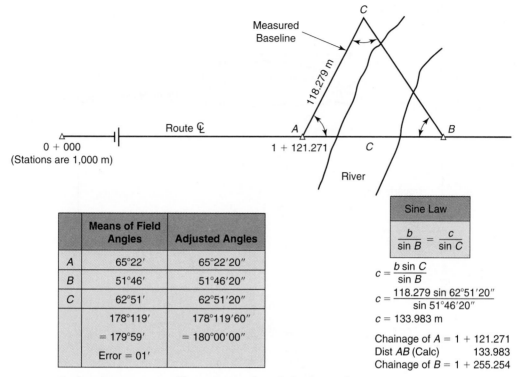

FIGURE 5-12 Prolonging a measured line over an obstacle by triangulation

The ideal (strongest) triangle is one having angles close to 60° (equilateral), although angles as small as 20° may be acceptable. The presence of rugged terrain and heavy tree cover adjacent to the river often results in a less than optimal geometric figure. The baseline and a minimum of two angles are measured so that the missing distance can be calculated. The third angle (on the far side of the river) should also be measured to check for mistakes and to reduce errors as shown by the correction method in Figure 5-12.

5.12 PROLONGING A LINE PAST AN OBSTACLE

In property surveying, obstacles such as trees often block the path of the survey. In route surveying, it is customary for the surveyor to cut down the offending trees (later construction will require them to be removed in any case). In property surveying, however, the removal of trees may be impermissible. Therefore, the surveyor must find an alternative method of providing distances and locations for blocked survey lines.

Figure 5-13(a) illustrates the technique of right-angle offset. Boundary line *AF* cannot be run because of the wooded area. The survey continues normally to point *B* just clear of the wooded area. At *B*, a right angle is turned (and doubled), and point *C* is located a sufficient distance away from *B* to provide a clear parallel line to the boundary line. The instrument is set at *C* and sighted at *B* (great care must be exercised because of the short sighting distance). An angle of 90° is turned to locate point *D*. Point *E* is located on the boundary line using a right angle and the offset distance used for *BC*. The survey can then continue to *F*. If distance *CD* is measured, then the required boundary distance (*AF*) is *AB* + *CD* + *EF*.

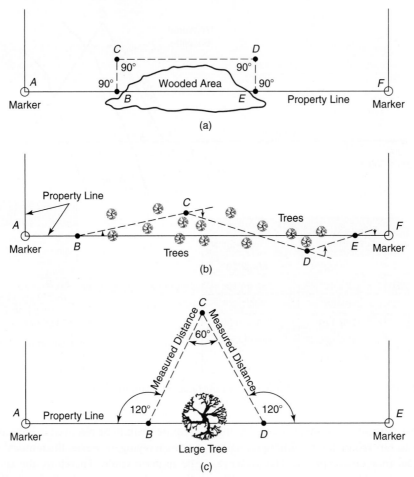

FIGURE 5-13 Prolonging a line past an obstacle. (a) Right-angle offset method. (b) Random-line method. (c) Triangulation method

If intermediate points are required on the boundary line between *B* and *E* (e.g., fencing layout), a right angle can be turned from a convenient location on *CD*, and the offset distance (*BC*) can be used to measure back to the boundary line. Use of a technique like this minimizes the destruction of trees and other obstructions. In Figure 5-13(b), trees are scattered over the area, preventing the establishment of a right-angle offset line. In this case, a random line (open traverse) is run (by deflection angles) through the scattered trees. The distance *AF* is the sum of *AB*, *EF*, and the resultant of *BC*, *CD*,

and *DE* (see Chapter 6 for appropriate computation techniques for "missing-course" problems such as this one).

In Figure 5-13(c), the line must be prolonged past an obstacle, a large tree. In this case, a triangle is constructed with the three angles and two distances measured as shown. As noted earlier, the closer the constructed triangle is to equilateral, the more accurate is the calculated distance (*BD*). Also, as noted earlier, the optimal equilateral figure cannot always be constructed due to topographic constraints, and angles as small as 20° are acceptable for many surveys.

We noted that the technique of right-angle offsets has a larger potential for error because of the weaknesses associated with several short sightings. At the same time, however, this technique gives a simple and direct method for establishing intermediate points on the boundary line. By contrast, the random-line and triangulation methods provide for stronger geometric solutions to the missing property line distances, but they also require less direct and much more cumbersome calculations for the placement of intermediate line points (e.g., fence layout).

5.13 TOTAL STATION FIELD TECHNIQUES

5.13.1 General Background

Total stations and their attached data collectors have been programmed to perform a wide variety of surveying functions. All total station programs require that the location of the instrument station and at least one reference station be identified so that all subsequent tied-in stations can be defined by their X (easting), Y (northing), and Z (elevation) coordinates. The instrument station's coordinates and elevation, together with the azimuth to the backsight reference station(s) (or its coordinates), can be entered in the field or uploaded prior to going out to the field, or can be computed through resection techniques; see Section 5.13.4. After setup, and before the instrument has been oriented for surveying as described above, the hi (height of instrument above the instrument station) and prism heights must be measured and recorded. The use of prismless total stations (see Section 3.24) will expedite many of the following field procedures.

Typical total station programs include point location, missing line measurement, resection, azimuth determination, determination of remote object elevation, offset measurements, layout or setting-out positions, and area computations. All these topics are discussed in the following sections.

5.13.2 Point Location

5.13.2.1 General Background After the instrument has been set up and properly oriented, the coordinates (northing, easting, and elevation) of any sighted point can be determined, displayed, and recorded in the following format: N.E.Z. (north, east, elevation) or E.N.Z. The format chosen reflects the design software program used to process the field data. At this time, the sighted point is numbered and coded for attribute data (point description) all of which is recorded with the location data. This program is used extensively in topographic surveys. See Figure 5-14.

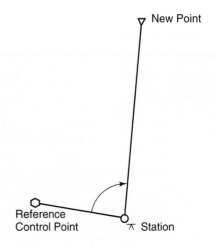

Known: N, E, and Z coordinates of the instrument station ($\bar{\lambda}$).
  N, E, and Z coordinates of reference control point, or at least the azimuth of the line joining the instrument station and the control point.

Measured: Angles, or azimuths, from the control point and distances to the new point from the instrument station.

Computed: N, E, and Z coordinates of new points. Azimuth of the line (and its distance) joining the instrument station to the new point.

FIGURE 5-14 Point location program inputs and solutions

5.13.2.2 Trigonometric Leveling   In Chapter 2, we saw how elevations were determined using automatic and digital levels. Section 2.13 briefly introduced the concept of trigonometric leveling. Field practice has shown that total station observations can produce elevations at a level of accuracy suitable for the majority of topographic, engineering, and project control surveying applications. For elevation work, the total station should be equipped with dual-axis compensation to ensure that angle errors remain tolerable. For more precise work, the total station should have the ability to measure to 1 second of arc.

In addition to the uncertainties caused by errors in the zenith angle, the surveyor should be mindful of the errors associated with telescope collimation; other instrument and prism imperfections; curvature, refraction, and other natural effects; and the errors associated with the use of a handheld prism pole. If the accuracy of a total station is unknown or suspect, and if the effects of the measuring conditions are unknown, the surveyor can compare her or his trigonometric leveling results with differential leveling results (using the same control points) under similar measuring conditions.

To improve precision, multiple readings can also be averaged; readings can be taken from both face 1 and face 2 of the instrument. Simultaneous reciprocal observations at both ends of the survey line may be the most reliable technique of determining differences in elevation using total stations. Table 5-1 shows the effect of angle uncertainty on various distances. The basis for the tabulated results is the cosine of $90° \pm$ the angle uncertainty or the sine of $0° \pm$ the angle uncertainty times the distance. Compare the errors shown in Table 5-1 with respect to the closure requirements shown in Tables 2-1 and 2-2 (see Figure 5-15).

5.13.3 Missing Line Measurement

This program enables the surveyor to determine the horizontal and slope distances between any two sighted points as well as the directions of the lines joining those sighted points. Onboard programs first determine the *N*, *E*, and *Z* coordinates of the sighted points and then compute (inverse) the joining distances and directions. See Figure 5-16.

5.13.4 Resection

This technique permits the surveyor to set up the total station at any convenient position (sometimes referred to as a **free station**) and then determine the coordinates and elevation of that instrument position by sighting previously coordinated reference stations. When sighting only two points of known position, it is necessary to measure and record both the distances and the angle between

Table 5-1   Elevation Errors in Feet or Meters

Distance (ft/m)	Vertical Angle Uncertainty				
	1 second	5 seconds	10 seconds	20 seconds	60 seconds
100	0.0005	0.0024	0.005	0.0097	0.029
200	0.0010	0.0048	0.010	0.0194	0.058
300	0.0015	0.0072	0.015	0.029	0.087
400	0.0019	0.0097	0.019	0.039	0.116
500	0.0024	0.0121	0.024	0.049	0.145
800	0.0039	0.0194	0.039	0.078	0.233
1,000	0.0048	0.0242	0.0485	0.097	0.291

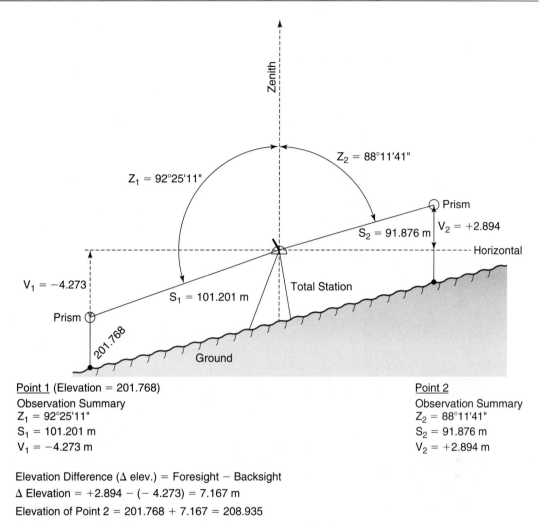

Point 1 (Elevation = 201.768)
Observation Summary
Z_1 = 92°25'11"
S_1 = 101.201 m
V_1 = −4.273 m

Point 2
Observation Summary
Z_2 = 88°11'41"
S_2 = 91.876 m
V_2 = +2.894 m

Elevation Difference (Δ elev.) = Foresight − Backsight
Δ Elevation = +2.894 − (− 4.273) = 7.167 m
Elevation of Point 2 = 201.768 + 7.167 = 208.935

FIGURE 5-15 Example of trigonometric leveling using a total station

the reference points; when sighting several points (three or more) of known position, it is necessary only to measure the angles between the points. It is important to stress that most surveyors take more readings than are minimally necessary to obtain a solution. These redundant measurements give the surveyor increased precision and a check on the accuracy of the results. Once the instrument station's coordinates have been determined and a known control point back-sighted, the instrument is now oriented, and the surveyor can continue to

survey using any of the other techniques described in this section (see Figure 5-17).

5.13.5 Azimuth Determination

When the coordinates of the instrument station and a backsight reference station have been entered into the instrument processor, the azimuth of a line joining any sighted points can be readily displayed and recorded (see Figure 5-18). When using a motorized total station equipped with automatic

Known: N, E, and Z coordinates of the instrument station (π)
N, E, and Z coordinates of a reference control point, or at least the azimuth of the line joining the instrument station and the control point

Measured: Angles, or azimuths, from the control point and distances to the new points from the instrument station

Computed: N, E, and Z coordinates of new points
Azimuths and horizontal, slope and vertical distances between the sighted points
Slopes of the lines joining any sighted points

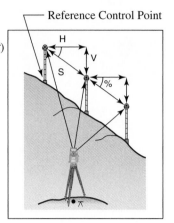

FIGURE 5-16 Missing line measurement program inputs and solutions

(Sketch courtesy of Sokkia Corporation, Overland Park, KS)

Known: N, E, and Z coordinates of control point #1
N, E, and Z coordinates of control point #2
N, E, and Z coordinates of additional sighted control stations (up to a total of 10 control points). These can be entered manually or uploaded from the computer depending on the instrument's capabilities.

Measured: Angles between the sighted control points
Distances are also required from the instrument station to the control points, if only two control points are sighted
The more measurements (angles and distances), the better the resection solution

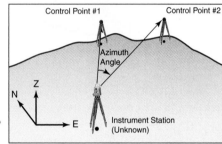

FIGURE 5-17 Resection program inputs and solutions

(Sketch courtesy of Sokkia Corporation, Overland Park, KS)

Known: N, E, and Z coordinates of the instrument station (π)
N, E, and Z coordinates of a reference control point, or at least the azimuth of the line joining the instrument station and the control point

Measured: Angles from the control point and distances to the new points from the instrument station

Computed: Azimuth of the lines joining the new points to the instrument station
Slopes of the lines joining any sighted points

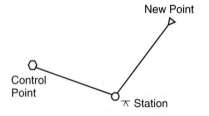

FIGURE 5-18 Azimuth program inputs and solutions

Known: N, E, and Z coordinates of the instrument station (⊼)
N, E, and Z coordinates of a reference control point (not shown), or at least the azimuth of the line joining the instrument station and the control point (optional) Using a prismless total station can expedite this field procedure

Measured: Horizontal angle, or azimuth from the reference control point (optional) and distance from the instrument station to the prism being held directly below (or above) the target point
Vertical angles to the prism and to the target point hi and height of the prism

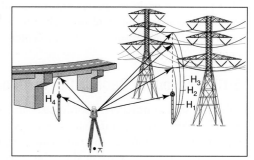

Computed: Distance from the ground to the target point (and its coordinates if required)

FIGURE 5-19 Remote object elevation program inputs and solutions
(Sketch courtesy of Sokkia Corporation, Overland Park, KS)

target recognition (ATR), the automatic angle measurement program permits you to repeat angles automatically to up to 10 previously sighted stations (including on face 1 and face 2). Just specify the required accuracy (standard deviation) and the sets of angles to be measured (up to 20 angle sets at each point), and the motorized instrument performs the necessary operations automatically.

5.13.6 Remote Object Elevation

The surveyor can determine the heights of inaccessible points (e.g., electricity conductors, bridge components), while using a total station that is already set up, simply by sighting the pole-mounted prism as it is being held directly under the object. When the object itself is then sighted, the object height can be promptly displayed (the prism height must first be entered into the total station). See Figure 5-19.

5.13.7 Offset Measurements

Hidden Object Offsets When an object is hidden from the total station, a measurement can be taken to the prism held in view of the total station, and then the offset distance can be measured. The angle (usually 90°) or direction to the hidden object along with the measured distance is entered into the total station, enabling it to compute the position of the hidden object (see Figure 5-20).

Known: N, E, and Z coordinates of the instrument station (⊼)
N, E, and Z coordinates (or the azimuth) for the reference control point (not shown)

Measured: Azimuth from the control point to the offset point
Distance from the instrument station to the offset point
Distance (*l*) from the offset point (at a right angle to the instrument line-of-sight) to the measuring point

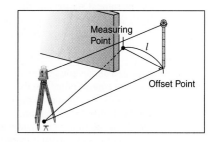

Computed: N, E, and Z coordinates of the "hidden" measuring points
Azimuths and distance from the instrument station to the "hidden" measuring point

FIGURE 5-20 Offset measurements (distance) programs solutions and inputs
(Sketch courtesy of Sokkia Corporation, Overland Park, KS)

Object Center Offsets To locate a solid object's center position, angles are measured to the prism as it is being held on each side of the object, equidistant from the object's center (see Figure 5-21).

5.13.8 Layout or Setting-Out Positions

After the station numbers, coordinates, and elevations of the layout points have been uploaded into the total station (or entered manually), the layout/setting-out software enables the surveyor to locate any layout point simply by entering that point's station number when prompted by the layout software. The instrument's display shows the left/right, forward/back, and up/down movements needed to place the prism in each of the desired position locations. This capability is a great aid in property and construction layouts (see Figure 5-22).

5.13.9 Area Computation

When this program has been selected, the processor computes the area enclosed by a series of measured points. First, the processor determines the coordinates of each station as described earlier and then, it uses those coordinates to compute the area in a manner similar to that described in Section 6.14 (see Figure 5-23).

Known: N, E, and Z coordinates of the instrument station (π)
N, E, and Z coordinates (or the azimuth) of the reference control point (not shown)

Measured: Angles from the control point to prism being held on each side of the measuring point (The prism must be held such that both readings are the same distance from the instrument station—if both sides are measured)

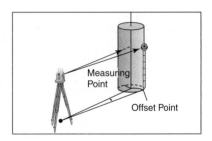

Computed: N, E, and Z coordinates of the "hidden" measuring points
Azimuths and distance from the instrument station to the "hidden" measuring point

FIGURE 5-21 Offset measurements (angles) program inputs and solutions

(Sketch courtesy of Sokkia Corporation, Overland Park, KS)

Known: N, E, and Z coordinates of the instrument station (π)
N, E, and Z coordinates of a reference control point (not shown), or at least the azimuth of the line joining the instrument station and the control point
N, E, and Z coordinates of the proposed layout points (Entered manually, or previously uploaded from the computer)

Measured: Indicated angles (on the instrument display), or azimuths, from the control point and distances to each layout point
The angles may be turned manually or automatically if a servo motor-driven instrument is being used.
The distances (horizontal and vertical) are continually remeasured as the prism is eventually moved to the required layout position

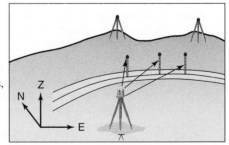

FIGURE 5-22 Laying out, or setting out, program inputs and solutions

(Sketch courtesy of Sokkia Corporation, Overland Park, KS)

Known: N, E, and Z coordinates of the instrument station (⋏)
 N, E, and Z coordinates of a reference control point (not shown), or at least the azimuth of the line joining the instrument station and the control point
 N, E, and Z coordinates of the proposed layout points (Entered manually, or previously uploaded from the computer)

Measured: Angles, or azimuths, from the control point and distances to the new points, from the instrument station.

Computed: N, E, and Z coordinates of the area boundary points
 Area enclosed by the coordinated points

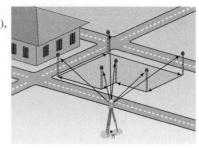

FIGURE 5-23 Area computation program inputs and solutions

(Sketch courtesy of Sokkia Corporation, Overland Park, KS)

5.14 SUMMARY OF TYPICAL TOTAL STATION CHARACTERISTICS

A *Parameter input selection*

 1. Angle units: degrees or gon

 2. Distance units: feet or meters

 3. Pressure units: inches Hg or mm Hg*

 4. Temperature units: °F or °C*

 5. Prism constant (usually—0.03 m). This constant is normally entered at the factory. When measuring to reflective paper (or when using reflectorless techniques), this prism constant must be corrected.

 6. Offset distance (used when an object is hidden from the total station)

 7. Face 1 or face 2 selection

 8. Automatic point number incrementation

 9. Height of instrument (hi)

 10. Height of reflector (HR)

 11. Point numbers and code numbers for occupied and sighted stations

 12. Date and time settings—for total stations with onboard clocks

B *Capabilities* (common to many total stations)

 1. Monitor: battery status, signal attenuation, horizontal and vertical axes status, collimation factors

 2. Compute coordinates: northing, easting, and elevation, or easting, northing, and elevation.

 3. Traverse closure and adjustment, and areas

 4. Topography reductions

 5. Remote object elevation, object heights

 6. Distances between remote points (missing line measurement)

 7. Inversing

 8. Resection

 9. Layout (setting out)

 10. Horizontal and vertical collimation corrections

 11. Vertical circle indexing and other error calibration programs

 12. Records search and review

 13. Programmable features; that is the loading of external programs

 14. Transfer of data to the computer (downloading)

 15. Transfer of computer files to the data collector (uploading) for layout and reference purposes.

---

*Some newer instruments have built-in sensors that detect atmospheric effects and automatically correct readings for these natural errors.

5.15 FIELD PROCEDURES FOR TOTAL STATIONS

Total stations can be used in any type of preliminary survey, control survey, or layout survey. They are particularly well suited for topographic surveys, in which the surveyor can capture the northings, eastings, and elevations of 700–1,000 points per day. This significant increase in productivity means that in some medium-density areas, ground surveys are once again competitive in cost compared to aerial surveys. Although the increase in efficiency in the field survey is notable, an even more significant increase in efficiency occurs when the total station is part of a computerized surveying system, including data collection, data processing (reductions and adjustments), and plotting; see Figure 1-7.

One of the notable advantages of electronic surveying techniques is that measurement data are recorded in the electronic field book (EFB); this cuts down considerably on the time required to record data and on the opportunity to make transcription mistakes. However, this does not mean that manual field notes are a thing of the past. Even in electronic surveys, there is a need for neat, comprehensive field notes. At the very least, manual field notes will contain the project title, field personnel, equipment description, date, weather, instrument setup station identification, and back-sight station(s) identification.

In addition, for topographic surveys, many surveyors include in the manual notes a sketch showing the area's selected (in some cases, all) details, such as trees, catch basins, and poles, and linear details, such as water's edge, curbs, walks, and building outlines. As the survey proceeds, the surveyor will place all or selected point identification numbers directly on the sketch feature, thus clearly showing the type of detail being recorded.

A topographic survey crew usually consists of the instrument operator and the prism holder, but if the topographic entities being surveyed are spread out such that it takes the prism-holder more than 30–50 seconds to walk from one tie-in point to the next, it might be more efficient to employ two prism-holders during the course of the survey; time is wasted if the instrument operator has to continually wait for the prism to be placed at the next pick up point.

After the data have been transferred to the computer, and as the graphics features are edited, the presence of manual field notes is invaluable for clearing up any ambiguous labeling or numbering of survey points.

5.15.1 Initial Data Entry

Most data collectors are designed to prompt for, or accept, some or all of the following initial configuration data:

- Project description
- Date and crew members' names
- Temperature
- Pressure [some data collectors require a parts per million (ppm) correction input, which is read from a temperature/pressure graph; see Figure 3-19]
- Prism constant (0.03 m is a typical value; check the specifications)
- Curvature and refraction settings (see Chapter 2)
- Sea-level corrections (see Chapter 10)
- Number of measurement repetitions: angle or distance (the average value is computed)
- Choice of face 1 and face 2 positions
- Choice of automatic point number incrementation (yes/no), and the value of the first point in a series of measurements
- Choice of Imperial units or metric units for all data

Many of these prompts can be bypassed, which causes the microprocessor to use previously selected default values or settings, in its computations. After the initial data have been entered, and the operation mode has been selected, most data collectors prompt the operator for all station and measurement entries. The following sections describe typical survey procedures.

5.15.2 Survey Station Descriptors

Each survey station, or shot location (point), must be described with respect to surveying activity (e.g., backsight, intermediate sight, foresight), station identification, and other attribute data.

Once a total station program or operation has been selected, some will prompt for the data entry [e.g., identification of occupied station, and backsight reference station(s)] and then automatically prompt for appropriate labels for intermediate sightings. Point description data can be entered as alpha or numeric codes (see Figure 5-24). This descriptive data will also show up on the printout and can (if desired) be tagged to show up on the plotted drawing.

Point Identification Codes
(shown is part of Seneca dictionary)

Survey Points

01	BM	Bench Mark
02	CM	Concrete Monument
03	SIB	Standard Iron Bar
04	IB	Iron Bar
05	RIB	Round Iron Bar
06	IP	Iron Pipe
07	WS	Wooden Stake
08	MTR	Coordinate Monument
09	CC	Cut Cross
10	N&W	Nail and Washer
11	ROA	Roadway
12	SL	Street Line
13	EL	Easement Line
14	ROW	Right of Way
15	CL	Centerline

Topography

16	EW	Edge Walk
17	ESHLD	Edge Shoulder
18	C&G	Curb and Gutter
19	EWAT	Edge of Water
20	EP	Edge of Pavement
21	RD	CL Road
22	TS	Top of Slope
23	BS	Bottom of Slope
24	CSW	Concrete Sidewalk
25	ASW	Asphalt Sidewalk
26	RW	Retaining Wall
27	DECT	Deciduous Tree
28	CONT	Coniferous Tree
29	HDGE	Hedge
30	GDR	Guide Rail
31	DW	Driveway
32	CLF	Chain Link Fence
33	PWF	Post and Wire Fence
34	WDF	Wooden Fence

Code Sheet for Field Use				
1 BM	2 CM	3 SIB	4 IB	5 RIB
6 IP	7 WS	8 MTR	9 CC	10 N&W
11 ROA	12 SL	13 EL	14 ROW	15 CL
16 EW	17 ESHL	18 C&G	19 EWAT	20 EP
21 RD	22 TS	23 BS	24 CSW	25 ASW
26 RW	27 DECT	28 CONT	29 HDGE	30 GDR
31 DW	32 CLF	33 PWF	34 WDF	35 SIGN
36 MB	37 STM	38 HDW	39 CULV	40 SWLE
41 PSTA	42 SAN	43 BRTH	44 CB	45 DCB
46 HYD	47 V	48 V CH	49 M CH	50 ARV
51 WKEY	52 HP	53 UTV	54 LS	55 TP
56 PED	57 TMH	58 TB	59 BCM	60 GUY
61 TLG	62 BLDG	63 GAR	64 FDN	65 RWYX
66 RAIL	67 GASV	68 GSMH	69 G	70 GMRK
71 TL	72 PKMR	73 TSS	74 SCT	75 BR
76 ABUT	77 PIER	78 FTG	79 EDB	80 POR
81 SLS	82 WTT	83 STR	84 BUS	85 PLY
86 TEN	0	0	0	0
0	0	0	0	0
0	0	0	0	0

(a)   (b)

FIGURE 5-24 (a) Some alphanumeric codes for sighted-point descriptions. (b) Code sheet for field use

Some data collectors (see Figure 5-25) are designed to work with all (or most) total stations on the market. These data collectors have their own routines and coding requirements. As this technology continues to evolve and take over many surveying functions, more standardized procedures should develop. Newer data collectors (and computer programs) permit the surveyor to enter not only the point code, but also various levels of attribute data for each coded point. For example, a tie-in to a utility pole can also tag the pole number, the use (e.g., electricity, telephone), the material (e.g., wood, concrete, steel), connecting poles, year of installation,

FIGURE 5-25 Trimble's TSCe data collector for use with robotic total stations and GPS receivers. When used with TDS Survey Pro Robotics software, this collector features real-time maps, 3D design stakeout, and interactive DTM with real-time cut-and-fill computations. When used with TDS Survey pro GPS software, this collector can be used for general GPS work, as well as for RTK measurements at centimeter-level accuracy

and so on. One obvious benefit of having street hardware inventories in a digital format is that maintenance work orders can be automatically generated by the system. This type of expanded attribute data is typical of the data collected for a GIS. See Chapter 9.

5.15.2.1 AASHTO's Survey Data Management System (SDMS)

The American Association of State Highway and Transportation Officials (AASHTO) has developed a survey data management system (SDMS) in 1991 to aid in the nationwide standardization of field coding and computer-processing procedures used in highway surveys. The field data can be captured automatically by the data collector (EFB), as in the case of total stations, or the data can be entered manually in the data collector for a wide variety of theodolite and level surveys. AASHTO's software is compatible with most recently designed total stations and some third-party data collectors, and data processing can be accomplished using most MS-DOS–based computers. For additional information on these coding standards, contact AASHTO at 444 N. Capitol St. N.W., Suite 225, Washington, D.C. 20001.

5.15.2.2 Occupied-Point (Instrument Station) Entries

Before the data collection can begin, the instrument station must be defined so that the processor can compute the positions of all sighted intermediate sight (IS) points:

- Height of instrument [this entry is the measurement from the survey station mark to the optical/electronic center mark of the instrument; see the cross (just below the logo) on the total station shown in Figure 5-31(b)]

- Station number, for example, 111 (see the example in Figure 5-26)

- Station identification code (see the code sheet, Figure 5-24)

- Coordinates of occupied station; the coordinates can be assumed, state plane, or Universal Transverse Mercator (UTM).

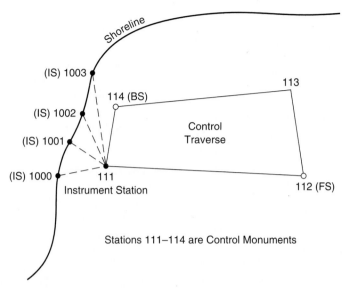

FIGURE 5-26 Sketch showing intermediate shoreline ties to a control traverse

- Coordinates of backsight (BS) station, or the reference azimuth to BS station

Note: With some data collectors, the coordinates of the above stations may instead be uploaded from computer files. Once the surveyor identifies (by point numbers) specific points as being instrument station, and backsight reference station(s), the coordinates of those stations then become active in the processor.

5.15.2.3 Sighted Point Entries

- Height of prism/reflector (HR) above base of pole; measured value is entered
- Station number, for example, 114 (BS); see Figure 5-26
- Station identification code (see Figure 5-24)

5.15.3 Typical Procedures for the Example Shown in Figure 5-26

5.15.3.1 Data Collection

1. Enter the initial data and instrument station data, as shown in Sections 5.15.1 and 5.15.2. Measure the height of the instrument, or adjust the height of the reflector to equal the height of instrument.

2. Sight at station 114. Zero the horizontal circle (any other value can be set instead of zero, if desired). Most total stations have a zero-set button.

3. Enter a code (e.g., BS), or respond to the data collector prompt.

4. Measure and enter the height of the prism/reflector (HR). If the height of the reflector has been adjusted to equal the height of the instrument, the value of 1.000 is often entered for both values because these hi and HR values usually cancel out in computations.

5. Press the appropriate "measure" buttons, for example, slope distance, horizontal angle, vertical angle.

6. Press the record button after each measurement; most instruments measure and record slope, horizontal, and vertical data after the pressing of just one button (when they are in the automatic mode).

7. After the station measurements have been recorded, the data collector prompts for the

station point number (e.g., 114) and the station identification code (e.g., 02 from Figure 5-24, which identifies "concrete monument").

8. If appropriate, as in traverse surveys, the next sight is to the FS station (112). Repeat steps 4, 5, 6, and 7, using correct data.

9. While at station 111, any number of intermediate sights (IS) can be taken to define the topographic features being surveyed (the shoreline, in this example). Most instruments have the option of speeding up the work by employing *automatic point number incrementation*. For example, if the topographic reading is to begin with point number 1000, the surveyor will be prompted to accept the next number in sequence (e.g., 1001, 1002, 1003) at each new reading. If the prism is later held at a previously numbered point (e.g., control point 107), the prompted value can be temporarily overridden by entering 107. The prism/reflector is usually mounted on an adjustable-length prism pole, and sometimes the height of the prism (HR) set to the height of the total station (hi). The prism pole can be steadied by using a bipod, as shown in Figure 3-21, to improve the accuracy for more precise sightings. Some software permits the surveyor to identify, by attribute name or further code number (see Figures 5-27 and 5-28), the stringed points that will be connected on the resultant plan (e.g., shoreline points in this example). This connect (on and off) feature permits the field surveyor to virtually prepare the plan (for a graphics terminal or plotter) while performing the actual field survey. The surveyor also can connect the points later while in edit mode on the computer depending on the software in use. Clear field notes are essential for this activity. Data collectors with touch-screen graphics can have points joined by touching the screen with a stylus or finger without any need for special stringing coding.

10. When all the topographic detail in the area of the occupied station (111) has been collected, the total station can be moved to the next traverse station (e.g., 112 in this example). When the total station is to be moved to another setup station, the instrument is always removed from the tripod and carried separately by its handle or in its case. The data collection can proceed in a similar manner as before, that is, BS at STA. 111, FS at STA. 113, with all relevant IS readings.

5.15.3.2 Data Transfer and Data Processing
In the example shown in Figure 5-24, the collected data now must be downloaded to a computer. The download computer program is normally supplied by the total station manufacturer, and the actual transfer can be sent through a USB cable. Once the data are in the computer, the data must be translated into a format that is compatible with the computer program that will process the data. This translation program is usually written or purchased separately by the surveyor.

Many modern total stations have data stored onboard, thus eliminating the handheld data collectors. Some instruments store data on a module that can be transferred to a computer-connected reading device (see Figure 1-4). Other total stations, including the Geodimeter (see Figure 5-34), can be downloaded by connecting the instrument (or its detachable keyboard) directly to the computer.

If the topographic data have been tied to a closed traverse, the traverse closure is calculated, and then all adjusted values for northings, eastings, and elevations (Y, X, and Z) are computed. Some total stations have sophisticated data collectors that can perform preliminary analysis, adjustments, and coordinate computations; others require the computer program to perform these functions.

Once the field data have been stored in coordinate files, the data required for plotting by digital plotters can be assembled, and the survey can be quickly plotted at any desired scale. The survey can also be plotted at an interactive graphics terminal for graphics editing, using one of the many available computer-aided design (CAD) programs.

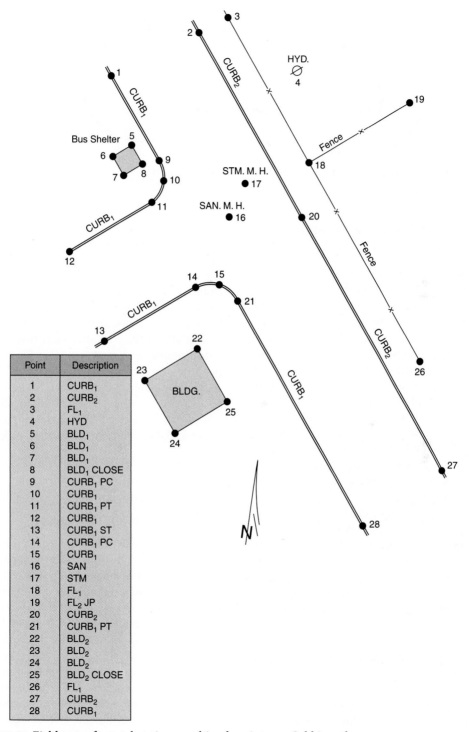

Point	Description
1	$CURB_1$
2	$CURB_2$
3	FL_1
4	HYD
5	BLD_1
6	BLD_1
7	BLD_1
8	BLD_1 CLOSE
9	$CURB_1$ PC
10	$CURB_1$
11	$CURB_1$ PT
12	$CURB_1$
13	$CURB_1$ ST
14	$CURB_1$ PC
15	$CURB_1$
16	SAN
17	STM
18	FL_1
19	FL_2 JP
20	$CURB_2$
21	$CURB_1$ PT
22	BLD_2
23	BLD_2
24	BLD_2
25	BLD_2 CLOSE
26	FL_1
27	$CURB_2$
28	$CURB_1$

FIGURE 5-27 Field notes for total station graphics descriptors—Sokkia codes

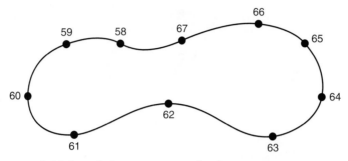

FIGURE 5-28 In-sequence field shots defining a topographic feature

5.15.4 Field-Generated Graphics Coding

Many surveying software programs permit the field surveyor to identify field data shots so that subsequent processing will produce appropriate computer graphics. This section includes some examples that may be regarded as typical. As noted in the previous section, newer total stations and their controllers have touch-screen graphics which permit the surveyor to edit the screen graphics field plot (including generating lines between selected points) simply by touching the screen (with stylus or finger) at the appropriate survey points. Older total stations require special coding to achieve similar results (see Figure 5-27).

Sokkia software gives the code itself a stringing capability (e.g., fence1, CURB1, CURB2, ℄), which the surveyor can easily turn on and off (see Figure 5-27). Several companies are now producing data collectors that permit the surveyor to create graphics in the field by simply identifying the appropriate graphics commands from the screen menus and then by tapping on the selected points shown on the data collector's graphics screen.

In some cases, it may be more efficient to assign the point descriptors from the computer keyboard after the survey has been completed. For example, the entry of descriptors is time consuming on some EFBs, particularly in automatic mode. In addition, some topographic features (e.g., the edge of the water in a pond or lake) can be captured in sequence, thus permitting the surveyor to edit in these descriptors efficiently from the computer in the processing stage instead of repeatedly entering dozens or even hundreds of identical attribute codes. Some data collectors prompt for the last entry thus avoiding the rekeying of identical attributes or descriptors. If the point descriptors are to be added at the computer, clear field notes are indispensable. See Figure 5-28 for an illustration of this stringing technique. The pond edge has been picked up (defined) by 10 shots, beginning with 58 and ending with 67. The point description edit feature can be selected from a typical pull-down menu, and the points can be described as follows:

1. "Points to be described?" 58..66 (*enter*)
2. "Description?" POND1 (*enter*)
3. "Points to be described?" 67 (*enter*)
4. "Description?" .POND10 (*enter*)

After the point descriptions have been suitably coded (either by direct field coding or by the editing technique shown here), a second command is accessed from another pull-down menu that simply (in one operation) converts the coded point description file so that the shape of the pond is produced in graphics. The four descriptor operations listed above require less work than would be required to describe each point using field entries for the 10 points shown in this example. Larger features requiring many more field shots would be even more conducive to this type of post-survey editing of descriptors.

It is safe to say that most projects requiring graphics to be developed from total station surveys utilize a combination of point description field coding and post-survey point description editing. Note that the success of some of these modern surveys still depends to a significant degree on old-fashioned, reliable survey field notes. The drafting of the plan of survey is increasingly becoming the responsibility of the surveyor, either through direct field-coding techniques or through post-survey data processing. All recently introduced surveying software programs enable the surveyor to produce a complete plan of survey on a digital plotter. Data collector software is now capable of exporting files in .dxf format for AutoCad post processing or shape files for further processing in ESRI GIS programs. Also, alignments and cross sections are created on desktop computers and then exported to data collectors in the recently established standard Land.xml file format.

5.15.5 Layout Surveys Using Total Stations

We saw in the previous section that total stations are particularly well suited for collecting data in topographic surveys. We also noted that the collected data can be readily downloaded to a computer and processed into point coordinates—northing, easting, elevation (Y, X, Z)—along with point attribute data. The significant increases in efficiency made possible with total station preliminary surveys can also be realized in layout surveys when the original point coordinates exist in computer memory together with the coordinates of all the key design points. To illustrate, consider the example of a road construction project. First, the topographic detail is collected using total stations set up at various control points (preliminary survey). The detail is then transferred to the computer; adjusted, if necessary; and converted into $Y, X,$ and Z coordinates. Various coordinate geometry and road-design programs can then be used to design the proposed road. When the proposed horizontal, cross-section, and profile alignments have been established, the proposed

coordinates ($Y, X,$ and Z) for all key horizontal and vertical (elevation) features can be computed and stored in computer files. The points coordinated will include top-of-curb and centerline positions at regular stations as well as all changes in direction or slope. Catch basins, traffic islands, and the like are also included, as are all curved or irregular road components.

These 3D data files now include coordinates of all control stations, all topographic detail, and all design component points. Control point and layout point coordinates ($Y, X,$ and Z) can then be uploaded into the total station. A layout is accomplished as follows. First, set up at an identified control point, lock onto another identified control point, and then set the instrument to layout mode. Second, while still in layout mode, enter the first layout point number (in response to a screen prompt) and note that the required layout angle or azimuth and layout distance are displayed on the screen. To set the first (and all subsequent) layout point(s) from the instrument station, the layout angle (or azimuth) is turned (automatically by motorized total stations) and the distance is set by following the display instructions (backward/forward, left/right, and up/down) to locate the desired layout point's position. When the total station is set to tracking mode, first set the prism close to the target layout point by rapid trial-and-error measurements and then switch the instrument back to precise measurement mode for final positioning.

Figure 5-29 illustrates a computer printout for a road construction project. The total station is set at control monument CX-80 with a reference BS on RAP (reference azimuth point) 2, which is point 957 on the printout. The surveyor has a choice of (1) setting the actual azimuth (213°57′01″, line 957) on the BS and then turning the horizontal circle to the printed azimuths of the desired layout points, or (2) setting 0 degrees for the BS and then turning the clockwise angle listed for each desired layout point. As a safeguard, after sighting the reference BS, the surveyor usually sights a second or third control monument (see the first 12 points on the printout) to check the azimuth or angle setting. The computer

MIS # E 152G

THE MUNICIPALITY OF METROPOLITAN TORONTO - DEPARTMENT OF ROADS AND TRAFFIC

W. R. ALLEN ROAD FROM SHEPPARD AVENUE TO STANSTEAD DRIVE

ENGINEERING STAKEOUT

FROM STATION 269+80.00   TO STATION 271+30.00

BASE LINE

INSTRUMENT ON CONST CONTROL MON CX-80
AZIMUTH 213-57- 1

SIGHTING   RAP #2 - ANTENNA C.F.B.

POINT NO.	STATION	DESCRIPTION	OFFSET FROM BASELINE		AZIMUTH DEG-MIN-SEC	DISTANCE	CLOCKWISE TURN ANGLE	ELEVATION	COORDINATES	
									NORTH	EAST
876	269+89.355	CONST CONTROL MON CX-76	22.862	LEFT	170-49-16	80.430	316-52-15	0.0	4845374.460	307710.370
885	271+29.785	CONST CONTROL MON CX-85	22.857	LEFT	350-49- 8	60.060	136-52- 7	0.0	4845513.091	307687.967
877	269+95.164	CONST CONTROL MON CX-77	22.861	RIGHT	139-19-24	87.513	285-52-24	0.0	4845387.490	307754.580
878	269+97.098	CONST CONTROL MON CX-78	38.095	RIGHT	130-50-11	94.861	276-53-11	0.0	4845391.830	307769.310
879	270+27.530	CONST CONTROL MON CX-79	38.081	RIGHT	115-33-22	74.155	261-36-21	0.0	4845421.870	307764.440
881	270+69.932	CONST CONTROL MON CX-81	22.862	RIGHT	80-38- 4	45.719	226-41- 3	0.0	4845461.300	307742.650
958	290+51.899	RAP #3 - RADIO TOWER	294.749	LEFT	344-19-40	1990.850	130-22-40	0.0	4847370.697	307159.747
884	271+29.932	CONST CONTROL MON CX-84	22.862	RIGHT	28- 3-30	75.551	174- 6-29	0.0	4845520.531	307733.077
959	0+00.000	RAP #4 - CN TOWER	0.0		152-46-14	**********	298-49-14	0.0	4833410.793	311894.638
933	270+16.535	CONST CONTROL MON CX-133	61.272	RIGHT	113- 9- 1	99.566	259-12- 0	0.0	4845414.717	307789.088
956	271+72.134	RAP #1 - BILLBOARD FRAME	471.682	RIGHT	69- 7-32	505.019	215-10-31	0.0	4845633.810	308169.411
960	269+67.759	CONTROL MON MTR77-619	9.329	RIGHT	53-18-33	106.983	299-11-32	196.768	4845358.277	307545.594
957	268+98.575	RAP #2 - ANTENNA C.F.B.	183.253	LEFT	213-57- 1	234.606	0- 0- 0	0.0	4845259.249	307566.519
483	269+83.555	BC CORNER ROUND	25.637	LEFT	172-39-51	86.275	318-42- 0	0.0	4845368.291	307708.556
753	269+83.622	CATCH BASIN GUTTER	25.142	LEFT	172-20-12	86.193	318-23-11	0.0	4845368.437	307709.034
485	269+84.988	PI CORNER ROUND	13.500	LEFT	164-33-15	85.312	310-34-15	0.0	4845371.643	307720.309
486	269+88.075	MP CORNER ROUND	16.973	LEFT	166-41-56	81.922	312-44-55	0.0	4845374.136	307716.388
446	269+88.654	PI CORNER ROUND	13.500	RIGHT	146-50-47	88.906	292-43-46	0.0	4845379.569	307716.378
2103	269+90.000	C/L OF CONSTRUCTION	0.0		154-49-55	82.995	300-52-54	196.352	4845378.744	307732.836
444	269+90.625	BC CORNER ROUND	28.986	RIGHT	137-35-48	94.626	283-38-47	0.0	4845383.986	307761.351
754	269+91.351	CATCH BASIN GUTTER	34.690	RIGHT	134-33- 3	97.281	280-36- 2	0.0	4845385.613	307766.866
461	269+91.604	BC CORNER ROUND	36.674	RIGHT	133-31-51	98.267	279-34-50	0.0	4845386.179	307768.784
434	269+92.355	BULLNOSE TOP OF CURB	1.500	RIGHT	153-21-22	81.172	299-24-21	196.425	4845381.308	307733.941
437	269+93.105	BC BULLNOSE TOP OF CURB	2.250	RIGHT	152-41-18	80.687	298-44-17	196.395	4845382.168	307734.562
435	269+93.105	CP BULLNOSE TOP ISLAND	1.500	RIGHT	153-11-45	80.456	299-14-44	196.410	4845382.048	307733.821
436	269+93.105	EC BULLNOSE TOP OF CURB	0.750	RIGHT	153-42-23	80.233	299-45-22	196.425	4845381.929	307733.081
447	269+93.947	MP CORNER ROUND	18.162	RIGHT	142-24-37	86.221	288-27-36	0.0	4845385.538	307750.135
482	269+97.210	CENTER PT CORNER ROUND	27.250	RIGHT	174-16-54	72.708	320-19-53	0.0	4845581.513	307704.785
484	269+97.210	EC CORNER ROUND-TOP CURB	13.500	LEFT	163-28-17	73.176	309-31-16	196.290	4845383.707	307718.358
755	269+98.000	CATCH BASIN TOP OF CURB	13.500	LEFT	163-29-28	72.392	309-26-28	196.270	4845384.488	307718.232
2107	270+00.000	TOP OF CURB	13.500	LEFT	163-10-51	70.410	309-13-51	196.036	4845386.462	307717.913
2106	270+00.000	C/L OF CONSTRUCTION	0.0		152-40-57	73.433	298-43-56	196.156	4845388.616	307731.240
443	270+04.265	CENTER PT CORNER ROUND	27.250	RIGHT	133-24-39	82.485	279-27-38	0.0	4845397.175	307757.460
445	270+04.265	EC CORNER ROUND-TOP CURB	13.500	RIGHT	141-41-32	74.932	287-50-31	196.050	4845394.981	307743.887
756	270+05.265	CATCH BASIN TOP OF CURB	13.500	RIGHT	141-25- 0	74.059	287-28- 0	196.110	4845395.968	307743.727
2111	270+10.000	TOP OF CURB	13.500	RIGHT	139-30-47	69.973	285-33-46	195.854	4845400.642	307742.971
2110	270+10.000	TOP OF CURB	13.500	LEFT	161-55-21	60.513	307-58-20	195.854	4845396.334	307716.317
2109	270+10.000	C/L OF CONSTRUCTION	0.0		149-53-42	64.006	295-56-41	195.974	4845398.488	307729.644
2113	270+20.000	TOP OF CURB	13.500	LEFT	160-10-24	50.657	306-13-23	195.686	4845406.206	307714.722

FIGURE 5-29 Computer printout of layout data for a road construction project

(Courtesy of Department of Roads and Traffic, City of Toronto)

CITY OF TORONTO

MANAGEMENT SERVICES DEPARTMENT DATA PROCESSING CENTRE

printout also lists the point coordinates and baseline offsets for each layout point.

Figure 5-30 shows a portion of the construction drawing that accompanies the computer printout shown in Figure 5-29. The drawing (usually drawn by digital plotter from computer files) is an aid to the surveyor in the field in laying out the works correctly. All the layout points listed on the

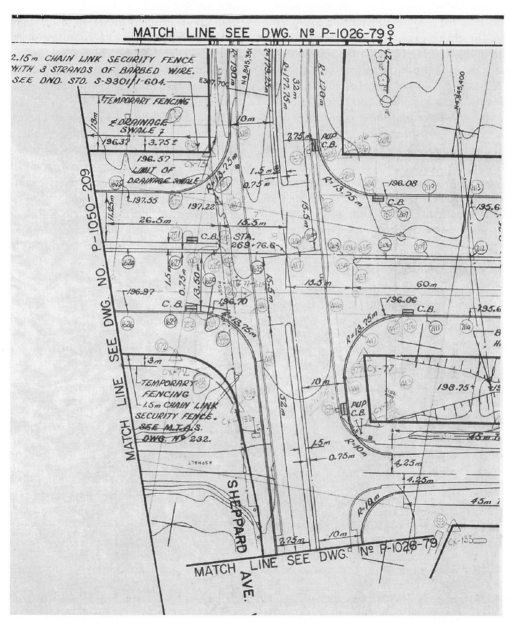

FIGURE 5-30 Portion of a construction plan showing layout points

(Courtesy of Department of Roads and Traffic, City of Toronto)

printout are also shown on the drawing, together with curve data and other explanatory notes.

More modern total stations offer an even more efficient technique. Instead of having the layout data only on a printout similar to that shown in Figure 5-29, the coordinates for all layout points can be uploaded into the total station microprocessor. The surveyor can then, in the field, identify the occupied control point and the reference BS point(s), thus orienting the total station.

The desired layout point number is then entered, with the required layout angle and distance inversed from the stored coordinates and displayed on the instrument's graphics screen. The layout can proceed by setting the correct azimuth, and then, with trial-and-error techniques, the prism is moved to the layout distance (the total station is set to tracking mode for all but the final measurements). With some total stations, the prism is simply tracked with the remaining left/right (±) distance being displayed alongside

the remaining near/far (±) distance. When the correct location has been reached, both displays show 0.000 m (0.00 ft). When a motorized total station (see Section 5.16) is used, the instrument itself turns the appropriate angle once the layout point number has been entered.

If the instrument is set up at an unknown position (free station), its coordinates can be determined by sighting control stations whose coordinates have been previously uploaded into the total station microprocessor. This technique, known as **resection**, is available on all modern total stations. Sightings on two control points can locate the instrument station, although additional sightings (up to a total of four) are recommended to provide a stronger solution and an indication of the accuracy level achieved (see Section 5.13.4).

Some theodolites and total stations come equipped with a guide light, which can help locate the prism holder very quickly (see Figures 5-31 and 5-32). The TC 800 is a total station that can be

(a)

FIGURE 5-31 (a) Leica TC 800 total station with EGLI® guide light. (Courtesy of Leica Geosystems Inc.) (b) Trimble's 3600 total station, featuring clamp-free operation (endless slow motion), can be upgraded to include reflectorless measurement, Tracklight™, laser plummet, and TDS onboard software

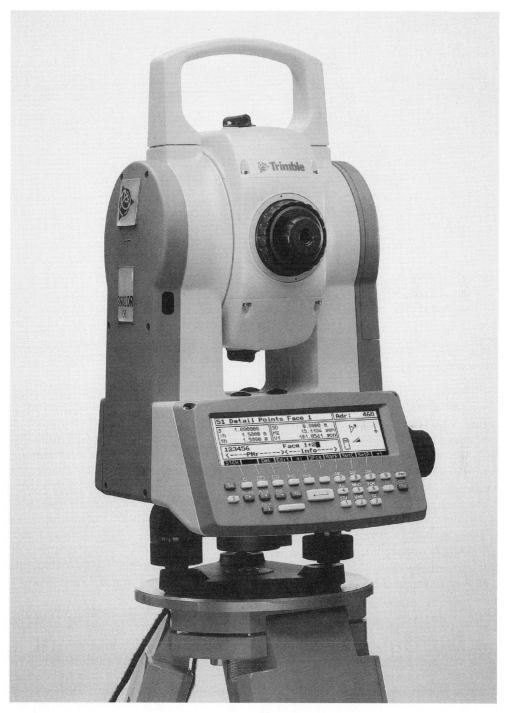

(b)

FIGURE 5-31 (*continued*)

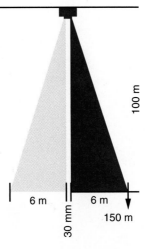

FIGURE 5-32 EGLI® guide light
(Courtesy of Leica Geosystems Inc.)

turned on and immediately used (no initialization procedure required) and comes equipped with internal storage for 2,000 points and a guide light, which is very useful in layout surveys, as prism holders can quickly place themselves on line by noting the colored lights sent from the total station. The flashing lights (yellow on the left and red on the right—as viewed by the prism holder), which are 12 m wide at a distance of 100 m, enable the prism holder to place the prism on line with final adjustments as given by the instrument operator. With ATR (see Section 5.16.1), the sighting-in process is completed automatically.

5.16 MOTORIZED TOTAL STATIONS

One extremely effective adaptation of the total station has been the addition of servomotors to drive both the horizontal and the vertical motions of these instruments. Motorized instruments have

been designed to search automatically for prism targets and then lock onto them precisely, to turn angles to designated points automatically using the uploaded coordinates of those points, and to repeat angles by automatically double centering. These instruments, when combined with a remote controller held by the prism surveyor, enable the survey to proceed with a reduced need for field personnel.

5.16.1 Automatic Target Recognition (ATR)

Some instruments are designed with ATR, which utilizes an infrared or laser beam that is sent coaxially through the telescope to a prism. The return is received by an internal charge-coupled device (CCD) camera. ATR is also referred to as autolock and autotracking. First, the telescope must be pointed roughly at the target prism, either manually or under software control. Then the motorized instrument places the crosshairs almost on the prism center (within 2 seconds of arc). Any residual offset error is automatically measured and applied to the horizontal and vertical angles. The ATR module is a digital camera that notes the offset of the reflected laser beam, permitting the instrument then to move automatically until the crosshairs have been electronically set on the point. After the point has thus been precisely sighted, the instrument can then read and record the angle and distance. Reports indicate that the time required for this process (which eliminates manual sighting and focusing) is only one-third to one-half of the time required to obtain the same results using conventional total station sighting techniques.

ATR comes with a lock-on mode, where the instrument, once sighted at the prism, continues to follow the prism as it is moved from station to station. To ensure that the prism is always pointed to the instrument, a 360° prism (see Figure 5-33) assists the surveyor in keeping the lock-on for a period of time. If lock-on is lost due to intervening obstacles, it is reestablished after manually roughly pointing at the prism, or by initiating a sweeping search until lock-on occurs again. ATR recognizes targets up to 1,000 m or 3,300 ft away, will function in darkness, requires no focusing or fine pointing, works with all types of prisms, and maintains a lock on prisms moving up to speeds of 11 mph or 5 mps (at a distance of 100 m).

5.16.2 Remote Controlled Surveying

Geodimeter, the company that first introduced EDM equipment in the early 1950s, marketed in the late 1980s a survey system in which the total station (Geodimeter 4000, see Figures 5-34 and 5-35) has been equipped with motors to control both the horizontal and the vertical movements. This total station can be used as a conventional instrument, but when interfaced to a controller located with the prism, the station instrument can be remotely controlled by the surveyor at the prism station by means of radio telemetry.

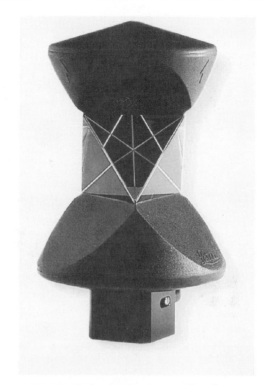

FIGURE 5-33 Leica 360° prism used with remote-controlled total stations and with automatic target recognition (ATR) total stations. ATR eliminates fine pointing and focusing. The 360° feature means that the prism is always facing the instrument (Courtesy of Leica Geosystems Inc.)

When the remote control feature button on the total station is activated, control of the station instrument is transferred to the remote controller, called the remote positioning unit (see Figure 5-35). The remote unit consists of the pole (with circular bubble), the prism, a data collector (capable of storing up to 10,000 points), telemetry equipment for communicating with the station instrument, and a sighting telescope that, when aimed back at the station instrument, permits a sensing of the angle of inclination, which is then transmitted to the station instrument via radio communication. As a result, the instrument can automatically move its telescope to the proper angle of inclination, thus enabling the instrument

(a)

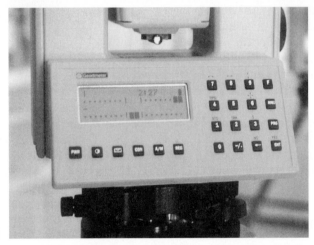

(b)

FIGURE 5-34 (a) Geodimeter 4400 Base Station. A total station equipped with servomotors controlling both the horizontal and vertical circle movements. It can be used alone as a conventional total station or as a robotic base station controlled by the remote positioning unit (RPU) operator. (b) Geodimeter keyboard showing in-process electronic leveling. Upper cursor can also be centered now by adjusting the third leveling screw

to commence an automatic horizontal sweeping that results in the station instrument being locked on the prism precisely.

A typical operation requires that the station unit be placed over a control station or over a free station whose coordinates can be determined using resection techniques (see Section 5.13.4) and that a BS be taken to another control point, thus fixing the location and orientation of the total station. The operation then begins with

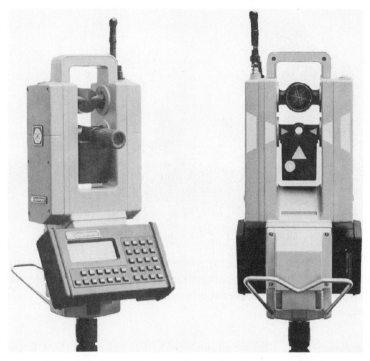

FIGURE 5-35 Remote positioning unit (RPU), a combination of prism, data collector, and radio communicator (with the base station) that permits the operator to engage in one-person surveys

both units being activated at the remote unit. The remote unit sighting telescope is aimed at the station unit, and the sensed vertical angle is sent via telemetry to the station unit. The station unit then automatically sets its telescope at the correct angle in the vertical plane and begins a horizontal search for the remote unit; the search area can be limited to a specific sector (e.g., 70°), thus reducing search time. The limiting range of this instrument is about 700 m. When the measurements (angle and distance) have been completed, the point number and attribute data codes can be entered into the data collector attached to the remote unit prism pole. One feature of a motorized total station with ATR permits the surveyor to accumulate and average a defined number of face 1 and face 2 angles to a point by simply sighting the point once; the instrument itself completes all the requested following operations.

When used for setting out, the desired point number is entered at the remote unit. The total station instrument then automatically turns the required angle, which it computes from previously uploaded coordinates held in storage (both the total station and the remote unit have the points in storage). The remote unit operator can position the prism roughly on line by noting the *guide light*, which shows as red or green (for this instrument), depending on whether the operator is left or right of the line, and as white when the operator is on the line (see also the guide light shown in Figure 5-32).

Distance and angle readouts are then observed by the operator to position the prism pole precisely at the layout point location. Because the unit can fast-track (0.4 second) precise measurements and because it is also capable of averaging multiple measurement readings, precise results can be obtained when using the prism pole by slightly waving the pole left and right and back

FIGURE 5-36 Leica TPS System 1000, used for roadway stakeout. Surveyor is controlling the remote-controlled total station (TCA 1100) at the prism pole using the RCS 1000 controller together with a radio modem. Assistant is placing steel bar marker at previous set-out point

(Courtesy of Leica Geosystems Inc.)

and forth in a deliberate pattern. Because all but the BS reference are obtained using infrared and telemetry, the system can be used effectively after dark, permitting night-time layouts for next-day construction and surveys in high-volume traffic areas that can be accomplished efficiently only in low-volume time periods.

Figure 5-36 shows a remotely controlled total station that utilizes ATR and the EGL1® to search for and then position the prism on the correct layout line, where the operator then notes the angle and distance readouts to determine the precise layout location. Figure 5-37 shows a motorized total station that has many of the features listed here, and also has wireless controllers.

5.17 HANDHELD TOTAL STATIONS

Several number of instrument manufacturers produce lower-precision reflectorless total stations that can be handheld, like a camcorder, or pole/tripod mounted like typical surveying instruments. These instruments have three integral components:

1. The pulse laser (FDA Class I*) distance meter measures reflectorlessly and to reflective papers and prisms. When used reflectorlessly, the distance range of 300–1,200 m varies with different manufacturers and model types; the type of sighted surface (masonry, trees, bushes, etc.) determines the distance range because some surfaces are more reflective than others. Masonry and concrete surfaces reflect light well, while trees, bushes, sod, etc. reflect light to lesser degrees. When used with prisms, the distance range increases to 5–8 km. Some instruments can be limited to an expected distance range envelope. This measuring restriction instructs the instrument not to measure and record distances outside the envelope; this feature permits the surveyor to sight more distant points through nearby clutter such as intervening electrical wires, branches, and other foliage. Distance accuracies are in the range of 1–10 cm, and ±5 cm ±20 ppm is

---

*FDA class 1 lasers are considered safe for the eyes. Class 2 lasers should be used with caution and should not be operated at eye level; classes 3A, 3B, and 4 lasers should only be used with appropriate eye protection.

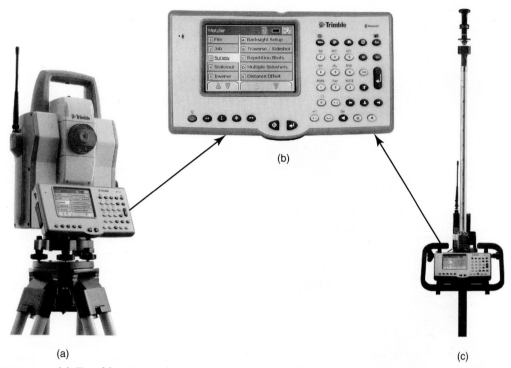

FIGURE 5-37 (a) Trimble 5600 Robotic Total Station. (b) Trimble ACU Controller, featuring wireless communications and a choice of software for all robotic total station applications as well as all GPS applications. (c) Remote pole unit containing an ACU Controller, 360° prism and radio
(Courtesy of Trimble)

typical. Power is supplied by two to six AA batteries, or two C cell batteries; battery type depends on the model. See Figure 5-38.

2. Angle encoders (and built-in inclinometers) determine the horizontal and vertical angles, much like the total stations described earlier in this chapter, and can produce accuracies up to greater than 1 minute of arc; the angle accuracy range of instruments on the market is 0.01–0.2°. Obviously, mounted instruments produce better accuracies than do handheld units.

3. Data collection can be provided by many of the collectors on the market. These handheld total stations can be used in a wide variety of mapping and GIS surveys, or they can be used in GPS surveys. For example, GPS surveys can be extending into areas under tree canopy or structural obstructions (where GPS signal are blocked), with the positional data collected directly into the GPS data collector.

5.18 SUMMARY OF MODERN TOTAL STATION CHARACTERISTICS

Modern total stations typically have the capabilities listed below:

- Some instruments require that horizontal and vertical circles be revolved through 360° to initialize angle-measuring operations, whereas some newer instruments require no such initialization.

FIGURE 5-38 The Reflectorless Total Station is equipped with integrating distance measurement, angle encoding, and data collection

(Courtesy of Riegl USA Inc., Orlando, FL)

- They can be equipped with servomotors to drive the horizontal and vertical motions (the basis for robotic operation).

- A robotic instrument can be controlled at a distance of up to 1,200 m from the prism pole.

- Telescope magnification: Most total stations are at 26× or 30×.

- The minimum focus distance: ranges of 0.5–1.7 m.

- Distance measurement to a single prism is in the range of 2–5 km (10 km for a triple prism).

- A coaxial visible laser has a reflectorless measuring range of 250–700 m.

- Angle accuracies are in the range of 1–10 seconds. Accuracy is defined as the standard deviation, direct and reverse DIN Spec. 18723.

- Distance accuracies, using reflectors: normal, from ±(2 mm +2 ppm); tracking, from ±(5 mm + 2 ppm); reflectorless: from ±(3 mm + 2 ppm); longer range reflectorless: from ±(5 mm + 2 ppm).

- Eye-safe lasers are class 1, and class 2 requires some caution.

- For many total stations, reflectorless ranges are up to 800 m to a 90 percent (reflection) Kodak gray card, and up to 300 m to a 18 percent (reflection) Kodak gray card; as noted earlier, Topcon total station reflectorless measurements can range up to 2,000 m from the instrument.

- Autofocusing.

- Automatic environment sensing (temperature and pressure) with ppm corrections applied to measured distances; or manual entry for measured corrections.

- Wireless Bluetooth communications, or cable connections.

- Standard data collectors (onboard or attached), or collectors with full graphics (four to eight lines of text) with touch-screen command capabilities for quick editing and line-work creation, in addition to keyboard commands.

- Onboard calibration programs for line-of-sight errors, tilting axis errors, compensator index errors, vertical index errors, and automatic target lock-on calibration. Corrections for sensed errors can be applied automatically to all measurements.

- Endless drives for horizontal and vertical motions (nonmotorized total stations).

- Laser plummets.

- Guide lights for prism placements, with a range of 5–150 m, used in layout surveys.

- Automatic target recognition (ATR)—also known as *autolock and autotracking*, with an upper range of 500–2,200 m.
- Single-axis or dual-axis compensation.

The capabilities of total stations are constantly being improved. Thus, this list is not meant to be definitive.

5.19 GROUND-BASED LiDAR IMAGING

In 2007, Trimble introduced the *VX Spatial Station*, a robotic total station to which 3D spatial scanning and live video on the controller screen have been added. See Figure 5-39. This combination permits the surveyor to identify and capture data points by tapping the images on the controller's touch screen to initiate total station measurements and scanning measurements; having all field points on the graphics screen greatly reduces the need for conventional field survey notes. This instrument combines millimeter-accuracy total station measurements that are typical distances from the instrument, with 10 mm accuracy in scanning measurements for points up to about 150 m away from the instrument. Trimble uses its extraction software *Realworks* to determine feature positions and descriptions. With its 2D and 3D capabilities, this instrument could also establish 3D works files for machine guidance and control projects. It is expected that other manufactures in the surveying field will soon join in these exciting new developments, thus fostering further advances in this technology.

LiDAR (light detection and ranging) scanning utilizes laser scanners that send out thousands of laser pulses per second to map "scenes" that would be hard to access using normal ground surveying techniques. The laser impulses (point clouds) are reflected by encountered surfaces and returned to the scanner; using time-of-flight data the distances to the reflected points can be determined, thus producing a mapping image of the scanned scene. Horizontal and vertical control of the captured data can be established by first knowing the 3D coordinates of the instrument station, by knowing the coordinates of any back-sighted control points, and by establishing control targets around the perimeter of the scene to be captured.

Several manufacturers have introduced ground-based LiDAR imaging instruments without the total station capabilities noted here. The VX Spatial Station has a scanning range of about 150 m while collecting only 5–15 points per second. In comparison, the ground-based scanners have typical scanning ranges of 350 m while collecting 5,000 distances per second; distances are determined using time-of-flight (TOF) technology similar to that used in Direct Reflex (DR)—equipped total stations.

See Figure 5-40. This unit, in addition to LiDAR scanners, comes with a high-resolution camera for added spatial data information; a GPS receiver and an inertial measuring unit (IMU) can be added to the instrument array to compensate for motion when these units are truck-mounted. This system is of great value when surveying inaccessible scenes, such as building facades, structural components, and complex industrial infrastructure networks (e.g., piping or electrical routings). The scene to be captured is first equipped with clearly identifiable control point targets (four is ideal; three is the minimum for georeferencing) near the limits of the proposed capture area. Although unwanted intrusions (people, cyclists, autos, animals, etc.) can be edited out in postprocessing, the fieldwork is scheduled when these intrusions will be at a minimum. Figure 11-4 shows the characteristics of reflected energy. If a portion of the scene is both highly reflective and angled away from the instrument, the LiDAR pulses may all be reflected away from the instrument, resulting in missed detail; as well, building windows may not reflect the LiDAR pulses, resulting in indoor data captures that are both unwanted and perhaps unexpected. These problems can be rectified by a pre-survey site inspection and remediation (including window coverings and the toning down of highly reflective surfaces), or in postprocessing editing.

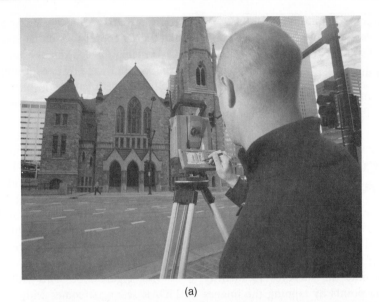

(a)

(b)

FIGURE 5-39 (a) VX Spatial station. A ground-based LiDAR imager providing 3D results from ground-level scanning. (b) 3D of an overhead expressway image created using the Spatial Station and Trimble software (Courtesy of Trimble, CA)

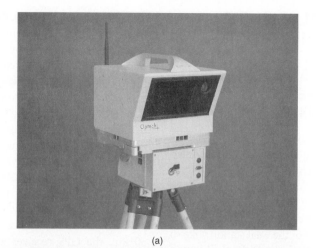

(a)

(b)

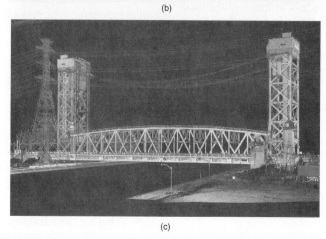

(c)

FIGURE 5-40 (a) ILRIS-3DMCER scanner; (b) bridge photo; and (c) 3D LiDAR image of same bridge captured by the LiDAR scanner

(Courtesy of Optech Incorporated, Toronto)

5.20 INSTRUMENTS COMBINING TOTAL STATION CAPABILITIES AND GPS RECEIVER CAPABILITIES

Leica Geosystems Inc. was the first, in early 2005, to market a total station equipped with an integrated dual-channel GPS receiver (see Figure 5-4). All TPS1200 total stations can be upgraded to achieve total station/GPS receiver capability. The GPS receiver, together with a communications device, is mounted directly on top of the total station yoke so that they are on the vertical axis of the total station and thus correctly positioned over the station point. The total station keyboard can be used to control all total station and GPS receiver operations. RTK (real-time kinematic) accuracies are said to be 10 mm + 1 ppm horizontal, and 20 mm + 1 ppm vertical. Also, see Chapter 7. The GPS receiver can identify the instrument's first uncoordinated position to high accuracy when working differentially with a GPS base station (located within 50 km). Once the instrument station has been coordinated, a second station can be similarly established, and the total station (GPS portion) can then be oriented while you are back-sighting the first established station. All necessary GPS software is included in the total station processor. Early comparative trials indicate that, for some applications, this new technology can be 30–40 percent faster than traditional total station surveying. Because nearby ground control points may not have to be occupied, the surveyor saves the time it takes to set up precisely over a point and to bring control into the project area from control stations that are too far away to be of immediate use. If the instrument is set up in a convenient unco-ordinated location allowing sight of the maximum number of potential survey points, the surveyor need only to turn on the GPS receiver and then level and orient the total station to a BS reference. By the time the surveyor is ready to begin the survey and assuming a GPS reference base station is within 50 km, the GPS receiver is functioning in RTK mode. The survey can commence with the points requiring positioning being determined using total station techniques and/or by using the GPS receiver after it has been transferred from the total station and placed on a rover antenna pole for roving positioning. The GPS data, collected at the total station, are stored along with all the data collected using total station techniques on compact memory flash cards. Data can be transferred to cell phones using wireless (e.g., *Bluetooth*) technology before being transmitted to the project computer for downloading. This new technology represents a new era in surveying, and has changed the way surveyors perform traditional traverses and control surveying, as well as topographic surveys and layout surveys.

Questions

5.1 How does an electronic theodolite differ from a total station?

5.2 What is the best method of setting up a theodolite?

5.3 What are optical and laser plummets used for?

5.4 Describe, in detail, how you would set up a total station, or theodolite on a steep slope.

5.5 Describe how you can use a theodolite or total station to prolong a straight line *AB* to a new point *C*.

5.6 How can you check to see that the plate bubbles of a theodolite or total station are adjusted correctly?

5.7 How can you determine the exact point of intersection of the centerlines of two cross streets?

5.8 What are some techniques of surveying a property line when trees block the line of sight?

5.9 What are the advantages of using a total station rather than using an electronic theodolite and steel tape?

5.10 What impact did the creation of electronic angle measurement have on surveying procedures?

5.11 With the ability to record field measurements and point descriptions in an instrument controller or electronic field book, why are manual field notes still important?

5.12 Explain the importance of electronic surveying in the extended field of surveying and data processing, now often referred to as the science of geomatics.

5.13 Using a programmed total station, how would you tie in (locate) a water valve "hidden" behind a building corner?

5.14 Some total station settings can be entered once and then seldom changed, while other settings must be changed each time the instrument is set up. Prepare two typical lists of instrument settings, one for each instance.

5.15 After a total station has been set up over a control station, describe what actions and entries must then be completed before beginning a topographic survey.

5.16 On a busy construction site, most control stations are blocked from a proposed area of layout. Describe how to create a convenient free station control point using resection techniques.

5.17 Describe a technique for identifying survey sightings on a curb line so that a series of curb line shots are joined graphically when they are transferred to the computer (or data collector graphics).

5.18 What are the advantages of using LiDAR scanning techniques for ground-based surveys?

5.19 Describe a total station technique for laying out pre-coordinated construction points.

5.20 What are the advantages of motorized total stations?

5.21 What are the advantages of using automatic target recognition (ATR)?

CHAPTER SIX

TRAVERSE SURVEYS

6.1 GENERAL BACKGROUND

A *traverse* is usually a control survey and is employed in all forms of legal, mapping, and engineering surveys. Essentially, traverses are a series of established stations that are tied together by angle and distance. The angles are measured using theodolites, or total stations, whereas the distances can be measured using total stations, electronic distance measurement (EDM) instruments, or steel tapes. Traverses can be open, as in route surveys, or closed, as in a closed geometric figure (see Figures 6-1 and 6-2). Boundary surveys, which constitute a closed consecutive series of established (or laid out) stations, are often described as being traverse surveys (e.g., retracement of the distances and angles of a section of property).

In engineering work, traverses are used as control surveys to (1) locate topographic detail for the preparation of plans, (2) lay out (locate)

engineering works, and (3) process earthwork and other engineering quantities. Traverses can also provide horizontal control for aerial surveys in the preparation of aerial imaging (see Chapter 12). Most control surveys in this more modern age now utilize satellite positioning techniques to provide survey control (see Chapter 7). The end product of satellite positioning surveys is the creation of the coordinates and elevations of all points surveyed. Similarly, one of the products of total station use are the coordinates and elevations of all points occupied and sighted.

Traverse computations include the following: balance field angles, compute latitudes and departures, compute traverse error, balance latitudes and departures, adjust original distances and directions, compute coordinates of the traverse stations, and compute the area enclosed by a closed traverse. In modern practice these computations are routinely performed on computers and on total stations, or their electronic field books and controllers. In this chapter we will

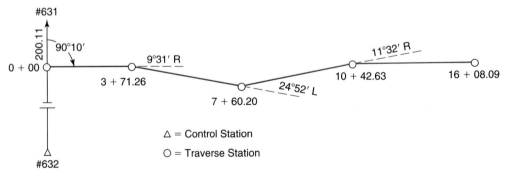

FIGURE 6-1 Open traverse

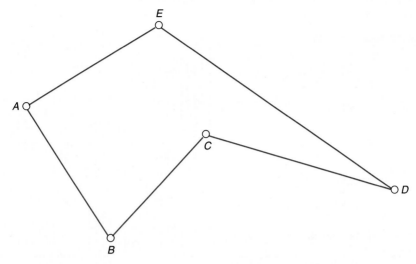

FIGURE 6-2 Closed traverse (loop)

perform traverse computations manually (using calculators) to demonstrate and reinforce the mathematical concepts underlying each stage of these computations. Without this knowledge, the student becomes just a button pusher working with a "black box" and little understanding of the surveying process.

6.2 OPEN TRAVERSE

Simply put, an *open traverse* is a series of measured straight lines (and angles) that do not geometrically close. This lack of geometric closure means that there is no geometric verification possible with respect to the actual positioning of the traverse stations. Accordingly, the measuring technique must be refined to provide for field verification. At a minimum, distances are measured twice (sometimes once in each direction) and angles are doubled.

In route surveys, open traverse stations can be verified by computation from available tied-in field markers as shown on property plans, by scale from existing topographic plans or through the use of satellite positioning (see Chapter 7). Directions can be verified by scale from existing plans, by observation on the sun or Polaris, or by

tie-ins to pre-set coordinated monuments. It is now often possible to tie in the initial and terminal stations of a route survey to coordinate grid monuments whose positions have been accurately determined using precise techniques either by previous surveys or by on-the-spot satellite positioning observations. In this case, the route survey becomes a *closed traverse* and is subject to geometric verification and analysis.

As was noted in Section 4.4, open traverses are tied together angularly by deflection angles, and distances are shown in the form of stations that are cumulative measurements referenced to the initial point of the survey (0 + 00) (see Figure 6-3).

6.3 CLOSED TRAVERSE

A *closed traverse* is either one that begins and ends at the same point or one that begins and ends at points whose positions have been previously determined (Section 6.2). In both cases the angles can be closed geometrically, and the position closure can be determined mathematically.

A closed traverse that begins and ends at the same point is known as a *loop traverse*

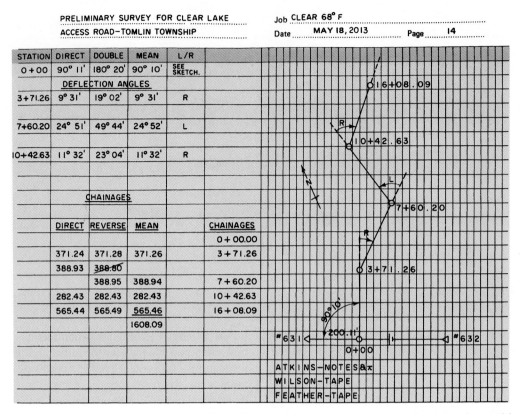

FIGURE 6-3 Field notes for open traverse (Note: The terms chainages and stations are interchangeable.)

(see Figure 6-2). In this case the distances are measured from one station to the next (and verified) by using a steel tape or EDM; the interior angle is measured at each station (and doubled). The loop distances and angles can be obtained by proceeding consecutively around the loop in a clockwise (CW) or counterclockwise (CCW) manner; in fact, the data can be collected in any order convenient to the surveyor. However, as noted in the previous chapter, the angles themselves (by convention) are always measured CW.

Example 6-1: In the following sections Example 6-1 will be used to illustrate error computations and error adjustments for a closed traverse. Data for this example are shown in Tables 6-1, 6-2, 6-4, 6-5, 6-8–6-10; and in Figures 6-4, 6-8–6-11, and 6-15. Section 6.4 shows how the field angles are

balanced; Section 6.5 introduces latitudes and departures and Section 6.6 shows the computations of latitudes and departures, the error of closure, and the resultant precision ratio; Section 6.9 shows how the errors in latitudes and departures are balanced (adjusted); Section 6.10 shows how the traverse adjustments affect the original distances and directions; Section 6.12 introduces rectangular coordinates and shows the computations for coordinates of the traverse stations shown in Example 6-1; Section 6.13 shows a summary of steps in the traverse problem computations; and Example 6-6, in Section 6.14, shows the computation of the area enclosed by the closed traverse introduced in Example 6-1.

Note that the data chosen for the examples used in this chapter result in relatively large errors, errors typically found when using steel tape and theodolite techniques. This approach was chosen so that the examples would produce errors large enough to illustrate the analysis/adjustment (calculator approach) techniques used here. More precise data typical of total station surveys have resultant errors so small that the time spent on analyses and adjustments could not be warranted for the lower order of accuracy often specified in some engineering surveys. Refer to Section 6.6 and Chapter 10 for more on survey accuracy requirements.

6.3.1 Summary of Traverse Computations

1. Balance the field angles.
2. Correct (if necessary) the field distances (e.g., for temperature).
3. Compute the bearings or azimuths.
4. Compute the linear error of closure and the precision ratio of the traverse.
5. Compute the balanced latitudes and balanced departures.
6. Compute the coordinates.
7. Compute the area.

Related topics, such as plotting by rectangular coordinates and the computation of land areas by other methods, are covered in Chapter 8.

6.4 BALANCING ANGLES

In Section 4.4, it was noted that the geometric sum of the interior angles in an n-sided closed figure is $(n - 2)180°$. For example, a five-sided figure would have $(5 - 2)180° = 540°$; a seven-sided figure would have $(7 - 2)180° = 900°$.

When all the interior angles of a closed field traverse are summed, they may not precisely equal the number of degrees required for geometric closure. This is due to systematic and random errors associated with setting the instrument over a point and with making sightings. Before mathematical analysis of the traverse can begin, even before the bearings can be calculated, the field angles must be adjusted so that their sum equals the correct geometric total. The angles can be balanced by distributing the angular error equally to each angle (if all angles were measured with the same precision), or one or more angles can be adjusted to force the closure. The acceptable total error of angular closure is usually quite small (i.e., <03′); otherwise, the fieldwork will have to be repeated. The actual size of the allowable angular error is governed by the specifications being used for that particular traverse. The angles for the traverse example, Example 6-1 (Section 6.3), are shown in Table 6-1 and Figure 6-4. If one of the traverse stations had been in a

Table 6-1   Example 6.1 Two Methods of Adjusting Field Angles

Station	Field Angle	Arbitrarily Balanced	Equally Balanced
A	101° 24′00″	101° 24′00″	101° 24′ 12″
B	149° 13′00″	149° 13′00″	149° 13′ 12″
C	80° 58′30″	80° 59′00″	80° 58′ 42″
D	116° 19′00″	116° 19′00″	116° 19′ 12″
E	92° 04′30″	92° 05′00″	92° 04′ 42″
	538° 119′00″	538° 120′00″	538°118′120″
	= 539° 59′00″	= 540° 00′00″	= 540° 00′ 00″
	Error = 01′	Balanced	Balanced
			Correction/angle $= \dfrac{60}{5} = 12″$

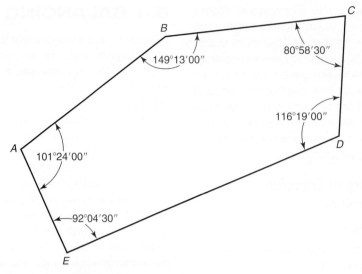

FIGURE 6-4 Example 6.1 (closed traverse problem) field angles

particularly suspect location (e.g., a swamp), a larger proportion of the angle correction could be assigned to that one station. In all balancing operations, however, a certain amount of guesswork is involved, because we do not know with certainty if any of the balancing procedures give us values closer to the true value than did the original field angle. The important point is that the overall angular closure is not larger than that specified.

6.5 LATITUDES AND DEPARTURES

In Section 1.14 it was noted that a point could be located by polar ties (direction and distance) or by rectangular ties (two distances at 90°). In Figure 6-5 (a), point A is located, with respect to point D, by direction (bearing) and distance. In Figure 6-5(b), point A is located, with respect to point D, by a distance north (Δy) and a distance east (Δx).

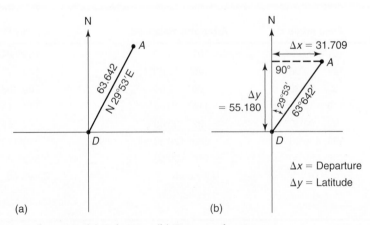

FIGURE 6-5 Location of a point. (a) Polar ties. (b) Rectangular ties

By definition, *latitude is the north/south rectangular component of a line*, and to differentiate direction, north is considered plus, and south is considered minus. Similarly, *departure is the east/west rectangular component of a line*, and to differentiate direction, east is considered plus, and west is considered minus. When working with azimuths, the plus/minus designation is directly given by the appropriate trigonometric function:

$$\text{Latitude } (\Delta y) = \text{distance } (H) \times \cos \alpha \quad (6\text{-}1)$$

$$\text{Departure } (\Delta x) = \text{distance } (H) \times \sin \alpha \quad (6\text{-}2)$$

where α is the bearing or azimuth of the traverse course, and distance (H) is the horizontal distance of the traverse course.

Latitudes (lats) and departures (deps) can be used to calculate the precision of a traverse by noting the plus/minus closure of both latitudes and departures. If the survey has been perfectly performed (angles and distances), the sum of the plus latitudes will equal the sum of the minus latitudes, and the sum of the plus departures will equal the sum of the minus departures.

In Figure 6-6, the computation direction has been performed in a CCW sense (all signs would simply be reversed for a CW approach). Latitudes *CD*, *DA*, and *AB* are all positive and should precisely (if the survey were perfect) sum to the latitude of *BC*, which is negative. Similarly, the departures of *CD* and *DA* are positive, and their sum should equal the sum of departures of *AB* and *BC*, which are both negative.

As noted earlier, when using bearings for directions, the north and east directions are positive, whereas the south and west directions are negative. Figure 6-7 shows these relationships. The figure also shows that the signs are those of the corresponding trigonometric function (cos or sin).

6.6 COMPUTATION OF LATITUDES AND DEPARTURES TO DETERMINE THE ERROR OF CLOSURE AND THE PRECISION OF A TRAVERSE

EXAMPLE 6-1

Step 1: Balance the angles Using the data introduced in Section 6.4 for this computation, and the angles resulting from equal distribution of the error among the field angles, balanced angles are shown in Figure 6-8.

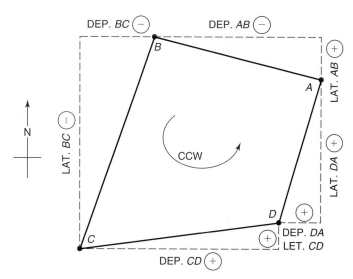

FIGURE 6-6 Closure of latitudes and departures (counterclockwise approach)

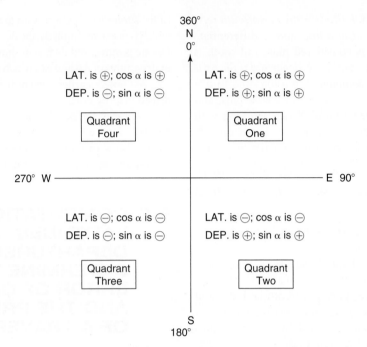

FIGURE 6-7 Algebraic signs of latitudes and departures by trigonometric functions (α is the azimuth)

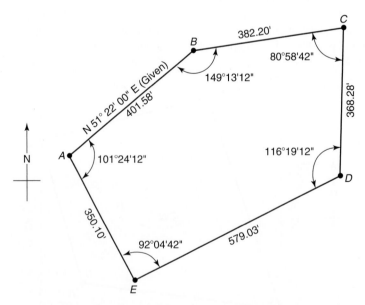

FIGURE 6-8 Distances and balanced angles for Example 6.1 (see Section 6.6, step 1)

Step 2: Compute the bearings (azimuths) Starting with the given direction of *AB* (N 51°22′00″ E), the directions of the remaining sides are computed. The computation can be solved going CCW (see Figure 6-9) or CW (see Figure 6-10) around the figure. Use the techniques developed in Section 4.9 (CCW approach) for this example.

Refer to Figure 6-9:

$$
\begin{aligned}
\text{Bg}\,AB &= \text{N}\,51°22′00″\text{E} \quad \text{(Given)} \\
\text{Az}\,AB &= 51°22′00″ \\
+\,\angle A &= 101°24′12″ \\
\text{Az}\,AE &= 152°46′12″ \\
&\quad +180° \\
\text{Az}\,EA &= 332°46′12″ \\
+\,\angle E &= 92°04′42″
\end{aligned}
$$

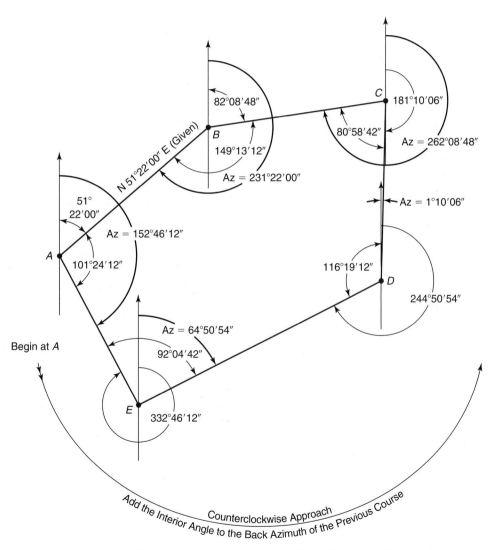

FIGURE 6-9 Azimuth computation, counterclockwise approach (see Section 6.6)

Az ED = 424°50'54"
      −360°
Az ED = 64°50'54"
      +180°
Az DE = 244°50'54"
+ ∠D 116°19'12"
Az DC = 361°10'06"
      −360°

Az DC = 1°10'06"
      +180°
Az CD = 181°10'06"
+ ∠C 80°58'42"
Az CB = 262°08'48"
      −180°
Az BC = 82°08'48"
+ ∠B 149°13'12"

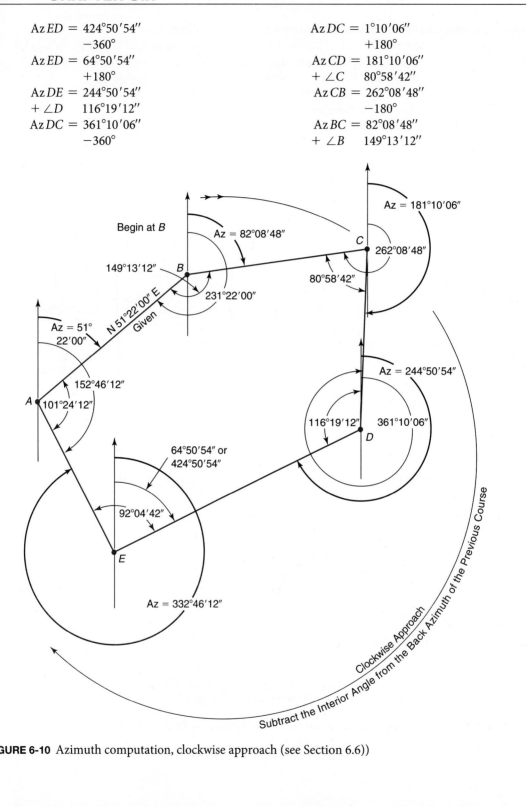

FIGURE 6-10 Azimuth computation, clockwise approach (see Section 6.6))

AzBA = 231°22′00″
−180°
AzAB = 51°22′00″

BgAB = N51°22′00″ E   Check

For comparative purposes, the CW approach is also shown (refer to Figure 6-10):

BgAB = N 51°22′00″ E
AzAB = 51°22′00″
+180°
AzBA = 231°22′00″
− ∠B   149°13′12″
AzBC = 82°08′48″
+180°
AzCB = 262°08′48″
− ∠C   80°58′42″
AzCD = 181°10′06″
+180°
AzDC = 361°10′06″
− ∠D   116°19′12″
AzDE = 244°50′54″
−180°
AzED = 64°50′54″
+ 360°00′00″*
AzED = 424°50′54″
− ∠E   92°04′42″

AzEA = 332°46′12″
−180°
AzAE = 152°46′12″
− ∠A   101°24′12″
AzAB = 51°22′00″   [Check]

Step 3: Compute the latitudes (Δy) and departures (Δx)

Reference to Table 6-2 will show the format for a typical traverse computation. The table shows both the azimuth and the bearing for each course (usually only the azimuth or the bearing is included); it is noted that the algebraic sign for both latitude and departure is given directly by the calculator (computer) for each azimuth angle. In the case of the bearings, as noted earlier, latitudes are plus if the bearing is north and minus if the bearing is south; similarly, departures are positive if the bearing is east and negative if the bearing is west. For example, course AE has an azimuth of 152°46′12″ (cos is negative; sin is positive; Figure 6-7), meaning that the latitude will be negative and the departure positive. Similarly, AE has a bearing of S 27°13′48″ E, which, as expected, results in a negative latitude (south) and a positive departure (east).

Table 6-2 shows that the latitudes fail to close by +0.06 (Σ lat) and the departures fail

Table 6-2 Latitudes and Departure Computations, Counterclockwise Approach

Course	Distance (ft)	Azimuth	Bearing	Latitude	Departure
AE	350.10	152°46′12″	S 27°13′48″ E	−311.30	+160.19
ED	579.03	64°50′54″	N 64°50′54″ E	+246.10	+524.13
DC	368.28	1°10′06″	N 1°10′06″ E	+368.20	+7.51
CB	382.20	262°08′48″	S 82°08′48″ W	−52.22	−378.62
BA	401.58	231°22′00″	S 51°22′00″ W	−250.72	−313.70
	P = 2081.19			Σ lat = +0.06	Σ dep = −0.49

$$E = \sqrt{(\Sigma \text{ lat})^2 + (\Sigma \text{ dep})^2} = \sqrt{0.06^2 + 0.49^2} = 0.49$$

$$\text{Precision ratio} = \frac{E}{P} = \frac{0.49}{2081.19} = \frac{1}{4247} \approx \frac{1}{4200}$$

*To permit subtraction of the next interior angle <E.

to close by -0.49 (Σ dep). Remember (as noted in Section 6.3) that these large errors (typical of steel tape and theodolite surveys) were used for illustrative purposes. If a total station had been used in the survey in Example 6-1, the resultant errors would have been significantly smaller and have much less of an impact on a typical engineering survey.

Figure 6-11 shows graphically the relationship between the C lat and C dep*. C lat and C dep are opposite in sign to Σ lat and Σ dep and reflect the direction consistent (in this example) with the CCW approach to the problem.

The traverse computation began at A and concluded at A'. The *linear error of closure* is given by the line $A'A$, and the correction closure is given by AA'; C lat and C dep are in fact the latitude and departure of the required correction course.

The length of AA' is the square root of the sum of the squares of the C lat and C dep:

$$AA' = \sqrt{(C\,\text{lat})^2 + (C\,\text{dep}^2)}$$
$$= \sqrt{0.06^2 + 0.49^2} = 0.494'$$

It is sometimes advantageous to know the bearing of the closure correction (misclosure). Reference to Figure 6-11 shows that C dep/C lat = tan bearing: So bearing $AA' = $ S 83°01' E.

The *error of closure* (linear error of closure) is the net accumulation of the random errors associated with the measurement of the traverse angles and traverse distances. In this example the total error is showing up at A simply because the computation began at A. If the computation had commenced at any other station, the identical linear error of closure would have shown up at *that* station.

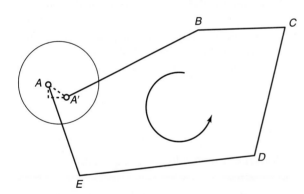

Closure Error $= A'A$
Closure Correction $= AA'$
Solution Proceeds Counterclockwise
Around the Traverse beginning at A.

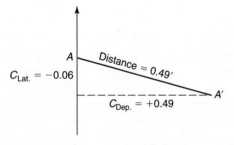

$$AA' = \sqrt{C_{\text{Lat.}}{}^2 + C_{\text{Dep.}}{}^2} = 0.494'$$

Bearing of AA' Can Be Computed from the Relationship:

$$\tan \text{Bearing} = \frac{C_{\text{Dep.}}}{C_{\text{Lat.}}} = \frac{0.49}{-0.06}$$

Bearing Angle $= 83.0189° = 83°01'$

Bearing AA' $= $ S 83°01' E

FIGURE 6-11 Closure error and closure corrections

---

*Σ lat is the error in latitudes; Σ dep is the error in departures; C lat is the required correction in latitudes; C dep is the required correction in departures.

The error of closure is compared to the perimeter (*P*) of the traverse to determine the *precision ratio*.

In the example of Table 6-2, the precision ratio is $E/P = 0.49/2,081$ (rounded).

$$\text{Precision ratio} = 1/4,247 \approx 1/4,200$$

The fraction of *E/P* is always expressed so that the numerator is 1, and the denominator is rounded to the closest 100 units. In the example shown, both the numerator and denominator are divided by the numerator (0.49).

The concept of an accuracy ratio was introduced in Section 1.15. Many states and provinces have accuracy ratios legislated for minimally acceptable surveys for boundaries. The values vary from one area to another, but usually the minimal values are 1/5,000–1/7,500. It is logical to assign more accurate values (e.g., 1/10,000) to high-cost urban areas. Engineering surveys are performed at levels of 1/3,000–1/10,000 depending on the importance of the work and the types of materials being used. For example, a gravel highway could well be surveyed at a 1/3,000 level of accuracy, whereas overhead rails for a monorail facility could well require levels of 1/7,500–1/10,000. Section 10.1.2 notes that the 2005 ALTA/ACSM Standards refer to relative positional accuracy (RPA). This is the value expressed in feet or meters that represents the uncertainty due to random errors in measurements in the location of any point on a survey relative to any other point on the same survey at the 95 percent confidence level. The allowable RPA for measurements controlling land boundaries on an ALTA/ACSM land title surveys are 0.07 ft (or 20 mm) + 50 ppm. (Note that 50 ppm = 50/1,000,000 = 1/20,000.)

As noted earlier, control surveyors for both legal and engineering projects must locate their control points at a much higher level of precision and accuracy than is necessary for the actual location of the legal or engineering project markers that are surveyed in from those control points.

If the precision using latitudes and departures computed from field measurements is not acceptable as determined by the survey specifications (see Chapter 10), additional fieldwork must be performed to improve the level of precision. (Fieldwork is never undertaken until all calculations have been double-checked.) When fieldwork is to be checked, usually the distances are checked first, as the angles have already been verified [i.e., (*n* – 2)180°].

If a large error (or mistake) has been made on one side, it will significantly affect the bearing of the linear error of closure. Accordingly, if a check on fieldwork is necessary, the surveyor first computes the bearing of the linear error of closure and checks that bearing against the course bearings. If a similarity exists (±5°), that course is the first course re-measured in the field. If a difference is found in that course measurement (or any other course), the new course distance is immediately substituted into the computation to check whether the required precision level has then been achieved.

A quick summary of initial traverse computations follows:

1. Balance the angles.
2. Compute the bearings and/or the azimuths.
3. Compute the latitudes and departures, the linear error of closure, and the precision ratio of the traverse.

If the precision ratio is satisfactory, further treatment of the data is possible (e.g., coordinates and area computations).

If the precision ratio is unsatisfactory (e.g., a precision ratio of only 1/4,000 when a ratio of 1/5,000 was specified), complete the following steps:

1. Double-check all computations.
2. Double-check all field book entries.
3. Compute the bearing of the linear error of closure and check to see if it is similar to a course bearing (±5°).
4. Re-measure the sides of the traverse, beginning with a course having a bearing similar

to the linear error of closure bearing (if there is one).

5. When a correction is found for a measured side, try that value in the latitude-departure computation to determine the new level of precision.

6.7 TRAVERSE PRECISION AND ACCURACY

The actual accuracy of a survey as suggested by the precision ratio of a closed traverse can be misleading. The opportunity exists for significant errors to cancel out; this can result in apparent high-precision closures from relatively inaccurate field techniques. In the case of a closed-loop traverse, untreated systematic errors will have been largely balanced during the computation of latitudes and departures. For example, in a square-shaped traverse, untreated systematic taping errors (long or short tape) will be completely balanced and beyond mathematical detection. Therefore, in

order for high precision to also reflect favorably on accuracy, control of specifications, field practices, and techniques is essential.

If a traverse has been closed to better than 1/5,000, the taping should have conformed to the specifications shown in Table 3-2. Reference to Figure 6-12 will show that, for consistency, the survey should be designed so that the maximum allowable error in angle (E_a) should be roughly equal to the maximum allowable error in distance (E_d). If the linear error is 1/5,000, the angular error should be consistent:

$$1/5,000 = \tan \theta$$
$$\theta = 0°00'41''$$

See Table 6-3 for additional linear and angular error relationships. The overall allowable angular error in an n-angled closed traverse would be $E_a \sqrt{n}$. (Random errors accumulate as the square root of the number of observations.) Thus, for a five-sided traverse with a specification for precision of 1/3,000, the maximum angular misclosure would be $01' \sqrt{5} = 02'$ (to the closest arc minute), and for a five-sided traverse with a specification

Point Y is to Be Set Out from Fixed Points X and Z.

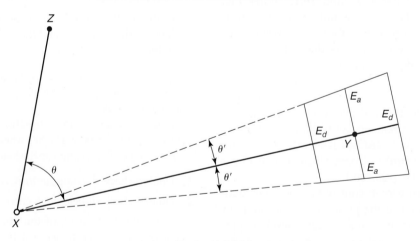

For the Line XY

E_d is the Possible Error in Distance Measurement and E_a is the Position Error Resulting from a Possible Angle Error of θ' in an Angle of θ.

FIGURE 6-12 Relationship between errors in linear and angular measurements

Table 6-3   Linear and Angular Error Relationships

Linear Accuracy Ratio	Maximum Angular Error, E_a	Least Count of Total Station or Theodolite Scale or Readout
1/1,000	0°03′26″	01′
1/3,000	0°01′09″	01′
1/5,000	0°00′41″	30″
1/7,500	0°00′28″	20″
1/10,000	0°00′21″	20″
1/20,000	0°00′10″	10″

for precision of 1/5,000, the maximum misclosure of the field angles would be $30″\sqrt{5} = 01′$ (to the closest 30″).

6.8 TRAVERSE ADJUSTMENTS

Latitudes and departures can be used for the computation of coordinates or the computation of the area enclosed by the traverse. In addition, the station coordinates can then be used to establish control for further survey layout. However, before any further use can be made of these latitudes and departures, they must be adjusted so that their errors are suitably distributed, and the algebraic sums of the latitudes and departures are each zero. These adjustments, if properly done, will ensure that the final position of each traverse station, as given by the station coordinates, is optimal with respect to the true station location.

Current surveying practice favors either the compass rule (Bowditch) adjustment or the least squares adjustment. The compass rule (an approximate method) is applied in most calculator solutions, with the least squares method being reserved for large, high-precision traverses (e.g., extension of a state or province coordinate grid) and for most computer software solutions. Although the least squares method requires more extensive computations, advances in computer technology are now enabling the surveyor to use the least squares technique with the help of a desktop computer or even

a handheld calculator. The 2005 ALTA/ACSM accuracy standards call for survey analysis and adjustments to be performed using least squares techniques, but with the very small errors and adjustments now encountered using modern electronic field equipment, the results of analysis and adjustments using the compass rule and the least squares techniques are nearly identical.

6.9 COMPASS RULE ADJUSTMENT

The *compass rule* is used in many survey computations to distribute the errors in latitudes and departures. The **compass rule** distributes the errors in latitude and departure for each traverse course *in the same proportion as the course distance is to the traverse perimeter.* That is, generally,

$$\frac{C \operatorname{lat} AB}{\Sigma \operatorname{lat}} = \frac{AB}{P} \quad \text{or} \quad C \operatorname{lat} AB = \Sigma \operatorname{lat} \times \frac{AB}{P} \quad (6\text{-}3)$$

where $C \operatorname{lat} AB = $ *correction in latitude AB*

$\Sigma \operatorname{lat} = $ error of closure in latitude

$AB = $ distance AB

$P = $ perimeter of traverse

and

$$\frac{C \operatorname{dep} AB}{\Sigma \operatorname{dep}} = \frac{AB}{P} \quad \text{or} \quad C \operatorname{dep} AB$$
$$= \Sigma \operatorname{dep} \times \frac{AB}{P} \quad (6\text{-}4)$$

where C dep $AB = $ *correction in departure AB*

$$\Sigma \text{ dep} = \text{error in closure in departure}$$

$$AB = \text{distance } AB$$

$$P = \text{perimeter of traverse}$$

In both cases, the sign of the correction is opposite from that of the error.

Referring to the traverse example, Example 6-1, Table 6-2 has been expanded in Table 6-4 to provide space for traverse adjustments. The magnitudes of the individual corrections are shown next:

$$C \text{ lat } AE = \frac{0.06 \times 350.10}{2,081.19} = 0.01$$

$$C \text{ lat } CB = \frac{0.06 \times 383.20}{2,081.19} = 0.01$$

$$C \text{ lat } ED = \frac{0.06 \times 579.03}{2,081.19} = 0.02$$

$$C \text{ lat } BA = \frac{0.06 \times 401.58}{2,081.19} = 0.01$$

$$C \text{ lat } DC = \frac{0.06 \times 368.28}{2,081.19} = 0.01$$

$$\text{Check: } \Sigma C \text{ lat} = 0.06$$

$$C \text{ dep } AE = \frac{0.49 \times 350.10}{2,081.19} = 0.08$$

$$C \text{ dep } CB = \frac{0.49 \times 382.20}{2,081.19} = 0.09$$

$$C \text{ dep } ED = \frac{0.49 \times 579.03}{2,081.19} = 0.14$$

$$C \text{ dep } BA = \frac{0.49 \times 401.58}{2,081.19} = 0.09$$

$$C \text{ dep } DC = \frac{0.49 \times 368.28}{2,081.19} = 0.09$$

$$\text{Check: } \Sigma C \text{ dep} = 0.49$$

When these computations are performed on a handheld calculator, the proportions 0.06/2,081.19 and 0.49/2,081.19 can be set up as constant multipliers to allow faster computations.

It remains now only for the algebraic sign to be determined. The rule is simple: The corrections are opposite in sign to the errors. Therefore, for this example, the latitude corrections are negative

and the departure corrections are positive. The corrections are now added algebraically to arrive at the balanced values. For example, in Table 6-4, the correction for latitude AE is –0.01, which is to be added to the latitude AE –311.30 algebraically to get –311.31. For the next course ED, the latitude correction (–0.02) and the latitude (+246.10) are added to get +246.08.

To check the work, the balanced latitudes and balanced departures are totaled to see if their respective sums are zero. It sometimes happens that the balanced latitude or balanced departure totals fail to equal zero by one or two last place units (0.01 in this example). This discrepancy is caused by rounding off and is normally of no consequence. The discrepancy could be is removed by arbitrarily changing one of the values to force the total to zero.

Note that when the error (in latitude or departure) to be distributed is quite small, the corrections can be arbitrarily assigned to appropriate courses. For example, if the error in latitude (or departure) were only 0.03 ft in a five-sided traverse, it would be appropriate to apply corrections of 0.01 ft to the latitude of each of the three longest courses. Similarly, in Example 6-1, where the error in latitude for that five-sided traverse is +0.06, it would have been appropriate to apply a correction of –0.02 to the longest course latitude and a correction of –0.01 to each of the remaining four latitudes, the same solution provided by the compass rule. See Appendix A for further treatment of error adjustments.

6.10 EFFECTS OF TRAVERSE ADJUSTMENTS ON THE ORIGINAL DATA

Once the latitudes and departures have been adjusted, the original polar coordinates (distance and direction) will no longer be valid. In most cases, the adjustment required for the polar coordinates is too small to warrant consideration; but

Table 6-4 Traverse Adjustments: Compass Rule, Section 6.6

Course	Distance (ft)	Bearing	Latitude	Departure	C lat	C dep	Balanced Latitudes	Balanced Departures
AE	350.10	S 27°13'48" E	−311.30	+160.19	−0.01	+0.08	−311.31	+160.27
ED	579.03	N 64°50'54" E	+246.10	+524.13	−0.02	+0.14	+246.08	+524.27
DC	368.28	N 1°10'06" E	+368.20	+7.51	−0.01	+0.09	+368.19	+7.60
CB	382.20	S 82°08'48" W	−52.22	−378.62	−0.01	+0.09	−52.23	−378.53
BA	401.58	S 51°22'00" W	−250.72	−313.70	−0.01	+0.09	−250.73	−313.61
	P = 2081.19		Σ lat = +0.06	Σ dep = −0.49	Σ C$_{lat}$ = −0.06	Σ C$_{dep}$ = +0.49	0.00	0.00

Table 6-5 Adjustment of Bearings and Distances Using Balanced Latitudes and Departures: (Section 6.6)

Course	Balanced Latitude	Balanced Departure	Adjusted Distance (ft)	Adjusted Bearing	Original Distance (ft)	Original Bearing
AE	−311.31	+160.27	350.14	S 27°14′26″ E	350.10	S 27°13′48″ E
ED	+246.08	+524.27	579.15	N 64°51′21″ E	579.03	N 64°50′54″ E
DC	+368.19	+7.60	368.27	N 1°10′57″ E	368.28	N 1°10′06″ E
CB	−52.23	−378.53	382.12	S 82°08′38″ W	382.20	S 82°08′48″ W
BA	−250.73	−313.61	401.52	S 51°21′28″ W	401.58	S 51°22′00″ W
	0.00	0.00	P = 2081.20		P = 2081.19	

if the data are to be used for layout purposes, the *corrected* distances and directions should be used.

The following relationships are inferred from Figures 6-5 and 6-11:

$$\text{Distance}\,AD = \sqrt{\text{lat}\,AD^2 + \text{dep}\,AD^2} \quad (6\text{-}5)$$

$$\tan \text{bearing}\,AD = \frac{\text{dep}\,AD}{\text{lat}\,AD} \quad (6\text{-}6)$$

Next, we use the values from Table 6-4. The solution for the course AE is shown in the following equations:

$$\text{Distance}\,AE = \sqrt{311.31^2 + 160.27^2} = 350.14\,\text{ft}$$

$$\tan \text{bearing}\,AE = \frac{+160.27}{-311.31}$$

$$\text{Bearing}\,AE = S27°14′26″\,E$$

The remaining corrected bearings and distances are shown in Table 6-5.

6.11 OMITTED MEASUREMENTS

The techniques developed in the computation of latitudes and departures can be used to supply missing course information on a closed traverse and can be used to solve any surveying problem that can be arranged in the form of a closed traverse.

Example 6-2: A missing course in a closed traverse is illustrated in Figure 6-13 and tabulated in Table 6-6. The data can be treated in the same manner as in a closed traverse. When the latitudes and departures are totaled, they will not balance. Both the latitudes and departures will fail to close by the amount of the latitude and departure of the missing course, DA.

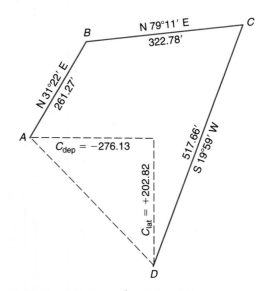

FIGURE 6-13 Example 6.2: Missing course computation

Table 6-6   Missing Course: Example 6.2

Course	Distance	Bearing	Latitude	Departure
AB	261.27	N 31°22′ E	+223.09	+135.99
BC	322.78	N 79°11′ E	+60.58	+317.05
CD	517.66	S 19°59′ W	−486.49	−176.91
			Σ lat = −202.82	Σ dep = +276.13
DA			C lat = +202.82	C dep = −276.13

The length and direction of *DA* can be simply computed by using Equations 6-5 and 6-6:

$$\text{Distance } DA = \sqrt{(\text{lat } DA)^2 + (\text{dep } DA)^2}$$
$$= \sqrt{202.82^2 + 276.13^2}$$
$$= 342.61 \text{ ft}$$

$$\tan \text{ bearing } DA = \frac{\text{dep } AD}{\text{lat } AD}$$
$$= \frac{-276.13}{+202.82}$$

$$\text{Bearing } DA = \text{N } 53°42′ \text{ W (rounded to closest minute)}$$

Note that this technique does not permit a check on the accuracy ratio of the fieldwork. Because this is the closure course, the computed value will also contain all accumulated errors.

Example 6-3: Figure 6-14 illustrates an intersection jog elimination problem that occurs in many municipalities. For a variety of reasons, some streets do not directly intersect other streets. They jog a few feet to a few hundred feet before continuing. In Figure 6-14, the sketch indicates that, in this case, the entire jog will be taken out on the north side of the intersection. The designer has determined that if the McCowan Road ℄ (South of Finch) is produced 300 ft northerly of the ℄ of Finch Avenue E, it can then be joined to the existing McCowan Road ℄ at a distance of

1,100 ft northerly from the ℄ of Finch Avenue E. Presumably, these distances will allow for the insertion of curves that will satisfy the geometric requirements for the design speed and traffic volumes. The problem here is to compute the length of *AD* and the defection angles at *D* and *A* [see Figure 6-14(b) and Table 6-7].

Solution: To simplify computations, assume a bearing for line *DC* of due south. The bearing for *CB* is therefore due west, and obviously the bearing of *BA* is N 0°29′ E. The problem is then set up as a missing course problem, and the corrections in latitude and departure are the latitude and departure of *AD* (see Table 6-7).

$$\text{Distance } AD = \sqrt{799.96^2 + 177.05^2} = 819.32 \text{ ft}$$

$$\tan \text{ bearing } AD = \frac{177.05}{-799.96}$$

$$\text{Bearing } AD = \text{S } 12°29′00″ \text{ E} \quad \text{(to the closest 30″)}$$

The bearings are shown in Figure 6-14(c), which leads to the calculation of the deflection angles as shown.

There are obviously many other situations where the missing course techniques can be used. If one bearing and one distance (not necessarily on the same course) are omitted from a traverse, the missing data can be solved. In some cases it may be necessary to compute intermediate cutoff lines and use cosine and sine laws in conjunction with the latitude/departure solution.

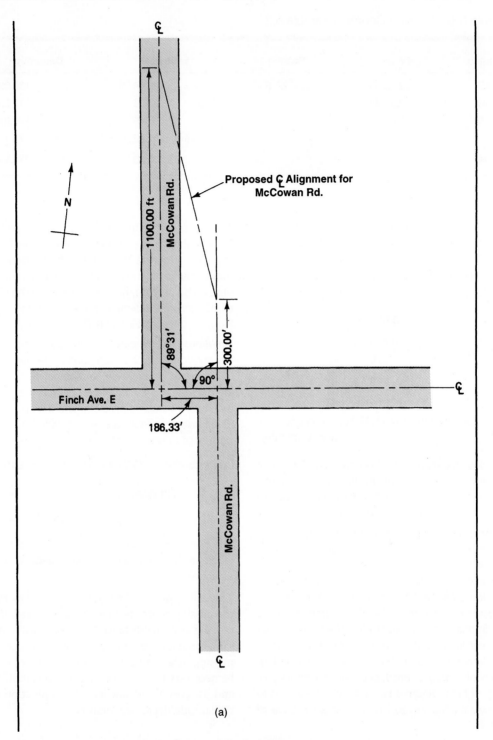

N

1100.00 ft

McCowan Rd.

Proposed ℄ Alignment for
McCowan Rd.

89°31'

300.00'

90°

Finch Ave. E

℄

186.33'

McCowan Rd.

℄

(a)

FIGURE 6-14 (a) Example 6.3: Missing course problem

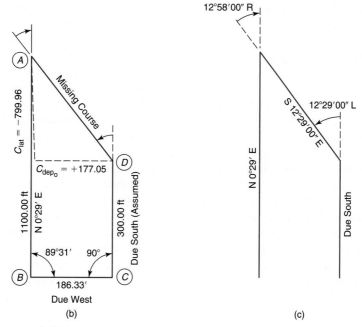

FIGURE 6-14 (*Continued*) (b) Distances and bearings. (c) Determination of deflection angles

Table 6-7   Missing Course: Example 6.3

Course	Distance	Bearing	Latitude	Departure
DC	300.00	S 0°00' E(W)	−300.00	0.00
CB	186.33	S 90°00' W	0.00	−186.33
BA	1100.00	N 0°29' E	+1099.96	+9.28
			Σ lat = 799.96	Σ dep = −177.05
AD			C lat = −799.96	C dep = +177.05

6.12 RECTANGULAR COORDINATES OF TRAVERSE STATIONS

6.12.1 Coordinates Computed from Balanced Latitudes and Departures

Rectangular coordinates define the position of a point with respect to two perpendicular axes.

Analytic geometry uses the concepts of a y axis (north–south) and an x axis (east–west), concepts that are obviously quite useful in surveying applications.

In universal transverse Mercator (UTM) coordinate grid systems, the X axis is often the latitude of 0° (equator), and the Y axis is a central meridian through the middle of the zone in which the grid is located (see Chapter 10); uppercase X and Y are used for assigned axes. For surveys of a limited nature, where a coordinate grid has not been established, the coordinate axes can be assumed. If the axes are to be assumed, values are chosen such that the coordinates of all stations will be positive (i.e., all stations will be in the northeast quadrant).

The traverse tabulated in Table 6-4 will be used for illustrative purposes. Values for the coordinates of station A are assumed to be 1,000.00 ft north and 1,000.00 ft east.

The coordinates of the other traverse stations are calculated simply by adding the *balanced* latitudes and departures to the previously calculated coordinates. In Figure 6-15 the balanced latitude and departure of course AE are applied to the assumed coordinates of station A to determine the coordinates of station E, and so on. These simple computations are shown in Table 6-8. A check on the computation is possible by using

the last latitude and departure (BA) to recalculate the coordinates of station A.

The use of coordinates to define the positions of boundary markers has been steadily increasing over the years (see Chapter 10). The storage of property-corner coordinates in large-memory civic computers will, in the not-so-distant future (already available in some jurisdictions), permit lawyers, municipal authorities, and others to have instant retrieval of current land registration information, assessment, and other municipal information, such as census and the level of municipal services. Although such use is important, the truly impressive impact of coordinate use will result from (1) the coordination of topographic detail (digitization) so that plans can be prepared by computer-assisted plotters, and (2) the coordination of all legal and

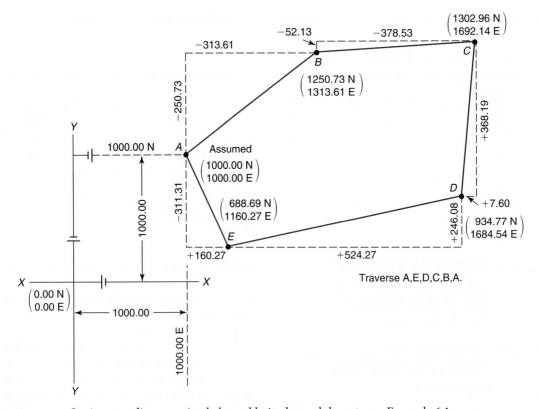

FIGURE 6-15 Station coordinates, using balanced latitudes and departures, Example 6.1

Table 6-8   Computation of Coordinates Using Balanced Latitudes and Departures

Course	Balanced Latitude	Balanced Departure	Station	North		East	
			A	1000.00	(assumed)	1000.00	(assumed)
AE	−311.31	+160.27		−311.31		+160.27	
			E	688.69		1160.27	
ED	+246.08	+524.27		+246.08		+524.27	
			D	934.77		1684.54	
DC	+368.19	+7.60		+368.19		+7.60	
			C	1302.96		1692.14	
CB	−52.23	−378.53		−52.23		−378.53	
			B	1250.73		1313.61	
BA	−250.73	−313.61		−250.73		−313.61	
			A	1000.00	Check	1000.00	Check

engineering details so that the plans be produced by computer-assisted plotters. The survey layout will be accomplished using sets of computer-generated coordinates (rectangular and polar) fed either manually or automatically into total stations (see Chapter 5). Complex layouts can then be accomplished quickly, by a few surveyors from one or two centrally located control points, with a higher level of precision and a lower incidence of mistakes.

6.12.2 Adjusted Coordinates Computed from Raw-Data Coordinates

Section 6.9 demonstrated the adjustment of traverse errors by the adjustment of the individual Δy's (northings) and Δx's (eastings) for each traverse course using the compass rule. Although this traditional technique has been favored for many years, lately, because of the widespread use of the computer in surveying solutions, coordinates are now often first computed from raw (unadjusted) bearing/distance data and then adjusted using the compass rule or least squares technique. Because we are now working with coordinates and not individual Δy's and Δx's, the distance factor to be used in the compass rule must be cumulative. This technique will be illustrated using the same field data from Section 6.6.

Corrections (C) to raw-data coordinates (see Tables 6-9 and 6-10): $C\Delta y$ = correction in northing (latitude), and $C\Delta x$ = correction in easting (departure).

Station E

$$C\Delta y \text{ is } (AE/P) \, \Sigma \Delta y = (350.10/2081.19)\,0.06 = 0.01$$
$$C\Delta x \text{ is } (AE/P) \, \Sigma \Delta x = (350.10/2081.19)\,0.49 = 0.08$$

Station D

$$C\Delta y \text{ is } [(AE + ED)/p] \, \Sigma \Delta y = (350.10 + 579.03)/2081.19 \times 0.06 = 0.03$$
$$C\Delta x \text{ is } [(AE + ED)/p] \, \Sigma \Delta x = (350.10 + 579.03)/2081.19 \times 0.49 = 0.22$$

Table 6-9   Computation of Raw-Data Coordinates

Station/Course	Azimuth	Distance	ΔY	ΔX	Coordinates (Raw Data)	
					Northing	Easting
A					1,000.00	1,000.00
AE	152°46′12″	350.10	−311.30	160.19		
E					688.70	1,160.19
ED	64°50′54″	579.03	246.10	524.13		
D					934.80	1,684.32
DC	1°10′06″	368.28	368.20	7.51		
C					1,303.00	1,691.83
CB	262°08′48″	382.20	−52.22	−378.62		
B					1,250.78	1,313.21
BA	231°22′00″	401.58	−250.72	−313.70		
A			ΣΔY = 0.06	ΣΔX = −0.49	1,000.06	999.51

Station C

$C\Delta y$ is $[(AE + ED + DC)/p] \Sigma \Delta y = [(350.10 + 579.03 + 368.28)/2081.19] \times 0.06 = 0.04$
$C\Delta x$ is $[(AE + ED + DC)/p] \Sigma \Delta x = [(350.10 + 579.03 + 368.28)/2081.19] \times 0.49 = 0.31$

Station B

$C\Delta y$ is $[(AE + ED + DC + CB)/p] \Sigma \Delta y = [(350.10 + 579.03 +$
$$368.28 + 382.20)/2081.19] \times 0.06 = 0.05$$
$C\Delta x$ is $[(AE + ED + DC + CB)/p] \Sigma \Delta x = [(350.10 + 579.03 +$
$$368.28 + 382.20)/2081.19] \times 0.49 = 0.40$$

Station A

$C\Delta y$ is $[(AE + ED + DC + CB + BA)/p] \Sigma \Delta y = [(350.10 + 579.03 + 368.28 +$
$$382.20 + 401.50)/2081.19] \times 0.06 = 0.06$$

$C\Delta x$ is $[(AE + ED + DC + CB + BA)/p] \Sigma \Delta x = [(350.10 + 579.03 + 368.28 +$
$$382.20 + 401.50)/2081.19] \times 0.49 = 0.49$$

6.13 AREA OF A CLOSED TRAVERSE BY THE COORDINATE METHOD

When the coordinates of the stations of a closed traverse are known, it is a simple matter to then compute the area within the traverse, either by computer or handheld calculator. Figure 6-16(a) shows a closed traverse 1, 2, 3, 4 with appropriate x and y coordinate distances. Figure 6-16(b) illustrates the technique used to compute the traverse area.

Figure 6-16(b) shows that *the desired area of the traverse is area 2 minus area 1,* where area 2 is the sum of the areas of trapezoids 4′433′

Table 6-10 Computation of Adjusted Coordinates, Distances, and Directions

Station/Course	Coordinates (Raw Data)				Adjusted Coordinates		Adjusted ΔY	Adjusted ΔX	Adjusted Distance	Adjusted Bearing
	Northing	Easting	CΔY*	CΔX*	Northing	Easting				
A	1,000.00	1,000.00			1,000.00	1,000.00				
AE				+0.08			−311.31	160.27	350.14	S 27°14'26" E
E	688.70	1,160.19	−0.01		688.69	1,160.27				
ED				+0.22			246.08	524.27	579.15	N 64°51'31" E
D	934.80	1,684.32	−0.03		934.77	1,684.54				
DC				+0.31			368.19	7.60	368.27	N 1°10'57" E
C	1,303.00	1,691.83	−0.04		1,302.96	1,692.14				
CB				+0.40			−52.23	−378.53	382.12	S 82°08'38" W
B	1,250.78	1,313.21	−0.05		1,250.73	1,313.61				
BA				+0.49			−250.73	−313.61	401.52	S 51°21'28" W
A	1,000.06	999.51	−0.06		1,000.00	1,000.00				

*Correction is opposite in sign to the error.

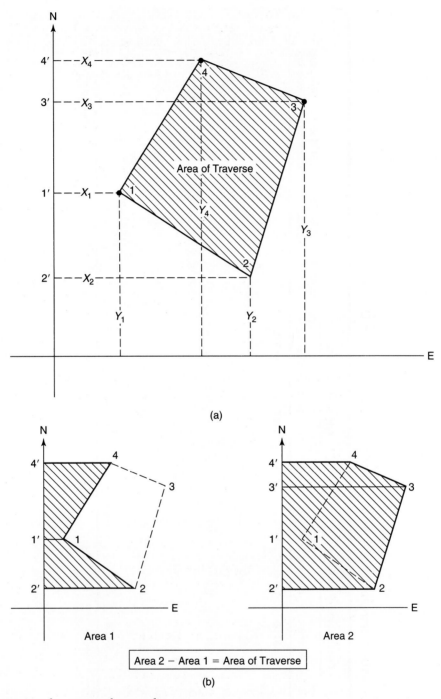

(a)

Area 2 − Area 1 = Area of Traverse

(b)

FIGURE 6-16 Area by rectangular coordinates

and 3'322' and area 1 is the sum of trapezoids 4'411' and 1'122'.

$$\text{Area } 2 = \frac{1}{2}(x_4 + x_3)(y_4 - y_3)$$
$$+ \frac{1}{2}(x_3 + x_2)(y - y_2)$$

$$\text{Area } 1 = \frac{1}{2}(x_4 + x_1)(y_4 - y_1)$$
$$+ \frac{1}{2}(x_1 + x_2)(y_1 - y_2)$$

$$2A = \left[(x_4 + x_3)(y_4 - y_3) + (x_3 + x_2)(y_3 - y_2)\right]$$
$$- \left[(x_4 + x_1)(y_4 - y_1) + (x_1 + x_2)(y_1 - y_2)\right]$$

Expanding this expression and collecting the remaining terms:

$$2A = x_1(y_2 - y_4) + x_2(y_3 - y_1) + x_3(y_4 - y_2)$$
$$+ x_4(y_1 - y_3) \qquad (6\text{-}7)$$

Stated simply, *the double area of a closed traverse is the algebraic sum of each x coordinate multiplied by the difference between the y coordinates of the adjacent stations.*

The double area is divided by 2 to determine the final area. This rule applies to a traverse with any number of sides. The final area can result in a positive or a negative number, reflecting only the direction of computation (CW or CCW). The physical area is, of course, positive; there is no such thing as a negative area.

Example 6-4: *Area Computation by Coordinates*
Referring to the traverse example in Section 6.6 (Example 6-1) and to Figure 6-15:

Station	North	East
A	1,000.00 ft	1,000.00 ft
B	1,250.73	1,313.61
C	1,302.96	1,692.14
D	934.77	1,684.54
E	688.69	1,160.27

The double area computation (to the closest ft^2) using the relationships shown in Equation 6-7 is as follows:

$$2A = x_1(y_2 - y_4) + x_2(y_3 - y_1) + x_3(y_4 - y_2) + x_4(y_1 - y_3)$$

$$X_A(Y_B - Y_E) = 1,000.00(1,250.73 - 688.69) = \quad +562,040$$
$$X_B(Y_C - Y_A) = 1,313.61(1,302.96 - 1,000.00) = +397,971$$
$$X_C(Y_D - Y_B) = 1,692.14(934.77 - 1,250.73) = \quad -534,649$$
$$X_D(Y_E - Y_C) = 1,684.54(688.69 - 1,302.96) = -1,034,762$$
$$X_E(Y_A - Y_D) = 1,160.27(1,000.00 - 934.77) = \quad + 75,684$$
$$ -533,716\,\text{ft}^2$$
$$2A = \quad 533,716\,\text{ft}^2$$
$$\text{Area} = 266,858\,\text{ft}^2$$

Also:

$$\text{Area} = \frac{246,858}{43,560} = 6.126\,\text{acres} \quad (1\,\text{acre} = 43,560\,\text{ft}^2)$$

6.14 REVIEW PROBLEM

Example 6-5: Figure 6-17 shows the field data for a five-sided closed traverse.

a. Balance the angles.

b. Compute the azimuths and bearings.

c. Compute the linear error of closure and the precision ratio of the traverse.

d. If the precision ratio is equal to or better than 1/4,000, balance the latitudes and departures using the compass rule.

e. Assuming that the coordinates of station B are (1,000.000 N, 1,000.000 E), compute the coordinates of the remaining stations.

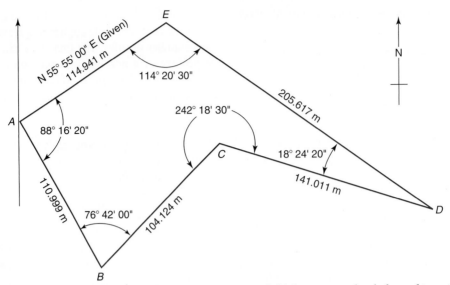

FIGURE 6-17 Traverse sketch for Example 6.5 showing mean field distances and unbalanced interior angles

Solution:

a.

Field Angles	Balanced Angles
A 88°16′20″	A 88°16′00″
B 76°42′00″	B 76°41′40″
C 242°18′30″	C 242°18′10″
D 18°24′20″	D 18°24′00″
E 114°20′30″	E 114°20′10″

538°120′00″ = 540°00′100″   538°119′60″ = 540°00′00″

$$(n - 2)180 = 540°$$
Error in angular closure = 100″
Correction per angle = −20″

b. See Figure 6-18 (use balanced interior angles)

(See Figure 6-18 and Table 6-11)

$$
\begin{aligned}
Az\,AE &= 55°55′00″ \\
+ \angle A &\quad 88°16′00″ \\
Az\,AB &= 144°11′00″ \\
+180° \\
Az\,BA &= 324°11′00″ \\
+ \angle B &\quad 76°41′40″ \\
Az\,BC &= 400°52′40″ \\
-360°
\end{aligned}
$$

$$
\begin{aligned}
Az\,BC &= 40°52′40″ \\
+180°\,E \\
Az\,CB &= 220°52′40″ \\
+ \angle C &\quad 242°18′10″ \\
Az\,CD &= 463°10′50″ \\
-360° \\
Az\,CD &= 103°10′50″ \\
+180° \\
Az\,DC &= 283°10′50″ \\
+ \angle D &\quad 18°24′00″ \\
Az\,DE &= 301°34′00″ \\
-180° \\
Az\,DE &= 121°34′50″ \\
+ \angle E &\quad 114°20′10″ \\
Az\,EA &= 235°55′00″ \\
-180° \\
Az\,AE &= 55°55′00″ \quad \text{Check}
\end{aligned}
$$

c. Referring to Table 6-12, the latitudes and departures are computed by using either the azimuths or the bearings and the distances. See Figure 6-18 for bearing computations. (Both azimuths and bearings are shown here only for comparative purposes; it is not necessary to use both.) Since the precision ratio (1/4,100) meets the specifications noted, the computations can continue.

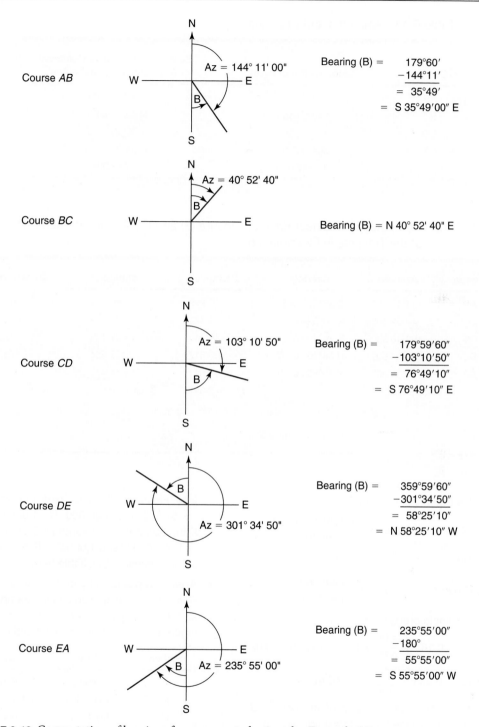

Course *AB*   Bearing (B) =   179°60′
                              −144°11′
                          =   35°49′
                          = S 35°49′00″ E

Course *BC*   Bearing (B) = N 40° 52′ 40″ E

Course *CD*   Bearing (B) =   179°59′60″
                              −103°10′50″
                          =   76°49′10″
                          = S 76°49′10″ E

Course *DE*   Bearing (B) =   359°59′60″
                              −301°34′50″
                          =   58°25′10″
                          = N 58°25′10″ W

Course *EA*   Bearing (B) =   235°55′00″
                              −180°
                          =   55°55′00″
                          = S 55°55′00″ W

FIGURE 6-18 Computation of bearings from computed azimuths, Example 6.5

Table 6-11   Azimuths and Bearings

Course	Azimuths (Clockwise)	Bearings (Clockwise) Computed from Clockwise Azimuths
AB	144°11'00"	S 35°49'00" E
BC	40°52'40"	N 40°52'40" E
CD	103°10'50"	S 76°49'10" E
DE	301°34'50"	N 58°25'10" W
EA	235°55'00"	S 55°55'00" W

Table 6-12   Computation of Linear Error of Closure (E) and the Accuracy Ratio E/P of the Traverse in Example 6.5

Course	Azimuth	Bearing	Distance	Latitude	Departure
AB	144°11'00"	S 35°49'00" E	110.999	−90.008	+64.956
BC	40°52'40"	N 40°52'40" E	104.124	+78.729	+68.144
CD	103°10'50"	S 76°49'10" E	141.011	−32.153	+137.296
DE	301°34'50"	N 58°25'10" W	205.617	+107.681	−175.166
EA	235°55'00"	S 55°55'00" W	114.941	−64.413	−95.197
			$P = 676.692$	$\Sigma\ \text{lat} = -0.164$	$\Sigma\ \text{dep} = +0.033$

$$E = \sqrt{(\Sigma\ \text{lat})^2 + (\Sigma\ \text{dep})^2} = \sqrt{0.164^2 + 0.033^2} = 0.167\ \text{m}$$

$$\text{Precision ratio of traverse} = \frac{0.167}{676.692} = \frac{1}{4052} = \frac{1}{4100}$$

d. Use Equations 6-3 and 6-4:

$$C\ \text{lat}\ AB = \Sigma\ \text{lat} \times \frac{\text{distance}\ AB}{\text{perimeter}}$$

and

$$C\ \text{dep}\ AB = \Sigma\ \text{dep} \times \frac{\text{distance}\ AB}{\text{perimeter}}$$

For the example problem:

$$C\ \text{lat}\ AB = 0.164 \times \frac{111}{677}\ (\text{values rounded}) = 0.027\ \text{m}$$

$$C\ \text{dep}\ AB = 0.033 \times \frac{111}{677} = 0.005\ \text{m}$$

As a check, the algebraic sum of the balanced latitudes (and balanced departures) will be zero.

e. The *balanced* latitudes and departures (Table 6-13) are used to compute the coordinates

(see Figure 6-19). The axes are selected so that B is 1,000.000 N, 1,000.000 E. If the computation returns to B with values of 1,000.000 N and 1,000.000 E, the computation is verified. See Table 6-14.

f. Area Computation by coordinates
Refer to the traverse shown in Example 6-5, and the computed coordinates shown in Figure 6-19, which are summarized next:

Station	North	East
A	1,089.981	935.049
B	1,000.000	1,000.000
C	1,078.754	1,068.139
D	1,046.635	1,205.498
E	1,154.366	1,030.252

Table 6-13   Computation of Balanced Latitudes and Departures Using the Compass Rule

Course	Latitude	Departure	C lat	C dep	Balanced Latitude	Balanced Departure
AB	−90.008	+64.956	+0.027	−0.005	−89.981	+64.951
BC	+78.729	+68.144	+0.025	−0.005	+78.754	+68.139
CD	−32.153	+137.296	+0.034	−0.007	−32.119	+137.289
DE	+107.681	−175.166	+0.050	−0.010	+107.731	−175.176
EA	−64.413	−95.197	+0.028	−0.006	−64.385	−95.203
	Σ lat = −0.164	Σ dep = +0.033	+0.164	−0.033	0.00	0.00

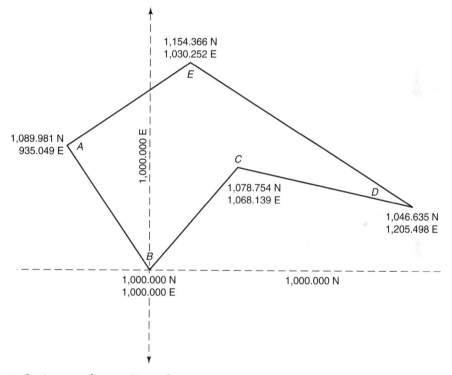

FIGURE 6-19 Station coordinates, Example 6.5

The *double area computation* (to the closest m²), using the relationships shown in Equation 6-7, is:

$$2A = x_1(y_2 - y_4) + x_2(y_3 - y_1) + x_3(y_4 - y_2) + x_4(y_1 - y_3)$$
$$X_A(Y_B - Y_E) = 935.049(1,000.000 - 1,154.366) = -144,340$$
$$X_B(Y_C - Y_A) = 1,000.000(1,078.754 - 1,089.981) = -11,227$$
$$X_C(Y_D - Y_B) = 1,068.139(1,046.635 - 1,000.000) = +49,813$$
$$X_D(Y_E - Y_C) = 1,205.498(1,154.366 - 1,078.754) = +91,150$$
$$X_E(Y_A - Y_D) = 1,030.252(1,089.981 - 1,046.635) = \underline{+44,657}$$
$$2A = +30,053 \, m^2$$
$$\text{Area} = 15,027 \, m^2$$
$$= 1.503 \, \text{hectares}$$

Table 6-14   Coordinates for Example 6.5

Station	North	East
B	1,000.000 +78.754 (lat BC)	1,000.000 +68.139 (dep BC)
C	1,078.754 −32.119 (lat CD)	1,068.139 +137.289 (dep CD)
D	1,046.635 +107.731 (lat DE)	1,205.428 − 175.176 (dep DE)
E	1,154.366 −64.385 (lat EA)	1,030.252 −95.203 (dep EA)
A	1,089.981 −89.981 (lat AB)	935.049 +64.951 (dep AB)
B	1,000.000 Check	1,000.000 Check

6.15 GEOMETRY OF RECTANGULAR COORDINATES

Figure 6-20 shows two points $P_1(x_1, y_1)$ and $P_2(x_2, y_2)$ and their rectangular relationships to the x and y axes.

$$\text{Length } P_1P_2 = \sqrt{(x_2 - x_1)^2 + (y_2 - y_1)^2} \quad (6\text{-}8)$$

$$\tan \alpha = \frac{x_2 - x_1}{y_2 - y_1} \quad (6\text{-}9)$$

where α is the bearing or azimuth of P_1P_2.

The length can be computed from either of the following:

$$\text{Length } P_1P_2 = \frac{x_2 - x_1}{\sin \alpha} \quad (6\text{-}10)$$

$$\text{Length } P_1P_2 = \frac{y_2 - y_1}{\cos \alpha} \quad (6\text{-}11)$$

For better accuracy, use the equation having the larger numerical value of $x_2 - x_1$ or $y_2 - y_1$ should be used.

It is clear from Figure 6-16 that $x - x_1$ is the departure of P_1P_2 and that $y_2 - y_1$ is the latitude of P_1P_2. In survey work, the y value (latitude) is usually known as the northing, and the x value (departure) is known as the easting. From analytic geometry, the slope of a straight line is $M = \tan(90° - \alpha)$, where $(90° - \alpha)$ is the angle of the straight line with x axis (see Figure 6-20); that is,

$$m = \cot \alpha$$

From coordinate geometry, the equation of straight line P_1P_2, when the coordinates of P_1 and P_2 are known, is

$$\frac{y - y_1}{y_2 - y_1} = \frac{x - x_1}{x_2 - x_1} \quad (6\text{-}12)$$

Equation 6-12 can also be written as

$$y - y_1 = \frac{y_2 - y_1}{x_2 - x_1}(x - x_1)$$

where $(y_2 - y_1)/(x_2 - x_1) = \cot \alpha = m$ (from analytic geometry), the slope of the line.

When the coordinates of one point P_1 and the bearing or azimuth of a line are known, the equation becomes

$$y - y_1 = \cot \alpha(x - x_1) \quad (6\text{-}13)$$

where α is the azimuth or bearing of the line through $P_1(x_1, y_1)$.

Also, from analytical geometry,

$$y - y_1 = \frac{1}{\cot \alpha}(x - x_1) \quad (6\text{-}14)$$

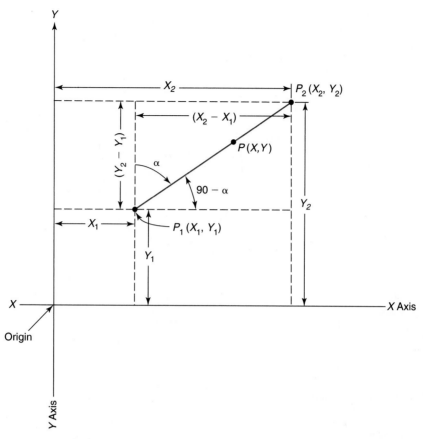

FIGURE 6-20 Geometry of rectangular coordinates

represents a line through P_1 *perpendicular* to the line represented by Equation 6-13; that is, the slopes of perpendicular lines are negative reciprocals.

Equations for circular curves are quadratics in the form

$$(x - H)^2 + (y - K)^2 = r^2 \qquad (6\text{-}15)$$

where r is the curve radius, (H, K) are the coordinates of the center, and (X, Y) are the coordinates of point P, which locates the circle (see Figure 6-21). When the circle center is at the origin, the equation becomes

$$x^2 - y^2 = r^2 \qquad (6\text{-}16)$$

6.16 ILLUSTRATIVE PROBLEMS IN RECTANGULAR COORDINATES

Several examples are provided in this section to give practice in both writing line equations and in solving for intersection coordinates.

Example 6-6: From the information shown in Figure 6-22, calculate the coordinates of the point of intersection K_1 of lines EC and DB.

Solution: From Equation 6-12, the equation of EC is

$$y - 688.69 = \frac{1{,}302.96 - 688.69}{1{,}692.14 - 1{,}160.27}(x - 1{,}160.27)$$

$$(1)$$

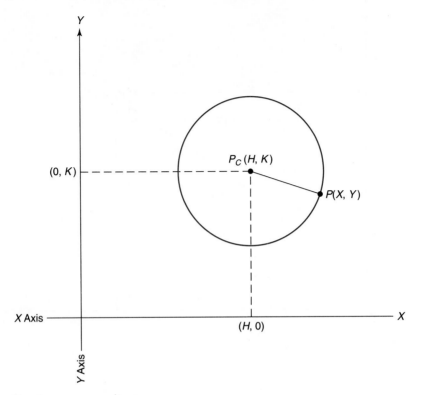

FIGURE 6-21 Circular curve coordinates

The equation of *DB* is

$$y - 934.77 = \frac{1,250.73 - 934.77}{1,313.61 - 1,684.54}(x - 1,684.54)$$

$$(2)$$

Simplifying, these equations become

$$EC: y = 1.154925x - 651.3355 \quad (1A)$$
$$DB: y = -0.8518049x + 2369.669 \quad (2A)$$
$$2.0067299x = 3021.004 \quad (1A-2A)$$
$$x = 1,505.436$$

Substitute the value of *x* in Equation 2A and check the results in Equation 1A:

$$y = 1,087.331$$

Therefore, the coordinates of point of intersection K_1 are (1,087.33 N, 1,505.44 E).

Example 6-7: From the information shown in Figure 6-22, calculate (a) the coordinates K_2, the point of intersection of line *ED* and a line

perpendicular to ED running through station *B*, and (b) distances K_2D and K_2E.

Solution:

a. From Equation 6-12, the equation of *ED* is

$$\frac{y - 688.69}{934.77 - 688.69} = \frac{x - 1,160.27}{1,684.54 - 1,60.27} \quad (1)$$

$$y - 688.69 = \frac{242.08}{524.27}(x - 1,160.27) \quad (1A)$$

From Equation 6-12, the equation of BK_2 is

$$y - 1,250.73 = \frac{524.27}{246.08}(x - 1,313.61) \quad (2)$$

Simplifying, these equations become (use five decimals to avoid rounding errors)

$$ED: 0.46938x - y = 144.09 \quad (1B)$$
$$BK_2: 2.13049x + y = +4,049.36 \quad (2B)$$
$$2.59987x = 3,905.27 \quad (1B + 2B)$$
$$x = 1,502.102$$

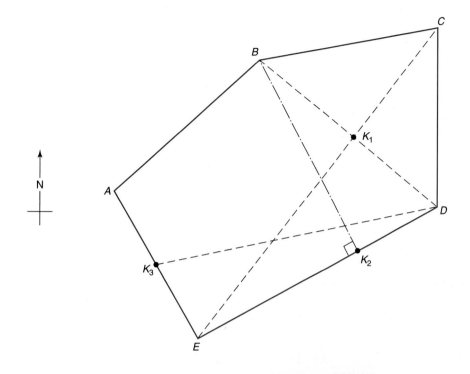

FIGURE 6-22 Coordinates for traverse problem, Section 6.12.2 (see Table 6-10)

	Coordinates	
Station	North (Y)	East (X)
A	1,000.00	1,000.00
B	1,250.73	1,313.61
C	1,302.96	1,692.14
D	934.77	1,684.54
E	688.69	1,160.27

Substitute the value of x in Equation 1A and check the results in Equation 2:

$$y = 849.15$$

Therefore, the coordinates of K_2 are (849.15 N, 1,502.10 E).

b. Figure 6-23 shows the coordinates for stations E and D and intermediate point K_2 from Equation 6-8:

$$\text{Length } K_2D = \sqrt{85.62^2 + 182.44^2} = 201.53$$

Length $K_2E = \sqrt{160.46^2 + 341.83^2} = 377.62$
$K_2D + K_2E = ED = 579.15$

Check:

$$\text{Length } ED = \sqrt{246.08^2 + 524.27^2} = 579.15$$

Example 6-8: From the information shown in Figure 6-22, calculate the coordinates of the point of intersection (K_3) of a line parallel to CB running from station D to line EA.

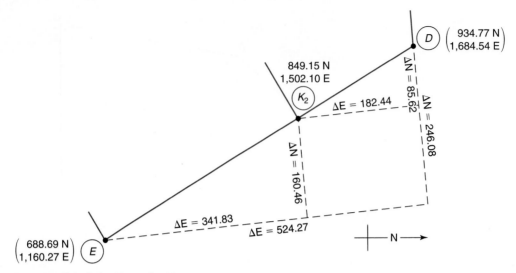

FIGURE 6-23 Sketch for Example 6.8

Solution: From Equation 6-12, the equation of CB is:

$$\frac{y - 1,302.96}{1,250.73 - 1,302.96} = \frac{x - 1,692.14}{1.313.61 - 1,692.14}$$

$$y - 1,302.96 = \frac{-52.23}{-378.53}(x - 1,692.14)$$

$$\text{Slope}(\cot \alpha)\text{ of } CB = \frac{52.23}{378.53}$$

Since DK_3 is parallel to BC,

$$\text{Slope}(\cot \alpha)\text{ of } DK_3 = \frac{52.23}{378.53}$$

$$DK_3: y - 934.77 = \frac{52.23}{378.53}(x - 1,684.54) \quad (1)$$

From Equation 6-12, the equation of EA is

$$\frac{y - 688.69}{1,000.00 - 688.69} = \frac{x - 1.160.17}{1,000.00 - 1,160.27} \quad (2)$$

$$DK_3: 0.13798x - y = 702.34 \quad (1A)$$
$$EA: 1.94241x - y = -702.34 \quad (2A)$$
$$2.08039x = +2,240.07 \quad (1A + 2A)$$
$$x = 1,076.7547$$

Substitute the value of x in Equation 1A and check the results in Equation 2A:

$$y = 850.91$$

Therefore, the coordinates of K_3 are (850.91 N, 1,076.75 E).

Example 6-9: From the information shown in Figure 6-24, calculate the coordinates of the point of intersection L of the centerline of Fisher Road and the centerline of Elm Parkway.

Solution: The coordinates of station M on Fisher Road are (4,850,277.01 N, 316,909.433 E) and the bearing of the Fisher Road centerline (ML) is S 75°10′30″ E. The coordinates of the center of the 350 m radius highway curve are (4,850,317.313 N, 317,112.656 E).

The coordinates here are referred to a coordinate grid system having 0.000 m North at the equator, and 304,800.000 m East at longitude 79°30′ W.

The coordinate values are, of necessity, very large and would cause significant rounding errors if they were used in the computations when using a calculator. Accordingly, an auxiliary set of coordinate axes will be used, allowing the values of the given coordinates to be greatly reduced for the computations; the amount reduced will later be added to give the final coordinates. The summary of coordinates is shown in Table 6-15.

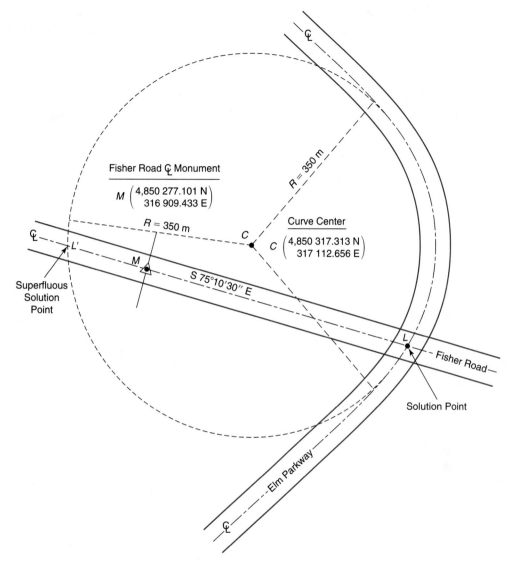

FIGURE 6-24 Intersection of a straight line with a circular curve, Example 6.9

Table 6-15   Grid Coordinates Reduced for Computations Using a Calculator

Station	Grid Coordinates		Reduced Coordinates	
	y	*x*	*y'* = (*y* – 4,850,000)	*x'* = (*x* – 316,500)
M	4,850,277.101	316,909.433	277.101	409.433
C	4,850,317.313	317,112.656	317.313	612.656

Table 6-16   Reduced Coordinates Restored to Grid Values

| Station | Reduced Coordinates | | Grid Coordinates | |
	y'	x'	y = (y' + 4,850,000)	x = (x' + 316,500)
L	142.984	916.151	4,850,142.984	317,416.151
L'	315.949	262.658	4,850,315.949	316,762.658

From Equation 6-13, the equation of Fisher Road centerline (*ML*) is

$$y' - 277.101 = \cot 75°10'30''(x' - 409.433) \quad (1)$$

From Equation 6-15, the equation of Elm Parkway centerline is

$$(x' - 612.656)^2 + (y' - 317.313)^2 = 350.000^2 \quad (2)$$

Simplify Equation 1 to

$$y' - 277.101 = -0.2646782x' + 108.368$$
$$y' = 0.2646782x' + 385.469 \quad (1A)$$

Substitute the value of y' into Equation 2:

$$(x' - 612.656)^2 = (0.2646782x' + 385.469$$
$$- 317.313)^2 - 350.000^2 = 0$$
$$(x' - 612.656)^2 + (-0.2646782x'$$
$$+ 68.156)^2 - 350.000^2 = 0$$
$$1.0700545\,x^2 - 1,261.391\,x' + 257,492.61 = 0$$

This quadratic of the form $ax^2 + bx + c = 0$ has roots

$$x = \frac{-b \pm \sqrt{b^2 - 4ac}}{2a}$$

$$x = \frac{1,261.3908 \pm \sqrt{1,591,107.30 - 1,102,124.50}}{2.140109}$$

$$x' = \frac{1,261.3908 \pm 699.27305}{2.140109}$$

$$x' = 916.1514 \quad \text{or} \quad x' = 262.658$$

Solve for y' by substituting in Equation 1A:

$$y' = 142.984 \quad \text{or} \quad y' = 315.949$$

When these coordinates are now shifted by the amount of the original axes reduction, the values shown in Table 6-16 are obtained.

Analysis of Figure 6-24 is required in order to determine which of the two solutions is the correct one. The sketch shows that the desired intersection point *L* is south and east of station *M*. That is, *L*(4,850,142.984 N, 317,416.151 E) is the set of coordinates for the intersection of the centerlines of Elm Parkway and Fisher Road. The other intersection point (*L'*) is superfluous.

Problems

6.1 A five-sided closed field traverse has the following angles: $A = 106°28'30''$, $B = 114°20'00''$; $C = 90°32'30''$; $D = 109°33'30''$; $E = 119°07'00''$. Determine the angular error of closure and balance the angles by applying equal corrections to each angle.

6.2 A four-sided closed field traverse has the following angles: $A = 81°53'30''$; $B = 70°28'30''$; $C = 86°09'30''$; $D = 121°30'30''$. The lengths of the sides are as follows: $AB = 636.45$ ft; $BC = 654.45$ ft; $CD = 382.65$ ft; $DA = 469.38$ ft.

The bearing of AB is S 17°17' W. BC is in the N.E. quadrant.

a. Balance the field angles.
b. Compute the bearings or the azimuths.

c. Compute the latitudes and departures.

d. Determine the linear error of closure and the accuracy ratio.

6.3 Using the data from Problem 6.2,

a. Balance the latitudes and departures by use of the compass rule.

b. Compute the coordinates of stations A, C, and D if the coordinates of station B are 1,000.00 N, 1,000.00 E.

6.4 Using the data from Problems 6.2 and 6.3, compute the area enclosed by the traverse using the coordinate method.

6.5 A five-sided closed field traverse has the following distances in meters: $AB = 51.766$; $BC = 26.947$; $CD = 37.070$; $DE = 35.292$; $EA = 19.192$. The adjusted angles are as follows: $A = 101°03'19''$; $B = 101°41'49''$; $C = 102°22'03''$; $D = 115°57'20''$; $E = 118°55'29''$. The bearing of AB is N 73°00'00'' E. BC is in the S.E. quadrant.

a. Compute the bearings or the azimuths.

b. Compute the latitudes and departures.

c. Determine the linear error of closure and the accuracy ratio.

6.6 Using the data from Problem 6.5:

a. Balance the latitudes and departures by use of the compass rule.

b. Compute the coordinates of the traverse stations using coordinates of station A as 1,000.000 N, 1,000.000 E.

6.7 Using the data from Problems 6.5 and 6.6, compute the area enclosed by the traverse using the coordinate method.

6.8 The two frontage corners of a large tract of land were joined by the following open traverse:

Course	Distance (ft)	Bearing
AB	100.00	N 70°40'05'' E
BC	953.83	N 74°29'00'' E
CD	818.49	N 70°22'45'' E

Compute the distance and bearing of the property frontage AD.

6.9 Given the following data for a closed property traverse:

Course	Bearing	Distance (m)
AB	N 37°10'49'' E	537.144
BC	N 79°29'49'' E	1109.301
CD	S 18°56'31'' W	?
DE	?	953.829
EA	N 21°10'08'' W	477.705

Compute the missing data (i.e., distance CD and bearing DE), and verify the results by summing the latitudes and departures of the property traverse.

6.10 A six-sided traverse has the following station coordinates: A (559.319 N, 207.453 E); B (738.562 N, 666.737 E); C (541.742 N, 688.350 E); D (379.861 N, 839.008 E); E (296.099 N, 604.048 E); F (218.330 N, 323.936 E). Compute the distance and bearing of each side.

6.11 Using the data from Problem 6.10, compute the area enclosed by the traverse.

6.12 Use the data from Problem 6.10 to solve the following. If the intersection point of lines AD and BF is K, and if the intersection point of lines AC and BE is L, compute the distance and bearing of line KL.

6.13 A five-sided field traverse has the following balanced angles: $A = 101°28'26''$; $B = 102°10'42''$; $C = 104°42'06''$; $D = 113°04'42''$; $E = 118°34'04''$. The lengths of the sides are as follows: $AB = 50.276$ m; $BC = 26.947$ m; $CD = 37.090$ m; $DE = 35.292$ m; $EA = 20.845$ m. The bearing of EA is N 20°20'20'' W, and AB is oriented northeasterly.

a. Compute the bearings.

b. Compute the latitudes and departures.

c. Compute the linear error of closure and the accuracy ratio.

6.14 From the data in Problem 6.13, balance the latitudes and departures employing the compass rule.

6.15 From the data in Problems 6.13 and 6.14, if the coordinates of station A are (1,000.000 N,

1,000.000 E), compute the coordinates of the other stations.

6.16 From the data in Problems 6.14 and 6.15, use the coordinate method to compute the area of the enclosed figure.

6.17 See Figure 6-25. A total station was set up at control station K that is within the limits of a five-sided property. Coordinates of control station K are 1,990.000 N, 2,033.000 E. Azimuth angles and distances to the five property corners were determined as follows:

Direction	Azimuth	Horizontal Distance (m)
KA	286°51′00″	34.482
KB	37°35′28″	31.892
KC	90°27′56″	38.286
KD	166°26′49″	30.916
KE	247°28′43″	32.585

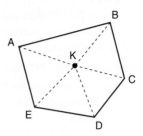

FIGURE 6-25 Sketch for Problem 6.13

Compute the coordinates of the property corners A, B, C, D, and E.

6.18 Using the data from Problem 6.17, compute the area of the property.

6.19 Using the data from Problem 6.17, compute the bearings and distances of the five sides of the property.

CHAPTER SEVEN

SATELLITE POSITIONING SYSTEMS

7.1 GENERAL BACKGROUND

In the United States and Canada, when people think of satellite navigation they think of global positioning system (GPS). However, GPS is only one satellite system. GNSS (Global Navigation Satellite System) is a term to describe all satellite navigation systems, current and planned. Currently there are two operational systems. The oldest, and the only system with a full constellation at present (2010), is the U.S. Navigation Satellite Timing and Ranging (NAVSTAR) GPS—generally referred to as the GPS. The Russian system, GLONASS, is also fully operational, but does not have a full constellation of satellites at present. See Table 7-2. Planned systems include the following:

1. European Union (EU) 30-satellite system, *Galileo*
2. China's *Compass*
3. Japan's *QZSS*
4. India's *IRNSS*

Galileo hopes to have its first four operational satellites launched in 2011. See http://www.esa.int/esaNA/galileo.html for current information on Galileo's deployment. China's Compass Satellite System (also known as BeiDou) will be comprised of 35 satellites when fully operational. Currently (January 2011) China has seven operational satellites allowing for testing and limited navigation. Japan's QZSS and India's IRNSS are both regional systems covering their areas of interest. QZSS is an enhancement to GPS, meaning it is used, among other tasks, to increase the precision of GPS positioning, while IRNSS is an independent GNSS.

This text will discuss primarily the NAVSTAR (GPS) system, which all survey-grade satellite receivers work with. However, the newer receivers also work with GLONASS and will work with Galileo when that system becomes operational.

7.2 UNITED STATES' GLOBAL SATELLITE POSITIONING SYSTEM (GPS)

The NAVSTAR system was developed by the U.S. Department of Defense (DoD) as a replacement for their current navigation systems such as TRANSIT and LORAN. The Air Force and Navy were the two divisions specifically interested in having precise navigation. The concept was approved by the DoD in 1973 and the first satellite was launched in 1978. By 1980 the system became operational and in 1994 the system had a complete constellation of 24 satellites (see Figure 7-1). A complete constellation provides enough satellites to determine position anywhere in the world at any time (see Figure 7-2). There are three components to the NAVSTAR system: the space component consisting of the satellites; the user component consisting of satellite receivers, of which there are two types, military and civilian; and finally the control component, which consists of control and monitoring stations for the satellites.

FIGURE 7-1 GPS Satellite
(Courtesy of Leica Geosystems Inc.)

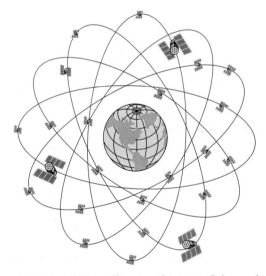

FIGURE 7-2 GPS satellites in orbit around the earth

The space segment consists of the satellite constellation. The original satellites were known as Block I satellites, which were considered experimental. Eleven Block I satellites were launched between 1978 and 1985. Block II and IIA production satellites were first launched in 1989. There were a total of 28 Block II and IIA satellites launched. The next set of satellites is Block IIR, the R standing for replenishment. The first Block IIR satellites were launched in 1997. Next Block IIR-M (M for modernization) were launched, starting in 2005. These satellites had additional signals to be used for the "modernization of GPS." The first Block IIF satellite was launched in May 2010. The Block IIF satellites provide an additional frequency (L5). Block III satellites are planned to be launched in the beginning of 2014. The reason for modernization of the GPS system is to accommodate the ever-expanding uses and reliance on GPS by the civilian clients as well as maintaining the needs of the military clients.

The fully operational GPS constellation consists of 24 operational satellites (plus spares) deployed in six orbital planes 20,200 km above the earth, with two orbit revolutions per sidereal day. The satellites appear over the same position by 4 min earlier each day. Spacing is such that at least six satellites are always visible anywhere on earth.

The control segment consists of the Master Control Station in Colorado Springs, Colorado, and monitoring stations throughout the world. Originally there were only Air Force Monitoring Stations, but because of times where satellites could not be monitored and redundancy concerns, National Geospatial Intelligence Agency (NGA) and the International GNSS Service (IGS) monitoring stations have been added. Currently (2010), there are 18 monitoring stations. Some of the monitoring stations also have the capability to upload the needed information to the satellites.

7.3 GPS CODES, SIGNALS, AND FREQUENCIES

In a very basic way, GPS works much like a radio in a car, in that information is being sent to the car from a transmitting tower. The tower is not directly broadcasting music, but sending signals on radio waves that can be interpreted by radios. The transmitter has no way of determining who, if anyone, is receiving the information, and therefore can be considered a passive system. The GPS codes, signals, and frequencies although related cannot be used interchangeably. The codes are the information a receiver needs to determine position, while the signal is that actual information that is being received by the receivers and the frequencies are how that information is being transported from the satellite to the receivers.

The satellites currently are transmitting on 2L band frequencies, known as the L1 and L2 frequencies. The L1 frequency is set at 1,575.42 MHz, which gives it a wavelength of 0.190 m or about 19 cm (see Equation 3.8). The L2 frequency is set at 1,227.60 MHz, with a wavelength of 0.244 m or about 24 cm. As mentioned before, the Block IIR-M transmits a third frequency known as L5, at 1,176.45 MHz (wavelength at about 25.5 cm).

Originally three codes were being sent by the satellites. The first is called the navigation code and it contains information about the satellite orbit (ephemeris), health, and clock correction. In addition, it provides the time, the GPS week, and anticipated orbits for all the satellites in the system (almanac). The navigation code takes about 12.5 min to receive, and is sent on both the L1 and L2 frequencies.

The C/A code, often called the Coarse/Acquisition code, is the code used by civilian receivers to determine distance from the satellite. It is a pseudo random code, meaning it looks like random noise but actually is a complicated code defining which satellite is sending the code. It repeats the code every millisecond.

The P code, often known as the Precise code, is the military code allowing for higher positioning precision than that of the C/A code. The P code is encrypted for security reasons; the encrypted code is called the Y code.

In addition to these three codes, with the advent of the Block IIR-M satellites, a fourth code, the L2C, is being sent on the L2 frequency. This is a civilian code that provides additional capabilities to locking onto the L2 frequency with "dual" frequency civilian receivers.

The L5 frequency, when fully deployed is to be used for aviation safety services, primarily to provide aircraft with signal redundancy. However, it is expected that this frequency will also increase the positional accuracy of any GPS receiver that can pick up the frequency.

7.4 RECEIVERS

There are generally three types of GPS receivers:

- Receivers with no augmentation
- Receivers with differential corrections
- Receivers with carrier phase differencing (CPD) capabilities

Nowadays, receivers with no augmentation cost less than $100, and often are accessories for other equipment, such as cell phones. The cost of receivers with differential correction can run from $100 to $5,000, depending on the hardware of the unit, the software, and the differential correction type. Receivers with CPD capabilities can range from $3,000 to $20,000 and up. One major factor in CPD units is if they are single- or dual-frequency receivers (if they can use only the L1 carrier or both the L1 and L2 carriers). See Figures 7-3–7-5.

FIGURE 7-3 Leica SYS 300. System includes SR 9400 GPS single-frequency receiver, with AT 201 antenna, and CR333 controller which collects data and provides software for real-time GPS. It can provide the following accuracies: 10–20 mm + 2 ppm using differential carrier techniques; and 0.30–0.50 m using differential code measurements (Courtesy of Leica Geosystems Inc.)

7.5 GPS POSITION MEASUREMENTS

Position measurements generally fall into one of two categories: code measurement and carrier phase measurement. Civilian code measurement is presently restricted to the C/A code, which can provide accuracies only in the range of 10–15 m when used in point positioning, and accuracies in the submeter to 5 m range, when used in various differential positioning techniques. P code measurements can apparently provide the military

with much better accuracies. Point positioning is the technique that employs one GPS receiver to track satellite code signals so that it can directly determine the coordinates of the receiver station.

Positional accuracies shown for GPS receivers relate to the positional horizontal accuracies; as a rule vertical positions are two to three times worse than the horizontal position. The errors causing the low accuracies cause the position to "drift," or not show much change over a short period of time. For this reason data collected over a short period of time may show a high precision but it still has a low accuracy.

The GPS system uses the World Geodetic System 1984 (WGS84) datum as is basis for its positions. The current adjustment used is WGS84 (1154). See Chapter 10 for more discussion of the relation of datums.

7.5.1 GPS Code Measurements

As previously noted, the military can utilize both the P code and the C/A code. The C/A code is used by the military to access the P code quickly. Until the L2C signal becomes operative on more satellites, the civilian user must be content with using only the C/A code. Both codes are digital codes comprised of zeros and ones [see Figure 7-6(a)], and each satellite transmits codes unique to that satellite. Although both codes are carefully structured, because the codes sound like random electronic "noise," they have been given the name pseudo random noise (PRN). The PRN code number (see Table 7-1) indicates which of the 37 seven-day segments of the P code PRN signal is presently being used by each satellite (it takes the P code PRN signal 267 days to transmit). Each 1-week segment of the P code is unique to each satellite and is reassigned each week (see Table 7-1). Receivers have replicas of all satellite codes in the onboard memory, which they use to identify the satellite and then to measure the time difference between the signals from the satellite to the receiver. Time is measured as the receiver moves the replica code (retrieved from memory) until a match between the transmitted code and the replica code is

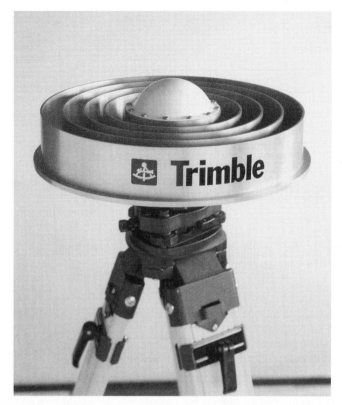

FIGURE 7-4 Geodetic-quality dual-frequency GPS receiver with a choke-ring antenna
(Courtesy of Trimble)

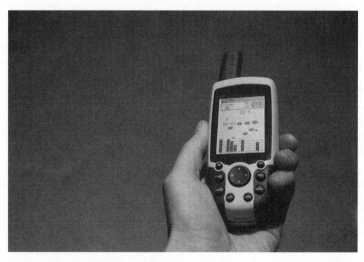

FIGURE 7-5 Garmin eMap handheld GPS receiver (accuracy: 15 m), used in navigation and GIS-type data location. It can be used with a DGPS radio beacon receiver to improve accuracy to within a few meters or less
(Eimantas Buzas/Shutterstock)

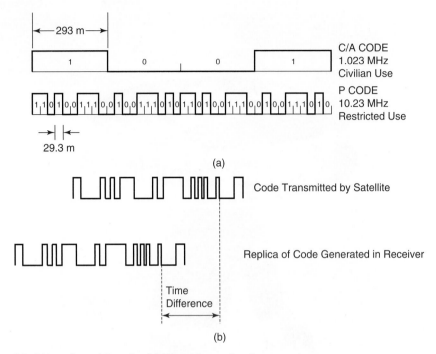

FIGURE 7-6 (a) C/A code and P code. (b) Time determination

achieved. See Figure 7-6(b). Errors caused by the slowing effects of the atmosphere on the transmission of satellite radio waves can be corrected by simultaneously performing position measurements utilizing two different wavelengths, such as L1 and L2.

One key dimension in positioning is the parameter of time. Time is kept onboard the satellites by so-called atomic clocks with a precision of 1 nanosecond (0.0000000001 s). The ground receivers are equipped with less precise quartz clocks. Uncertainties caused by these less precise clocks are resolved when observing the signals from four satellites instead of the basic three-satellite configuration required for x, y, and z positioning. Some predict that future increases in spatial positioning accuracy will result from improvements in satellite onboard clocks planned for future satellite constellations.

The distance, called pseudorange, is determined by multiplying the time factor by the speed of light:

$$\rho = t(300,000,000)$$

where

ρ (lowercase Greek letter rho) = the pseudorange

t = travel time of the satellite signal in seconds

$300,000,000$ = velocity of light in meters per second (actually 299,792,458 m/s)

When the pseudorange (ρ) is corrected for clock errors, atmospheric delay errors, multipath errors, and the like, it is then called the range and is abbreviated by the uppercase Greek letter rho, P; that is, P = (ρ + error corrections).

Figure 7-7 shows the geometry involved in point positioning. Computing three pseudoranges (ρ_{AR}, ρ_{BR}, ρ_{CR}) is enough to solve the intersection of three developed spheres, although this computation gives two points of intersection. One point (a superfluous point) will be obviously irrelevant; that is, it probably will not even fall on the surface of the earth. The fourth pseudorange (ρ_{DR}) is required to

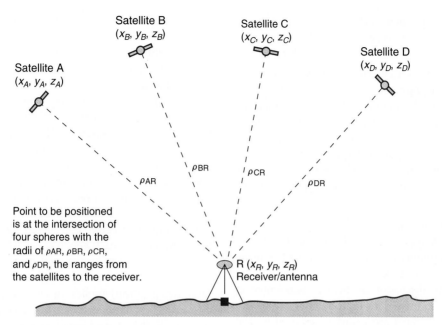

FIGURE 7-7 Geometry of point positioning

remove the fourth unknown—the receiver clock error. This provides an absolute position on the WGS84 datum.

7.5.2 Errors

There are a number of factors that limit the accuracy of GPS; the chief sources are described below: It should be remembered that although 10–15 m is not considered accurate enough for almost all survey measurements, these positions are absolute all over the world, which is actually very high precision.

1. Clock errors of the receivers.
2. Ionospheric refraction (occurring 50–1,000 km above earth) and tropospheric refraction (occurring from the earth's surface to 80 km above earth). Signals are slowed as they travel through these earth-centered layers. The errors worsen as satellite signals move from directly overhead to down near the horizon. These errors can be reduced by scheduling nighttime observations, by gathering sufficient redundant data and using reduced

baseline lengths (1–5 km), or by collecting data on both frequencies over long distances (20 km or more). Most surveying agencies do not record observations from satellites below 10–15° of elevation above the horizon.

3. Multipath interference, which is similar to the ghosting effect seen on early TVs. Some signals are received directly and others are received after they have been reflected from adjacent features such as tall buildings and steel fences. Recent improvements in antenna design have significantly reduced these errors.

4. A weak geometric figure that results from poorly located, four-satellite signal intersections. This consideration is called the dilution of precision (DOP). DOP can be optimized if many satellites (beyond the minimum of four) are tracked; the additional data strengthen the position solution. Most survey-level receivers are now capable of tracking up to 12 satellites simultaneously (see Section 7.5.2.1).

5. Errors associated with the satellite orbital data. The locations of the satellites are based

on their predicted orbit data, which are being sent via the navigation code. The predicated orbit may not reflect the actual orbit.

6. Setup errors. Centering errors can be reduced if the equipment is checked to ensure that the optical plummet is true, and hi measuring errors can be reduced by utilizing equipment that provides a built-in (or accessory) measuring capability to precisely measure the antenna reference height (ARH) (directly or indirectly) or by using fixed-length tripods and bipods.

Many of the effects of the above errors, including denial of accuracy by the DoD (if it were to be re-introduced), can be surmounted by using differential positioning surveying techniques. Most of the discussion in this text is oriented to relatively short baselines; for long lines (>150 km), more sophisticated processing is required to deal with natural and human-made errors.

7.5.2.1 Dilution of Precision   The quality of the solution is dramatically affected by the geometry of the satellite positions. Dilution of precision (DOP) is the indicator used to measure the geometry quality. There are a number of types of DOP measurements: horizontal (HDOP), vertical (VDOP), time (TDOP), relative (RDOP), and the two most commonly used in surveying—geometric (GDOP) and position (PDOP).

The better the geometry of the satellites, the lower the DOP value will be, providing a higher certainty of the solution. For PDOP, usually a value of six or less is desired when obtaining a position. Nowadays, with so many available satellites, the PDOP will usually be around two in open skies. It is when obstructions are blocking a portion of the sky where there may be sufficient satellites for a solution, but they are all in one area of the sky providing a very weak solution and high DOP.

7.5.3 Differential GPS

Differential GPS (DGPS) is a relative positioning technique that employs two GPS receivers to track satellite code signals to determine the baseline

vector (*X*, *Y*, and *Z*) between the two receiver stations. The two receivers must collect data from the same satellites simultaneously (same epoch) and their observations are then combined to produce results that are superior to those achieved with just a single-point positioning. This is providing relative positioning between the two receivers. One receiver is considered a "base" station, or a receiver with a known position. This position may be a precise absolute position, or it may be just an approximate position. The relative positions between the two receivers can achieve submeter accuracies. The concept is fairly simple: The "base" station (receiver considered on a known position) looks at each epoch to see what the difference between the known and computed position is, and then the "rover" (receiver with the unknown position) applies the correction for that difference for that epoch. Because this difference reflects all the measurement errors (except for multipath) for nearby receivers, this "difference" can be included in the nearby roving GPS receivers' computations to remove the common errors in their measurements. This can be done in real time, if the two receivers have communication between them, or it can be postprocessed if the receivers store the collected information.

Differential GPS can be achieved by using two GPS receivers in the field, or there are a number of services that will provide the base station and broadcast the information, so that a user only needs one receiver and access to the broadcast information. One advantage to using a service is the position of the base station is precisely known.

7.5.4 GPS Augmentation Services— Differential GPS

There are commercial services, such as OmiSTAR (www.omniSTAR.com), that collect GPS data at several sites and weight them to produce an "optimized" set of corrections that are then uplinked to geostationary satellites, which, in turn, broadcast the corrections to subscribers who pay an annual fee for this service. This service provides real-time measurements because the corrections are automatically applied to data collected by the subscribers' GPS roving receivers. In addition there are two differential services provided by the U.S. government, NDGPS (an extension of the U.S. Coast Guard DGPS system) and WAAS.

7.5.4.1 U.S. Coast Guard's DGPS System

The U.S. Coast Guard Maritime Differential GPS Service has created a differential GPS (DGPS), consisting of more than 80 remote broadcast sites and two control centers. It was originally designed to provide navigation data in coastal areas, the Great Lakes, and major river sites. This system received full operational capability (FOC) in 1999. The system includes continuously operating GPS receivers, at locations of known coordinates, that determine pseudorange values and corrections and then radio-transmit the differential corrections to working navigators and surveyors at distances ranging from 100 to 400 km. The system uses international standards for its broadcasts, published by the Radio Technical Commission for Maritime (RTCM) services. Effectively the surveyor or navigator, using the DGPS broadcasts along with his/her own receiver, has the advantages normally found when using two receivers. Since the corrections for many of the errors (orbital and atmospheric) are very similar for nearby GPS receivers, once the pseudoranges have been corrected, the accuracies thus become much improved.

Information on DGPS and on individual broadcast sites can be obtained on the Internet at the U.S. Coast Guard Navigation Center website: www.navcen.uscg.gov/. Effective April 2004, the U.S. Coast Guard, together with the Canadian Coast Guard, implemented a seamless positioning service in the vicinity of their common border.

7.5.4.2 Nationwide DGPS (NDGPS)

The success of the U.S. Coast Guard's DGPS prompted the U.S. Department of Transportation (DOT), in 1997, to design a terrestrial expansion over the land surfaces of the conterminous or continental United States (CONUS), and major transportation routes in Alaska and in Hawaii. The U.S. Coast Guard (USCG) was given the task of being the lead agency in this venture which essentially consisted of expanding the maritime DGPS across all land areas. By 2007, about 37 NDGPS stations were operating. In January 2007, work on NDGPS was halted pending congressional review of future project funding. In 2008, the program was approved to continue, with USCG still responsible for managing the system. Currently the NDGPS system has 38 operational sites, with an emphasis being given to update the receivers and an eventual goal of dual coverage across the continental United States. See the website at www.navcen.uscg.gov/ for updates.

7.5.4.3 Wide Area Augmentation System

Wide Area Augmentation System (WAAS) was designed and built by the Federal Aviation Administration (FAA) and uses 25 ground stations in the conterminous United States, with additional stations in Alaska, Canada, and Mexico either operating or planned; the northing, easting, and elevation coordinates of these ground stations have been precisely determined. Signals from available GPS satellites are collected and analyzed by all stations in the network to determine errors (orbit, clock, atmosphere, etc.). The differential corrections are then forwarded to one of the two master stations (one located on each U.S. coast). The master stations create differential correction messages that are then relayed to one of three geostationary satellites located near the equator. The WAAS geostationary satellites then rebroadcast the differential correction messages on the same frequency as that used by GPS—that is, L1 at 1,575.42 MHz—to receivers located on the ground, at sea, or in the air. This positional service is free and many modern GPS receivers have the built-in capability of receiving WAAS corrections.

Accuracy specifications call for a position accuracy of 7.6 m (horizontal and vertical) more than 95 percent of the time, and current field trials consistently produce results in the 1.5 m range for horizontal position and 1.5 m in the vertical position.

Although this service is free and most GPS receivers have access to the correction, because the satellites are located near the equator, they are low on the horizon in the United States. This makes maintaining signal lock difficult, especially for the inexpensive receivers. The base stations are currently all in the United States, which restricts the accuracies to primarily U.S. locations.

7.6 GPS CARRIER PHASE MEASUREMENT

GPS codes, which are modulations of the carrier frequencies, are comparatively lengthy. Compare the C/A code at 293 m and the P code at 29.3 m (Figure 7-6) with the wavelengths of L1 and L2 at 0.19 and 0.24 m, respectively. It follows that carrier phase measurements have the potential for much higher accuracies than do code measurements.

We first encountered phase measurements in Chapter 3, when we observed how the electronic distance measurement (EDM) equipment measured distances (see Figure 3-18). Essentially EDM distances are determined by measuring the phase delay required to match up the transmitted carrier wave signal with the return signal (two-way signaling). Equation 3.10 (repeated below) calculates this distance:

$$L = \frac{n\lambda + \varphi}{2}$$

where φ is the partial wavelength determined by measuring the phase delay (through comparison with an onboard reference), n is the number of complete wavelengths (from the EDM instrument to the prism and back to the EDM instrument), and λ is the wavelength. The integer number of wavelengths is determined as the EDM instrument successively sends out (and receives back) signals at different frequencies, thus permitting the creation of equations allowing for the computation of the unknown n.

Because GPS ranging involves only one-way signaling, other techniques must be used to determine the number of full wavelengths. GPS receivers can measure the phase delay (through comparison with onboard carrier replicas) and count the full wavelengths after lock-on to the satellite has occurred, but more complex treatment is required to determine N, the initial cycle ambiguity. That is, N is the number of full wavelengths sent by the satellite prior to lock-on. Since a carrier signal is comprised of a continuous transmission of sine-like waves with no distinguishing features, the wave count cannot be accomplished directly.

$$P = \varphi + N\lambda + \text{errors}$$

where P = satellite − receiver range

φ = measured carrier phase

λ = wavelength

N = initial ambiguity (the number of full wavelengths at lock-on)

Once the cycle ambiguity between a receiver and a satellite has been resolved, it does not have to be addressed further unless a loss of lock occurs between the receiver and the satellite. When loss of lock occurs, the ambiguity must be resolved again. Loss of lock results in a loss of the integer number of cycles and is called a cycle slip. As an example, loss of lock can occur when a roving receiver passes under a bridge, a tree canopy, or any other obstruction that blocks all or some of the satellite signals.

Cycle ambiguity can be determined through the process of differencing. GPS measurements can be differenced between two satellites, between two receivers, and between two epochs. An **epoch** is an event in time—a short observation interval in a longer series of observations. After the initial epoch has been observed, later epochs will reflect the fact that the constellation has moved relative to the ground station and, as such, presents a new geometrical pattern and thus new intersection solutions.

7.6.1 Differencing

Relative positioning occurs when two receivers are used to simultaneously observe satellite signals and to compute the vectors (known as a baseline) joining the two receivers. Relative positioning can

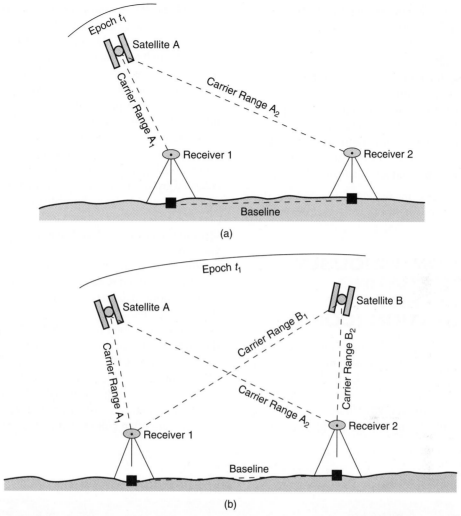

FIGURE 7-8 Differencing. (a) Single difference: two receivers observing the same satellite simultaneously (between receivers difference). (b) Double difference: two receivers observing two satellites simultaneously (between satellite difference)

provide better accuracies because of the correlation possible between measurements simultaneously made over time by two or more different satellite receivers. Differencing is the technique of simultaneous baseline measurements and falls into the categories of single difference, double difference, and triple difference.

- *Single difference*: When two receivers simultaneously observe the same satellite, it is possible

to correct for most of the effects of satellite clock errors, orbit errors, and atmospheric delay. See Figure 7-8(a).

- *Double difference*: When one receiver observes two (or more) satellites, the measurements can be freed of receiver clock error, and atmospheric delay errors can also be eliminated. Further, when using both differences (between the satellites and between the receivers), a double difference occurs. Clock errors, atmospheric delay errors, and orbit errors can all be eliminated. See Figure 7-8(b).

- *Triple difference*: The difference between two double differences is the triple difference. That is, the double differences are compared over two (or more) successive epochs. This procedure is also effective in detecting and correcting cycle slips and for computing first-step approximate solutions that are used in double difference techniques. See Figure 7-8(c).

The references, in Section 7.15, provide more information on the theory of signal observations and ambiguity resolution.

7.7 CONTINUOUSLY OPERATING REFERENCE STATION (CORS)

The continuously operating reference station (CORS) system is a differential measurement system developed by the National Geodetic Survey (NGS) that has now become nationwide (see Figure 7-9). As of May 2010 there were 1,450 operational CORS throughout the world. The goal is to expand the network until all points within the conterminous United States (CONUS) will be within 200 km of, at least, one operational CORS site. The GPS satellite signals observed at each site are used to compute the base station position, which is then compared to the correct position coordinates (which have been previously and precisely determined). The *difference* (thus differential) between the correct position and the computed position is then made available over the Internet for use in the postprocessing of the field observations which are collected by the roving receivers used by a wide variety of government and private surveyors.

The official CORS network includes stations set up by the NGS, the U.S. Coast Guard, the U.S. Army Corps of Engineers (USACE), and, more recently, stations set up by other federal and local agencies.

The coordinates at each site are computed from 24-hour data sets collected over a 10–15 day period. These highly accurate coordinates are then transformed into the NAD83 horizontal datum for use by local surveyors. Each CORS continuously tracks GPS satellite signals and creates files of both carrier measurements and code range measurements, which are available to the public via the Internet, to assist in positioning surveys. The data are often collected at a 30-s epoch rate,

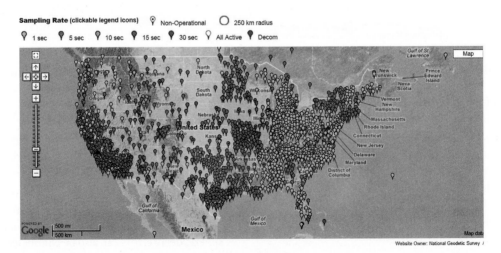

FIGURE 7-9 CORS sites in continental United States from NGS

(National Geodetic Survey/NOAA)

although some sites collect data at 1-s, 5-s, 10-s, and 15-s epoch rates. NGS converts all receiver data and individual site meteorological data to receiver independent exchange (RINEX) format. The files can be accessed via the Internet at NGS website: http://www.ngs.noaa.gov. Select CORS from the menu to access CORS data. The data may be accessed by directly accessing an FTP site and downloading the necessary files, or by using the User Friendly CORS (UFCORS), or you may have NGS's OPUS determine the position of your receiver (see Section 7.7.1). The NGS stores data from all sites from the first day of operation of the site; however, after 30 days the data are only stored in 30-s epochs. The local surveyor, armed with data sets in his or her locality at the time of an ongoing survey and having entered these data into his or her GPS program, has the equivalent of an additional dual-frequency GPS receiver. A surveyor with just one receiver can proceed as if two receivers were being used in the differential positioning mode. Data from the CORS stations form the foundation of the national spatial reference system (NSRS). At the time of this writing, positions are given in both NAD83 coordinates (used in the U.S. for surveying and civil applications) and ITRF (ITRF00) position coordinates. Future CORS receivers will be capable of receiving signals from GLONASS and Galileo satellites—in addition to GPS satellites. Some future CORS sites will be established at additional U.S. tide gauge stations to help track sea level fluctuations.

There are thousands of other Continuous GPS (CGPS) stations that are not a part of the official NGS CORS site, but still have high-accuracy positions. Many of these sites can be accessed through the University NAVSTAR Consortium (UNAVCO www.unavco.org) website under their Plate Boundary Observatory (PBO) page.

7.7.1 Online Positioning User Service (OPUS)

Online Position User Service (OPUS) operated by NGS provides position solutions for GPS data. OPUS uses only NGS CORS to determine the position of a point. OPUS is not the only online positioning service, but it is the one most commonly used by surveyors. Its limitation is that it uses only NGS CORS sites, and they may be a significant distance away (50 km or more). The claimed accuracy of the position is a centimeter (0.03 ft), under ideal conditions. This service uses three CORS to derive a solution. It has two options for processing: Rapid Static for time spans of 15 minutes to 2 hours and Static for 2 hours to 48 hours. OPUS requires data from a dual-frequency receiver. The surveyor must know the antenna model used and the antenna height to the Antenna Reference Point (ARP) in meters. The surveyor sends the receiver data files—as collected, or in receiver independent exchange (RINEX) format to NGS via the Web. The results are emailed back to the surveyor within minutes, containing positions (latitude and longitude and Earth Centered Earth Fixed) on two datums; NAD 83 (CORS96 Adjustment) with an epoch of 2002.00 and ITRF00 with an epoch of the date of the survey. In addition, UTM coordinates and State Plane Coordinates are provided (in meters). The State Plane Coordinates are based on the NAD83 CORS96 values. The Rapid Static report will provide standard deviations for each value, while the Static report will provide peak to peak errors. Elevations are given as both ellipsoidal heights and as orthometric heights (using GEOID09 model; see Section 7.12.1.2). See www.ngs.noaa.gov/OPUS/ for further information.

7.8 CANADIAN ACTIVE CONTROL SYSTEM

The Geodetic Survey Division (GSD) of Geomatics Canada has combined with the Geological Survey of Canada to establish a network of active control points (ACP) in the Canadian Active Control System. The system includes 10 unattended dual-frequency tracking stations (ACPs) that continuously measure and record carrier phase and pseudorange measurements for all satellites in view at a 30-s sampling interval. A master ACP in Ottawa coordinates and controls the system. The data are archived in RINEX format and available online 4 hours after the end of the day. Precise ephemeris data, computed with input from 24 globally distributed core GPS tracking stations of the International GPS Service for Geodynamics (IGS), are available online within 2 to 5 days after the observations; precise clock corrections are also available in the 2- to 5-day time frame. The Canadian Active Control System is complemented by the Canadian Base Network, which provides 200-km coverage in Canada's southern latitudes for high accuracy control (centimeter accuracy). These control stations are used to evaluate and to complement a wide variety of control stations established by various government agencies over the years. These products, which are available for a subscription fee, enable a surveyor to position any point in the country with a precision ranging from a centimeter to a few meters. Code observation positioning at the meter level, without the use of a base station, is possible using precise satellite corrections. Real-time service at the meter level is also available. The data can be accessed on the Web at http://www.geod.nrcan.gc.ca.

7.9 SURVEY PLANNING

As with any survey field operation, preplanning is important for GPS surveys in order to be successful and efficient. Certain GPS survey types require significant planning, while other types require only minimal planning. One typical planning task for GPS is to determine the number of available satellites and the geometric strength of the solution. Typically, only satellites that are at least 10° above the horizon are used for GPS solutions. In addition there can be obstructions limiting satellite signal. Most GPS processing software has mission planning functions. Figures 7-10 through 7-12 shows the results provided by a Web-based mission planning program from Ashtech (http://www.ashtech.com). Continual checks on the operating status of the GPS and GLONASS constellations (see Tables 7-1 and 7-2) will keep the surveyor aware of instances where satellites are temporarily set "unhealthy" to effect necessary adjustments and recalibrations.

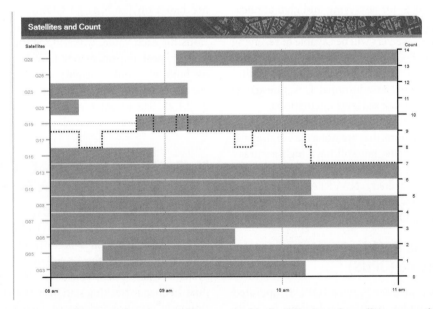

FIGURE 7-10 Mission planning showing satellite count (dashed line) and satellites over the duration of the mission using Ashtech Web Mission Planning application (http://asp.ashtech.com/wmp/)

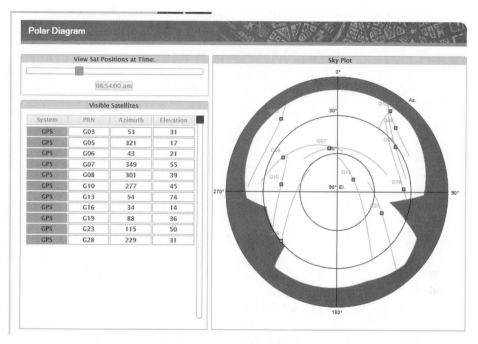

FIGURE 7-11 Mission planning showing satellite positions relative to observer (0° is north, 90° elevation is directly above observer); gray area is area blocked out by obstructions and low altitudes

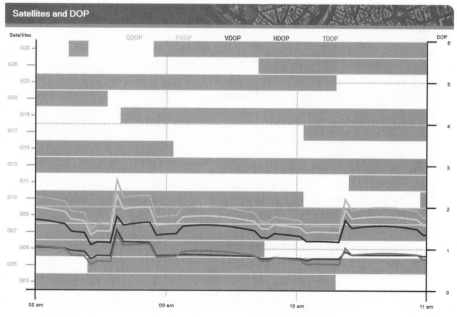

FIGURE 7-12 Mission planning showing dilution of precision (DOP) values over the duration of the mission

Table 7-1   Status of GPS Operational Block II Satellites as of January 2011

			Current Block II/IIA/IIR/IIR-M Satellites			
Launch Order	PRN	SVN	Launch Date	Freq STD	Plane	U.S. Space Command**
*II-1		14	Feb 14, 1989			19802
*II-2		13	Jun 10, 1989			20061
*II-3		16	Aug 18, 1989			20185
*II-4		19	Oct 21, 1989			20302
*II-5		17	Dec 11, 1989			20361
*II-6		18	Jan 24, 1990			20452
*II-7		20	Mar 26, 1990			20533
*II-8		21	Aug 02, 1990			20724
*II-9		15	Oct 01, 1990			20830
IIA-10	32	23	Nov 26, 1990	Rb	E5	20959
IIA-11	24	24	Jul 04, 1991	Cs	D5	21552
*IIA-12		25	Feb 23, 1992	Rb	A5	21890
*IIA-13		28	Apr 10, 1992			21930
IIA-14	26	26	Jul 07, 1992	Rb	F5	22014
IIA-15	27	27	Sep 09, 1992	Cs	A4	22108
*IIA-16		32	Nov 22, 1992		F6	22231
*IIA-17		29	Dec 18, 1992			22275
*IIA-18		22	Feb 03, 1993			22446
*IIA-19		31	Mar 30, 1993			22581
*IIA-20		37	May 13, 1993			22657
IIA-21	09	39	Jun 26, 1993	Cs	A1	22700
*IIA-22		35	Aug 30, 1993	Rb		22779
IIA-23	04	34	Oct 26, 1993	Rb	D4	22877
IIA-24	06	36	Mar 10, 1994	Rb	C5	23027
IIA-25	03	33	Mar 28, 1996	Cs	C2	23833
IIA-26	10	40	Jul 16, 1996	Cs	E3	23953
IIA-27	30	30	Sep 12, 1996	Cs	B2	24320
IIA-28	08	38	Nov 06, 1997	Cs	A3	25030
***IIR-1		42	Jan 17, 1997			
IIR-2	13	43	Jul 23, 1997	Rb	F3	24876
IIR-3	11	46	Oct 07, 1999	Rb	D2	25933
IIR-4	20	51	May 11, 2000	Rb	E1	26360
IIR-5	28	44	Jul 16, 2000	Rb	B3	26407
IIR-6	14	41	Nov 10, 2000	Rb	F1	26605
IIR-7	18	54	Jan 30, 2001	Rb	E4	26690
IIR-8	16	56	Jan 29, 2003	Rb	B1	27663
IIR-9	21	45	Mar 31, 2003	Rb	D3	27704

Table 7-1   (*Continued*)

Current Block II/IIA/IIR/IIR-M Satellites						
Launch Order	PRN	SVN	Launch Date	Freq STD	Plane	U.S. Space Command**
IIR-10	22	47	Dec 21, 2003	Rb	E2	28129
IIR-11	19	59	Mar 20, 2004	Rb	C3	28190
IIR-12	23	60	Jun 23, 2004	Rb	F4	28361
IIR-13	02	61	Nov 06, 2004	Rb	D1	28474
IIR-14M	17	53	Sep 26, 2005	Rb	C4	28874
IIR-15M	31	52	Sep 25, 2006	Rb	A2	29486
IIR-16M	12	58	Nov 17, 2006	Rb	B4	29601
IIR-17M	15	55	Oct 17, 2007	Rb	F2	32260
IIR-18M	29	57	Dec 20, 2007	Rb	C1	32384
IIR-19M	07	48	Mar 15, 2008	Rb	A6	32711
IIR-20M	01	49	Mar 24, 2009	Rb	B6	34661
IIR-21M	05	50	Aug 17, 2009	Rb	E6	35752
IIF-1	25	62	May 28, 2010	Rb	B2	36585

Source: Taken from Department of the Navy's Time Service Depart ment website at ftp://tycho.usno.navy.mil/pub/gps/gpsb2.txt

Table 7-2   GLONASS Constellation Status as of January 10, 2011

Orb. Pl.	Orb. Slot	RF Chnl	# GC	Launched	Operation Begins	Operation Ends	Life-Time (months)	Satellite Health Status		Comments
								In Almanac	In Ephemeris (UTC)	
	1	01	730	14.12.09	30.01.10		12.9	+	+ 19:00 10.01.11	In operation
	2	−4	728	25.12.08	20.01.09		24.5	+	+ 19:00 10.01.11	In operation
	3	05	727	25.12.08	17.01.09	08.09.10	24.5			Maintenance
I	5	01	734	14.12.09	10.01.10		12.9	+	+ 16:59 10.01.11	In operation
	6	−4	733	14.12.09	24.01.10		12.9	+	+ 16:59 10.01.11	In operation
	7	05	712	26.12.04	07.10.05		72.5	+	+ 17:15 10.01.11	In operation
	8	06	729	25.12.08	12.02.09		24.5	+	+ 18:45 10.01.11	In operation

(*Continued*)

Table 7-2 (*Continued*)

Orb. Pl.	Orb. Slot	RF Chnl	# GC	Launched	Operation Begins	Operation Ends	Life-Time (months)	Satellite Health Status		Comments
								In Almanac	In Ephemeris (UTC)	
II	9	−2	736	02.09.10	04.10.10		4.3	+	+ 17:30 10.01.11	In operation
	10	−7	717	25.12.06	03.04.07		48.6	+	+ 18:59 10.01.11	In operation
	11	00	723	25.12.07	22.01.08		36.6	+	+ 19:00 10.01.11	In operation
	12	−1	737	02.09.10	12.10.10		4.3	+	+ 19:00 10.01.11	In operation
	13	−2	721	25.12.07	08.02.08		36.6	+	+ 19:00 10.01.11	In operation
	14	−7	722	25.12.07	25.01.08		36.6	+	+ 16:59 10.01.11	In operation
		−7	715	25.12.06	03.04.07	24.10.10	48.6			Maintenance
	15	00	716	25.12.06	12.10.07		48.6	+	+ 16:59 10.01.11	In operation
	16	−1	738	02.09.10	11.10.10		4.3	+	+ 16:59 10.01.11	In operation
III	17	−5	714	25.12.05	31.08.06	08.01.11	60.6	−	−05:00 08.01.11	Maintenance
		04	718	26.10.07	04.12.07	29.11.10	38.5			Maintenance
	18	−3	724	25.09.08	26.10.08		27.5	+	+ 18:30 10.01.11	In operation
	19	03	720	26.10.07	25.11.07		38.5	+	+ 19:00 10.01.11	In operation
	20	02	719	26.10.07	27.11.07		38.5	+	+ 19:00 10.01.11	In operation
	21	04	725	25.09.08	05.11.08		27.5	+	+ 19:00 10.01.11	In operation
	22	−3	731	02.03.10	28.03.10		10.3	+	+ 16:59 10.01.11	In operation
			726	25.09.08	13.11.08	31.08.09	27.5			Maintenance
	23	03	732	02.03.10	28.03.10		10.3	+	+ 16:59 10.01.11	In operation
	24	02	735	02.03.10	28.03.10		10.3	+	+ 16:59 10.01.11	In operation

Source: From Russian Federal Space Agency website at http://www.glonass-ianc.rsa.ru/pls/htmldb/f?p=202:20:43196687266 63003::NO

7.9.1 Static Surveys

For Static surveys (see Section 7.10.2 for a definition), survey planning includes a visit to the field in order to locate and inspect existing stations, to place monuments for new stations, and to determine the best routes and access for all stations. A compass and clinometer (Figure 3-7) are handy in sketching the location and elevation of potential obstructions at each station on a visibility (obstruction) diagram (see Figure 7-13)—these obstructions are entered into the software for later display. The coordinates (latitude and longitude) of stations should be scaled from a topographic map—scaled coordinates can help some receivers to lock onto the satellites more quickly.

Mission planning displays include the number of satellites available (Figure 7-10); satellite orbits and obstructions—showing orbits of satellites as viewed at a specific station on a specific day, at a specific time (Figure 7-11); and a plot of DOP values. Generally surveyors look at the geometric DOP (GDOP) or the positional DOP (PDOP). These DOP values are critical in determining when it is best to be collecting data and when it is inadvisable to be collection data.

For Static surveys, attention must be given the sampling rate (epoch) of all the receivers; faster rates require more storage, but can be helpful in detecting cycle slips—particularly on longer lines (>50 km). In addition an occupation schedule needs to be established, showing the approximate start times and durations of occupation for all sites.

7.9.2 Kinematic Surveys

Much of the discussion above, for planning Static surveys, also applies to kinematic surveys (see Section 7.10.3 for definitions). For kinematic surveys, over long distances, the route also must be planned for each roving receiver so that best use is made of control points, crews, and equipment. The base station receiver must be compatible with the survey mission and the roving receivers; one base station can support any number of roving receivers. The technique to be used for initial ambiguity resolution must be determined; that is, will it be antenna swap, or known station occupation, or on-the-fly (OTF)? And at which stations will this resolution take place?

7.10 GPS FIELD PROCEDURES

7.10.1 Tripod- and Pole-Mounted Antenna Considerations

GPS antennas can be mounted directly to tripods via optical plummet tribrachs or use fixed height tripods, which are fixed height rods with an adjustable tripod attached. Take care to center the antenna precisely and to measure the antenna height precisely. The antenna height can be measured to an Antenna Reference Point (ARP), which is usually at the base of the antenna, or can be measured to a tape measure station, which is usually along the outside of the antenna, where it is easy to measure to. The tape measure stations are slant measurements that require a correction for the slant and the offset to the ARP. Earlier antennas have a direction mark (often an N or arrow, or a series of numbered notches along the outside perimeter of the antenna), which enabled the surveyor to align the roving antenna in the same direction (usually north) as the base station antenna. This helps to eliminate any bias in the antennas.

GPS receiver antennas can also be mounted on adjustable-length poles (similar to prism poles) or bipods. Adjustable-length poles are not recommended for Static surveys. As with most surveys, field notes are important, both as backup and as confirmation of entered data. It is critical for Static surveys to be precise in measurement and complete in the notes about each occupation. Figure 7-14 shows a sample field log that is to be completed during the occupation.

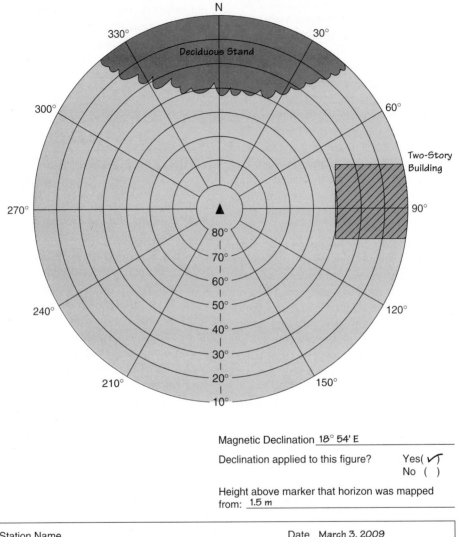

GPS Station Obstruction Diagram

Magnetic Declination _18° 54' E_

Declination applied to this figure?   Yes(✓)
                                       No (   )

Height above marker that horizon was mapped
from: _1.5 m_

Station Name _____   Date _March 3, 2009_

Station Number _1079_   Operator _RJB_

Latitude _46° 36' 30" N_   Longitude _122° 18' 00" W_

FIGURE 7-13 Station visibility diagram

GPS FIELD LOG

Page 1 of __

Project Name _____　　　　*Project Number* _____

Receiver Model/No. _____　　Station Name _____
Receiver Software Version _____　　Station Number _____
Data Logger Type/No. _____　　4-Character ID _____
Antenna Model/No. _____　　Date _____
Cable Length _____　　　　Obs. Session _____
Ground Plane Extensions Yes ( ) No ( )　　Operator _____

Data Collection　　　　　　　　　　*Receiver Position*
Collection Rate _____　　　Latitude _____
Start Day/Time _____　　　Longitude _____
End Day/Time _____　　　　Height _____

Obstruction or possible interference sources _____
_____
General weather conditions _____
_____
Detailed meteorological observations recorded: Yes ( ) No ( )

Antenna Height Measurement

Show on sketch measurements taken to derive the antenna height. If slant measurements are taken, make measurement on two opposite sides of the antenna. Make measurements before and after observing session.

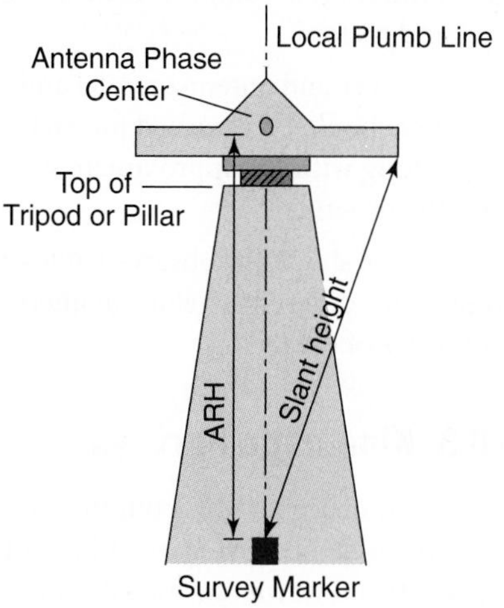

Vertical measurements   ( )

Slant measurements   ( ) : radius _____ m

BEFORE	AFTER
_____ m _____ in.	_____ m _____ in.
_____ m _____ in.	_____ m _____ in.

Mean _____
Corrected to vertical
if slant measurement _____
Vertical offset to
phase center _____
Other offset
(indicate on sketch) _____
TOTAL HEIGHT _____

Verified by: _____

Antenna Phase Center
Local Plumb Line
Top of Tripod or Pillar
ARH
Slant height
Survey Marker

FIGURE 7-14 GPS field log

(Courtesy of Geomatics, Canada)

7.10.2 Static Surveys

7.10.2.1 Traditional Static   In this technique of GPS positioning, two, or more, receivers collect data from the same satellites during the same epochs. Accuracy can be improved by using the differential techniques of relative positioning whereby one base receiver antenna (single or dual frequency) is placed over a point of known coordinates (*X*, *Y*, *Z*) on a tripod, while other antennas are placed, also on tripods, over permanent stations that are to be positioned. Observation times are 2 hours or more, depending on the receiver, the accuracy requirements, the length of the baseline, the satellites' geometric configuration, and atmospheric conditions. This technique is used for long lines in geodetic control, control densification, and photogrammetric control for aerial surveys and precise engineering surveys; it is also used as a fallback technique when the available geometric array of satellites is not compatible with other GPS techniques (see the sections describing GPS measurement methods). This process, nowadays, is not used often, with most surveyors performing Rapid Static surveys (See Section 7.10.2.2) for control. The field procedures for Static surveys are identical as those for Rapid Static, except for the occupation times.

7.10.2.2 Rapid Static   This technique, which was developed in the early 1990s, can be employed over short (up to 15 km) lines. Accuracies of a few millimeters are possible using this technique. Observation times for dual-frequency receivers of 5–20 min are typical—depending on the length of the baseline. Single-frequency receivers can perform Rapid Static surveys, requiring longer observation times. When performing a GPS survey, observation times indicate the time duration that both receivers are simultaneously collecting data. Therefore, the time starts when the second receiver starts collecting data.

There are two approaches to a Rapid Static survey: The first would be similar to a conventional traverse, where the two GPS receivers measure a line, or leg of a traverse, and then measure the next leg, until the traverse is completed; the second method is similar to a radial survey, where there is a "base" station that does not move during the survey, and constantly collects data, while a "rover" moves from station to station. The radial method is the most common method, which allows the use of CORS or ACP stations as the base station.

Solutions for a Rapid Static survey require postprocessing, meaning the data between all the receivers must be processed through GPS software after the field collection.

Rapid Static surveys are done for high-order control-type surveys generally, and certain procedures should be in place.

- Multiple (a minimum of two) horizontal control points with known positions should be observed. This will fix the horizontal system.

- Multiple (a minimum of three) vertical control points with known positions should be observed if orthometric heights are desired. Ideally the vertical control would wrap around the boundary of the entire project.

- The antenna height measurement should be measured at the beginning and end of the observation. It is common to measure in both feet and meters.

- The receiver and antenna model and serial number should be recorded for each observation, along with the approximate starting time and the observer.

- All control should be observed at least two times with different satellite geometry (different times of the day).

7.10.3 Kinematic Surveys

This technique begins with both the base unit and the roving unit occupying a 10-km (or shorter) baseline (two known positions) until ambiguities are resolved. Alternatively, short baselines with one known position can be used where the distance between the stations is short enough to permit antenna swapping. Here, the base station and a nearby (within reach of the antenna cable) undefined station are occupied for a short period of time (say, 2 min) in the static mode, after which the antennas are swapped (while still receiving the satellite signals—but now in the rove mode) for a further few minutes of readings in the static mode (techniques may vary with different manufacturers). Receivers having wireless technology (e.g., Bluetooth wireless technology) may not have the baselines restricted by the length of antenna cable. After the antennas have been returned to their original tripods and after an additional short period of observations in the static mode, the roving receiver—in rove mode—then moves (on a pole, backpack, truck, boat, etc.) to position all required detail points—keeping a lock on the satellite signals. If the lock is lost, the receiver is held stationary for a few seconds until the ambiguities are once again resolved so that the survey can continue. The base station stays in the static mode unless it is time to leapfrog the base and rover stations.

A third technique, called on-the-fly (OTF) ambiguity resolution, occurs as the base station remains fixed at a position of known coordinates and the rover receiver is able to determine the integer number of cycles using software designed for that purpose—even as the receiver is on the move.

7.10.4 Real-Time Kinematic (RTK)

The real-time combination of GPS receivers, mobile data communications, onboard data processing, and onboard applications makes RTK the most common method used with GPS receivers for surveying. As with the motorized total stations described in Section 6.7, real-time positioning offers the potential of one-person capability in positioning, mapping, and quantity surveys (the base station receiver can be unattended). RTK can achieve an accuracy of 2 cm (0.06 ft) horizontally and 3 cm (0.09 ft) vertically. This precision limits the types of survey that can be done with RTK to lower precision work, such as topographic surveys and layouts of rough grade for construction.

RTK requires a base station for receiving the satellite signals and then retransmitting them to the roving surveyor's receiver, which is simultaneously tracking the same satellites. The roving surveyor's receiver can then compare the base station signals with the signals received by the roving receiver to process baseline corrections and thus determine accurate positions in real time. The epoch collection rate is usually set to 1 s. Messages to the rover are updated every 0.5–2 s, and the baseline processing can be done in 0.5–1 s. Figure 7-15 shows a typical radio and amplifier used in code and carrier differential surveys. Radio reception requires a line of sight to the base station. Today's RTK systems also have the capability of communicating using cell phones. The use of cell phones (in areas of cellular coverage) can overcome the interference on radio channels that sometimes frustrates and delays surveyors. Third generation (G3) mobile phone technology has created the prospect of transferring much more that voice communications and small data-packets; this evolving technology will enable the surveyor to transfer much larger data files from one phone to another, and as well, will enable the surveyor to access computer files via the Internet—both with data uploads and downloads. This new wireless technology will greatly impact both the collection of field data and the layout of surveying and engineering works.

The rover must first solve for a precise vector from the base station to the rover. This is often called a fixed solution. Once the solution is fixed, as long as the rover and base are both collecting from the same satellites, this precise relationship will be maintained, even as the rover moves. Dual-frequency receivers can resolve the vector, without being on a point with a known position and can even resolve it when the receiver is on the move, and can usually resolve it in seconds. This allows the surveyor to be less concerned about maintaining constant lock on all the satellites while moving.

Additionally the rover often needs to get onto a coordinate system, other than the base station's system. This is often called transformation or localization. This requires the rover to occupy stations with known coordinates, so that the GPS software can establish the transformation parameters.

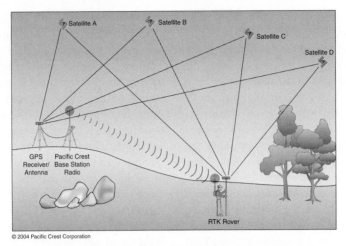

© 2004 Pacific Crest Corporation

(a)

(b)

FIGURE 7-15 (a) A base station GPS receiver and base station radio, as well as roving GPS receiver equipped with a radio receiver that allows for real-time positioning. (b) Base station radio

(Tony Hertz/Alamy)

Once this is done, any other station will be transformed to the desired coordinate system.

As mentioned before, using RTK does not provide high accuracy; however, there are certain procedures to ensure that the positions are within the accepted accuracies.

- The base station must be set in an area that provides the maximum open sky.

- The rover should not go beyond 10 km from the base station, as the accuracy of RTK is dependent on the distance from the rover to the base station.

- Maximum acceptable PDOP should be set at six on the rover. The software will not automatically accept a position when the PDOP value is exceeded.

- When transforming onto a coordinate system, at least one known position not used in the solution should be checked to verify the quality of the solution.

- Pay attention to large obstructions, as they will decrease the quality of the measurement. With GPS, satellites do not go over the North Pole, so keeping obstructions to the north of the rover will minimize problems. Data Collection software will have methods for locating points using offsets, which can minimize obstructions.

RTK surveys can collect positions in 1 s, so extremely high production rates can be achieved with RTK. Units can be mounted to ATVs or trucks to perform topographic surveys of a large area in a short time. Even with walking, a large area can be surveyed quickly. It is important with these methods to pay attention to the actual antenna height above the ground; keeping the rods level and the point of the rod near the ground will give the best results.

RTK surveys are the most demanding GPS survey on the equipment. The base station and rover have to be collecting the same satellites; they must be communicating with each other and the data collection software must be continually doing computations. Surveyors keep all of this equipment in top shape in order to avoid equipment failure in the field and understand that no matter what they do, from time to time the system will fail. Proper planning requires consideration of what to do when failure occurs.

7.10.5 Real-Time Networks (RTN)

RTK is the positioning technique that has seen the greatest technological advances and greatest surveyor acceptance over the past few years. Private companies, municipalities, and even states are now installing (or planning to install) RTK base station networks to service their constituent private and public surveyors. With the large number of CORS stations now established, the need for additional RTK base stations has been reduced because the existing CORS can also be used in RTK operations. Multibase RTK networks can greatly expand the area in which a wide variety of surveyors can operate. It is reported that multibase RTK base stations, using cellular communications, can cover a much larger area (up to four times) than can radio-equipped RTK base stations.

In 2004, Trimble Inc. introduced a variation on the multibase network concept called virtual reference station (VRS) technology. This real-time network (RTN) requires a minimum of three base stations up to 60 km apart (this is apparently the maximum spacing for postprocessing in CORS), with communications to rovers and to a processing center via cell phones. This process requires that the rover surveyor dial into the system and provide his or her approximate location. The network processor determines measurements similar to what would have been received had a base station been at the rover's position and transmits those measurements to the rover receiver. The accuracies and initialization times in this system (using 30-km baselines) are said to be comparable to results obtained in very short (1–2 km) baseline situations. Accuracies of 1 cm in northing and easting and 2 cm in elevation are expected.

RTN was obviously an idea whose time had come. Since 2004 there has been a flurry of work in establishing RTNs across North America, parts of Asia, and Europe. Trimble's main competitors (Topcon, Leica, and Sokkia) have all joined in the expansion of RTNs using their own hardware and software in network design. Once the network of base stations has been established, GPS signals, continuously received at each base station, can be transmitted via the Internet to a central Internet server. Server software can model measurement errors and then provide appropriate corrections to any number of roving receivers via Internet protocol–equipped cell phones—which are working within allowable distances of the base stations. A roving GPS surveyor now has the same capabilities as roving GPS surveyors working with their own dedicated base stations.

Some RTNs have been established by state DOT agencies for their own use, but who also make the service freely available to the public. Other RTNs have been established by entrepreneurs who license (monthly or annual) use of their network to working surveyors.

It is expected that all these continuously operating base station receivers will be added to the CORS network, thus greatly expanding that system.

7.11 GPS APPLICATIONS

Although GPS was originally devised to assist in military navigation, guidance, and positioning, civil applications continue to evolve at a rapid rate. From its earliest days, GPS was welcomed by the surveying community and recognized as an important tool in precise positioning. Prior to GPS, published surveying accuracy standards, then tied to terrestrial techniques (e.g., triangulation, trilateration, and precise traversing), had an upper accuracy limit of 1:100,000. With GPS, accuracy standards have risen as follows: AA (global), 1:100,000,000; A (national—primary network), 1:10,000,000; and B (national—secondary network), 1:1,000,000. In addition to continental control, GPS has now become a widely- used technique for establishing and verifying state/provincial and municipal horizontal control, as well as for horizontal control for large-scale engineering and mapping projects.

7.11.1 Control Surveys

Control surveys are usually done using Rapid Static methods, although some low-order control surveys are done using RTK. Most GPS systems allow the surveyor to enter all the pertinent information in when starting the Static survey. This would include the station name and ID, and the antenna height, as well as what the antenna height was measured to. The software will recognize when there are sufficient satellites and start collecting the information. The GPS receiver will either store the data onboard, on some type of memory card, or in a data collector. Once enough data is collected, the receiver can be shut down and moved to the next position.

When the field collection is completed, the postprocessing of the data begins. The first process is to determine if an acceptable solution for each vector has been achieved. The software will look at every possible receiver pair combination to determine vectors. Often, if using more than two receivers, there are redundant vectors and unresolved vectors that must be removed. Once all the vector solutions are acceptable, an adjustment of the network will take place. First, usually a minimally constrained adjustment will be performed, where only one known horizontal and vertical position will be held. The adjusted solution will then be compared with other known values to see if there are problems in the solution. If orthometric heights are desired, often a geoid model will be applied (see Section 7.12.1). Any positions that do not meet statistical precision will require additional observations.

A survey report is mandatory for control surveys, showing what was used as a basis, what the final positions were, and what their precision is estimated to be, along with a description of the work done and the equipment used.

7.11.2 Topographic Surveys

RTK is generally used for topographic surveys, locating details with short occupation times and described with the input of appropriate coding. Input may be accomplished by keying in, by screen-tapping, by using bar-code readers, or by keying in prepared library codes. Line work may require no special coding (see Field Generated Coding, Section 5.15.4), as entities (curbs, fences, etc.) can be joined by their specific codes (curb2, fence3, etc., or by tapping the display screen to activate the appropriate stringing feature). Some software will display the accuracy of each observation for horizontal and vertical position, giving the surveyor the opportunity to take additional observations if the displayed accuracy does not yet meet job specifications. In addition to positioning random detail, this GPS technique permits the collection of data on specified profile, cross section, and boundary locations—utilizing the navigation functions; contours may be readily plotted from the collected data. Data captured using these techniques can be added to a mapping or GIS database or directly plotted to scale using a digital plotter (see Chapter 9). Although it is widely reported that GPS topographic surveys can be completed more quickly than total station topographic surveys, they do suffer from one major impediment: A GPS receiver must have line of sight to four (preferably five) satellites to determine position. When topographic features are located under tree canopy or hidden by other obstructions, a GPS receiver cannot be used to directly determine position. In this case offset procedures are often used.

7.11.3 Layout Surveys

For layout work, the coordinates of all relevant control points and layout points are uploaded from computer files before going out to the field. Often, instead of layout points, horizontal and vertical alignments are loaded into the data collector. This allows the surveyor to lay out any point shown on the plans relative to these alignments. As each layout point number is keyed into the collector, the azimuth and distance to the required position are displayed on the screen. The surveyor, guided by these directions, eventually moves to the desired point—which is then staked. Once the point is staked, a last observation is taken, and all the pertinent information for that point is presented on the screen and stored. One base receiver can support any number of rover receivers, permitting the instantaneous layout of large-project pipelines, roads, and building locations by several surveyors, each working only on a specific type of facility or by all roving surveyors working on all proposed facilities—but on selected geographic sections of the project.

As with topographic applications, the precision of the proposed location is displayed on the receiver as the antenna pole is held on the grade stake or other marker to confirm that layout specifications have been met; if the displayed precision is below specifications, the surveyor simply waits at the location until the processing of data from additional epochs provides the surveyor with the necessary precision. On road layouts, both line and grade can be given directly to the builder (by marking grade stakes), and progress in cut and fill can be monitored. Slope stakes can be located without any need for inter-visibility. The next logical stage was to mount GPS antennas directly on various construction excavating equipment to directly provide line and grade control.

As with the accuracy/precision display previously mentioned, not only are cut and fill, grades, and the like displayed at each step along the way, but also a permanent record is kept on all of these data in case a review is required. Accuracy can also be confirmed by re-occupying selected layout stations and noting and recording the displayed measurements—an inexpensive, yet effective, method of quality control.

When used for material inventory measurements, GPS techniques are particularly useful in open-pit mining, where original, in-progress, and final surveys can be easily performed for quantity and payment purposes. As well, material stockpiles can be surveyed quickly and volumes computed using appropriate onboard software.

For both GPS topographic and layout work, there is no way to get around the fact that existing and proposed stations must be occupied by the antenna. If some of these specific locations are such that satellite visibility is impossible, perhaps because obstructions are blocking the satellites' signals (even when using receivers capable of tracking both constellations), then ancillary surveying techniques (e.g., total stations, Chapter 5) must be used.

As noted in Section 5.20, one manufacturer has produced a combined prismless total station/GPS receiver (see Figure 5-4) so that most needed measurements can be performed by one instrument. In cases where millimeter accuracy is needed in vertical dimensions (e.g., some structural layouts), one GPS manufacturer has incorporated a precise rotating laser level into the GPS receiver–controller instrumentation package.

7.11.4 Additional Applications

GPS is ideal for the precise type of measurements needed in deformation studies—whether they are for geological events (e.g., plate slippage) or for structure stability studies such as bridges and dams monitoring. In both cases, measurements from permanently- established remote sites can be transmitted to more central control offices for immediate analysis.

In addition to the Static survey control as described above, GPS can also be utilized in dynamic applications of aerial surveying and hydrographic surveying where onboard GPS receivers can be used to supplement existing ground or shore control or where they can now be used in conjunction with inertial guidance equipment [Inertial Navigation System (INS)] for control purposes, without the need for external (shore or ground) GPS receivers. Navigation has always been one of the chief uses made of GPS. Civilian use in this area has really taken off. Commercial and pleasure boating now have an accurate and relatively inexpensive navigation device. With the cessation of Selective Availability (SA), the precision of low-cost receivers has improved to the <10 m range and that can be further improved to the submeter level using differential (e.g., DGPS radio beacon) techniques. Using low-cost GPS receivers, we can now navigate to the correct harbor, and we can navigate to the correct mooring within that harbor. As well, GPS, together with

onboard inertial systems (INS), is rapidly becoming the norm for aircraft navigation during airborne remote-sensing missions.

7.12 VERTICAL POSITIONING

Until recently, most surveyors have been able to ignore the implications of geodesy for normal engineering plane surveys. The distances encountered are so relatively short that global implications are negligible. However, the elevation coordinate (h) given by GPS solutions refers to the height from the surface of the reference ellipsoid to the ground station, whereas the surveyor needs the orthometric height (H). The ellipsoid is referenced to a spatial Cartesian coordinate system (Figure 7-16), called the World Geodetic System of 1984 (WGS84), in which the center (0, 0, 0) is the center of the mass of the earth, and the X axis is a line drawn from the origin through the equatorial plane to the Greenwich meridian. The Y axis is in the equatorial plane perpendicular to the X axis, and the Z axis is drawn from the origin perpendicular to the equatorial plane, as shown in Figure 7-16.

Essentially, GPS observations permit the computation of X, Y, and Z Cartesian coordinates of a geocentric ellipsoid. These Cartesian coordinates can then be transformed to geodetic coordinates: latitude (φ), longitude (λ), and ellipsoidal height (h). The geodetic coordinates, together with geoid corrections, can be transformed to UTM, state plane, or other grids to provide working coordinates (northing, easting, and elevation) for the field surveyor.

Traditionally surveyors are used to working with spirit levels and reference orthometric heights (H) to the average surface of the earth, as depicted by mean sea level (MSL). The surface of MSL can be approximated by the equipotential surface of the earth's gravity field, called the **geoid**.

The density of adjacent landmasses at any particular survey station influences the geoid,

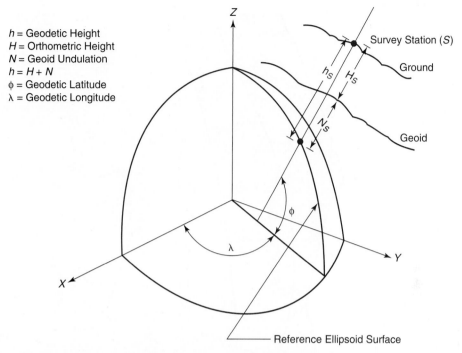

h = Geodetic Height
H = Orthometric Height
N = Geoid Undulation
h = H + N
ϕ = Geodetic Latitude
λ = Geodetic Longitude

FIGURE 7-16 Relationship of geodetic height (h) and orthometric height (H)

which has an irregular surface. Thus, its surface does not follow the surface of the ellipsoid; sometimes it is below the ellipsoid and other times above it. Wherever the mass of the earth's crust changes, the geoid's gravitational potential also changes, resulting in a nonuniform and unpredictable geoid surface. Because the geoid does not lend itself to mathematical expression, as does the ellipsoid, geoid undulation (the difference between the geoid surface and the ellipsoid surface), often called the geoid separation, must be measured at specific sites to determine the local geoid undulation value (see Figures 7-17 and 7-18).

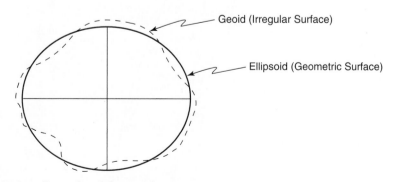

FIGURE 7-17 GRS80 ellipsoid and the geoid

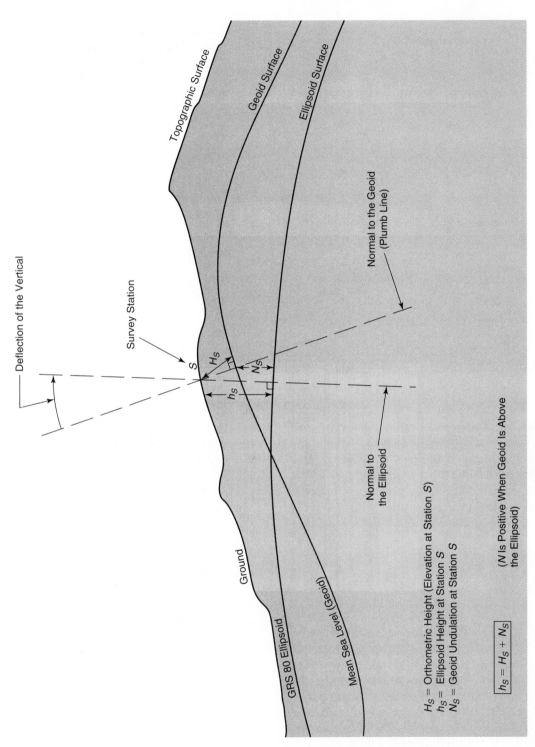

H_S = Orthometric Height (Elevation at Station S)
h_S = Ellipsoid Height at Station S
N_S = Geoid Undulation at Station S

(N Is Positive When Geoid Is Above
the Ellipsoid)

$$h_S = H_S + N_S$$

FIGURE 7-18 The three surfaces of geodesy (undulations greatly exaggerated)

7.12.1 Geoid Modeling

Geoid undulations can be determined both by gravimetric surveys and by the inclusion of points of known elevation in GPS surveys. When the average undulation of an area has been determined, the residual undulations over the surveyed area must still be determined. While residual undulations are usually less than 0.020 m over areas of 50 km^2, the earth's undulation itself ranges from +75 m at New Guinea to –104 m at the south tip of India.

After all the known geoid separations have been plotted, the geoid undulations (N) at any given survey station can be interpolated; the orthometric height (H) can be determined from the relationship $H = h - N$, where h is the ellipsoid height (N is positive when the geoid is above the ellipsoid and negative when below the ellipsoid)—see Figure 7-17. Geoid modeling data can be obtained from government agencies, and in many cases, GPS receiver suppliers provide these data as part of their onboard software.

7.12.1.1 CGG2000 GEOID (Canada)

Natural Resources Canada (NRCan) has developed an improved geoid model—the Canadian Gravimetric Geoid model (CGG2000), which is a refinement of previous models (GSD95 and GSD91). It which takes into account about 700,000 surface gravity observations in Canada, with the addition of about 1,477,000 observations taken in the United States and 117,100 observations taken in Denmark. Although the model covers most of North America, it was designed for use in Canada. Geoid data (the GPS.H package includes the new geoid model and GPS.Hv2.1, which is the latest software) are available at the ministry's website at www.geod.nrcan.gc.ca (under "Earth Sciences → Mapping").

7.12.1.2 GEOID09 (United States)

GEOID09 (a refinement of GEOID06, GEOID03, GEOID99, GEOID96, GEOID93, and GEOID90) is a geoid-elevation estimation model of the conterminous United States (CONUS), referenced to the GRS80 ellipsoid. Thus, it enables mainland surveyors to convert ellipsoidal heights (NAD 83/GRS80) reliably to the more useful orthometric heights (NAVD88) at the centimeter levels. GEOID09 was created from over 20,000 points-and used updated GPSBMs that were not available for the previous geoid model (GEOID03). GEOID09 files are among the many services to be found at the National Geodetic Survey's Geodetic tool kit website at www.ngs.noaa.gov/TOOLS/.

7.13 CONCLUSION

Table 7-3 summarizes the GPS positioning techniques described in this chapter. GPS techniques hold such promise that most future horizontal and vertical control could be coordinated using these techniques. It is also likely that many engineering, mapping, and GIS surveying applications will be developed using emerging advances in real-time GPS data collection. In the future, the collection of GPS data will be enhanced as more and more North American local and federal agencies install continuously operating receivers and transmitters, which provide positioning solutions for a wide variety of private and government agencies involved in surveying, mapping, planning, GIS-related surveying, and navigation.

7.14 GPS GLOSSARY

absolute positioning   The direct determination of a station's coordinates by receiving positioning signals from a minimum of four GPS satellites. Also known as point positioning.

active control station (ACS)   See CORS.

ambiguity   The integer number of carrier cycles between the GPS receiver and a satellite.

CORS   Continuously Operating Reference Station (GPS). CORS-transmitted data can be used by single-receiver surveyors or navigators to permit higher precision differential positioning through postprocessing computations.

Table 7-3   GPS measurements summary*

Two basic modes

Code-based measurements: Satellite-to-receiver pseudorange is measured and then corrected to provide the range; four satellite ranges are required to determine position by removing the uncertainties in X, Y, Z, and receiver clocks. Military can access both the P code and the C/A code; civilians can access only the C/A code.

Carrier-based measurements: The carrier waves themselves are used to compute the satellite(s)-to-receiver range, similar to EDM. Most carrier receivers utilize both code measurements and carrier measurements to compute positions.

Two basic techniques

Point positioning: Code measurements are used to directly compute the position of the receiver. Only one receiver required.

Relative positioning: Code and/or carrier measurements are used to compute the baseline vector (ΔX, ΔY, and ΔZ) from a point of known position to a point of unknown position, thus enabling the computation of the coordinates of the new position.

Relative positioning

Two receivers required—simultaneously taking measurements on the same satellites.

Static: Accuracy—5 mm + 1 ppm. Observation times—1 hour to many hours. Use—control surveys; standard method when lines longer than 20 km. Uses dual- or single-frequency receivers.

Rapid Static: Accuracy—5–10 mm + 1 ppm. Observation times—5–15 min. Initialization time of 1 min for dual-frequency and about 3–5 min for single-frequency receivers. Receiver must have specialized Rapid Static observation capability. The roving receiver does not have to maintain lock on satellites (useful feature in areas with many obstructions). Used for control surveys, including photogrammetric control for lines 10 km or less. The receiver program determines the total length of the sessions, and the receiver screen displays "time remaining" at each station session.

Kinematic: Accuracy—10 mm + 2 ppm. Observation times—1–4 epochs (1–2 minutes on control points); the faster the rover speed, the quicker must be the observation (shorter epochs). Sampling rate is usually between 0.5 and 5 s. Initialization by occupying two known points—2–5 min, or by antenna swap—5–15 min. Lock must be maintained on four satellites (five satellites are better in case one of them moves close to the horizon). Good technique for open areas (especially hydrographic surveys) and where large amounts of data are required quickly.

DGPS: The U.S. Coast Guard's system of providing differential code measurement surveys. Accuracy—submeter to 10 m. Roving receivers are equipped with radio receivers capable of receiving base station broadcasts of pseudorange corrections, using Radio Technical Commission for Maritime (RTCM) standards. For use by individual surveyors working within range of the transmitters (100–400 km). Positions can be determined in real time. Surveyors using just one receiver have the equivalent of two receivers.

Real-time differential surveys: Also known as real-time kinematic, RTK. Accuracies—1–2 cm. Requires a base receiver occupying a known station, which then radio-transmits error corrections to any number of roving receivers, thus permitting them to perform data gathering and layout surveys in real time. All required software is onboard the roving receivers. Dual-frequency receivers permit on-the-fly (OTF) reinitialization after loss of lock. Baselines are restricted to about 10 km. Five satellites are required. This, or similar techniques, is without doubt the future for many engineering surveys.

*Observation times and accuracies are affected by the quality and capability of the GPS receivers, by signal errors, and by the geometric strength of the visible satellite array (GDOP). Vertical accuracies are about half the horizontal accuracies.

cycle slip   A temporary loss of lock on satellite carrier signals causing a miscount in carrier cycles; lock must be reestablished to continue positioning solutions.

differential positioning   Obtaining satellite measurements at a known base station in order to correct simultaneous same-satellite measurements made at rover receiving stations. Corrections can be postprocessed, or corrections can be real time (RTK) as when they are broadcast directly to the roving receiver.

DOP (dilution of precision)   A value that indicates the relative uncertainty in position, based on the geometry of the satellites, relative to the GPS receiver. DOPS can be referenced as horizontal (HDOP), vertical (VDOP), time (TDOP), relative (RDOP), and then the two most commonly used in surveying—geometric (GDOP) and position (PDOP).

epoch   An observational event in time that forms part of a series of GPS observations.

geodetic height (*h*)   The distance from the ellipsoid surface to the ground surface.

geoid surface   A surface that is approximately represented by mean sea level (MSL), and is, in fact, the equipotential surface of the earth's gravity field.

geoid undulation (*N*)   The difference between the geoid surface and the ellipsoid surface. *N* is negative if the geoid surface is below the ellipsoid surface. Also known as *geoid height*.

GNSS (Global Navigation Satellite System)   The term used to describe all satellite navigation systems, current and planned.

global positioning system (GPS)   A ground positioning (*Y*, *X*, and *Z*) technique based on the reception and analysis of NAVSTAR satellite signals.

ionosphere   That section of the earth's atmosphere that is about 50 km to 1,000 km above the earth's surface.

ionospheric refraction   The impedance in the velocity of signals (GPS) as they pass through the ionosphere.

NAVSTAR   A set of orbiting satellites used in navigation and positioning, also known as GPS.

orthometric height (*H*)   The distance from the geoid surface to the ground surface. Also known as *elevation*.

pseudorange   The uncorrected distance from a GPS satellite to a GPS ground receiver determined by comparing the code transmitted from the satellite to the replica code residing in the GPS receiver. When corrections are made for clock and other errors, the pseudorange becomes the range.

real-time positioning (real-time kinematic—RTK)   RTK requires a base station to measure the satellites' signals, process the baseline corrections, and then broadcast the corrections (differences) to any number of roving receivers that are simultaneously tracking the same satellites.

relative positioning   The determination of position through the combined computations of two or more receivers simultaneously tracking the same satellites, resulting in the determination of the baseline vector (*X*, *Y*, *Z*) joining two receivers.

troposphere   That part of the earth's atmosphere which stretches from the surface to about 80 km upward (includes the stratosphere as its upper portion).

7.15 RECOMMENDED READINGS

Books and articles

Geomatics Canada. 1993. *GPS Positioning Guide.* Ottawa: Natural Resources Canada.

Hofman-Wellenhof, et al., 1997. *GPS Theory and Practice,* 4th ed. New York: Springer-Verlag Wien.

Hurn, Jeff. 1993. *Differential GPS Explained.* Sunnyvale, CA: Trimble Navigation Co.

Leick, Alfred. 1995. *GPS Satellite Surveying,* 2nd ed. New York: John Wiley & Sons.

Reilly, James P. "The GPS Observer" (ongoing columns), *Point of Beginning* (POB).

Spofford, Paul, and Neil Weston. 1998. "CORS— The National Geodetic Survey's Continuously Operating Reference Station Project," *ACSM Bulletin*, March/April.

Snay, Richard A., and Soler, Tomás. 2008. "Continuously Operating Reference Station (CORS): History, Applications, and Future Enhancements," *Journal of Surveying Engineering ASCE*, November, 134(4), 95–104.

Trimble Navigation Co. 1989. GPS, *Surveyor's Field Guide.* Sunnyvale, CA. (Information on these and other Trimble publications is available at the Trimble website shown on the following page.)

Trimble Navigation Co. 1992. GPS, *A Guide to the Next Utility.* Sunnyvale, CA.

Van Sickle, Jan. 2008. *GPS for Land Surveyors*, 3rd ed. Chelsea, MI: Ann Arbor Press Inc.

Wells, David, et al., 1986. *Guide to GPS Positioning.* Fredericton, NB: Canadian GPS Associates.

Magazines for general information (including archived articles)

ACSM Bulletin, American Congress on Surveying and Mapping, http://www.survmap.org/

GPS World, http://www.gpsworld.com/

Point of Beginning (POB), http://www.pobonline.com/

Professional Surveyor, http://www.profsurv.com

Websites for general information, reference, and Web links, and GPS receiver manufacturers (see the list of additional Internet references in Appendix E)

Canada-Wide real-Time DGPS Service, www.cdgps.com

DGPS, http://www.navcen.uscg.gov/ (U.S. Coast Guard Navigation Center)

Galileo, http://europa.eu.int/comm/dgs/energy_transport/galileo/index_en.htm

GLONASS, http://www.glonass-ianc.rsa.ru/en/ Open Directory Project, http://www.dmoz.org/Science/Earth_Sciences/Geomatics/

Land Surveyors' Reference Page, Stan Thompson, PLS, Huntington Technology Group, http://www.lsrp.com/

Leica, http://www.leica-geosystems.com

Online positioning service (OPUS), www.ngs.noaa.gov/OPUS/

NOAA, What is Geodesy?, http://www.nos.noaa.gov/education/kits/geodesy/geo01_intro.html

Natural Resources Canada, http://www.nrcan.gc.ca/

National Geodetic Survey (NGS), http://www.ngs.noaa.gov/

Sokkia, http://www.sokkia.com/

Topcon, http://www.topconpositioning.com/

Trimble, http://www.trimble.com/

Also see the list of Internet references in Appendix B.

Questions

7.1   Why is it necessary to observe a minimum of four GPS satellites to solve for position?

7.2   How does the United States' GPS constellation compare with the GLONASS constellation and with the proposed Galileo constellation?

7.3   How does differential positioning work?

7.4   What is the difference between range and pseudorange?

7.5   What are the chief sources of error in GPS measurements? How can you minimize or eliminate each of these errors?

7.6   What are the factors that must be analyzed in GPS planning?

7.7   Describe RTK techniques used for a layout survey.

7.8   Which positioning method is best for precise control surveys and why?

7.9   Explain why station visibility diagrams are used in survey planning.

7.10   Explain the difference between orthometric heights and ellipsoid heights.

7.11   Explain the CORS system.

7.12   For RTK to work, what do we need besides two or more receivers collecting data from a sufficient number of satellites simultaneously?

7.13   Assume a horizontal positional accuracy for the determination of a point using GPS to be 2 cm, and the vertical positional accuracy to be 3 cm. How many meters would two points have to be apart before I could be assured that the error in slope is less than 1 percent?

7.14   What type of GPS process gets positions within 1–2 m?

7.15   What is the geometric indicator of precision used by the GPS system?

CHAPTER EIGHT

TOPOGRAPHIC AND HYDROGRAPHIC SURVEYING AND MAPPING

8.1 GENERAL BACKGROUND

Topographic surveys are preliminary (also known as preengineering) surveys used to locate the horizontal and vertical positions of natural and constructed surface features in a defined geographic area. Hydrographic surveys are much the same as topographic surveys, except that the surveyed area is under water or on shore. Such features are located relative to one another in the field by tying them all in to the same control line or control grid, or through the direct positioning techniques common to satellite positioning field surveys (see Chapter 7). Additionally, topographic spatial detail positioning can be conducted using remote-sensing surveys; see Chapters 11 ("Satellite Imagery") and 12 ("Airborne Imagery").

Mapping and *drafting* are terms that cover a broad spectrum of scale graphics and related computations. With the rapid development of geographic information systems (GISs), which are discussed Chapter 9, mapping has now largely become one of the products of the geodata collection, management, and analyses associated with GIS; additionally, digital topographic data can be downloaded directly into computer-assisted drafting or design (CAD) programs, the most popular of which is AutoCad (ACAD). Onscreen editing and digital plotting are core capabilities of the software. Generally, *drafting* is a drawing term usually reserved for large- and intermediate-scale (see Table 8-1) graphics and is often encountered in surveying, engineering, and architecture applications in the preparation of plans; on the other hand, *mapping* is a drawing term usually reserved

for small-scale graphics, in the form of maps, often depicting topographic and boundary detail of relatively large areas of the earth's surface. As noted in Chapter 5, when topographic surveys are undertaken using total stations, survey drafting has now become the responsibility of the field surveyor who can prepare plot files using specific field-codes, or data collector graphics manipulation, which reflect graphics' connecting or *stringing* characteristics of topographic or built features (such as shoreline or a specific curb line).

The essential difference between maps and plans is their use. Maps portray, as in an inventory, the detail (e.g., topography, surface features, boundaries) for which they were designed. Maps can be of a general nature, such as the topographic maps compiled and published by the U.S. Geological Survey (scales normally ranging from 1:24,000 down to 1:1,000,000), or maps can be specific, showing only those data (e.g., cropland inventory) for which they were designed (see Chapter 9). Engineering-surveying plans, on the other hand, not only show existing terrain (or other) conditions, but can also depict proposed alterations (i.e., *designs*) to the existing landscape. Most plans are drawn to a large scale, although comprehensive functional planning or route design plans for state and provincial highways can be drawn to a small scale (e.g., 1:50,000) to give a bird's-eye view of a large study area.

The techniques of plan and map preparation have now become digital, with the maps and plans being composed and edited on computers and the drawings being produced on digital plotters. The mechanical techniques of mapping and quantity estimates are presented here to give you an

Table 8-1   Summary of Map and Plan Scales and Contour Intervals

	Metric Scale	Foot/Inch Scale Equivalents	Contour Interval for Average Terrain	Typical Uses
Large scale	1:10	$1'' = 1'$		Detail
	1:50	$\frac{1}{4}'' = 1', 1'' = 5'$		Detail
	1:100	$\frac{1}{4}'' = 1', 1'' = 8'$		Detail, profiles
	1:200	$1'' = 20'$		Profiles
	1:500	$1'' = 40', 1'' = 50'$	0.5 m, 1 ft	Municipal design plans
	1:1,000	$1'' = 80', 1'' = 100'$	1 m, 2 ft	Municipal services and site engineering
Intermediate scale	1:2,000	$1'' = 200'$	2 m, 5 ft	Engineering studies
	1:5,000	$1'' = 400'$	5 m, 10 ft	and planning
	1:10,000	$1'' = 800'$	10 m, 20 ft	(e.g., drainage areas, route planning)
Small scale	1:20,000			
	1:25,000	$1:25,000, \quad 2\frac{1}{2}'' = 1, mi$		
	1:50,000	$1:63,360, \quad 1'' = 1, mi$		Topographic maps, Canada and United States
	1:100,000	$1:126,720, \quad 2\frac{1}{2}'' = 1, mi$		Geological maps, Canada and United
	1:200,000			States
				Special-purpose maps and atlases (e.g., climate, minerals)
	1:250,000	$1:250,000, \quad \frac{1}{2}'' = 1, mi$		
	1:500,000	$1:625,000, \quad \frac{1}{2}'' = 1, mi$		
	1:1,000,000	$1:1,000,000, \quad \frac{1}{2}'' = 1, mi$		

*The contour interval chosen must reflect the scale of the plan or map, but the terrain (flat or steeply inclined) and intended use of the plan are also factors in choosing the appropriate contour interval.

introduction to the logic behind the computer programs now used in this process. Table 8-1 summarizes typical scales along with appropriate contour intervals; Table 8-2 shows standard drawing sizes.

8.2 MAPS AND PLANS

The traditional techniques for producing maps and plans are included to show how survey data, obtained from either ground or aerial surveys, are added by hand to scale drawings. The reproduction of maps traditionally involved photographing the finished inked or scribed map and preparing a printing plate from the negative. Lithographic offset printing was used to create the maps; multicolor maps required a separate plate for each color, although shading can be accomplished with screens. Modern techniques use digital plotters to convert computer data (vector or raster formats) to produce the desired graphics.

Scribing is a cartographic mapping technique in which the map details are directly cut onto drafting film that has a soft, opaque coating. Modern scribing techniques employ digital plotters where the pens are replaced with scribing tools. This scribed film takes the place of a photographic negative in the photolithography printing process. Scribing is preferred by many because of the sharp definition made possible by this cutting technique. Plans, on the other hand, are reproduced in an entirely different manner. The completed plan, in ink or pencil, can simply be run through a direct-contact

Table 8-2   Standard Drawing Sizes

| | International Standards Organization (ISO) | | | | | | ACSM* Recommendations | |
| | Inch Drawing Size | | Metric Drawing Sizes (mm) | | | | | |
Drawing Sizes	Border Size	Overall Paper Size	Drawing Size	Border Size	Overall Paper Size	Drawing Size	Paper Size
A	8.00 × 10.50	8.50 × 11.00	A4	195 × 282	210 × 297	–	150 × 200
B	10.50 × 16.50	11.00 × 17.00	A3	277 × 400	297 × 420	A4	200 × 300
C	16.00 × 21.00	17.00 × 22.00	A2	400 × 574	420 × 594	A3	300 × 400
D	21.00 × 33.00	22.00 × 34.00	A1	574 × 821	594 × 841	A2	400 × 600
E	33.00 × 43.00	34.00 × 44.00	A0	811 × 1,159	811 × 1,159	A1	600 × 800
						A0	600 × 1,200

*American Congress on Surveying and Mapping Metric Workshop, March 14, 1975. Paper sizes rounded off for simplicity, still have cut-in-half characteristic.

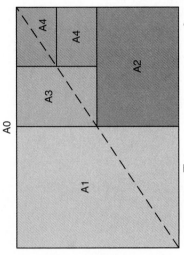

Ratio 1: √2   Area of A0 size = 1 m²

Metric Drawing Paper

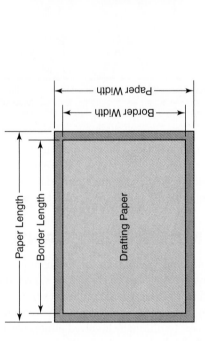

negative printing machine (blueprint) or a direct-contact positive printing machine (whiteprint). The whiteprint machine, now in use in most engineering drafting and design offices, uses paper sensitized with diazo compounds, which, when exposed to light and ammonia vapor, produce prints. The quality of whiteprints cannot be compared to map-quality reproductions; however, the relatively inexpensive whiteprints are widely used in surveying and engineering offices where they are used as working plans, customer copies, and contract plans. Although reproduction techniques are vastly different for maps and plans, the basic plotting procedures can be quite similar.

8.3 SCALES AND PRECISION

Maps and plans are drawn so that a distance on the map or plan conforms to a set distance on the ground. The ratio (called *scale*) between plan distance and ground distance is consistent throughout the plan. Scales can be stated as equivalencies, for example, $1'' = 50'$ or $1'' = 1,000'$, or the same scales as can be stated as 1:600 or 1:12,000. When the corresponding ratios are used, any units are valid; that is, 1:500 is the same scale for inches, feet, meters, and so on. Only ratios are used in the metric (SI) system. Table 8-1 shows recommended map and plan scales and the equivalent scales in the foot–inch system. Almost all surveying work required for the production of intermediate- and small-scale maps is done by aerial or satellite imaging, with the maps being produced photogrammetrically (see Chapter 12).

Even in municipal areas where services (roads, sewers, water) plans for housing developments can be drawn at 1:1,000 and plans for municipal streets are drawn at 1:500, it is not uncommon to have the surveys conducted aerially and the maps produced photogrammetrically. For street surveys, the surveying manager will have a good idea of the cost per kilometer or mile for various orders of urban density and will arrange for field or aerial surveys depending on which method is deemed cost-effective.

Before a field survey is undertaken, a clear understanding of the reason for the survey is necessary so that appropriately precise techniques can be employed. If the survey is required to locate points that will later be shown on a small-scale map, the precision of the survey will be of a very low order. Generally, points are located in the field with a precision that will at least be compatible with the plotting precision possible at the designated plan (map) scale. For example, if we can assume that points can be plotted to the closest 0.5 mm (1/50 in.) at a scale of 1:500, this represents a plotting capability to the closest ground distance of 0.25 m (i.e., 0.0005×500), whereas at a scale of 1:20,000 the plotting capability is $(0.0005 \times 20,000) = 10$ m of ground distance. In the former example, although plotting capabilities indicate that a point should be tied in to the closest 0.25 m, in reality the point probably would be tied in to a higher level of precision (e.g., 0.1 m).

The following points should be kept in mind:

1. Some detail (e.g., building corners, railway tracks, bridge beam seats, or other structural components) can be defined and located precisely.

2. Some detail (e.g., stream banks, edges of a gravel road, ₡ of ditches, limits of a wooded area, rock outcrops) cannot be defined or located precisely.

3. Some detail (e.g., large single trees, culverts, docks) can be located with only moderate precision, using normal techniques.

Usually, the detail that is fairly well defined is located with more precision than is required just for plotting. The reasons for this are as follows:

1. As in the preceding example, it takes little (if any) extra effort to locate detail to 0.1 m compared with 0.25 m.

2. By using the same techniques to locate all detail, the survey crew is able to develop uniform practices, which reduces mistakes and increases efficiency.

3. Some location measurements taken in the field, and design parameters that may be based on those field measurements, are also shown on the plan as layout dimensions (i.e., levels of

precision are required that greatly supersede the precision required simply for plotting).

4. Modern surveys using radial total station techniques or precise global positioning system (GPS) techniques give high-precision positioning for all intermediate sightings, and modern surveys using GIS-compatible GPS instruments routinely give submeter (or better) positioning results.

Most natural features are themselves not precisely defined. If a topographic survey is required in an area having only natural features (e.g., stream or watercourse surveys, site development surveys, large-scale mapping surveys), a relatively imprecise survey method such as aerial or satellite imaging can be employed.

All topographic surveys are tied into both horizontal and vertical (benchmarks) control. The horizontal control for topographic surveys can use closed-loop traverses, traverses from a coordinate grid monument closed to another coordinate grid monument, route centerline (℄), or some assumed baseline, and by GPS. The survey measurements used to establish the horizontal and vertical control are always taken more precisely than are the location ties.

Surveyors are conscious of the need for accurate and well-referenced survey control. If the control is inaccurate, the survey and resultant design will also be inaccurate; if the control is not well referenced, it will be costly (perhaps impossible) to precisely relocate the control in the field once it is lost. In addition to providing control for the original survey, the same survey control should be used if additional survey work is required to supplement the original survey, and of course the same survey control should be used for any construction layout resulting from designs based on the original survey.

8.4 PLAN PLOTTING

8.4.1 General Background

The size of drafting paper required can be determined by knowing the scale to be used and the area or length of the survey. Standard paper sizes are shown in Table 8-2. The title block is often of standard size and has a format similar to that shown in Figure 8-1. The block is often placed in the lower right corner of the plan but placement often depends on the filing system. Revisions to the plan are usually referenced immediately

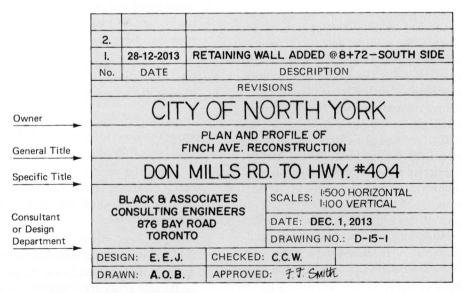

FIGURE 8-1 Typical title block

above the title block, showing the date and a brief description of the revision.

Many consulting firms and engineering departments attempt to limit the variety of their drawing sizes so that plan filing can be standardized. Some vertical-hold filing cabinets are designed such that title blocks in the upper right corner are more easily seen.

Manual plotting begins by first plotting the survey control (e.g., ₵, traverse line, coordinate grid) on the drawing. The control is plotted so that the data plot will be suitably centered on the available paper. Sometimes the data outline is roughly plotted first on tracing paper so that the plan's overall dimension requirements (for the given scale) can be properly oriented on the available drafting paper. It is customary to orient the data plot so that north is toward the top of the plan; a north arrow is included on all property survey plans (plats) and many engineering drawings. The north direction on maps is clearly indicated by lines of longitude or by N-S, E-W grid lines. The plan portion of the plan and profile does not usually have a north indication. Instead, local practice often dictates the direction of increasing stationing (e.g., stationing increasing left to right for west to east and south to north orientation).

Chapter 5 describes how field drafting is begun using graphic connections or "stringing" field-codes as the data are recorded in the total station; graphical data can be then edited and even created (from auxiliary data files) on an interactive graphics screen before being transferred to a computer-aided drafting (CAD) program for final presentation work and output on a digital plotter (see Section 8.4.2). Next we will describe how the field data can be manually plotted.

Once the control has been plotted and checked, the features can be plotted with either rectangular (X, Y coordinates) or polar [(radial) r, θ coordinates] methods. Pre-electronic rectangular plots (X, Y coordinates) can be laid out with a T-square and set square, although the parallel rule has now largely replaced the T-square. When using either the parallel rule or the T-square, the paper is first set square and then secured with masking tape to the drawing board. Once the paper is square and secure, the parallel rule, together with a set square and scale, can be used to lay out and measure rectangular dimensions.

However, hand-drawn polar plots are laid out with a protractor and a scale. The protractor can be a plastic graduated circle or half-circle of various diameters (the larger the circle or half-circle, the more precise it is), a paper full-circle protractor for use under or on the drafting paper, or a flexible-arm drafting machine complete with right-angle-mounted graduated scales. Field data that have been collected with polar techniques (e.g., total station) can be efficiently plotted with polar techniques. See Figure 8-2 for standard map and plan symbols.

The techniques described here are still used in surveying and engineering applications; however, computer-based techniques have mostly replaced manual techniques.

8.4.2 Digital Plotting

Once the plotting files have been established by working with CAD programs, data can be plotted in a variety of ways. Data can be plotted onto a high-resolution graphics screen. The plot can be checked for completeness and accuracy. If interactive graphics are available, the plotted features can be deleted, enhanced, corrected, crosshatched, labeled, dimensioned, and so on. At this stage, a hard copy of the screen display can be printed either on a simple printer or on an ink-jet color printer. Plot files can be plotted directly on a digital plotter similar to that shown in Figure 8-3. The resulting plan can be plotted to any desired scale, limited only by the paper size. Some plotters have only one or two pens, although plotters are available with four to eight pens; a variety of pens permit colored plotting or plotting using various line weights. Scribing tools can be used in place of pens. Plans, and plan and profile views, drawn on digital plotters are becoming more common on construction sites. Automatic plotting (with a digital plotter) can be used where the field data have been coordinated and stored in computer memory. Coordinated field data are a by-product of data collection by total stations,

Primary highway, hard surface	Boundaries: National	
Secondary highway, hard surface	State	
Light-duty road, hard or improved surface	County, parish, municipio	
Unimproved road	Civil township, precinct, town, barrio	
Road under construction, alignment known	Incorporated city, village, town, hamlet	
Proposed road	Reservation, National or State	
Dual highway, dividing strip 25 feet or less	Small park, cemetery, airport, etc.	
Dual highway, dividing strip exceeding 25 feet	Land grant	
Trail	Township or range line, United States land survey	

Township or range line, approximate location

Railroad: single track and multiple track	Section line, United States land survey
Railroads in juxtaposition	Section line, approximate location
Narrow gage: single track and multiple track	Township line, not United States land survey
Railroad in street and carline	Section line, not United States land survey
Bridge: road and railroad	Found corner: section and closing
Drawbridge: road and railroad	Boundary monument: land grant and other
Footbridge	Fence or field line
Tunnel: road and railroad	
Overpass and underpass	Index contour —— Intermediate contour
Small masonry or concrete dam	Supplementary contour Depression contours
Dam with lock	Fill Cut
Dam with road	Levee Levee with road
Canal with lock	Mine dump Wash
	Tailings Tailings pond
Buildings (dwelling, place of employment, etc.)	Shifting sand or dunes Intricate surface
School, church, and cemetery	Sand area Gravel beach
Buildings (barn, warehouse, etc.)	
Power transmission line with located metal tower	Perennial streams Intermittent streams
Telephone line, pipeline, etc. (labeled as to type)	Elevated aqueduct Aqueduct tunnel
Wells other than water (labeled as to type)	Water well and spring Glacier
Tanks: oil, water, etc. (labeled only if water)	Small rapids Small falls
Located or landmark object; windmill	Large rapids Large falls
Open pit, mine, or quarry; prospect	Intermittent lake Dry lake bed
Shaft and tunnel entrance	Foreshore flat Rock or coral reef
	Sounding, depth curve Piling or dolphin
Horizontal and vertical control station:	Exposed wreck Sunken wreck
Tablet, spirit level elevation BM△ 5653	Rock, bare or awash; dangerous to navigation
Other recoverable mark, spirit level elevation △ 5455	
Horizontal control station: tablet, vertical angle elevation VABM △95/9	Marsh (swamp) Submerged marsh
Any recoverable mark, vertical angle or checked elevation △3775	Wooded marsh Mangrove
Vertical control station: tablet, spirit level elevation BM × 957	Woods or brushwood Orchard
Other recoverable mark, spirit level elevation × 954	Vineyard Scrub
Spot elevation × 7369 × 7369	Land subject to controlled inundation Urban area
Water elevation 670 670	

(a)

FIGURE 8-2 (a) Topographic map symbols (*Continued*)

(Courtesy of U.S. Department of Interior, Geological Survey)

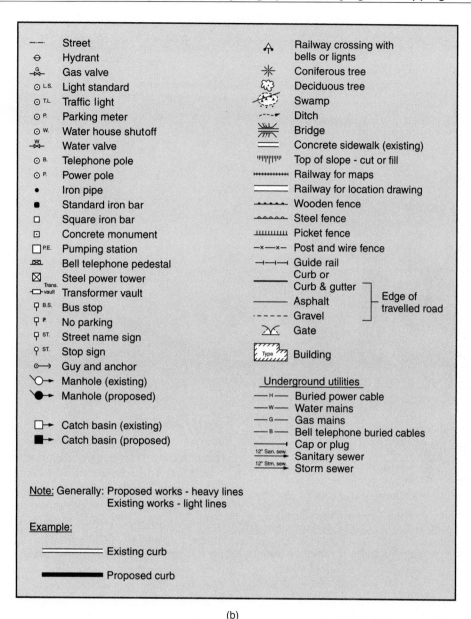

(b)

FIGURE 8-2 (b) Municipal works plan symbols

airborne imagery, satellite imagery, and digitized, or scanned, data from existing plans and maps (see Figure 1-7).

A plot file can include the following:

- Title
- Scale

- Limits (e.g., can be defined by the southwesterly coordinates, northerly range, and easterly range)
- Plotting of all points
- Connecting specific points (through feature coding or CAD commands)

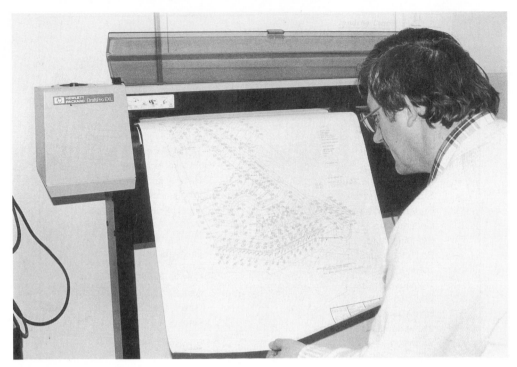

FIGURE 8-3 Land division plot on a Hewlett-Packard eight-pen digital plotter

- Pen number (various pens could have different line weights or colors—two to four pens are common)
- Symbols (symbols are pre-designed and stored by identification number in a symbol library)
- Height of characters (the heights of labels, text, coordinates, and symbols can be defined)

The actual plotting can be performed by simply keying in the plot command required by the specific computer program and then by keying in the name of the plot file that is to be plotted.

The coordinated field point files can be transferred to an interactive graphics terminal (see Figure 8-4), with the survey plot being created and edited graphically right on the high-resolution graphics screen. Some surveying software programs have this graphics capability, whereas others permit coordinate files to be transferred easily to an independent graphics program (i.e., CAD) using a .dxf-type format for later editing and plotting. Once the plot has been completed on the

graphics screen (and all the point coordinates have been stored in the computer), the plot can be transferred to a digital plotter for final presentation.

This latter technique is now being successfully used in a wide variety of applications. The savings in time and money are too great to be overlooked, especially when the actual engineering or construction design can be accomplished on the same graphics terminal, with all design elements' coordinates also being stored in the computer files and all construction drawings being produced on the digital plotter.

8.4.3 Computerized Surveying Computations and Drawing Preparation

As noted in Section 8.1, maps and plans are now being prepared routinely with the aid of computer programs. Survey data can be transferred to the computer either as discrete plot points (northing, easting, and elevation) or as already joined

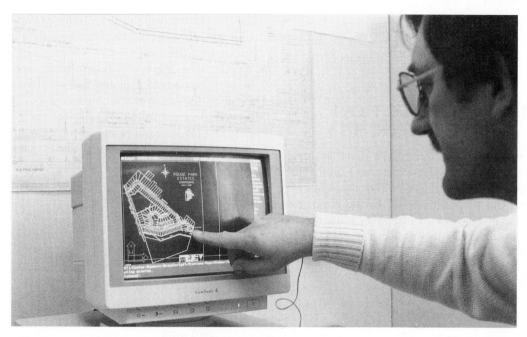

FIGURE 8-4 Land division design and editing on a desktop computer

graphic entities. If the data have been referenced to a control traverse for horizontal control, various software programs can quickly determine the acceptability of the traverse closure. Feature coding can produce graphics labels, or they can be created right in the CAD programs.

Computer-generated models of existing elevations are called digital elevation models (DEMs) or digital terrain models (DTMs). Some agencies use these two terms interchangeably while others regard a DTM as a DEM that includes the location of break lines. **Break lines** are joined coordinated surface points that define changes in slope, such as valley lines, ridge lines, the tops and bottoms of slopes, ditch lines, and the tops and bottoms of curbs.

Computer programs are also available to help create designs in digital terrain modeling (together with contours production), land division and road layout, highway layout, and other projects. The designer can quickly assemble a database of coordinated points reflecting both the existing surveyed ground points and the proposed key points created through the various design

programs. Some surveying software includes drawing capabilities, whereas others create .dxf or .dwg files designed for transfer to CAD programs; Autocad (ACAD) is the most widely used. Surveying and drawing programs have some or all of the following capabilities:

- Survey data import.
- Project definition with respect to map projection, horizontal and vertical datums, and ellipsoid and coordinate system.
- Coordinate geometry (COGO) routines for accuracy determination and for the creation of auxiliary points.
- Graphics creation.
- Feature coding and labeling.
- Digital terrain modeling and contouring, including break-line identification and contour smoothing.
- Earthwork computations.
- Design of land division, road (and other alignment) design, design of horizontal and vertical curves, site grading, etc.

- Creation of plot files.
- File exports in .dxf, .dwg, and .xml; compatibility with GIS through ESRI (e.g., Arcinfo) files and shape files.

8.5 INTRODUCTION TO CONTOURS

Contours are lines drawn on a plan that connect points having similar elevations. Contour lines represent an even elevation value (see Table 8-1), with the contour interval selected for terrain, scale, and intended use of the plan. It is commonly accepted that elevations can be determined to half

the contour interval; this permits, for example, a 10-ft contour interval on a plan where elevations are known to the closest 5 ft.

Contours are manually plotted by scaling between two adjacent plotted points of known elevation (assuming that a uniform slope exists between these points). In Figure 8-5(a), the scaled distance (any scale can be used) between points 1 and 2 is 0.75 units, and the difference in elevation is 5.4 ft. The difference in elevation between point 1 and contour line 565 is 2.7 ft; therefore, the distance from point 1 to contour line 565 is

$$\frac{2.7}{5.4} \times 0.75 = 0.38 \, \text{units}$$

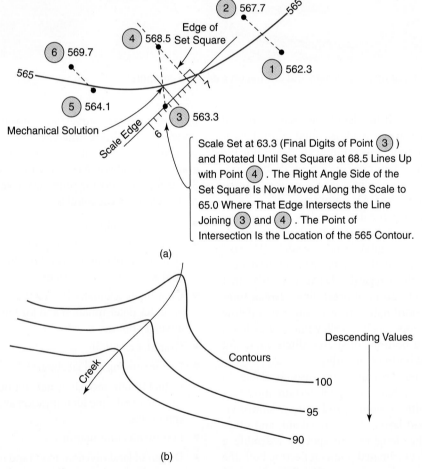

FIGURE 8-5 Contours. (a) Plotting contours by interpolation. (b) Valley line. (*Continued*)

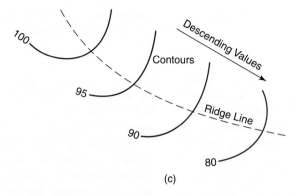

(c)

CONTOUR LINES

These are drawn through points having the same elevation. They
show the height of ground above sea level (M.S.L.) in either feet
or meters and can be drawn at any desired interval.

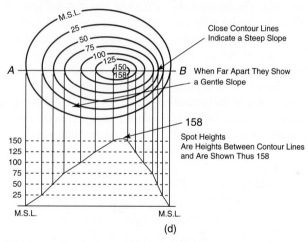

(d)

FIGURE 8-5 (c) Ridge line. (d) Contour plan with derived profile (line AB).

(Courtesy of Department of Energy, Mines, and Resources, Canada)

To verify this computation, the distance from contour line 565 to point 2 is

$$\frac{2.7}{5.4} \times 0.75 = 0.38 \, \text{units}$$

$$0.38 + 0.38 \approx 0.75 \quad \text{Check}$$

The scaled distance between points 3 and 4 is 0.86 units, and their difference in elevation is 5.2 ft. The difference in elevation between point 3 and contour line 565 is 1.7 ft; therefore, the distance from point 3 to contour line 565 is

$$\frac{1.7}{5.2} \times 0.86 = 0.28 \, \text{units}$$

This can be verified by computing the distance from contour line 565 to point 4:

$$\frac{3.5}{5.2} \times 0.86 = 0.58 \, \text{units}$$

$$0.58 + 0.28 = 0.86 \quad \text{Check}$$

The scaled distance between points 5 and 6 is 0.49 units, and the difference in elevation is 5.6 ft. The difference in elevation between point 5 and contour line 565 is 0.9 ft; therefore, the distance from point 5 to contour line 565 is

$$\frac{0.9}{5.6} \times 0.49 = 0.08 \, \text{units}$$

and from line 565 to point 6 the distance is

$$\frac{4.7}{5.6} \times 0.49 = 0.41 \text{ units}$$

In addition to the foregoing arithmetic solution, contours can be interpolated with mechanical techniques. It is possible to scale off units on a barely taut elastic band and then stretch the elastic so that the marked-off units fit the interval being analyzed. Alternately, the problem can be solved by rotating a scale while using a set square to line up the appropriate divisions with the field points. In Figure 8-5(a), a scale is set at 63.3 on point 3 and then rotated until the 68.5 mark lines up with point 4 using a set square on the scale. The set square is then slid along the scale until it lines up with 65.0; the intersection of the set square edge (90° to the scale) with the straight line joining points 3 and 4 yields the solution (i.e., the location of elevation at 565 ft). This technique is faster than the arithmetic technique.

Because contours are plotted by analyzing adjacent field points, it is essential that the ground slope be uniform between those points. An experienced survey crew will ensure that enough rod readings or total station sightings are taken to suitably define the ground surface. The survey crew can further define the terrain if care is taken in identifying and tying in valley lines and ridge lines, ditch lines, top and bottom of slopes lines, that is, any linear series of elevations (*break lines*) that define a change in slope of a natural ground surface or a built surface.

Figure 8-5(b) shows how contour lines bend uphill as they cross a valley; the steeper the valley, the more the line diverges uphill. Figure 8-5(c) shows how contour lines bend downhill as they cross ridge lines. Figure 8-6 shows the plot of control, elevations, and valley and ridge lines. Figure 8-7 shows contours interpolated from the data in Figure 8-6. Figure 8-8(b) shows the completed plan, with additional detail (roads and buildings) also shown. Figure 8-8(a) shows typical field notes for a stadia survey (also see Appendix H). Figure 8-8(c) shows a contour plan together with a derived profile line (*AB*).

Contours are now almost exclusively produced using any of the current software programs. Most programs generate a triangulated irregular

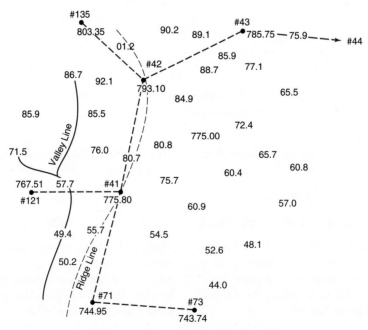

FIGURE 8-6 Plot of survey control, ridge and valley lines, and spot elevations (note only final digits shown)

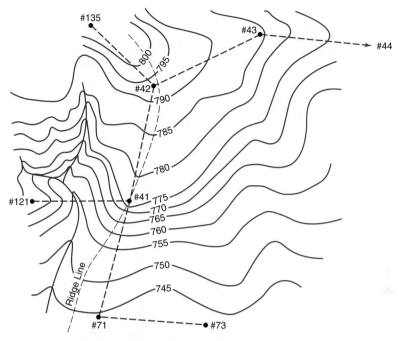

FIGURE 8-7 Contours plotted by interpolating between spot elevations, with additional plotting information given when the locations of ridge and valley lines are known

network (TIN); the sides of the triangles are analyzed so that contour crossings can be interpolated. The field surveyor must also note and mark *break lines* that define significant changes in topography (e.g., bottom/top of hills, channels, depressions, creeks) so that the software program can generate contours that accurately reflect actual field conditions.

Figure 8-8 shows the steps in a typical contour production process. Two basic methods have been used: the uniform grid approach and the more popular TIN approach (used here). First, the TIN is created from the plotted points and defined break lines [Figure 8-9(a)] and the raw contours [Figure 8-9(b)] are computer-generated from those data. To make the contours pleasing to the eye, as well as representative of the ground surface, suitable smoothing techniques must be used to soften the sharp angles occurring when contours are generated from TINs (as opposed to the less angular lines resulting in contours

derived from a uniform grid approach). For more on this topic, refer to Christensen, A. H. J., "Contour Smoothing by an Eclectic Procedure," *Photogrammetric Engineering and Remote Sensing (RE & RS)*, April, 2001: 516.

8.6 SUMMARY OF CONTOUR CHARACTERISTICS

1. Closely spaced contours indicate steep slopes.

2. Widely spaced contours indicate moderate slopes (spacing here is a relative relationship).

3. Contours must be labeled to give the elevation value. Either each line is labeled or every fifth line is drawn darker (wider) and it is labeled.

4. Contours are not shown going through buildings.

Point	Horizontal Angle	Horizontal Distance	Difference in Elevation	Elevation
42				793.10
41	00.0	197.80	− 17.30	775.80
43	232 25.2	199.10	− 7.35	785.75
135	120 05.2	145.20	+ 10.25	803.35
a	234 50	76.10	− 2.60	790.50
b	247 10	76.30		793.10
c	277 22	85.30	− 8.20	784.90
d	322 10	100.50	− 7.60	785.50
etc.				

(a)

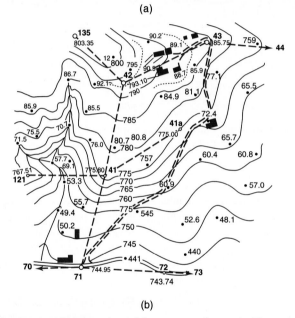

(b)

FIGURE 8-8 (a) Survey field notes. (b) Plan plotted from notes shown in Figures 8-6–8-8(a). (Courtesy of Leica Geosystems Inc.)

5. Contours crossing a built horizontal surface (roads, railroads) will be straight parallel lines as they cross the facility.

6. Because contours join points of equal elevation, contour lines cannot cross. (Caves present an exception.)

7. Contour lines cannot begin or end on the plan.

8. Depressions and hills look the same; the reader must note the contour value to distinguish the terrain (some agencies use hachures or shading to identify depressions).

9. Contours deflect uphill at valley lines and downhill at ridge lines. Line crossings are perpendicular: U-shaped for ridge crossings; V-shaped for valley crossings.

10. Contour lines must close on themselves, either on the plan or in locations off the plan.

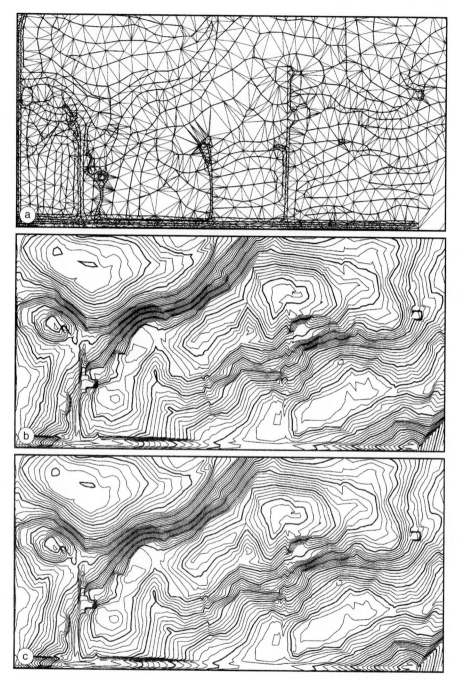

FIGURE 8-9 Contouring. (a) Break lines and TIN. (b) Raw contours. (c) Contour smoothing with the eclectic procedure

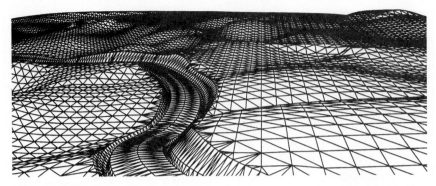

FIGURE 8-10 QuickSurfTGRID showing smoothed, rolling terrain coupled with road definition
(Courtesy of MicroSurvey Software Inc., Westbank, BC, Canada)

11. The ground slope between contour lines is uniform. Had the ground slope not been uniform between the points, additional elevation readings would have been taken at the time of the survey.

12. Important points can be further defined by including a "spot" elevation (height elevation).

13. Contour lines tend to parallel each other on uniform slopes.

In addition to contours, some 3D software programs provide 3D perspective plots to portray the landscape as viewed from any azimuth position and from various altitudes. See Figures 8-10 and Figure F-4 for a plot of LiDAR-collected points color coded to display elevation layers.

8.7 TOPOGRAPHIC (PLANIMETRIC) SURVEYS

8.7.1 General

Topographic surveys are performed in order to determine the position of natural and built features (e.g., trees, shorelines, roads sewers, buildings). These features can then be drawn to scale on a plan or map. In addition, topographic surveys include the determination of ground elevations, which can later be drawn on plans or maps for the construction of contours, or plotted in the form of cross sections and profiles.

The vast majority of topographic/planimetric surveys are now performed using aerial surveying techniques with the plans and DEMs being constructed using modern computerized *photogrammetric*, or laser imaging, techniques. Smaller surveys are often performed using electronic equipment, such as *total stations*. The horizontal location (x and y) and the vertical (elevation) location (z) can be easily captured with one sighting, with point descriptions and other attribute data being entered into electronic storage for later transfer to the computer. Electronic surveying techniques are discussed, in detail, in Chapter 5.

The focus here will be on the rectangular and polar surveying techniques, first introduced in Section 1.14, employing pre-electronic field techniques. The rectangular technique, discussed here, utilizes right-angle offsets for detail location and cross sections for elevations and profiles. The polar technique discussed here utilizes total station or theodolite/EDM techniques for both horizontal location and elevation capture and contours for elevation depiction.

8.7.2 Feature Locations by Right-Angle Offsets

Many traditional topographic surveys, excluding mapping surveys, but including most preengineering surveys, used the right-angle offset technique to locate detail. This technique not only provides the location of plan detail but also provides location for area elevations taken by cross sections.

Plan detail is located by measuring the distance perpendicularly from the baseline to the object and, in addition, measuring along the baseline to the point of perpendicularity (see Figure 1-11). The baseline is laid out in the field with stakes (nails in pavement) placed at appropriate intervals, usually 100 ft or 20/30 m. A sketch is entered in the field book before the measuring commences. If the terrain is smooth, a tape can be laid on the ground between the station marks. This will permit the surveyor to move along the tape (toward the forward station), noting and booking the stations of the already-sketched detail on both sides of the baseline. The right angle for each location tie can be established using a pentaprism (see Figure 8-11), or a right angle can be approximately established in the following manner (swung-arm technique): The surveyor stands on the baseline facing the detail to be tied in and then points one arm down the baseline in one direction and then the other arm down the baseline in the opposite direction; after checking both arms (pointed index fingers) for proper alignment, the surveyor closes his or her eyes while swinging his or her arms together in

FIGURE 8-11 Double right-angle prism
(Courtesy of Keuffel & Esser Company, Morristown, NJ)

front, pointing (presumably) at the detail. If the surveyor is not pointing at the detail, the surveyor moves slightly along the baseline and repeats the procedure until the detail has been correctly sighted in. The station is then read off the tape and booked in the field notes. This approximate method is used a great deal in manual route surveys and municipal surveys. This technique provides good results over short offset distances (50 ft or 15 m). For longer offset distances or for very important detail, a pentaprism or even a theodolite can be used to determine the stationing.

Once all the stations have been booked for the interval (100 ft or 20–30 m), it only remains to measure the offsets left and right of the baseline. If the tape has been left lying on the ground during the determination of the stations, it is usually left in place to mark the baseline while the offsets are measured with another tape.

Figure 8-12(a) illustrates topographic field notes that have been booked using a single baseline and in Figure 8-12(b) using a split baseline. In Figure 8-12(a) the offset distances are shown on the dimension lines and the baseline stations are shown opposite the dimension line or as close as possible to the actual tie point on the baseline.

In Figure 8-12(b) the baseline has been "split"; that is, two lines are drawn representing the baseline, leaving a space of zero dimension between them for the inclusion of the stations. The split-baseline technique is particularly valuable in densely detailed areas where single-baseline notes would tend to become crowded and difficult to decipher. The earliest topographic surveyors in North America used the split-baseline method of note keeping (see Figure 14-11).

8.7.3 Feature Locations by Polar (Radial) measurement

Planimetric and topographic measurements can be efficiently captured using total stations or theodolites with attached EDMs. In this technique, survey control is first established on the survey site, for example, by using a verified traverse (open or closed), or a GPS station. The total station or

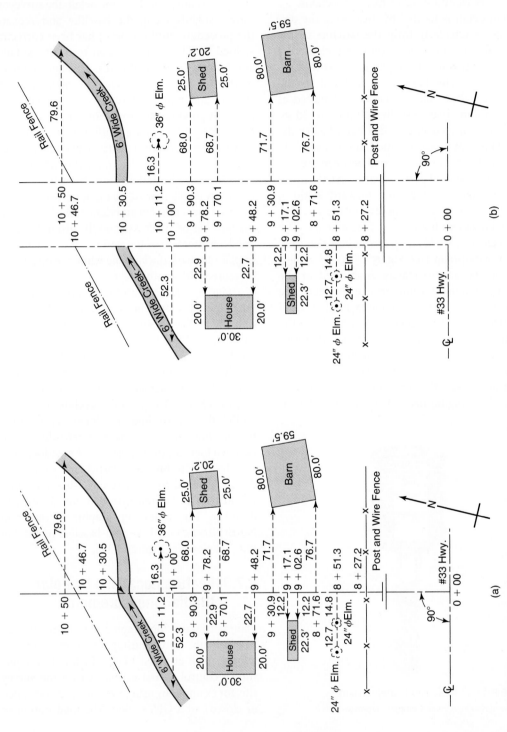

FIGURE 8-12 Topographic field notes. (a) Single baseline. (b) Split baseline

theodolite/EDM is set on a control station with a reference backsight to another control point; the horizontal angle can be set to zero or any azimuth value. Once the total station or theodolite/EDM has been field-oriented in this manner, the survey can commence with HCR (horizontal circle right), with vertical and horizontal distances being recorded in the data collector (onboard or attached) or in the field book (for manual surveys). Elevations are determined automatically (total stations) or later by adding/subtracting the vertical distances to the elevation of the instrument station; if the height of the prism (HR) was not made to equal the height of the instrument (hi), then (+hi − HR) must be factored into the computations for elevations (see Section 2.13).

The party chief usually prepares a sketch of the area to be surveyed and places the shot number directly on the sketch as the captured data are being field recorded. Complete field notes are invaluable when questions later arise concerning wrongly numbered or wrongly identified field data. See Appendix G for sample field notes for this type of survey.

8.7.4 Control Surveys

The horizontal control for many surveys consists of appropriate ties to the legal framework that is in the vicinity of the survey. For example, in municipal surveying, the horizontal control is usually given by the property frontage markers, from which a road centerline (℄) can be established. A significant portion of the surveyor's time is taken up in looking for, identifying, and verifying these property markers. In most cases, urban property markers are metallic in nature, for example, iron bars or iron pipes; sometimes concrete monuments are used to mark key points for government property (e.g., highways); and in wilderness areas, property and or legal markers are often 4 × 4 in cedar posts, with or without rock cairns. Several devices can be used to locate buried metallic markers (see one example in Figure 8-13), but the most useful tool in this regard is the old-fashioned shovel. Some jurisdictions now publish the coordinates of selected

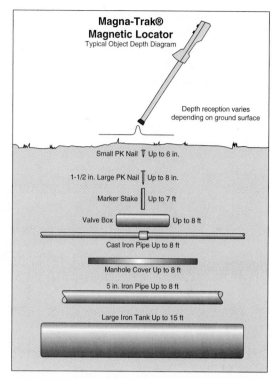

FIGURE 8-13 Magnetic Locator, used to find buried surveying markers and other metallic objects. The presence of buried metallic objects is indicated both on a digital bar graph display and by increasing/decreasing audio tone
(Courtesy of CST/Berger, IL)

property markers. In that case, a handheld GPS receiver is useful in guiding the surveyor to the theoretical location of the marker.

8.8 CROSS SECTIONS AND PROFILES

8.8.1 Cross Sections

Cross sections are a series of elevations taken at *right angles to a baseline* at specific stations, whereas *profiles* are a series of elevations taken *along a baseline* at some specified repetitive station interval. The elevations thus determined can be plotted on maps and plans either as spot elevations or as contours, or they can be plotted as end areas for construction quantity estimating. As in offset ties, the baseline interval is usually 100 ft (20–30 m), although in rapidly changing terrain the interval is usually smaller (e.g., 50 ft or 15–20 m). In addition to the regular intervals, cross

sections are also taken at each abrupt change in the terrain (top, bottom of slopes, etc.).

Figure 8-14 illustrates how the rod readings are used to define the ground surface. In Figure 8-14(a) the uniform slope permits a minimum (℄ and

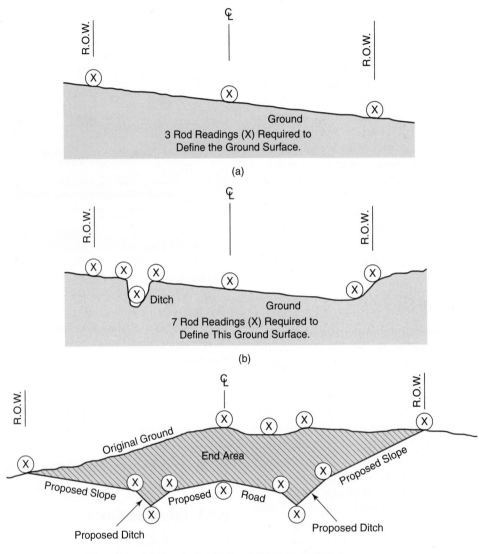

FIGURE 8-14 Cross sections used to define ground surface. (a) Uniform slope. (b) Varied slope. (c) Ground surface before and after construction

both limits of the survey) number of rod readings. In Figure 8-14(b) the varied slope requires several more (than the minimum) rod readings to adequately define the ground surface. Figure 8-14(c) illustrates how cross sections are taken before and after construction.

The profile consists of a series of elevations along the baseline. If cross sections have been taken, the necessary data for plotting a baseline profile will also have been taken. If cross sections are not planned for an area for which a profile is required, the profile elevations can be determined simply by taking rod readings along the baseline at regular station intervals and at all points where the ground slope changes (see Figure 2-17).

Typical field notes for profile leveling are shown in Figure 2-18. Cross sections are booked in two different formats. Figure 2-21 shows cross sections booked in standard level note format. All the rod readings for one station (that can be "seen" from the HI) are booked together. In Figure 2-22, the same data are entered in a format once popular with highway agencies. The latter format is more compact and thus takes less space in the field book; and space allows for a detailed description of each rod reading, an important consideration for municipal surveyors.

The assumption in this and the previous section is that the data are being collected using conventional offset ties and cross-section methods (i.e., tapes for the tie-ins and a level and rod for the cross sections). In municipal work, a crew of four surveyors can be used efficiently. While the party chief makes sketches for the detail tie-ins, the instrument operator (rod readings and bookings) and two surveyors (one on the tape, the other on the rod) could perform the cross sections. In theory, using the motorized total station described in Section 5.17.2, it is possible that this crew of four surveyors could be replaced by just one surveyor, a compelling argument for instrument modernization.

Chapter 5 described total stations. These instruments, often used in polar techniques, can also easily be used in the rectangular format

dictated by cross sections; they can measure distances and differences in elevation much more quickly and efficiently than pre-electronic techniques [nonrobotic total stations require only two surveyors in the crew, unless it is decided that a third crew member (second prism holder) can be utilized to improve crew efficiency]. Most of these instruments record the distance and elevation data automatically for future computer processing, but the surveyor must enter the attribute data for each shot (if the attribute is the same as for the previous shot, it can be entered more easily using a repeat *default* function).

8.8.2 Profiles Derived from Contours

Profiles establish ground elevations along a defined route [see line *AB* on Figures 8-8(c) and line 71-41-42 on Figure 8-15]. Profile data can be directly surveyed, as when a road ℄ has rod readings taken at specific intervals (e.g., 100 ft), as well as at all significant changes in slope; or profile data can be taken from contour drawings. For example, referring to Figure 8-8, the profile of line 71-41-42 can be determined as follows: The original scale was given as 1 in. = 150 ft, and this is used to scale the distance 71 to 41 (198 ft) and 41 to 42 (198 ft). The distances to the various contour crossings are as follows:

71 to 745 = 16′	71 to 775 = 188′
71 to 750 = 63′	41 to 780 = 51′
71 to 755 = 115′	41 to 785 = 118′
71 to 760 = 138′	41 to 790 = 165′
71 to 765 = 155′	41 to 795 = 225′
71 to 770 = 172′	

Figure 8-15 shows line 71-41-42 and the contour crossings replotted at 1 in. = 100 ft (for clarity). Directly below, the elevations are plotted vertically at $1'' = 30'$. The horizontal scale for both plan and profile is always identical; the vertical scale of the profile is usually exaggerated to display the terrain surface properly. In this example, a convenient datum line (740) is suitably placed on the available paper so that the profile is centered.

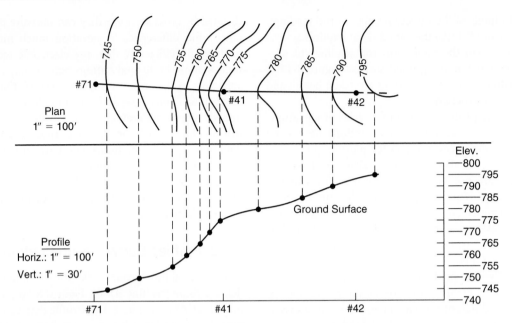

FIGURE 8-15 Plan and profile [from Figure 8.8(b)]

8.9 CROSS SECTIONS, END AREAS, AND VOLUMES

Cross sections establish ground elevations at right angles to a proposed route. Cross sections can be developed from a contour plan as were profiles in Section 8.8.2, although it is common to have cross sections taken by field surveys. See Chapter 2. Cross sections are useful in determining quantities of cut and fill in construction design. If the original ground cross section is plotted, and then the proposed design cross section is also plotted, the end area at that particular station can be computed. In Figure 8-16(a), the proposed road at station 7 + 00 is at an elevation below existing ground. This indicates a *cut* situation (i.e., the contractor will cut out that amount of soil shown between the proposed section and the original section). In Figure 8-16(b) the proposed road elevation at station 3 + 00 is above the existing ground, indicating a *fill* situation (i.e., the contractor will bring in or fill in that amount of soil shown). Figure 8-16(c) shows a transition section between cut and fill sections.

When the end areas of cut or fill have been computed for adjacent stations, the volume of cut or fill between those stations can be computed by simply averaging the end areas and multiplying the average end area by the distance between the end-area stations; Figure 8-17 illustrates this concept.

$$V = \frac{(A_1 + A_2)L}{2} \qquad (8\text{-}1)$$

Equation 8-1 gives the general case for volume computation, where A_1 and A_2 are the end areas of two adjacent stations and L is the distance (feet or meters) between the stations. The answer in cubic feet is divided by 27 to give the answer in cubic yards; when metric units are used, the answer is left in cubic meters.

The average end-area method of computing volumes is entirely valid only when the area of the midsection is, in fact, the average of the two end areas. This is seldom the case in actual earthwork computations; however, the error in volume resulting from this assumption is insignificant for the usual earthwork quantities of cut and fill. For special earthwork quantities (e.g., expensive structures excavation) or for higher-priced materials

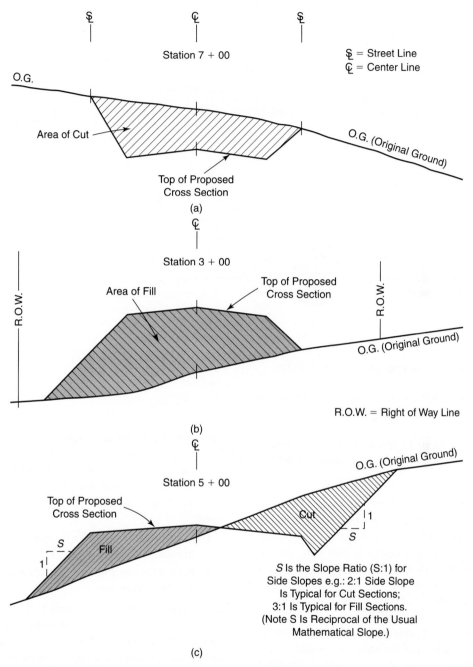

FIGURE 8-16 End areas. (a) Cut section. (b) Fill section. (c) Transition section (i.e., both cut and fill)

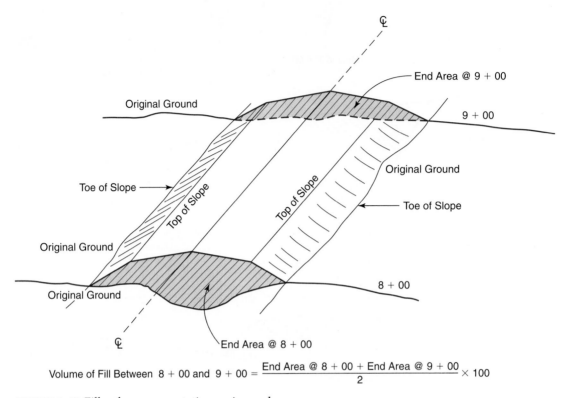

$$\text{Volume of Fill Between } 8 + 00 \text{ and } 9 + 00 = \frac{\text{End Area @ } 8 + 00 + \text{End Area @ } 9 + 00}{2} \times 100$$

FIGURE 8-17 Fill-volume computations using end areas

(e.g., concrete in place), a more precise method of volume computation, the prismoidal formula, must be used.

Example 8-1: Volume by End Areas, Metric Units
Figure 8-18(a) shows a pavement cross section for a proposed four-lane curbed road. As shown, the total pavement depth is 605 mm, the total width is 16.30 m, the subgrade is sloping to the side at 2 percent, and the top of the curb is 20 mm below the elevation of the ℄. The proposed cross section is shown in Figure 8-19(a) along with the existing ground cross section at station 0 + 340. All subgrade elevations were derived from the proposed cross section, together with the six design elevation of 221.43. The desired end area is the area shown below the original ground plot and above the subgrade plot.

Solution: At this point, an elevation datum line is arbitrarily chosen (220.00). The datum line

chosen can be any elevation value rounded to the closest foot, meter, or 5-ft value that is lower than the lowest elevation in the plotted cross section. Figure 8-19(a) illustrates that end-area computations involve the computation of two areas:

1. Area between the ground cross section and the datum line
2. Area between the subgrade cross section and the datum line

The desired end area (cut) is area 1 minus area 2. For fill situations, the desired end area would be area 2 minus area 1. The end-area computation can be determined as shown in Table 8-3. Each entry computes the area of a trapezoidal portion of the cross section by multiplying the average height (above the datum) by the width. This calculation is performed for both cross sections. Assuming that the end area at 0 + 300 has been computed to be 18.05 m², the volume of cut

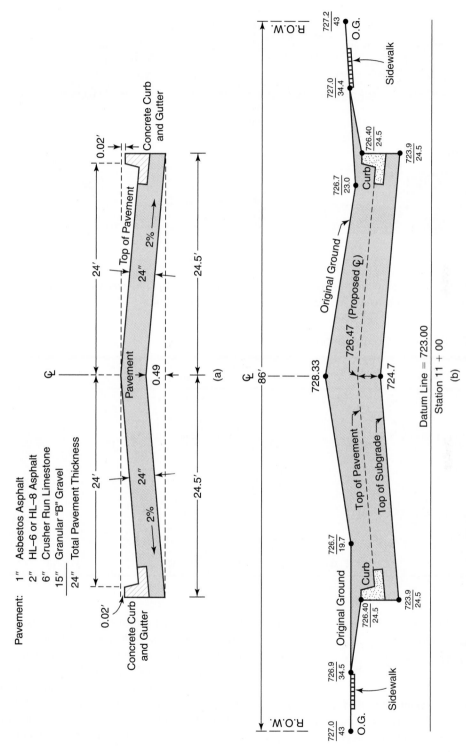

FIGURE 8-18 End-area computation, general (foot units). (a) Typical arterial street cross section—48 ft of roadway centered on an 86 ft row. (b) Survey and design data plotted to show both the original ground and the design cross sections

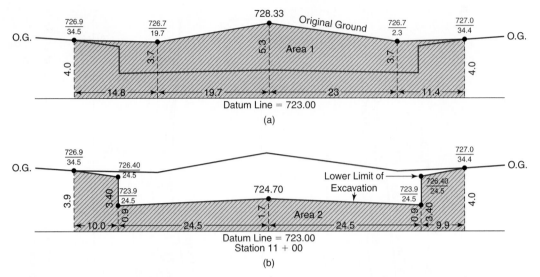

FIGURE 8-19 End-area computations for cut areas (foot units). (a) Area between ground cross section and datum line. (b) Area between subgrade cross section and datum line

Table 8-3   End-Area Computations

Station	Plus	Area 1	Minus	Area 2
0 + 340	$\dfrac{1.55 + 1.50}{2} \times 4.5 =$	6.86	$\dfrac{1.55 + 1.41}{2} \times 2.35 =$	3.48
	$\dfrac{1.50 + 2.00}{2} \times 6.0 =$	10.50	$\dfrac{0.66 + 0.82}{2} \times 8.15 =$	6.03
	$\dfrac{2.00 + 1.50}{2} \times 7.0 =$	12.25	$\dfrac{0.82 + 0.66}{2} \times 8.15 =$	6.03
	$\dfrac{1.50 + 1.59}{2} \times 3.5 =$	5.41	$\dfrac{1.14 + 1.59}{2} \times 2.35 =$	3.53
	Check: 21 m	35.02 m²	Check: 21 m	19.07 m²

End area = 35.02 − 1,907 = 15.95 m²

between 0 + 300 and 0 + 340 can now be computed using Equation 8-1:

$$V = \frac{(18.05 + 15.95)}{2} \times 40 = 680\,m^2$$

Example 8-2: *Volume by End Areas, Foot Units* Figure 8-18(b) shows a pavement cross section for a proposed four-lane curbed road. As

shown, the total pavement depth is 2.0 ft (24″), the total width is 49 ft, the subgrade is sloping to the side at 2 percent, and the top of the curb is 0.02 ft below the elevation of the 6. The proposed cross section is shown in Figure 8-19(b) along with the existing ground cross section at station 11 + 00. You can see that all subgrade elevations were derived from the proposed cross section, together with the six design elevation of 726.47. The desired end area is the area shown below the original ground plot and above the subgrade plot.

Solution: At this point, an elevation datum line is arbitrarily chosen (723.00). The datum line chosen can be any elevation value rounded to the closest foot, meter, or 5-ft value that is lower than the lowest elevation in the plotted cross section. Figure 8-19(b) illustrates that end-area computations involve the computation of two areas:

1. Area between the ground cross section and the datum line

2. Area between the subgrade cross section and the datum line

The desired end area (cut) is area 1 minus area 2. For fill situations, the desired end area would be area 2 minus area 1. The end-area computation can be determined as shown in Table 8-4. Assuming that the end area at 10 + 00 has been computed to be 200 ft^2, the volume of cut between 10 + 00 and 11 + 00 can now be computed using Equation 8-1:

$$V = \frac{(200 + 181)}{2} \times 100 = 19{,}050\,\text{ft}^3 \div 27$$

$$= 705.6\,\text{yd}^3$$

8.10 PRISMOIDAL FORMULA

If values more precise than those given by end-area volumes are required, or if the sections change from cut to fill to form linearly a prism shape, the prismoidal formula can be used. A **prismoid** is a solid with parallel ends joined by plane or warped surfaces. The prismoidal formula is

$$V = L\,\frac{(A_1 + 4A_m + A_2)}{6}\,\text{ft}^3 \text{or m}^3 \qquad (8\text{-}2)$$

where A_1 and A_2 are the two end areas, A_m is the area of a section midway between A_1 and A_2, and L is the distance from A_1 to A_2. A_m is not the average of A_1 and A_2, but is derived from cross-section dimensions that are the average of corresponding dimensions required for A_1 and A_2 computations.

The formula in Equation 8-2 is also used for other geometric solids (e.g., truncated prisms, cylinders, and cones). To justify its use, the surveyor must refine the field measurements to reflect the increase in precision sought. A typical application of the prismoidal formula is the computation of in-place volumes of concrete. The cost of a cubic yard or meter of concrete as opposed to that of a cubic yard or meter of earth cut or fill is sufficient reason to require increased precision.

Table 8-4 End-Area Computations

Station	Plus	Area 1	Minus	Area 2
11 + 00	$\dfrac{4.0 + 3.7}{2} \times 14.8 =$	57	$\dfrac{3.9 + 0.9}{2} \times 10.0 =$	24
	$\dfrac{3.7 + 5.33}{2} \times 19.7 =$	89	$\dfrac{0.9 + 1.7}{2} \times 24.5 =$	32
	$\dfrac{5.33 + 1.7}{2} \times 23.0 =$	104	$\dfrac{1.7 + 0.9}{2} \times 24.5 =$	32
	$\dfrac{3.7 + 4.0}{2} \times \dfrac{11.1}{68.9} =$	43	$\dfrac{0.9 + 4.0}{2} \times \dfrac{9.9}{68.9} =$	24
	Check	293	Check	112

End Area = 293 − 112 = 181 ft^2

8.11 CONSTRUCTION VOLUMES

In highway construction, for economic reasons, the designers try to balance cut and fill volumes optimally. Cut and fill cannot be balanced precisely because of geometric and esthetic design considerations, and because of the unpredictable effects of shrinkage and swell. **Shrinkage** occurs when a cubic yard (meter) is excavated and then placed while being compacted. The same material that formerly occupied 1 yd^3 (m^3) volume now occupies a smaller volume. Shrinkage reflects an increase in density of the material and is obviously greater for silts, clays, and loams than it is for granular materials such as sand and gravel. **Swell** is a term used to describe the change in volume

of shattered (blasted) rock. Obviously, 1 yd^3 (m^3) of solid rock will expand significantly when shattered. Swell is usually in the range of 15–20 percent, whereas shrinkage is in the range of 10–15 percent, although values as high as 40 percent are possible with organic material in wet areas.

To keep track of cumulative cuts and fills as the profile design proceeds, the cumulative cuts (plus) and fills (minus) are shown graphically in a mass diagram. The total cut-minus-fill is plotted at each station directly below the profile plot. Note the cumulative mass starts to decrease when a fill section is reached and starts to increase when a cut section is reached. The mass diagram is an excellent method of determining waste or borrow volumes, and can be adapted to show haul (transportation) considerations (Figure 8-20).

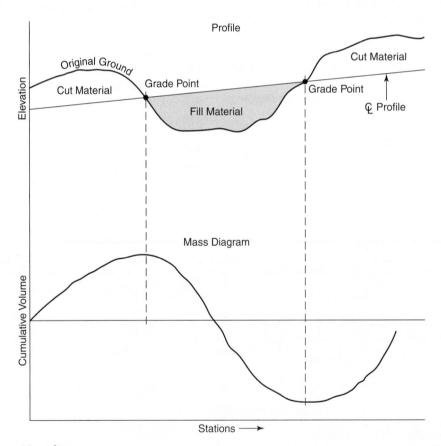

FIGURE 8-20 Mass diagram

Large fills require *borrow* material, usually taken from a nearby borrow pit. Borrow pit leveling procedures are described in Section 2.9 and Figure 2-23. The borrow pit in Figure 2-23 was laid out on a 50-ft grid. The volume of a grid square is the average height $(a + b + c + d)/4$ times the area of the base (50^2). The partial grid volumes (along the perimeter of the borrow pit) can be computed by forcing the perimeter volumes into regular geometric shapes (wedge shapes or quarter-cones).

When high precision is less important, volumes can be determined by analysis of contour plans; the smaller the contour interval, the more precise the result. The areas enclosed by a contour line can be taken off by planimeter; electronic planimeters are very useful for this purpose. Modern earthwork software programs will also compute quantities from DTMs before and after construction. The volume of material between elevations h_1 and h_2 can be estimated by

$$V = I \frac{(C_1 + C_2)}{2} \qquad (8\text{-}3)$$

where V is the volume (ft^3 or m^3) of earth or water, C_1 and C_2 are areas of contours at elevations h_1 and h_2, and $I = h_2 - h_1$ is the contour interval. The prismoidal formula can be used if m is an intervening contour (C_m) between C_1 and C_2.

This method is well suited for many water-storage volume computations.

Perhaps the most popular current volume-computation technique involves the use of computers with an earthwork software program, of which there are a large number of available. The computer programmer uses techniques similar to those described here, but the surveyor's duties may end with proper data entry into the computer.

8.12 AREA COMPUTATIONS

Areas enclosed by closed traverses can be computed using the coordinate method (see Section 6.13). Figure 8-21 illustrates two additional area computation techniques.

8.12.1 Trapezoidal Technique

The area in Figure 8-21 was measured using a fiberglass tape for the offset distances. A common interval of 15 ft was chosen to delineate the riverbank suitably. Had the riverbank been more uniform, a larger interval could have been used. The trapezoidal technique assumes that the lines joining the ends of each offset line are straight lines (the smaller the common interval, the more valid this assumption).

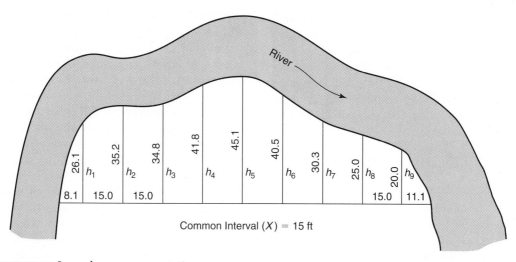

FIGURE 8-21 Irregular area computation

The end sections can be treated as triangles:

$$A_0 = \frac{8.1 \times 26.1}{2} = 106\,\text{ft}^2$$

and

$$A_9 = \frac{11.1 \times 20.0}{2} = 111\,\text{ft}^2$$
$$= 217\,\text{ft}^2 \text{ Subtotal}$$

The remaining areas can be treated as trapezoids. The trapezoidal rule is stated as follows:

$$\text{Area} = X\left(\frac{h_1 + h_n + h_2 + \ldots h_{n-1}}{2}\right) \quad (8\text{-}4)$$

where X = common interval between the offset lines

h = offset measurement

n = number of offset measurements

$$A_{1-8} = 15\left(\frac{26.1 + 20.0}{2} + 35.2 + 34.8 + 41.8\right.$$
$$\left. + 45.1 + 40.5 + 30.3 + 25.0\right)$$
$$= 4{,}136\,\text{ft}^2$$

$$\text{Total area} = 4{,}136 + 217 = 4{,}353\,\text{ft}^2$$

8.12.2 Simpson's One-Third Rule

This technique gives more precise results than the trapezoidal technique and is used where one boundary is irregular, like the one shown in Figure 8-21. The rule assumes that an odd number of offsets n is involved and that the lines joining the ends of three successive offset lines are parabolic in configuration. Simpson's one-third rule is stated as follows:

$$A = \frac{\text{interval}}{3}(h_1 + h_n + 2\Sigma h_{\text{odd}} + 4\Sigma h_{\text{even}})$$
$$(8\text{-}5)$$

That is, one-third of the common interval times the sum of the first and last offsets $(h_1 + h_n)$ plus twice the sum of the other odd numbered offsets (Σh_{odd}) plus four times the sum of the even-numbered offsets (Σh_{even}).

From Figure 8-21:

$$A_{1-8} = \frac{15}{3}\,[26.1 + 20.0 + 2(34.8 + 45.1 + 30.3)$$
$$+ 4(35.2 + 41.8 + 40.5 + 25.0)]$$
$$= 4{,}183\,\text{ft}^2$$

$$\text{Total area} = 4{,}183 + 217\,(\text{from preceding}$$
$$\textit{example})$$
$$= 4{,}400\,\text{ft}^2$$

If a problem has an even number of offsets, the area between the odd number of offsets is determined by Simpson's one-third rule, with the remaining area determined with the trapezoidal technique. The discrepancy between the trapezoidal technique and Simpson's one-third rule here is 47 ft^2 in a total of 4,400 ft^2 (about 1 percent in this case).

8.13 AREA BY GRAPHICAL ANALYSIS

We have seen that areas can be determined by using coordinates (Chapter 6) and, less precisely, by using the somewhat approximate methods illustrated by the trapezoidal rule and Simpson's one-third rule. Areas can also be determined by analyzing plotted data on plans and maps. For example, if a transparent sheet is marked off in grid squares to some known scale, an area outlined on a map can be determined by placing the squared paper over (sometimes under) the map and counting the number of squares and partial squares within the boundary limits shown on the map. The smaller the squares, the more precise will be the result.

Another method of graphic analysis involves the use of a planimeter (see Figures 8-22, 8-23, and 8-24). A **planimeter** consists of a graduated measuring drum attached to an adjustable or fixed tracing arm (which itself is attached to a pole arm), one end of which is anchored to the working surface by a needle. The graduated measuring drum gives partial revolution readings, while a disc keeps count of the number of full revolutions. Areas are determined by placing the pole-arm

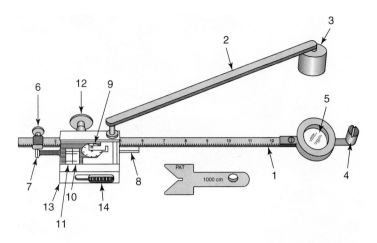

Part 1 Tracer Arm
Part 2 Pole Arm
Part 3 Pole Weight
Part 4 Hand Grip
Part 5 Tracing Magnifier
Part 6 Clamp Screw
Part 7 Fine Movement Screw

Part 8 Tracer Arm Vernier
Part 9 Revolution Recording Dial
Part 10 Measuring Wheel
Part 11 Measuring Wheel Vernier
Part 12 Idler Wheel
Part 13 Carriage
Part 14 Zero Setting Slide Bar

FIGURE 8-22 Polar planimeter

FIGURE 8-23 Area takeoff by polar planimeter

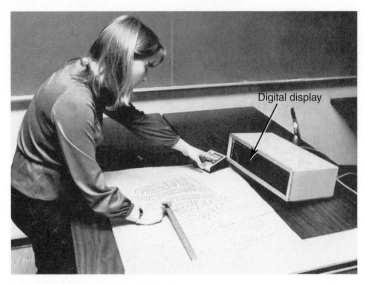

Digital display

FIGURE 8-24 Area takeoff by an electronic planimeter

needle in a convenient location, setting the measuring drum and revolution counter to zero (some planimeters require recording an initial reading), and then tracing (using the tracing pin) the outline of the area to be measured. As the tracing proceeds, the drum (which is also in contact with the working surface) revolves, thus measuring a value that is proportional to the area being measured.

Some planimeters measure directly in square inches, while others can be set to map scales. When in doubt, or as a check on planimeter operation, the surveyor can measure out a scale figure [e.g., 4 in. (100 mm) square] and then measure the area (16 in.^{2}) with a planimeter so that the planimeter area can be compared with the actual area laid off by scale. If the planimeter gives a result in square inches, say, 51.2 in.^{2}, and the map is at a scale of 1 in. = 100 ft, the actual ground area portrayed by 51.2 in.^{2} would be 51.2 100^2 = 512,000 ft^2 = 11.8 acres.

The planimeter is normally used with the pole-arm anchor point outside the area to be traced. If it is necessary to locate the pole-arm anchor point inside the area to be measured, as in the case of a relatively large area, the area of the zero circle of the planimeter must be added to the planimeter readings. This constant is supplied by the manufacturer or can be deduced by simply measuring a large area twice, once with the anchor point outside the area and once with the anchor point inside the area.

Planimeters are particularly useful in measuring end areas (Section 8.9) used in volume computations. Planimeters are also used effectively in measuring watershed areas, as a check on various construction quantities (e.g., areas of sod, asphalt), and as a check on areas determined by coordinates. Electronic planimeters (see Figure 8-24) measure larger areas in less time than do traditional polar planimeters. Computer software is available for highways and other earthwork applications (e.g., cross sections) and for drainage basin areas. The planimeter shown in Figure 8-24 has a 36 × 30 in. working area capability, with a measuring resolution of 0.01 in (0.02 in.^{2} accuracy).

Computer programs are now used to compute and display profiles, cross sections, and areas. Once the coordinates (N, E, and elevation) of field points are in the database, the quantity determinations and engineering design can be readily accomplished. Digitizing tablets or tables can also be used to digitize (i.e., determine X and Y positions) the location of many features on maps, plans, and cross sections, thus permitting the computation of areas.

8.14 HYDROGRAPHIC SURVEYS

8.14.1 General Background

Hydrographic surveys are used to define shoreline and underwater features. Why do we need hydrographic surveys? Offshore engineering and the shipping industry have continued to expand. Drilling rigs, located up to 125 mi (200 km) offshore, search for resources, particularly oil and gas. Offshore islands are constructed (sometimes under severe weather conditions) of dredged material to support marine structures. Containerization has become an efficient and preferred method of cargo handling and harbor depths up to 80 ft (25 m) are required to accommodate larger ships and tankers. The demand for recreational transportation ranges from large pleasure cruise ships to small sailboats.

Hydrographic surveys are made to acquire and present data on rivers, oceans, lakes, bays, or harbors. In addition to harbor construction and offshore drilling, these surveys are carried out for one or more of the following activities:

- Determining the water depths and locations of rocks, sandbars, and wrecks for navigation channel openings and salvage operations.

- Dredging for harbor deepening, maintenance, mineral recovery, and navigation channel access.

- Evaluating areas of sedimentation and erosion for coastline protection and offshore structures.

- Measuring areas subject to siltation or scouring to determine the effects on water quality and existing structures, such as bridge abutments and storm sewage outfalls.

- Providing recreational facilities such as beaches and marinas.

- Determining site locations for submarine cables and underwater pipelines and intakes.

- Determining pollution sources.

- Evaluating the effects of corrosion, particularly in salt water.

- Determining the extent of wetland areas.

- Determining profiles and cross sections of river channels to assess the potential for hydroelectric-generation dam sites.

These sections describe the procedures required to obtain the necessary data for hydrographic surveys, hydrographic plans, and electronic charts.

8.14.2 Objectives of Hydrographic Mapping and Electronic Charting

The primary requirement involves showing the topographic configuration of the underwater features, both natural and built. The resulting product is therefore similar to a topographic map of land areas. See Figure F-1. However, the methods used to obtain the information for hydrographic surveys are vastly different.

In the horizontal plane, the position of a survey vessel must be fixed to the required accuracy. Determining the accurate location of the vessel is complicated by weather conditions, particularly wind, waves, and fog. The depth of the seabed below the survey vessel, known as a **sounding**, is subject to variations caused by wave and tidal action. After the original sounding has been corrected, the resulting depth is called a reduced sounding. See Figure 8-25.

The surveyor must also be aware that the costs and accuracies of hydrographic surveys are normally not comparable to land surveys. While the surveyor on land can see the features, the hydrographic surveyor cannot, except in shallow, clear waters. This may result in the omission of important features, such as rocks or wrecks, that vitally affect the proposed undertaking.

The objective of the hydrographic survey must be considered in light of these issues. For example, cost overruns may be justified to locate important underwater features through an increased number of soundings. Therefore, the proper planning of a hydrographic survey, as discussed in the following section, is critical for the final product to satisfy the project requirements at a cost that is acceptable.

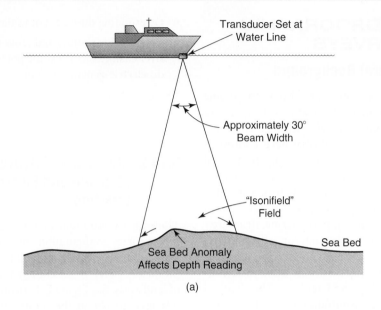

Transducer Set at
Water Line

Approximately 30°
Beam Width

"Isonifield"
Field

Sea Bed

Sea Bed Anomaly
Affects Depth Reading

(a)

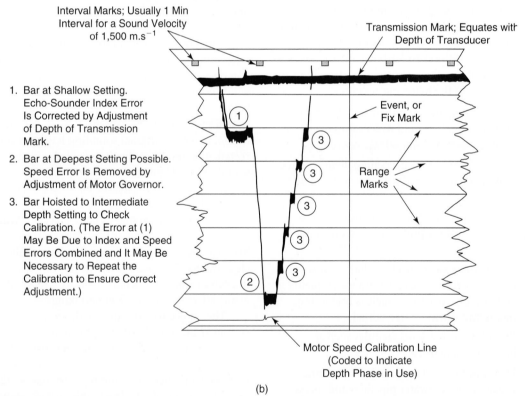

Interval Marks; Usually 1 Min
Interval for a Sound Velocity
of 1,500 m.s⁻¹

Transmission Mark; Equates with
Depth of Transducer

1. Bar at Shallow Setting.
Echo-Sounder Index Error
Is Corrected by Adjustment
of Depth of Transmission
Mark.

2. Bar at Deepest Setting Possible.
Speed Error Is Removed by
Adjustment of Motor Governor.

3. Bar Hoisted to Intermediate
Depth Setting to Check
Calibration. (The Error at (1)
May Be Due to Index and Speed
Errors Combined and It May Be
Necessary to Repeat the
Calibration to Ensure Correct
Adjustment.)

Event, or
Fix Mark

Range
Marks

Motor Speed Calibration Line
(Coded to Indicate
Depth Phase in Use)

(b)

FIGURE 8-25 Isonified area and bar check for echo sounder depths. Transducer and isonified area (a).
Typical bar check (b).

8.14.3 Planning

Careful planning and preparation are essential for any undertaking. Flexibility in the plan allows for delays due to weather and equipment breakdowns. In the preplanning stage, all available and applicable information should be examined, and copies should be obtained whenever possible. Hydrographic charts of the general area provide indications of the sounding depths and depth variations that are likely to be encountered. This information is useful for sounding equipment selection and the design of the sounding pattern.

Topographic maps indicate the configuration of the shoreline and natural or built features that may be used as control stations for horizontal position fixing. Aerial photographs viewed stereoscopically (in three dimensions) provide this information, as well as information about offshore bar and shore formations (see Chapter 12), in greater detail. Color aerial photographs and LiDAR images are particularly useful because of their high penetration capabilities through water. The most successful modern sounding techniques involve the use of satellite positioning coordinated with a sounding at the same instant in time.

Navigational directions and boating restrictions, including channel locations, normal and storm wave heights, prevailing wind directions, and areas of restricted vessel use, assist in sounding vessel selection and in identifying sources for obtaining permits in restricted areas. Tide tables are essential for the design of recording gauges in ocean work. If existing gauges are operational, their locations, frequency of water level recordings, base datum used, and the availability of the data to the surveyor should be known in advance. Horizontal control data from previous land and hydrographic surveys reduce the effort and cost of the survey measurements among shoreline control stations. Previous knowledge of existing coordinate control networks allows the survey data to be tied to the survey system, if required.

When the survey is to be controlled by a satellite positioning system, attention must be given to the availability of Coast Guard differential GPS (DGPS) radio beacons so that differential techniques can be utilized. Software that coordinates all data collection (soundings and positionings) and integrates with onboard electronics must be acquired. If necessary, a decision must be made about whether to buy or rent the equipment.

8.14.4 Survey Vessels

General considerations should include the following:

1. Overall purpose of the survey, particularly the need for geophysical survey equipment or additional survey requirements.

2. Weather conditions, such as wave heights.

3. Size of the survey team and whether team members are to live on the craft.

Specific conditions that always apply include the following:

1. Sufficient space for position fixing and plotting. The plotting board or electronic chart should be under cover and relatively free from engine vibration.

2. An all-around view for visual position-fixing techniques.

3. Sufficient electrical power at the required voltages for all equipment needs.

4. Compatibility of fuel capacity and storage for supplies within the range and operational requirements.

5. Stability and maneuverability at slow speeds (up to 6 km/h, or 4 knots).

6. Cruising speeds of at least 15 km/h (10 knots) to minimize time loss from the base to the survey area and to provide sufficient speed to return safely to port in the event of sudden storms.

8.14.5 Depth Measurements

The depth of the point below the water surface or sounding must be related to the desired datum or reference level. This normally involves corrections for seasonal water levels under nontidal conditions, and for tidal variations where tides are a factor.

Depths measurements can range from the manual techniques of weighted lines, graduated in meters, feet, or both—used only for projects involving a small number of soundings, and echo sounders which provide depth measurements by timing the interval between transmission and reception of an acoustic pulse, which travels to the bottom and back at a rate of approximately 1,500 m/s (5,000 ft/s). Separate transmissions are made at rates of up to six per second. The beamwidth emitted from the vessel is typically about 30°, as illustrated in Figure 8-25(a). The portion of the seabed within the beamwidth is termed the *isonified area*.

In depth measurement, the most significant point is directly below the transducer, which is the vibratory diaphragm that controls the frequency of transmission. However, the echo sounder records the earliest return from its transmission (that which has traveled the shortest distance). Within a beamwidth of 30°, this return may not be from a target directly beneath the transducer, but from seabed anomalies within the isonified area [see Figure 8-25(a)]. This will lead to anomalies in the soundings, which may be differentiated by an experienced observer. Highly reflective targets such as bare rock and shipwrecks, located near the edges of the beam, show as narrow, clearly defined bands on the readout, compared with thick, poorly defined bands over soft sediments, weedbeds, and the like. Constant monitoring of the transmission returns and notes of anomalies should be incorporated in the hydrographic survey.

Sound velocity in water is a function of temperature, salinity, and density. These factors vary daily and seasonally, and as a result of periodic occurrences such as heavy rainfalls and tidal streams. Attempts to correct the soundings for these variables are usually unsatisfactory and costly. As a result, it must be recognized that accuracies in acoustic measurements in seawater are not better than 1 part in 200.

Calibration of echo sounders is carried out either by comparison with direct measurements, using weighted lines, or by a bar check in depths less than 30 m, as illustrated in Figure 8-25(b). The latter involves setting a bar or disc horizontally beneath the transducer at various depths. The echo sounder recorder is adjusted to match the directly measured depths. If the sounder is not adjustable, the differences are recorded for regular depth intervals, and the resulting corrections are applied to each sounding. It is advisable to calibrate the echo sounder at the beginning and end of each day's use, particularly at 10 percent, 50 percent, and 80 percent of the maximum depths measured.

Recent improvements to depth measurements include the now widespread use of multibeam echo sounding in shallow waters. Because of the fast rate of data acquisition (900 depth points per second), manual editing is not feasible. This system requires an onboard computer (color monitor and plotter) for data logging, navigation, quality control, multibeam calibration, data editing, and plotting. The computer software can also integrate the data collected by the sensors on pitch, heave, and roll.

Weighted (hand) sounding lines are seldom used for depths over 30 m (100 ft). The lines may be small-linked steel chain, wire, cotton, hemp cord, or nylon rope. A weight, usually made of lead, is attached to one end. Markers are placed at intervals along the line for depth reading. Lines constructed of link chains are subject to wear through abrasion. Wire lines stretch moderately when suspended, depending on the size of the bottom weight. The weights vary from 2.3 kg (5 lb) to 32 kg (75 lb), although 4.5 kg (10 lb) is usually sufficient for moderate depths and low velocities. Cotton or hemp lines must be stretched before use and graduated when wet. They must be soaked in water for at least 1 hour before use to allow the rope to assume its working length. Nylon lines stretch appreciably and unpredictably, and they are not recommended for other than very approximate depth measurements. The hydrographic surveyor should calibrate the sounding lines against a steel survey tape regularly under the conditions most similar to actual usage, such as hemp lines after soaking.

8.14.6 Position-Fixing Techniques

The location of the survey vessel in the horizontal plane when a particular sounding has been measured is a fundamental requirement for the hydrographic survey. Directional control of the vessel along the sounding lines is an important factor for ensuring that the survey area is covered sufficiently to meet the specifications.

Before discussing each position-fixing technique, some generalities in this field should be recognized. Three overall methods of position-fixing techniques are described in this section: manual, electronic, and satellite positioning (GPS). The manual operations involve more basic equipment, such as theodolites and sextants, as well as larger field crews for taking and recording the large number of visual readings necessary. The electronic techniques involve more sophisticated equipment and correspondingly smaller field crews because many of the readings are recorded automatically. GPS techniques include differential GPS (DGPS) for submeter precision and real-time kinematic (RTK) for centimeter precision (useful in engineering works).

The factors governing the selection of the technique to be used relate primarily to the location of the site, the complexity of the site area, the volume of data to be collected, and the necessity of collecting similar data over the same area on a weekly, monthly, or seasonal basis. The electronic devices are rapidly becoming easier to operate, and they provide greater accuracies and are more easily available. Familiarity with the manual techniques is important for two main reasons. First, an understanding of these techniques is essential for comprehending the basic requirements for position fixing. Second, these techniques may still be used regularly for local projects that are limited in scope.

8.14.6.1 Intersection from Theodolite Stations on Shore

The position of the vessel can be determined by two simultaneous horizontal angles measured by total stations set up on shore stations (three shore stations provide for error analysis). The total stations are zeroed on any station that is part of the horizontal control network. As the boat proceeds along the sounding lines, the total station observers track the vessel's path, sighting a target on the boat mounted over the echo sounder. At the instant that a fix is required, the boat driver raises a prearranged signal or gives a radio signal, the sounding is recorded on the vessel, and each shore observer records the measured angle. Each fix is numbered consecutively in the field notes by both total station operators and the echo sounder reader in the vessel. Plotting of the fixes is undertaken after the data from the vessel and the shore station have been correlated.

8.14.6.2 Plotting of Data

The fix numbers and, if necessary, the recorded times should be compiled by the survey team after completion of a maximum of four range lines or 50 fixes. At this time, the farthest points offshore on each sounding line should be plotted to ensure that the spacing between the lines satisfies the specifications. The use of total stations permits the computation of sighted location coordinates.

8.14.6.3 Known Range Line and Single Angle

The range line markers control the course of the vessel, as shown in Figure 8-26. The fix is obtained by taking a horizontal angle from shore station (with a total station or theodolite), or from the vessel using a handheld total station. A three-person team is required: a boat driver using range poles for line, a total station angle observer, and an echo sounder reader. Two lines are normally fixed, and the targets are then moved as the work progresses.

This method is usually restricted to within 3 km (approximately 2 miles) of the shoreline, depending on the size of the targets and the size of the horizontal angles. The accuracy of the fix drops below normally acceptable standards when the angle from the vessel is below 10°. This should be considered in the location of the shore stations.

This method is used when it is necessary to repeat the soundings at the same points. The determinations of dredging quantities and of changes in the bottom due to scour of silt or sand are two common applications.

Fixed range lines are established on shore and are located to intersect as closely as possible

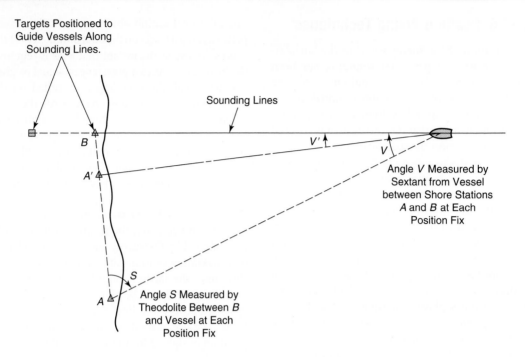

Targets Positioned to
Guide Vessels Along
Sounding Lines.

Sounding Lines

B

A'

Angle V Measured by
Sextant from Vessel
between Shore Stations
A and B at Each
Position Fix

V'

V

S

A

Angle S Measured by
Theodolite Between B
and Vessel at Each
Position Fix

Note: If Shore Station Located at A' Instead of A, Angle V' Is Less than 10°
Thus Reducing Accuracy of Fix Below Acceptable Survey Standards.

Only One of the Two Angles Shown Is Required for the Position Fix, That Is
Either S or V.

FIGURE 8-26 Range line and single angle

to right angles, as shown in Figures 8-27 and 8-28. The shore stations are permanently marked, usually by iron bars driven into the ground, mortared stone cairns, or painted crosses on bedrock. Targets are erected over the shore stations during the survey and are stored for reuse.

The boat proceeds to the intersection and takes the sounding, as required. A common difficulty is keeping the vessel stationary long enough to obtain an accurate sounding at the intersection point, particularly under strong wind conditions. It is therefore advisable to use a boat capable of good maneuverability at low speeds. The point of intersection should be approached into the direction of the wind.

Poor weather conditions must be anticipated using this method because the depth

measurements are taken at preselected intervals, monthly, weekly, and so on. Since the surveyor cannot wait for more suitable weather conditions and the soundings must be compared with previous readings, the following procedure is useful:

1. Swells caused by wave action can cause the sounding referred to the mean water level or mean low-water level to be in error by as much as half the wave height. This error can be minimized by recording at least three soundings at both the wave crest and the wave trough, and averaging the figures. The vessel may have to re-approach the intersection point each time to avoid rapid drifting.

2. Strong winds blowing in one direction for periods of over 24 hours cause the mean water level

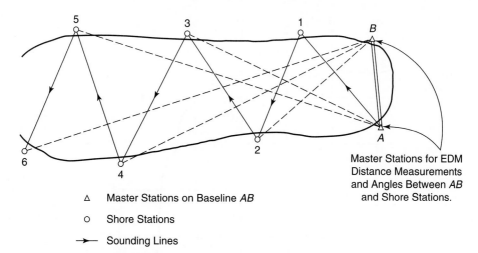

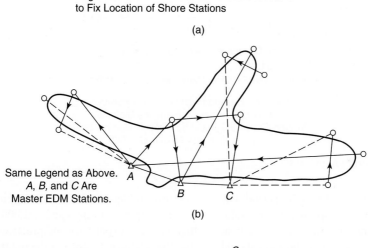

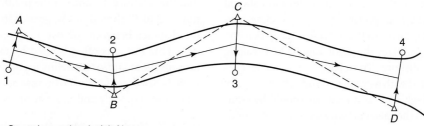

FIGURE 8-27 Layout for constant vessel velocity soundings. (a) Inland lake with regular shoreline. (b) Inland lake with irregular shoreline. (c) River

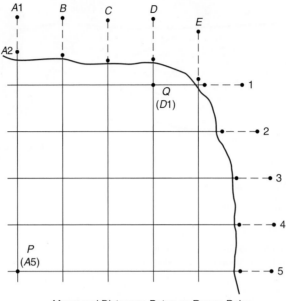

— —   Measured Distances Between Range Poles

——   Sounding Lines

•   Range Poles Set Over Permanent Shore Stations

Position of *Q* Located with Greater Accuracy than *P*
Due to Proximity of Shoreline Stations.
Position Fixes Designated by Letter + Number System,
For Example *P* Is A5 and *Q* Is "*D1*."

FIGURE 8-28 Intersecting range lines

to rise or lower during and after this period. The resulting error in all soundings taken under these conditions must be corrected. Permanent marking of the mean datum elevation should be established, preferably on a shoreline having a slope toward the water of over 25 percent. This minimizes the error caused by breaking waves offshore. The change in water level is measured with a graduated rod. The average of the trough and crest readings should be compared to the mean datum elevation. All soundings are corrected accordingly.

3. Tidal variations must be accounted for in addition to the preceding considerations.

The precision of this method depends primarily on the success of locating the intersection points. This, in turn, depends on the distance between the two range poles at each shore station,

as shown in Figure 8-28. The desired level of accuracy for surveys of this type is 1:1,000. To achieve this accuracy, experience and practice have shown that the offshore distance to the point of intersection (*A2* to *P*) should not be greater than 10 times the distance between the two range poles (*A1* to *A2*). This factor is often limited by existing land uses on shore and/or shoreline topography. When GPS is used to locate positions, soundings are taken more quickly and more accurately.

8.14.7 Hydrographic Surveying and the Global Positioning System (GPS)

As noted in Chapter 7, this remarkable positioning development is revolutionizing the way we determine our geographical location. Its impact on marine positioning and navigation will probably be even greater than on land-based surveys. Marine positioning involves few obstructions to the reception of the satellite signals, and the electronic nature of the technology permits simultaneous capture of position, headings, and sounding data.

Figure 8-29 shows a small craft using differential GPS techniques to record river-sounding positions. The same equipment used with newly developed handheld total stations (see Figure 8-30) permits simultaneous collection of positioning, sounding, and shoreline details. Various commercial data collectors can be used for this multitechnology data capture. A notebook computer with three serial ports would be particularly suitable. Onboard applications software can be expanded to include layout capabilities for marine construction, thus giving the user a very powerful surveying capability.

As we noted in Chapter 7, the U.S. Coast Guard has created the DGPS, which enables coastal, Great Lakes, and major river radio beacons to transmit differential GPS code corrections continuously to an unlimited number of working surveyors and sailors within ranges of 100–400 km. Canada has a similar service for its coastal regions. Surveyors working onboard ships can use the radio beacon transmissions [in the worldwide Radio Technical Commission for Maritime (RTCM) standard] with onboard GPS receivers to locate themselves to within submeter accuracy, and to provide a differential speed accuracy of about 0.1 knot. The RTCM Services, Special Committee 104 (RTCM SC-104), which created standards for DGPS transmission, has also created RTK messages; because their length is almost double that available from equipment manufacturers, the current trend is to utilize the manufacturers' standards for RTK radio transmission. Trimble Inc. has released a compact measurement record (CMR) for general

FIGURE 8-29 Leica GPS antenna and receiver on a small craft, providing positioning data for echo-soundings in a differential kinematic mode. The data collector here is a notebook computer
(Courtesy of Leica Geosystems Inc., Norcross, GA)

FIGURE 8-30 Prosurvey 1000 handheld survey laser: used (without prisms) to record distances and angles to survey stations. Shown here being used to determine river width in a hydrographic survey
(Courtesy of Laser Atlanta, Norcross, GA)

use to encourage more efficient base station data transfer.

Figure 8-31 shows an L1-frequency C/A code, six-channel GPS receiver. Other receivers have as many as 12 channels, and some dual-frequency receivers can also process carrier observations, which allow for RTK surveys (1 cm accuracy in position and 0.5 m/s accuracy in differential speed) when they are referenced to shore-based receivers radio-transmitting differential corrections. Current RTK radio transmitters have a range of only 10–20 km. Most receivers require five or more satellites to accomplish initialization quickly. After initialization, the survey can continue with just four satellites. Postprocessed solutions are also available by utilizing the differential data transmitted by the dual-frequency continuously operating reference station (CORS) stations (see Section 7.8).

For areas outside the DGPS range, manufacturers have developed base stations (single- and dual-frequency) capable of tracking 12 channels of continuous code and carrier data at a rate of 0.5 s, with accuracies of 10 cm for code and 0.005 m for carrier observations. These base stations can transmit differential corrections using RTCM standards (for code corrections) to onboard receivers. As noted earlier, when transmitting carrier data, radios have a range of 10 m to 20 km.

Onboard considerations for hydrographic surveying include the measurement from the mast-mounted GPS antenna to the GPS receiver on the bridge, and from there to the sounding transducer on the hull. If the distance from the antenna to the receiver is greater than 30 m, an amplifier may be needed to boost the signals. Another consideration is the applications software, which should be capable of performing all needed processing and the linking of all peripheral electronic equipment. For example, Trimble Inc. HYDRO software can link positioning with electronic sensors such as echo sounders, compasses,

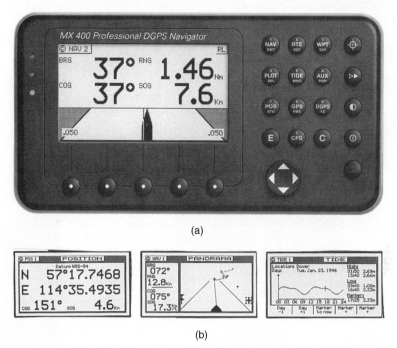

(a)

(b)

FIGURE 8-31 (a) Receiver. (b) Typical displays. Leica MX400 DGPS Navigator—a single-frequency, six-channel GPS receiver equipped with an optional DGPS beacon receiver. Positions can be displayed as latitude/longitude, Loran-C, Decca, and UTM
(Courtesy of Leica Geosystems Inc.)

tide gauges, sidescan sonar, and acoustic positioning. Additional modules include contouring, profiles, volumes, and digitizing. This software also displays an electronic chart of the area showing the planned range lines together with the location (distance in feet or meters right or left of the range line) and the heading of the vessel. Figure 8-32 is a screen display showing the vessel on sounding line CH00140. The vessel is 1.1 m off line to the right. Also shown are the point coordinates (latitude/longitude) and a displayed depth of roughly 26½ m.

It is possible to mount two separate GPS receivers on large survey vessels to reduce the effects of sounding errors caused by the roll, pitch, heave, and yaw of the vessel. Hydrographic maps are now giving way to onboard electronic charts. As new data on soundings, tides, currents, and

the like are collected into a central database, the upgraded data can then be transferred electronically to a vessel to update that vessel's electronic chart, giving it some real-time characteristics.

8.14.8 Sweeping

Sweeping is the term applied to taking additional soundings to locate underwater features in areas not covered by the original sounding plan. Figure 8-33 illustrates the problem. The isonified area of the seabed below the echo sounder is proportional to the water depth and since the cone beneath the transducer of the echo sounder is approximately 30°, the shallower depths along the sounding lines *A* and *B* in Figure 8-33 leave a gap between the two lines that has not been sounded. As illustrated, an important seabed feature can be missed during the

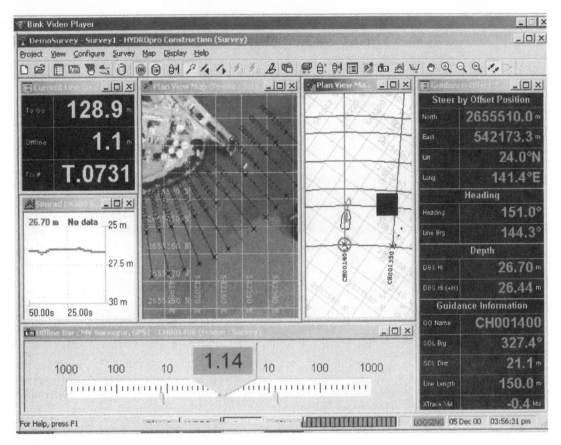

FIGURE 8-32 Hydrographic software display
(Courtesy of Trimble, Sunnyvale, CA)

survey. A gap exists between lines *B* and *C*, while the area between lines *C* and *D* is covered adequately.

The difficulty comes from the fact that the surveyor does not know the water depths until the survey is completed. Therefore, the sweeping operation is the last requirement of the project. Areas requiring sweeping must be identified. Using the cone of 30° beneath the echo sounder, depths less than 1.85 times the distance between the sounding lines may require sweeping. For example, if the sounding lines are 50 m (165 ft) apart, depths less than 93 m (305 ft) will have gaps. At the discretion of the surveyor, extra sounding lines may be used to sound the gaps. The same position-fixing techniques employed during the survey, as illustrated in solution (b) of

Figure 8-33, are used. A cost-effective compromise involves running the vessel at higher speeds parallel to the shoreline in sounding out lines to the depths where coverage is ensured. See solution (c) of Figure 8-33. The surveyor should note the location of any depth anomalies, and position-fix only these unusual occurrences.

Improvements to sweeping include (1) using an array of transducers and (2) using swath sounders. The array of transducers is lowered from booms projecting from the sides of the vessel; each transducer can obtain depth measurements every second, all of which are logged. The swath sounders give a fan-shaped beam; this technique uses a sidescan sonar that can detect the angle of arrival of incoming acoustic energy.

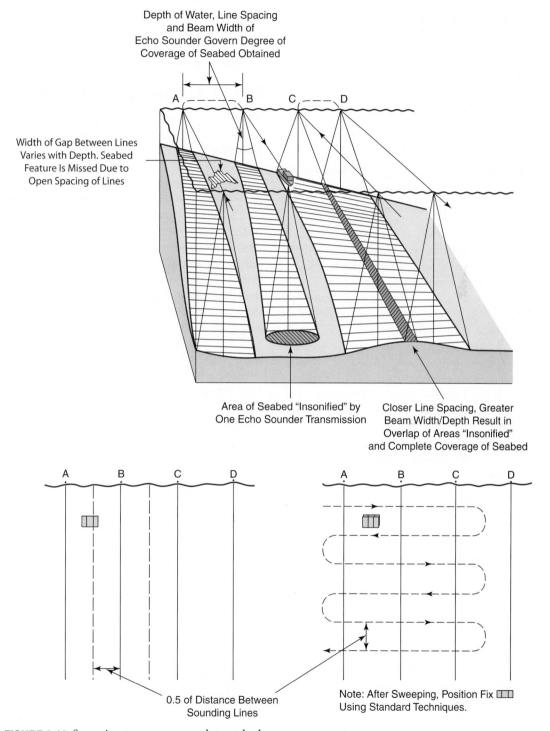

Depth of Water, Line Spacing
and Beam Width of
Echo Sounder Govern Degree of
Coverage of Seabed Obtained

Width of Gap Between Lines
Varies with Depth. Seabed
Feature Is Missed Due to
Open Spacing of Lines

Area of Seabed "Insonified" by
One Echo Sounder Transmission

Closer Line Spacing, Greater
Beam Width/Depth Result in
Overlap of Areas "Insonified"
and Complete Coverage of Seabed

0.5 of Distance Between
Sounding Lines

Note: After Sweeping, Position Fix
Using Standard Techniques.

FIGURE 8-33 Sweeping to ensure complete seabed coverage

8.14.9 Sounding Plan

The most economical coverage of the seabed is achieved through a series of equally spaced sounding lines over the survey area. Specific considerations are

- Appropriate scale of the survey.
- Spacing between the sounding lines and their orientation with the shoreline.
- Interval between fixes along a sounding line.
- Speed of the vessel.
- Direction in which the sounding lines are run.

The scale of the survey is set by the degree of thoroughness and the precision of the soundings. Therefore, the scale determines the number of sounding lines and fixes along each line. Depending on the preceding factors, scales for surveys within 5 km (approximately 3 miles) of the shoreline range between 1:1,000 (1 in. = ±83 ft) and 1:20,000 (1 in. = ±1,666 ft). The distance between the sounding lines is based on the rule that it should not exceed 10 mm or 1 cm (approximately 0.4 in.) on the drawing. Therefore, at a scale of 1:5,000, the lines should not be greater than 5,000 × 0.01 m, or 50 m apart.

The Canadian Hydrographic Service's line spacing is a half centimeter (0.005 m) at any scale, for example:

$$1:20,000 \text{ line spacing} = 100 \text{ m}$$
$$1:10,000 \text{ line spacing} = 50 \text{ m}$$
$$1:5,000 \text{ line spacing} = 25 \text{ m}$$
$$1:1,000 \text{ line spacing} = 5 \text{ m}$$

One can drop two zeros and divide the scale by 2 to calculate the line spacing.

The speed of the vessel during sounding is determined by the realistic assumption that the time interval between fixes will be a minimum of 1 min. For a scale of 1:5,000, the speed of the vessel would be 7 km/h (approximately 4 knots).

The sounding lines are run in a direction that is nearly at right angles with the direction of the depth contours. The effects of geological conditions, offshore bars, and the like should be considered in determining the most efficient and economical direction for the angle of the sounding lines with the shoreline.

8.14.10 Airborne Laser Bathymetry

In the past dozen, or so, years, airborne laser bathymetry (ALB) has become operational in the field of shallow-water hydrographic surveying. Most of these LiDAR systems employ two spectral bands: one to detect the water surface (1,064 nm infrared band) and the other to detect the bottom (532 nm blue/green band). The depth of the water is determined from the time difference of laser returns reflected from the surface and from the seabed. See Figure 8-34. Although this technology is still relatively new (since 1994), it is predicted that its short mobilization times (compared to shipboard echo sounding), efficient area coverage, lower costs, and relatively fast processing of data will make it the mainstay of shallow-water bathymetry.

The ability to penetrate water depends to a high degree on the clarity of the water. Very turbid water may permit penetration of only a few meters, while clear water may permit penetration

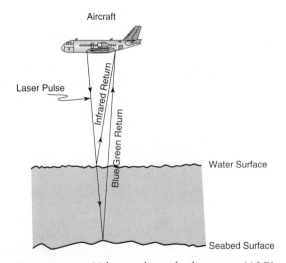

FIGURE 8-34 Airborne laser bathymetry (ALB) depth measurement

up to 60 m (2–3 times Secchi disk depth). Typical shallow-water LiDAR surveys are operated from helicopters or fixed-wing aircraft flying at 200–300 m altitude and at speeds of between 60 and 120 knots (110–220 km/h). They collect depth soundings on a 4-m horizontal grid, giving the position accurate to 0.5–2 m horizontally and 15 cm vertically (when controlled by GPS measurements).

ALB can produce more effective coverage of shallow water soundings because, given the altitude of the LiDAR instrument package, the cone of signal emissions has the room to spread and thus to cover consistently a much wider area of the seabed than conventional shipborne multibeam echo sounders, which operate on the surface of the water. The shallower the water, the more pronounced is this advantage. In very shallow water, the shipborne multibeam echo sounder can cover only a very narrow swath of seabed and runs the continual risk of hull-bottom damage.

Problems

A topographic survey was performed on a tract of land using total station techniques to locate the topographic detail. The sketch in Figure 8-35 shows the traverse (A to G) and the grid baseline (0 + 00 at A) used to control the survey. Also given are bearings and distances of the traverse sides (Table 8-4), grid elevations (Table 8-5), and angle and distance ties for the topographic detail

(Table 8-6). Problems 8.1 through 8.8 combine to form a comprehensive engineering project; the project can be covered completely by solving all the problems of this section, or individual problems can be selected to illustrate specific topics. Numerical values can be chosen as meters or feet. All field and design data for these problems are shown on this page and on the following pages.

8.1 Establish the grid, plot the elevations (the decimal point is the plot point from Table 8-5), and interpolate the data to establish contours at 1-m (1-ft) intervals. Scale at 1:500 for metric units, or 1 in. = 10 ft or 15 ft for foot units. Use pencil.

8.2 Compute the interior angles of the traverse and check for geometric closure, that is $(n - 2)180°$ (see Table 8-5).

8.3 Plot the traverse, using the interior angles and the given distances. Scale as in Problem 8.1 (see Table 8-7).

Table 8-5 Balanced Traverse Data

Course	Bearing	Distance (m or ft)
AB	N 3°30′ E	56.05
BC	N 0°30′ W	61.92
CD	N 88°40′ E	100.02
DE	S 23°30′ E	31.78
EF	S 28°53′ W	69.11
FG	South	39.73
GA	N 83°37′ W	82.67

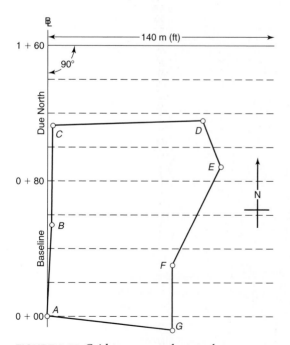

FIGURE 8-35 Grid traverse and control

Table 8-6 Elevation Data: Surveying Grid Elevations (Problem 8.1)

Station	Baseline	20 m (ft) E	40 m (ft) E	60 m (ft) E	80 m (ft) E	100 m (ft) E	120 m (ft) E	140 m (ft) E
1 + 60	68.97	69.51	70.05	70.53	70.32			
1 + 40	69.34	69.82	71.12	71.00	71.26	71.99		
1 + 20	69.29	70.75	69.98	71.24	72.07	72.53	72.61	
1 + 00	69.05	71.02	70.51	69.91	72.02	73.85	74.00	75.18
0 + 80	69.09	71.90	74.13	71.81	69.87	71.21	74.37	74.69
0 + 60	69.12	70.82	72.79	72.81	71.33	70.97	72.51	73.40
0 + 40	68.90	69.66	70.75	72.00	72.05	69.80	71.33	72.42
0 + 20	68.02	68.98	69.53	70.09	71.11	70.48	69.93	71.51
0 + 00	67.15	68.11	68.55	69.55	69.92	71.02		
@ Sta. A								

Table 8-7 Survey Notes (Problem 8.4)

	Horizontal Angle	Distance [m (ft)]	Description
		Sta. ⊼ @ Station B (sight C, 0°00′)	
1	8°15′	45.5	S. limit of treed area
2	17°00′	57.5	S. limit of treed area
3	33°30′	66.0	S. limit of treed area
4	37°20′	93.5	S. limit of treed area
5	45°35′	93.0	S. limit of treed area
6	49°30′	114.0	S. limit of treed area
		⊼ @ Station A (sight B, 0°00′)	
7	50°10′	73.5	₵ gravel road (8 m ± width)
8	50°10′	86.0	₵ gravel road (8 m ± width)
9	51°30′	97.5	₵ gravel road (8 m ± width)
10	53°50′	94.5	N. limit of treed area
11	53°50′	109.0	N. limit of treed area
12	55°00′	58.0	₵ gravel road
13	66°15′	32.0	N. limit of treed area
14	86°30′	19.0	N. limit of treed area
		⊼ @ Station D (sight E, 0°00′)	
15	0°00′	69.5	₵ gravel road
16	7°30′	90.0	N. limit of treed area
17	64°45′	38.8	N.E. corner of building
18	13°30′	75.0	N. limit of treed area
19	88°00′	39.4	N.W. corner of building
20	46°00′	85.0	N. limit of treed area

8.4 Plot the detail using the plotted traverse as control. Scale as in Problem 8.1. See Table 8-7.

8.5 Determine the area enclosed by the traverse in m² (ft²) by using one or more of the following methods.

    a. Use grid paper as an overlay or underlay. Count the squares and partial squares enclosed by the traverse, and determine the area represented by one square at the chosen scale. From that relationship, determine the area enclosed by the traverse.

    b. Use a planimeter to determine the area.

    c. Divide the traverse into regular-shaped figures (squares, rectangles, trapezoids, triangles) and use a scale to determine the figure dimensions. Calculate the areas of the individual figures and add them to produce the overall traverse area.

    d. Use the balanced traverse data in Table 8-4 and the technique of coordinates to compute the area enclosed by the traverse.

    e. Use the software program in the total station (or its controller) to compute the area.

8.6 Draw profile A to E showing both the existing ground and the proposed ₵ of highway. Use the following scales: metric units—horizontal is 1:500, vertical is 1:100; foot units—horizontal is 1 in. = 10 ft or 15 ft, vertical is 1 in. = 2 ft or 3 ft.

8.7 A highway is to be constructed to pass through points A and E of the traverse. The proposed highway ₵ grade is +2.30 percent rising from A to E (₵ elevation at $A = 68.95$). The proposed cut and fill sections are shown in Figure 8-36(a) and (b). Plot the highway ₵ and 16 m (ft) width on the plan (see Problems 8.1, 8.3, and 8.4). Show the limits of cut and fill on the plan.

8.8 Combine Problems 8.1, 8.3, 8.4, 8.6, and 8.7 on one sheet of drafting paper. Arrange the plan and profile together with the balanced traverse data and an appropriate title block. Use size A2 paper or the equivalent.

8.9 State which type of electromagnetic position-fixing system you would use for the following circumstances and give three main brief reasons for your selection.

    Accuracy requirements: ±0.25 m (0.8 ft)

    Maximum distance offshore: 10 km (6 mi)

8.10 Figure F-1 is a hydrographic map compilation of the lower Niagara River. See also Figures F-2 and F-3 (aerial images); Figure F-5 (satellite image); and Figure F-4 (LiDAR images), which show the same general geographic area. Compare and contrast the information available from this hydrographic map with the information available from the various remotely sensed images. List the general uses to which each of these same-area images can be put.

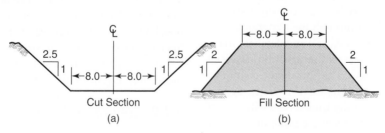

Cut Section
(a)

Fill Section
(b)

FIGURE 8-36 Proposed cross sections (Problem 8.7). (a) Cut section. (b) Fill section

CHAPTER NINE

GEOGRAPHIC INFORMATION SYSTEMS

9.1 BACKGROUND

9.1.1 The Evolution from Mapping to Geographic Information System (GIS)

Geographic information systems (GISs) over the last few years has had the same impact that many breakthrough technologies have had. In the 1980s it was the fax machine that became so dominant that businesses could not function without one. In the mid-1990s it was the Internet, allowing society immediate access to information, useful or otherwise. In the early 2000s the cell phone first allowed parents to keep in constant contact with their children and then allowed the children to text to all their friends about what they were telling their parents they were doing. Although society doesn't see GIS as one of those technologies, who hasn't got online to get directions from one location to another or looked a Google Earth to see where the nearest coffee house is or what their house looks like from above or at street view. Along with GPS, GIS has enabled cars to talk to their drivers and tell them how to get home. Agencies and companies make decisions impacting society based on the analysis provided by GIS on a daily basis.

GIS came about through the development of computer-aided design (CAD) and database management. GIS in its current incarnation has been around since the 1960s. The Canadian Geographic Information System, Harvard University, School of Design GRID, and the United States Bureau of the Census's TIGER system are some of the early adaptations of GIS. However, the concept of GIS goes back to the history of mapping. Since the earliest days of mapping, topographic features have routinely been portrayed on scaled maps and plans; these maps and plans provided an inventory of selected or general features which were found in a given geographic area.

In its most basic sense GIS is geographically displaying objects, often called "features," and information, often called "attributes," attached to the objects. The objects can be physical, such as a fire hydrant, sidewalk, or building, or they can be documented objects such as zoning boundaries, accident site, or proposed right of way. The introduction of topological techniques permitted the data to be connected in a relational sense to their spatial connections. It thus became possible not only to determine where a point (e.g., hydrant) or a line (e.g., road) or an area (park, neighborhood, etc.) was located, but also to analyze those features with respect to the adjacency of other spatial features, connectivity (network analysis), and direction of vectors. Adjacency, connectivity, and direction opened the database to a wider variety of analyzing and querying techniques. For example, given the enormous amount of data now stored in modern urban databases, it is possible in a GIS real-estate application to display, in municipal map form, all the industrially zoned parcels of land with rail-spur possibilities, within 1.5 miles of freeway access, in the range of 1.2–3.1 acres in size. Relational characteristics also permit the database to be used for routing from emergency vehicles to vacationers.

9.1.2 General Definition

The formal concept of GIS was born out of the integration of CAD and database management. It is an evolutionary process of automating the use of maps to catalog and analyze information relative to a geographic position. The ability to orient the spatial relationship of an object to the earth is called georeferencing. The concept is that any bit of information can be attached to a georeferenced object. This allows one to look at the information in relation to other georeferenced objects. GIS works on the concept of "layers" or themes in which each layer or theme relates to an object type, such as roads or vegetation. See Figure 9-1.

GIS allows analysis of data in a spatial relationship, such as "How many houses are within a 2 mile radius of the nuclear power plant" or "Which blocks have an average household income of at least $50,000 and are within two blocks from a coffee house?"

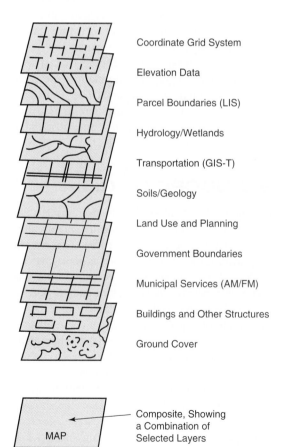

Coordinate Grid System

Elevation Data

Parcel Boundaries (LIS)

Hydrology/Wetlands

Transportation (GIS-T)

Soils/Geology

Land Use and Planning

Government Boundaries

Municipal Services (AM/FM)

Buildings and Other Structures

Ground Cover

MAP

Composite, Showing
a Combination of
Selected Layers

FIGURE 9-1 Illustration of thematic layers

Today GIS is applied to a number of studies and industries including but certainly not limited to the following:

- Cadastral information systems
- Natural resource management
- Wildfire management
- Facilities management
- Driving instructions
- Hydrology
- Planning
- Infrastructure maintenance

9.1.3 Specific Definition

The author's choice for a definition comes from the University of Edinborough GIS faculty, which states that "a geographic information system (GIS) is a computerized system for capturing, storing, checking, integrating, manipulating, analyzing, and displaying data related to positions on, or near, the earth's surface."

The United States Geological Survey defines GIS as "a computer system capable of capturing, storing, analyzing, and displaying geographically referenced information; that is, data identified

History of GIS

Dr. Roger Tomlinson is credited with coining the term *geographic information system* (GIS) when he headed the Canadian Geographic Information System (CGIS) in the early 1960s. CGIS is credited by many with initiating modern GIS, although clearly there were others involved with this concept in the early 1960s.

The concept of attaching information to maps clearly had been around for centuries. The cartographer for General Rochambeau's army, Louis-Alexandre Berthier, during the Revolutionary war is credited with creating overlay maps showing the troop movements which aided in the American-French allied victory in Yorktown. Dr. John Snow mapped the London cholera outbreak in 1854 by using bars to indicate deaths per households. These are just two famous examples, of which there are many day-to-day examples of extending the use of a map in order to analyze data.

according to location. Practitioners also define a GIS as including the procedures, operating personnel, and spatial data that go into the system" (http://erg.usgs.gov/isb/pubs/gis_poster/).

A more detailed definition of GIS can be found in the article "What Does GIS' Mean?" by Nicholas R. Chrisman (Chrisman, N. R. 1999, What Does "GIS" Mean? *Transactions in GIS* 3 (2), 175–186):

> Geographic information System (GIS)—The organized activity by which people
>
> - measure aspects of geographic phenomena and processes;
>
> - represent these measurements, usually in the form of a computer database, to emphasize spatial themes, entities, and relationships;
>
> - operate upon these representations to produce more measurements and to discover new relationships by integrating disparate sources; and
>
> - transform these representations to conform to other frameworks of entities and relationships.

It is clear from the definitions and applications that term and use of GIS today is very broad and crosses many disciplines. There is ongoing discussion about GIS being a science or a tool. Those discussions are well beyond the breadth of this book. We will look at the basics of GIS and see how surveyors benefit from GIS and can benefit a GIS.

We first want to discuss how surveying and GIS interact. The specifics of GIS are discussed at the end of this chapter. If you need a better understanding of GIS feel free to read those sections first or refer to those sections as needed.

9.1.4 Surveyor's Role in GIS

The surveyor has a unique perspective on GIS, as historically surveyors provided much of the information that has been used to establish the geographical reference of objects in GIS. However, there has often been resistance between the surveying community and the GIS community to work together. This has been caused by a lack of understanding on both sides caused by the differing terminology used by both technologies as well as issues with the cost of developing GIS. The relationship between the two professions has slowly developed over the last

20 years as the GIS community begins to understand the impact of showing ownership lines on a GIS and the surveying community develops cost-effective data collection methods. There is still a need for the surveyor to understand and convey to the GIS professional verifiable field data collection procedures that ensure precision requirements are being met.

Today many surveyors are involved in GIS by providing a high-precision geographically referenced base map from which all other layers are built. Some surveyors are also GIS managers, who coordinate all the data collection, storage, analysis, and reporting. In addition, much of the work done in surveying, although not primarily for GIS purposes, finds its way into a GIS data set. This may include orthophotos that are controlled by surveyors, tract maps that are mapped and monumented by surveyors, and infrastructure that is topographically mapped by surveyors.

The U.S. Department of Labor Occupational Outlook puts Geographic Information Specialist under the Surveyor category. The Canadian National Occupational Classification for Statistics defines a Geographic Information Specialist under Mapping Technicians.

9.2 SCOPE OF GIS

It is hard, if not impossible, to overestimate the scope of GIS. Almost every discipline incorporates GIS in at least some of their studies. GIS is used extensively for local analysis, such as managing city services, it is used extensively for regional analysis such as watersheds, it is used at state and providence level, and it is used as the national level and is used for analysis of impacts on the world. These analyses can provide data in a graphic format that is easy for the general public to understand. These features are both the strength and weakness of GIS.

As a user of GIS, it is important to question the process of the analysis. Does the analysis reflect the correct precision of the data collected? Is it appropriate to define building sites 50 feet from stream banks, when the stream banks were digitized in from a map with a scale of 1:24,000?

If someone was determining the sites on the original paper map, most people would question the accuracy of the site locations, but because it is shown on a graphic screen or plot at $1'' = 100$ ft it appears precise. Within GIS there are many analytical tools available, such as the determination of slope. This provides an adequate interpretation of slopes for general use; however, there are factors that must be considered if the slope analysis is critical. What is the basis for elevations, and how precise is it? What is the methodology used for slope determination? These are some of the issues that must be considered when using GIS, when its answers are critical to decision making.

9.3 DAY-TO-DAY GIS

To understand how universal GIS is these days, one just needs to get on the Internet. Mapquest (http://www.mapquest.com/) uses GIS to find directions for travel. Google Earth (http://earth.google.com/) lets you explore our world and also find the nearest cafe to you.

USGS has the national map (http://nationalmap.gov/) which is intended to replace their topographic map series that allows you to view and download a variety of the layers created by USGS and other federal agencies.

Many other local agencies provide online GIS for the use and benefit of their citizens. One example is the City of San Luis Obispo (http://www.slocity.org/publicworks/gishome.asp) in California (see Figure 9-2), which uses GIS to allow the public to check on zoning, points of interest, and heritage trees.

It is important to understand that GIS is becoming as invasive within our society as the cell phone. It is not within the scope of this book to discuss the social issues that come about with the increased use of this technology. There are real

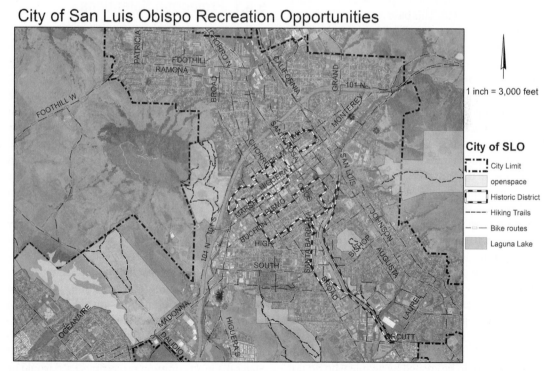

FIGURE 9-2 City of San Luis Obispo, California, with online GIS showing recreational opportunities within a portion of the city

concerns about how GIS can affect people's right to privacy. For more information, you can search Ogle Earth blog (http://ogleearth.com/) or other sites on privacy.

These national or world wide public GISs are generally using low-resolution information, and would not be considered adequate for regional GISs. However, because of the public acceptance of these large systems, we are starting to see some partial integration into local systems. It is not hard to believe that in the near future systems such as Google Earth will be used for the basis of a local GIS, much the same way the Google Search engine is used for localized searches on many websites.

9.4 WHAT SURVEYORS NEED TO KNOW

Surveyors are often involved in GIS on the data collection end of the projects. A solid GIS project will create a precise base map to which all other objects are related to. This base map may be a map of legal parcels for a city GIS, may be a precisely georeferenced orthophoto for a forest land use study, may be field map for the management of a farming enterprise, or may be high-precision control points for a control diagram.

It is important for surveyors to understand what the concept of GIS is in order to provide these base maps for GIS. In addition the surveyors must understand what the final product of the GIS is, so that the precision used to establish a base map is appropriate. Many surveyors feel that any base map should be established to the highest precision possible. There are valid arguments for this approach. First the profession of surveying is known for the ability to provide precise measurements; second surveyors argue that the actual use for any data is not known; therefore in order to be safe, all data should be collected at the highest precision possible. The arguments against measuring to the highest precision possible are as follows: First the cost can be prohibitive so that base maps would be created by people who do not understand measurement theory and error; and second we know from measurement theory that

providing precise measurement does not mean we know the absolute position. In surveying, we make a determination as to what the appropriate precision should be, and this should be the same in GIS.

What is important for a surveyor to do for any base map that they create is to document the information. In GIS the report about the information provided is known as "metadata," described as data about the data. Create metadata that clearly state what was done, how it was done, and what precision it was done at. This is what will limit the liability of a surveyor if the data are used for something beyond their intended use. These metadata should be put together in the form a report. The report should indicate the reference datum(s) used, the control used to get on the reference datums, the equipment and procedures used, the final results and the estimated precisions, and other information about the data.

9.4.1 Projection Selections

One of the most common problems that occur in setting up GIS is the establishment of the projections that are used. As discussed in Chapter 10, a projection is a two-dimensional representation of a three-dimensional object, and the three-dimensional object we usually use is a datum that is approximating the shape of the earth. Any projection generates loss of precision from the datum. It is possible to actually present the data in GIS without any projection. This would show all data positions as either geographic position (latitude and longitude) or ellipsoidal coordinates. This is an appropriate coordinate system for very large coverages such as the whole earth or all of North America. The problem with showing everything as its three-dimensional components, determining distances and angular relationships, can be complex. Also computations involving elevation such as slope may provide incorrect values.

Therefore, it is common to have GIS use projections for its coordinate system. In the United States the most common projections used would be State Plane Coordinates zones for high-precision, relatively small areas or UTM coordinates for lower precision but larger areas. However, there are a number of other projections that are used. The state of California created a low-precision Albers projection, known as the Teale Albers projection, which covers the entire state.

Remember, a projection is always associated with a datum. In North America the most common datums are NAD27 and NAD83. (See Chapter 10 for a thorough discussion of these two datums.) NAD83 has a number of significant adjustments associated with it, so often a decision must also be made as to what adjustment should be used.

Generally, if the base map is based on field information, the datum and projection used for that field information will be the basis for the GIS. Where the decision on projections and datums becomes an issue is when spatial data are being added to the GIS which does not have the projection and datum assigned to it. There is a specific order that should be used in making a determination of which system are the data based on.

1. Check the metadata.

   a. Often there will be metadata that have the information shown on them. Then it is just a matter of assigning the correct projection.

2. Check the source.

   a. First see if you can determine the source of the data. If the source is not directly credited, you may be able to determine the source by the file name. Generally, agencies that generate large data sets have a systematic way to name all the files.

      i. If you can find the source, you may be able to find the exact system used, or you may be able to find in general what system was used for that particular data set.

3. Look at the coordinate limits of the data.

   a. If the values are relatively small such as (−120.34567, 35.45678), then it may be latitudes and longitudes. If it is latitudes and longitudes, you need to determine which

datum was used. If the data set was probably collected using GPS, is it a good guess that the underlying datum is WGS84. However, if it has been processed it may be NAD83. If the data set is based on older information, the datum may be NAD27.

   b. If the values are large, then they are probably based on a projection. You can look for control within the area of interest and then look at the known values for those control points. This may give you the projection system you are looking for.

If you end up having to determine the system being used by a data set, you need to do some verification on your assumptions. It is best to field verify a position if possible. There are data sets that may contain correct spatial relationship amongst the data within the data set, but have no projection associated with it. This is very common on older surveyed information, as most older surveys were based on assumed coordinate systems.

Also it is important to remember that there are additional errors generated when projecting from one datum to another, as these are only approximations. These approximations may be acceptable depending on the anticipated precision of the project (see Table 9-1).

9.4.2 Field Collection Procedures

Although field collection for GIS can be similar to field collection for topographic surveys, generally there are better ways to collect data for GIS. It is important to remember that GIS looks at objects as points, lines, or polygons. When collecting field data, there must be given some consideration as to what type of object is going to be collected. Some examples are as follows: A field might be collected as a polygon as it has area. Fences, however, could be collected either as a polygon, with area, or as lines because fences do not necessarily always close. Often underground utilities, such as water lines, are mapped by locating surface features, such as water valves. General procedure would be to collect the valves as a point object, allowing critical attributes to be assigned to the valves, and

Table 9-1   Sample of Variation in Datums

Control Point Monterey (San Luis Obispo, California)				
			Variation from NSRS (2007)	
Datum	Latitude (North)	Longitude (West)	North (ft)	East (ft)
NAD83 (NSRS2007)	35°17′37.70262″	120°38′44.37447″	0.00	0.00
NAD83 (1992)	35°17′37.68347″	120°38′44.35562″	−1.98	1.51
NAD83 (1986)	35°17′37.67867″	120°38′44.34635″	−2.48	2.27
NAD27	35°17′37.76855″	120°38′40.72601″	−1.29	302.50
Control Point Afton (Sedgwick, Kansas)				
			Variation from NSRS (2007)	
Datum	Latitude (North)	Longitude (West)	North (ft)	East (ft)
NAD83 (NSRS2007)	37°37′11.57382″	97°38′17.94272″	0.00	0.00
NAD83 (1997)	37°37′11.57376″	97°38′17.94273″	−0.01	0.00
NAD83 (1986)	37°37′11.58313″	97°38′17.93887″	0.94	0.30
NAD27	37°37′11.53341″	97°38′16.76431″	−3.21	94.84

Note the variations shown were generated using the Corpcon program and setting all values as NAD83 HARN (High Accuracy Reference Network).

then map and attribute the water lines in the office using valve locations. Once the feature type is determined, the information that is to be collected with each feature must be determined. The non-spatial data that are collected with each feature are called attributes. Attributes differ from general descriptions that are applied in topographic surveys in that there can be multiple attributes associated with a single object and attributes are named, so that there is similar information going into each attribute. Attributes for a fence might be the type of fence, the number of strands, if it is barb wire, and the condition of the fence. Attributes for a water valve might be the size and type valve, the manufacturer, and the condition of the valve.

When collecting features as discussed above, it is important to remember that line and polygon features, once started, must be completed before another feature is collected. Therefore, there are times when all features might be collected as point features and then the line and polygon features would be compiled in the office. The decision made before the actual data collection is started is what will determine if the collection process goes smoothly or is plagued with problems. Therefore, it is a good idea to give plenty of consideration on how the data are going to be collected.

The use of GPS is the most common method for collecting data for GIS; this is not only because of the ease of use of GPS, but also because generally GISs are for large areas that make using other methods of data collection cumbersome. Almost always, some other methods must also be employed in areas where GPS either does not work, or is so sporadic that the use of GPS becomes inefficient. We will first discuss the use of GPS and then how to employ other systems.

The method for collecting the data will depend on the positional precision expected. If the precision is expected to be within a 0.1′ (3 cm), then standard surveying procedures would be applied. Note this is at a higher quality than are standard topographic surveys, so it can be expected that the collection of data at this level will take longer than normal topographic methods. As a rule most GIS precisions are not at this level, although one for public works within a city may require this precision. A more common precision might be 0.5′ (15 cm) for fixed works. Current technology would still require survey-grade GPS to achieve 0.5′ positional precision. Field procedures at these precisions would be similar to topographic survey procedures. Attention must be paid to maintaining lock, having low DOP, and having the rod vertical when collection positions.

Much data for GIS can be collected using GIS grade GPS, that is, using differential GPS to get positional accuracies of 1 m. For natural features, this precision is more than precise enough. The GIS grade GPS has the advantage over survey-grade GPS in that maintaining the differential correction is easier and can handle obstructions better. The downside of using GIS grade GPS is associated with how the antenna is positioned. Most of these GPSs either have an antenna coming out of a backpack or are held in the palm of the hand. This means that being directly over the position being collected only happens by chance. Therefore, another significant error is being introduced into the positional precision, the error of being over the point being collected. These systems do not use the same rod system as survey-grade GPS is because the underlying GPS is not as precise. See Section 9.5.1 for a discussion on data collection procedures. One option when collecting either a line or polygon feature is to continually collect data. The software will automatically collect a position either based on time span such as every 10 seconds or distance span such as every 10 ft. This is the same concept as continuous point collection in topographic surveys. There are times when this option is an appropriate method for data collection, such as a field or creek that has an irregular shape and there are times when it is not appropriate, such as a fence line which will have specific angle points. The time or distance span should be set so that it sufficiently represents the figure being located, but not so dense that it adds unneeded detail to the figure. When collecting positions, there are a number of times when you cannot directly collect the data with GPS. This might be because you are locating the corner of a building, which will block out the GPS signals, or you are working under dense vegetation canopy or maybe you are trying to locate wire heights. Most of the data collection software will have some sort of offset options that will allow you to collect two points or the position you want, and then use direction and distance from those points or just distance to determine the location of the missing point. A quick way in the field to collect buildings with GPS is to get in line with the face of the building and then take two points along that line, not concerning yourself with the distance off the corner. Do this for every face of the building, and then in the office you can connect the points to create the building. This process takes some office time but can drastically reduce the field time, if there are a number of buildings to locate, or if the building has a complex footprint.

Sometimes you will need to use a total station to pick up additional information. Points can be set using the GPS to measure from using the total station. When doing this, consideration must be given to the precision of the point being set with GPS. Assuming that two points will be set, one to set the total station on and a second for a backsight, each point will carry with it the random error associated with the GPS. Assuming each point was set at ±0.5′ means the relative error between them could be as much as 1 ft. This error is going to be carried through, as an angular error to any positions located with the total station. As a rule of thumb it is a good idea to set the backsight a distance at least as twice as far from the total station as any point that will be collected with the total station.

Another method of collecting additional data is with a range finder. Many GIS grade GPSs will allow a range finder to be linked into the GPS, so that data collected with the range finder are automatically used to determine the position (see Figure 9-3). Range finders are able to measure distance and direction (based on compass) in order to determine a position. Distance accuracy can be up to 0.02 ft, angular accuracy up to 0.1 degrees, and

FIGURE 9-3 "Advantage Total Station" range finder. It can be equipped to measure horizontal distances, vertical distances, and angles from a baseline so that the surveyor can extend a GIS survey into areas where GPS signals are blocked

(Courtesy of Laser Atlanta, Norcross, GA)

inclinometer accuracy to 0.4 degrees. Although these have high-accuracy distance measurement, range finders send out a beam that gets wider the further it is from the unit; therefore, irregular shaped objects will carry some uncertainty as to where the unit is measuring to. In the field these systems can be useful for low-precision positioning. Positional precision can be increased with these units if proper procedures are used. This includes establishing the relation of the compass readings to the coordinate system by first measuring direction along a known line. Also for higher precision and blunder detection, it is recommended to locate a position from two different GPS positions.

9.4.2.1 Attributes As discussed earlier, attributes are descriptions applied to the features being collected that provide specific information about the feature. Attributes are defined by a name describing the information, the information type, such as text, integer, real number, or true/false, and the length of the allowed answer. Often default values can be set, so that the most common answer is shown when entering the data.

These attribute fields are one of the primary sources for solving queries entered into a GIS. Because of this, it is important that there is consistency in the data being entered into each field. If you were looking up for all mailing addresses that were sent to a post office, you might search the mailing address for listing going to a post office box. If variations on the data entered in the address attribute looks like "Post Office Box 111," "P.O. Box 111," "PO Box 111," "Box 111," it will be harder to extract the information, and more probably that some of the information will be missed. Therefore, it is good to establish a list of acceptable answers for all those attributes that have only limited answers. This list can be set up in the GIS and in the field software to show up as a pull-down list, so that one can select the appropriate answer when filling out the attributes. For those attributes that do not have finite answers, it is good to establish specific procedures for filling in the attributes.

Finally one or more of these attributes may be used to label the features in the GIS so thought should be given to the length of the answers and how text should be entered: should it be all capitalized or not. Mapping in GIS has the same cartographic constraints that exist in all surveying mapping. Short concise descriptions will make the map more readable. Text within the map is generally more readable if it is all capitalized.

9.4.3 Understanding What Is Lacking

Currently GIS is not an efficient way to create standard surveying maps such as topographic surveys, boundary surveys, and control plans. It is not that they can't be done using GIS; it is just not nearly as efficient as doing it within a CAD environment. Bearings and distances can be shown on a GIS, details can be created, line types can be varied, coordinate calculations can be done, and notes can be added. However, the current strength in GIS is its ability to analyze data and present that analysis in a graphic format. Its strength is not in precise vector editing.

GIS topology rules can force buildings to be square and lot lines to always abut adjacent lines with common corners always meeting. However, this takes a significant amount of knowledge of the GIS software to ensure that this is happening. Currently, surveyors are more comfortable in verifying this by the use of coordinates and coordinate geometry.

GIS has the capability to create three-dimensional models. However, none of the software can easily create design plans for a road or subdivision. Although GIS can create profiles, it does not currently have the capability to automatically generate the precise information that needs to be shown on plan and profiles in order for the construction of the design to be completed.

9.4.4 Example Using GIS for Surveying

Often when performing a large control survey, the control must be based on national datums. In the United States the national control is managed by National Geodetic Survey (NGS). In this example control must be established for a 22,000-acre LiDAR survey along the California coast by rapid static procedures using dual-frequency GPS units. The project requires use of High Accuracy Reference Network (HARN) stations. In California these are known as High Precision Geodetic Network (HPGN) stations. In addition, a continuously operating reference station (CORS) will be used to verify the quality of the survey. With the survey being done by GPS and vertical precision important, benchmarks will be used to model the local geoid separation. The benchmarks are to be based on NAVD88. The control needs to be as close as possible or within the boundaries of the LiDAR survey, but also "surround" the project. In addition there was a previous control survey performed 15 years ago that must be transformed to the current project (see Figure 9-4).

9.4.4.1 Establishing the Site Location
The first step would be to find a map that would show the general area. A good general area map is a USGS 7½ minute quadrangle for the area that is georeferenced. These can easily be obtained from a variety of sources. Next the project area needs to be overlaid on the quadrangle map.

9.4.4.2 Finding National Control
The next process would be to determine the national control points in the area. The NGS website (http://www.ngs.noaa.gov/) has a link to the database with all the national control points. Each control point has a datasheet with all the information associated with the control point. Control points can be selected based on name, area, or a number of other options. Besides the datasheets, the control points can be downloaded in a shape file. A shape file is a GIS format for a single layer of information. The shape file from NGS is a point shape file of the control selected. One option is by quadrangle, which works well when using the quadrangle map as a backdrop. This brings up 172 control points in the area. This is good, but they needed to be separated by the type of control. The shape file not only provides the location, but provides attributes for each point. With the NGS data the attributes contain information on the type of control point it is; in fact there are 35 attributes associated with each point (see Figure 9-5).

9.4.4.3 Control Point Selection
First the HPGN and CORS points are isolated using the attribute Datum_Tag. All HPGN monuments will be suitable for GPS occupation, so that does not need to be checked (see Figure 9-6). Next the benchmarks must be isolated. The potential benchmark sites will be those that are on NAVD88, have good stability, and are GPS accessible, all determined from the attributes.

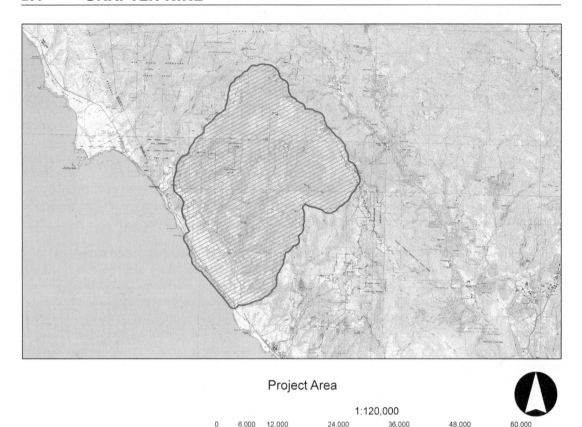

Project Area

1:120,000

0   6,000   12,000   24,000   36,000   48,000   60,000
                                                                       Feet

FIGURE 9-4 A project area overlaid on a mosaic of 7½ minute USGS quadrangle maps of the area

The GIS map shows that the only benchmarks that fit the criteria are along the coast, not surrounding the project (see Figure 9-7). This means that some considerations will have to be given to benchmarks along the northeast side of the project. All of these control points have the datasheet that contains all the information on the point. These will be important to access if a point is being considered for use.

One of the attributes is the website link to the datasheet for the particular control point. This information can be accessed by clicking on the control point and then clicking the link (see Figure 9-8). This will provide the datasheet for viewing or printing.

After careful review of all the control, a plan is established for the national control to be used. The final national control map shows the selected

control points for the surveyors to locate in the field (see Figure 9-9).

Next, the old local control survey must be considered. As with most local control, the information is limited to a point identifier (integer number); the northing, easting, and elevation of the point; and a brief description of the point. These data are in a standard text format. The process is to convert the text into a database, which can usually be done easily. The text files usually contain a separator between each value, such as a comma. This allows the text file to be imported into a database program. From there the database can be read into a GIS, with the easting defining the X coordinate and the northing defining the Y coordinate. This allows a plot on the map of the local coordinates. Often, if available, an orthophoto can be overlaid onto the GIS to aid in the

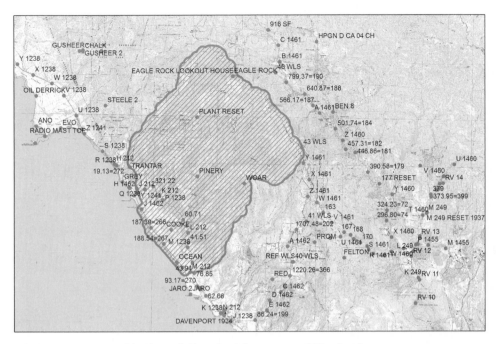

National Control in area of Project

1:120,000

Legend
• NGS_Control
▨ Limit of Project

0   2   4   8   12   16
Kilometers

FIGURE 9-5 All national control found in the general area of the project

determination of which control points have the best potential for being located with GPS.

9.4.4.4 Benefits Using This Example

This is a simple and yet very useful example of surveyors using GIS to their benefit. All of these data can then be loaded into a mobile GIS and attached to the cars GPS. This allows for the surveyors to get near the site and then pull up the datasheets when searching for the monuments.

This example indicates one of the advantages for a surveyor to maintain all of their surveys in a GIS. For a small cost in time up front, it would allow a large return in allowing the surveyor immediate access to all of their surveys based on location and long-term management of their field surveys.

9.5 CONSTRUCTION OF DATA

One of the largest costs in GIS is that of collecting the data. This is because great care has to be taken in order to ensure the integrity of the data is maintained so that it is useful in the GIS. Integrity of data is not the same as the precision of the data. The integrity relates to how objects were collected and what information was associated with the objects. A polygon object must close correctly, a point object must be over the object, and the information must be consistent and correct.

Many people tend to think that the precision of the data coming from GIS is absolute. As one taking surveying, you have come to realize that there is no absolute in field measurements. In

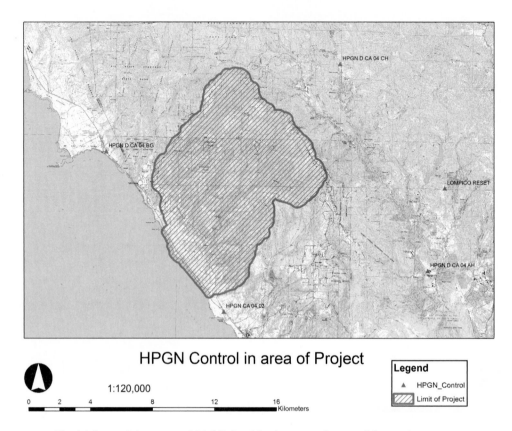

HPGN Control in area of Project

1:120,000

0   2   4   8   12   16
Kilometers

Legend

▲   HPGN_Control

▨   Limit of Project

FIGURE 9-6 The high-precision control highlighted in the general area of the project

fact there can be a large span in the precision of a position, which is affected by the equipment used, the measurement technique, and the ability of the person collecting the information.

9.5.1 Field Collections

Vector data are often collected in the field. The precision needed is often dependent on the task being done. Often field data are collected using GPS. However, the precision with GPS is dependent on the GPS and the method being used. In addition, many times GPS will not function in areas where the information is being collected. This requires either the use of a range finder or total station along with the GPS to pick up the additional information.

In consideration of the quality of the positional value, there are two factors that must be taken into account. First is the absolute position of the data collector, that is, what is the precision that is nominally being achieved by the GPS? Is it 2 cm, ½ m, 1 m, 10–15 m? The second is, what is the absolute position of the object itself? If you are using a backpack GPS with a correction that is providing a precision of 1 m 95 percent of the time, that precision is to the phase center of the GPS antenna. If you are using that for location of fence angle points, this means that in order for the position of the fence angle points to be within 1 m, you must have that GPS antenna directly over the base of the fence post. The reality is that with a backpack GPS unit, most

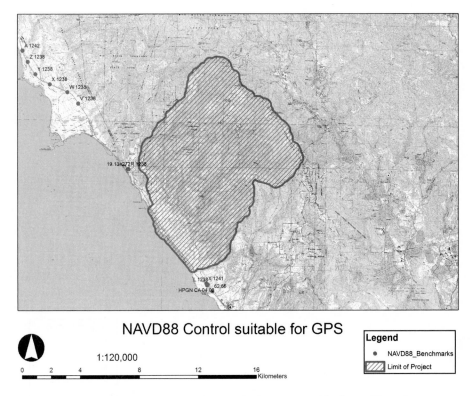

NAVD88 Control suitable for GPS

1:120,000

0   2   4   8   12   16
                                    Kilometers

Legend
● NAVD88_Benchmarks
▨ Limit of Project

FIGURE 9-7 Benchmarks that meet all the criteria including acceptable for satellite observation

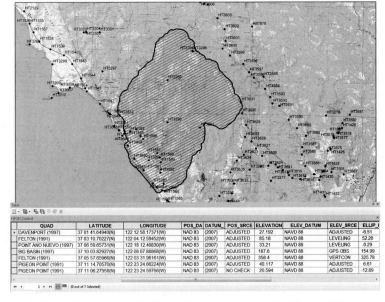

QUAD	LATITUDE	LONGITUDE	POS_DA	DATUM	POS_SRCE	ELEVATION	ELEV_DATUM	ELEV_SRCE	ELLIP_
DAVENPORT (1997)	37 01 41.64940(N)	122 12 58.17371(W)	NAD 83	(2007)	ADJUSTED	27.192	NAVD 88	ADJUSTED	-8.51
FELTON (1991)	37 03 10.70227(N)	122 04 12.59452(W)	NAD 83	(2007)	ADJUSTED	85.18	NAVD 88	LEVELING	52.20
POINT ANO NUEVO (1997)	37 06 59.65731(N)	122 18 12.40830(W)	NAD 83	(2007)	ADJUSTED	33.21	NAVD 88	LEVELING	-0.29
BIG BASIN (1997)	37 10 03.82927(N)	122 08 07.88868(W)	NAD 83	(2007)	ADJUSTED	187.6	NAVD 88	GPS OBS	154.99
FELTON (1991)	37 05 57.65966(N)	122 03 31.96161(W)	NAD 83	(2007)	ADJUSTED	358.4	NAVD 88	VERTCON	325.78
PIGEON POINT (1991)	37 11 14.70570(N)	122 23 24.66224(W)	NAD 83	(2007)	ADJUSTED	40.117	NAVD 88	ADJUSTED	6.61
PIGEON POINT (1991)	37 11 06.27958(N)	122 23 24.59766(W)	NAD 83	(2007)	NO CHECK	20.594	NAVD 88	ADJUSTED	-12.89

FIGURE 9-8 Some of the attributes available from the data downloaded from NGS

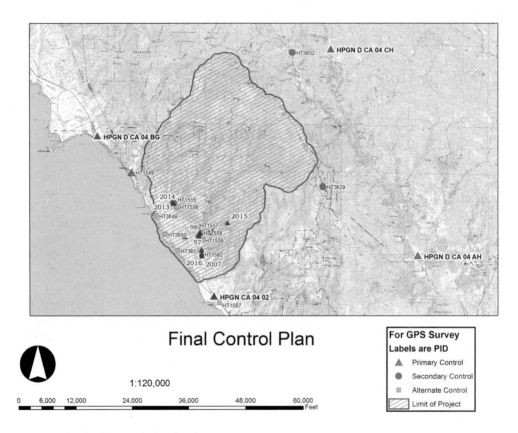

FIGURE 9-9 The final control plan for the project

people are going to stand next to the fence when locating the angle points. This adds an additional error into the position. This error may be acceptable for needs of the GIS, but it would be incorrect to indicate that the fence angle point location is precise to the nearest meter at the 95 percent level. This is true for any data collection that is using a backpack or handheld unit. Most survey-grade GPSs use a rod with a rod level. This will generally give the position to the level of the quality of the GPS assuming the rod level is adjusted.

In surveying, when using GPS or any other measurement system, specific procedures are devised to minimize the possibility of error and to maintain the degree of precision that

is desired. This is a trait that has been carried through the surveying profession, even as the technology advances. It is an expertise within the profession of surveying that is distinct from other professions. The process of collecting data for a GIS has much more of a chance of being successful if specific procedures are established for the equipment being used. The procedures should include methods for verifying the precision, assuring that all features are collected and all attributes are noted correctly. All too often, field collection consists of sending inexperienced field personnel out with an RTK or DGPS and an aerial photo with a list of points to collect. The data are entered into the GIS without any knowledge of the quality of the work.

9.5.2 Remote Sensing

Remote sensing, at field of study in and of itself, is more and more being seen as a part of GIS. Remote sensing is defined as "The measurement or acquisition of information of some property of an object or phenomenon, by a recording device that is not in physical or intimate contact with the object or phenomenon under study" (American Society for Photogrammetry and Remote Sensing). Remote sensing used to be known as photo interpretation, but the term *remote sensing* was coined by the Office of Navel Research with the advent of non-photographic sensors and the use of satellite data.

Remote sensing can provide information on the surface of the earth by the use of LiDAR or radar; it can provide information on vegetation types, soil types, and land use. Remote sensing can be very effective in classifying large areas. Remote sensing forsakes precision for the ability to analyze large areas quickly; we can very easily look at trends to see changes in land use over large area. You as a user of these data must understand the low quality of precision, however.

Remote sensing is generally using raster type of data. There are number of specialty programs that can manipulate the raster data so as to allow for the exploitation of the information contained. Raster data are a cell type format, where there is a single value within each cell. This value when collected with an electromagnetic sensor is an indication of brightness within a certain wavelength range. That is, the sensor is recording the electromagnetic energy within a certain wavelength (see Chapter 11 for more information).

9.5.3 Scanning

Scanning has been used extensively within the GIS realm to convert hard-copy data to digital data to be used within GIS. Scanning of aerial photographs is the only way to convert analog images to a raster data set. Professionally done scanning uses high-quality scanners with extremely high spatial resolution, are very quick in processing the images, have high radiometric performance giving them the ability to measure correct reflectance with very high quality lighting and software that will allow precise dodging and equalizing. There is always some information loss in transferring a photograph to a scanned image, but these high-quality scanners minimize those losses.

Scanning of vector data, such as maps and plans, is done as a cost-saving measure. Generally, this is the most cost-effective way to convert hard-copy maps. There is software that will then take the scanned image which is raster data and convert the scanned lines back to vector data. The early versions of the software would often create hundreds of lines for what was a single line on a map because of defects in either the original map or the scan of the map. Newer software can generally recognize this and combine all those lines into a single line.

It has been this author's experience that the use of this software provides a low-quality product for a number of reasons. The quality of the scan can affect the software's ability to interpret the data; the complexity of the data on the plan or map can cause misinterpretations by the software. The original line work, especially if hand drafted, can be not very precise. This vectorization process requires extensive editing in order to ensure that vectorized data correctly represent what is on the plans or maps. Still, there are times when the only way to recreate this information is either by scanning or digitizing.

9.5.4 Coordinate Geometry Entry

Another option that is not used much because of the perceived time to create is using coordinate geometry of the plans or maps to create the vector data. Most plans and maps have textual information indicating the distance and directions of lines and the radius, arc length, and central angles of arcs. In addition, sometimes there is a reference coordinate system shown on the plans. Using coordinate geometry, it is possible to precisely recreate these lines. To the inexperienced eye, converting a map showing 1,000 lots using coordinate geometry may seem a daunting task; however, often it can be done in the same time frame as scanning, vectorizing, and editing the same map.

By using CAD systems designed to work with survey data, these coordinate geometry calculations can be done quickly, with the end results showing up immediately on the screen. As with all measurement processes, procedures must be set up to minimize errors while maximizing efficiency.

9.6 BASIC ANALYSIS OF DATA

The reason that GIS has become so popular is not because of its ability to create maps but because of the ability to perform analysis on the map and the underlying information. The ability to analyze the data is almost limitless. There are many books on the subject. It is not within the concept of this text to discuss the details of analysis or database management. We will discuss some of the basic concepts so that a reader would be able to understand the basics of GIS analysis.

9.6.1 Queries

The first basic analysis is a query, or asking a question of the data. This can be as simple as which parcels are greater than 10,000 sq ft or they could be which parcels are greater than 10,000 sq ft, are vacant, and are zoned single-family residential. Once these queries are asked, a selection set is created. That is all those objects that match the query are grouped. Then additional analysis can be done on the group, or a layer of just the group can be created.

An often used type of query is a buffer type of query. Buffer queries ask a question in relation to location. We could take our previous query of those single-family residential vacant parcels greater than 10,000 sq ft, and add to find all those parcels that meet those criteria and are within 1 mile of a school. In this example we would want to create a buffer of 1 mile around all schools and find the parcels within our query that touched that buffer. When working with location queries, thought must be given to what is being asked, that is, do we want the whole parcel within a mile or just any portion of the parcel within a mile?

Of course when creating a query it will only work if the tabular data associated with the parcels have the information needed to answer those questions. Also it must be remembered that the selection set is only as good as the underlying information. Errors in the data will generate errors in the results.

9.6.2 Topology

One special area in GIS analysis is concept of topology. Topology as defined by "interface standard for vector product format" Mil Std 2407 "is the branch of mathematics concerned with geometric relationships unaltered by elastic deformation. In geographic applications, topology refers to any relationship between connected geometric primitives that is not altered by continuous transformation."

Topology maintains physical relationships between objects. Lots in a subdivision should always be abutting each other as shown in Figure 9-10. A common back lot corner should stay a common back lot corner, so if on one lot the common corner is moved, all the other lots will be adjusted accordingly. A pressurized water line should have water going in only one direction, depending on the valve configuration as shown in Figure 9-11. Therefore, the GIS should be able to determine the direction of flow based on the valve configuration.

Topology gives GIS the potential to model systems such as water, sewer, and traffic as well as to ensure correct geometrical relationships between objects.

9.7 COMPONENTS OF GIS

9.7.1 Hardware

9.7.1.1 Office Hardware Any GIS clearly needs a computer system. The size and type of computer system are dependent on how the GIS product is going to be deployed. Is it for a single user, multiple users, a Web-based application? What type of users will there be, power users, analysts, or viewers? Every type of deployment has a number of considerations to be made. The hardware

Martin County Utilities System Map Book

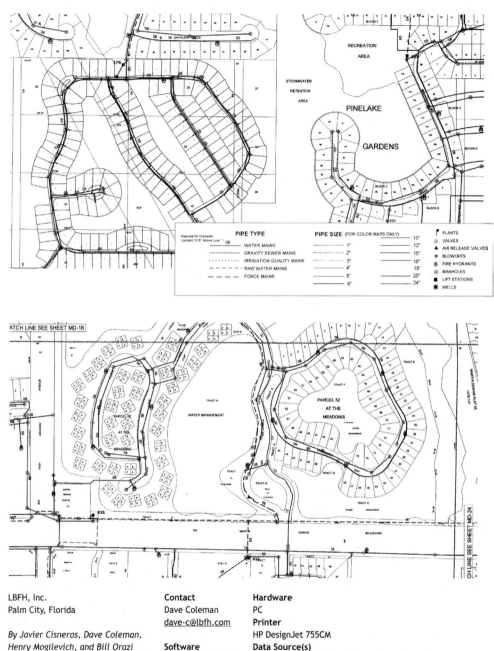

LBFH, Inc.
Palm City, Florida

By Javier Cisneros, Dave Coleman,
Henry Mogilevich, and Bill Orazi

Contact
Dave Coleman
dave-c@lbfh.com

Software
ArcView GIS 3.1

Hardware
PC
Printer
HP DesignJet 755CM
Data Source(s)
Martin County GIS and Martin County
Utilities as-built drawings

FIGURE 9-10 Typical GIS application—utilities as-built drawings
(Courtesy of David S. Coleman and Henry Mogilevich, LBFH Inc., Palm City, FL)

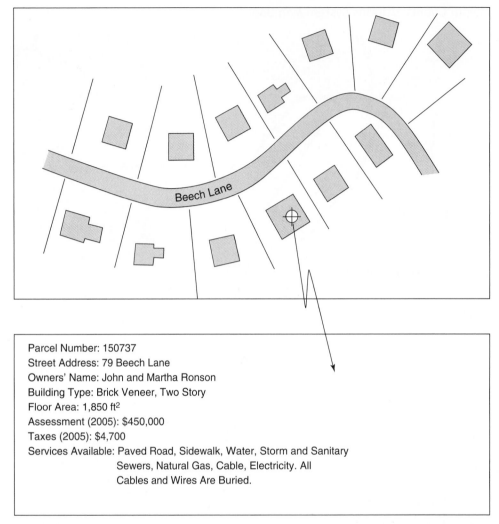

Parcel Number: 150737
Street Address: 79 Beech Lane
Owners' Name: John and Martha Ronson
Building Type: Brick Veneer, Two Story
Floor Area: 1,850 ft^2
Assessment (2005): $450,000
Taxes (2005): $4,700
Services Available: Paved Road, Sidewalk, Water, Storm and Sanitary
                    Sewers, Natural Gas, Cable, Electricity. All
                    Cables and Wires Are Buried.

FIGURE 9-11 Zoomed-in portion of GIS layer showing the type of feature data available with specific cursor selections

is also dependent on the type of data that will be used. Raster data can take much more computing than vector data. A GIS for a state may have much more data than a GIS for a city. As companies, consultants, or agencies become comfortable with GIS, the number of uses increases significantly; therefore, it is common to design the hardware for a much larger capacity than what is initially intended.

The speed and memory capabilities of computer systems are continually increasing, so there is no sense for this textbook to try and provide minimum requirements. It is important to understand that GIS is a computing and memory intensive process, much like CAD; the hardware systems should be at the high-end range to be productive in the application of GIS.

9.7.1.2 Field Hardware Field hardware used to collect GIS data is primarily GPS. However, the GPSs often are not survey-grade

GPS, but what is commonly called GIS grade GPS. These types of GPSs are ones that have differential corrections providing positional accuracies in the range of 0.5–3 m (horizontal). Along with the GPS unit, many of the handheld data collectors are low-cost, touch screen–based units as opposed to the ruggedized data collectors used by general surveying systems.

Beyond GPSs, range finders, handheld total stations, sensors, and digital cameras can be added to the system to collect additional information. Range finders are used to locate features that cannot be located with GPS alone. The range finders generally will have a reflectorless EDM for distance and a digital compass for direction. See Section 5.18 for a discussion on handheld total stations. Although a classical total station can be used and is more precise, they are generally not used in GIS data collection except when high precision is required (see Figure 9-3).

9.7.2 Software

9.7.2.1 Office Software There are a number of GIS software systems available. The most common commercial GIS software is the ArcGIS software collection from ESRI of Redlands, California. However, there are a number of other systems such as Autodesk Map series and ERDAS IMAGINE family. There are also free GIS software platforms such as GRASS (http://grass.osgeo.org/) and OpenGIS (http://www.opengeospatial.org/ogc). Originally, the GIS programs were developed by large agencies or universities. Many of these original programs are still being used in some updated version.

There are a variety of versions of the software, the idea being the user is able to tailor the software to meet their needs. In general there are desktop versions, small server versions, large server versions, Web-based versions, and combinations of all of them. Although this does give the user much more flexibility, it comes at the price of complexity.

9.7.2.2 Field Software There are a number of GIS software packages for field collection of

data. Software for field collection of data is sometimes called mobile software. A number of the mobile software packages are developed for a specific industry using GIS, such as the Utility industry. These software packages differ from the surveying field software, in that there is much emphasis in attribute data entry. This allows multiple notes about a single object to be entered. For instance, if the location of a power pole was being collected, the software might ask the field person to enter the date of the pole, the length, the number and type of wires, and the general condition of the pole. The GIS mobile software does not have the computing capabilities associated with surveying data collection software, but generally is easier to use for people not familiar with the software. Many of the developers of surveying data collection software are adding the capability of multiple attribute association with a single object. However, there will always be GIS specific field software that sacrifices capabilities for ease of use.

9.7.3 Database

Besides the GIS software, a database system is sometimes required. A database is the storage of data in an organized format. There are a number of formats used, but they are all created for the ease of searching through large set of data. Most of the GIS software will have some basic flat file type database associated with the software. This is a simple file format that works well for moderate size data sets with a single or just a few number of users. The database will store the information about the object location and shape as well as all the attributes associated with that object.

On large GIS implementations a large-scale relational database will be built. These relational databases are far more flexible in searching and indexing, can handle very large data sets, and can be accessed by a large number of users at one time. The downside of the relational data set is the complexity involved in building the database for the GIS.

When developing the database structure for a GIS, much consideration must be given to what the final product is intended to be.

A well-designed database will allow quick extraction of information needed for the analysis the GIS was designed for.

A newer development in database management concerns managing large data sets of images. Google Earth is probably the best known use of deployment of large image data sets. The database organizes all the images based on their geographic position and resolution quality. The system then provides the user a small snapshot of the area defined by the virtual location displayed on the screen and the apparent altitude of the viewer. The systems must do this using extremely large data sets with near instantaneous responses for large group of users at one time. Think about an Olympic-size pool full of ping pong balls, each with a specific address written on it. You need to select 100 specific ping pong balls out of the pool within a second and then another specific 100 the next second. This would be a small image database server.

9.7.4 GIS Manager

The larger the implementation, the more complex the management of the system becomes. It is important to understand that taking GIS from an individual user to a multiuser application requires a GIS manager who has a comprehensive understanding of the GIS software, information technology, and network systems. Failure of GIS implementations is often caused by a lack of understanding of the overall requirements by the GIS manager. The GIS manager is the person who will make the GIS a useful tool to the users of the system.

The advancements in the software have made the position of GIS manager a more specialized field requiring constant training on the newest capabilities. The GIS manager is the person who must make the critical decision of what is the most cost-effective approach to their GIS. In the years to come this will require the individual to be an expert in information technology, business administration, and public policy.

9.8 TYPES OF DATA

The data used in a GIS can be broken down into three main categories. There are vector based, raster based, and tabular data. There is one additional type of data called metadata. It is often referred to as "data about the data." Each one of these types of data has specific characteristics and uses. Understanding these characteristics and uses is the first step to understanding how to use GIS as a tool. Figure 9-12 shows various types of data sets.

CAD to GIS and GIS to CAD

When the CAD and GIS started, they were considered software for different disciplines that would not overlap. Soon it was clear that much of the mapping that would benefit GIS was being done in CAD. This made it of interest to the GIS community to be able to convert CAD files to GIS files. The structures of the two systems were not completely compatible. Two main issues had to be dealt with. First generally mapping in CAD was done on an assumed coordinate system, and often not based on real world dimensions, but just on mapping dimensions. Second CAD standards did not require or even recommend that layers be object specific. This means a single layer could contain polylines, lines, points, and text. In addition CAD did not enforce polygon objects (an object that starts and ends on the same point). This required quite a bit of cleanup on the GIS side when converting CAD files.

Soon the CAD industry noticed that CAD was being used for creating much of the line work for GIS systems. The CAD industry wanted to advance in this market so soon their software had the capability of converting their files to GIS file formats. In addition the CAD market started to develop GIS software packages. To compete with this, the GIS industry developed even better conversion tools, so that now the process is relatively simple. It has also been more common to develop CAD drawings on specific coordinate systems, such as state plane coordinates.

Today both areas are still completely distinct in the nature of their use, but have established the ability to use either's data with only minor cleanup being done by the users.

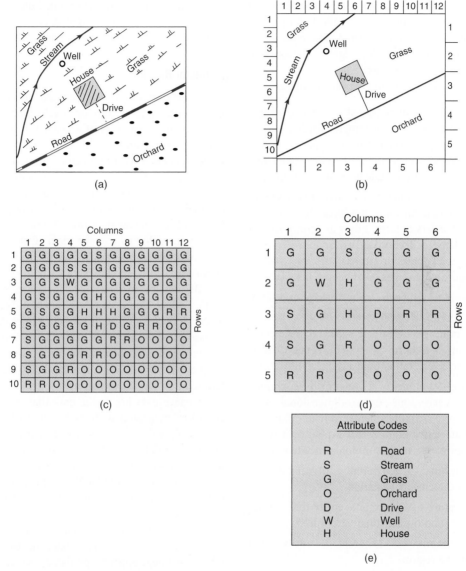

FIGURE 9-12 Different methods of displaying spatial data. (a) Topographic map. (b) Vector model. (c) Raster model (120 grid cells—moderate resolution). (d) Raster model (30 grid cells—coarse model). (e) Attribute codes

1. Vector based

Vector-based data are data that can be recreated by a mathematical formula. This is the type of data that is often thought of as CAD data. The common types used in GIS are points, lines, and polygons. However, in understanding how GIS manipulates

vector information, you need to understand how it sees this vector information. The National Geospatial-Intelligence Agency (http://www.nga.mil/portal/site/nga01/) has an "interface standard for vector product format" Mil Std 2407, which defines various vector types or what they call features. Features are composed of what they call primitives. A primitive is the smallest component of which all features are created. There are three geometric primitives (nodes, edges, faces) and one cartographic primitive (text).

Below are the definitions of each of these:

Node—A zero-dimensional geometric primitive that is composed of a single coordinate tuple (pair or triple). There are two types of nodes: entity nodes and connected nodes. Only one node can occupy a single geographic location.

Edge—A one-dimensional primitive used to represent the location of a linear feature and/or the borders of faces. Edges are composed of an ordered collection of two or more coordinate tuples (pairs or triplets). At least two of the coordinate tuples must be distinct.

Face—A region enclosed by an edge or a set of edges (a face has area).

Text primitive—Characters placed in specific locations in a coordinate system. Text is a cartographic object, rather than a geographic entity.

Feature—A model of a real-world geographic entity. A zero-, one-, or two-dimensional entity of uniform attribute scheme from an exhaustive attribute distribution across a plan or a set of such entities sharing common attribute values.

Point feature—A geographic entity that defines a zero-dimensional location.

Line feature—A geographic entity that defines a linear (one-dimensional) structure.

Area feature—A geographic entity that encloses a region.

Text feature—A cartographic entity that relates a textual description to a zero- or one-dimensional location.

Complex feature—A single feature that relates directly to other features rather than to a primitive.

Each object or feature then can have attributes associated with it. This is what is stored in the database, along with the feature geometry. Each layer has the same type of features so that the attribute list can be consistent. This provides solid database structure that allows systematic analysis of data.

2. Raster Based

Raster data are commonly thought of as images. Raster data have a cell- or pixel-based format, where each cell or pixel of a raster file has a specific size (length and width), location, and a single value associated with that cell or pixel. The array of cells or pixels is recorded in a specific format. There are a number of different formats that can be used for storage of raster data. Any program that reads the raster data must know the correct format in order to use the data. Color images have multiple layers which have the cells or pixel aligned but different values for each layer.

Besides image raster data there are also thematic and surface raster data. Thematic data are where there are a finite number of values that can be associated with any one cell. Thematic data are theme-based data showing spatial limits of a single theme. This could include land classifications, temperature ranges, or zoning information.

Surface data are raster data that contain an elevation associated with each cell. Surface data commonly will be stored in different formats as other raster data and often are displayed as shading or contours. There are also vector-based surfaces, such as triangulated irregular networks (TIN) or digital elevation models (DEM).

The difference between raster and vector data is that raster data have a fixed cell size. This can be seen when the data are scaled up to a point where the image appears "pixilated," which means the boundary of the cell is more defined than the image that it is trying to convey.

The smaller the cell size, the more detail there is and the larger the file size. Raster data tend to create significantly larger file sizes than vector data.

3. Tabular

Tabular data are just as important, if not more so, as the vector and raster data. However, they are not visual, and so they are not easily recognized.

Tabular data contain all the attributes associated with the features. Each feature is going to have a field that contains a unique identity for that object. Then it will have information about the object such as the type of object and the geometry of the object. It then can have any nonspatial information that is critical to the user. This might include some sort of identifier that can be used to link to other tabular data. A polygon type feature showing land parcels could have the street address associated with the parcel as an attribute. That street address could be used to access an assessor's database that would then allow for additional information to be linked to the parcel, such as the owner's name and the zoning of the property.

4. Metadata

Metadata give the information being used and collected in a GIS a sort of lineage. This allows a user to understand what the data are showing, when they were created, where they came from, and what the quality of the information is. All information has some value; the question becomes, does the value associated with a specific data set achieve the required needs of the task at hand?

In the United States, within the federal government there is the Federal Geographic Data Committee (FGDC). This group is an interagency committee that promotes the coordinated development, use, sharing, and dissemination of geospatial data on a national basis. They have established a structure for federal metadata in a 90-page report titled "Content Standard for Digital Geospatial Metadata" (http://www.fgdc.gov/). Although this only pertains to federal agencies, it is one of the most common formats for saving metadata.

9.9 GLOSSARY

Address matching   GIS geocoding software that can identify street addresses (such as those compiled in the US Census Bureau's TIGER data) by conversion from grid coordinates.

arc   An arc is a line entity, used in topology. See link.

attribute   This is a nongraphical descriptor of an geographic feature or entity; it can be qualitative or quantitative—such as numerical codes tied to qualitative descriptors, for example, the number 8 representing a commercial development@ on a land-use layer.

buffer   A specified zone around an entity which can be used to query a database.

cell   The basic rectangular element of a raster grid—it is called a pixel in digital images.

centroid   The centerpoint (or other identified point) of a polygon at which location attribute information for that polygon may be tagged. For example, for a polygon representing a specific neighborhood, population, assessment, voting preferences, etc., all may be stored at the designated centroid coordinates or grid cell.

chain   Directional and nonintersecting arcs or strings having nodes at each end are called chains in topology.

digitizing   This involves the conversion of analog data from existing map sheets into digital (co-ordinated) data by using a cursor on a digitizing table.

entity   Also called feature, this is a real-world object that can be geographically positioned or located.

layer   In a GIS, a layer is a collection of similar (thematic) data for a given geographic area, which is stored in a specific location in computer memory.

line   A one-dimensional entity that directly links two coordinated end points.

link   A one-dimensional topological entity that directly connects two nodes. A directed link defines direction from a from node to a to node. See arc.

network   An entity of interconnected lines that permits the analysis of route flow.

node   A zero-dimensional topological point entity which represents a beginning or ending of chains (representing arcs or lines)—including intersections.

overlay   The GIS techniques of selectively comparing and combining layers (categories) of entities and their variable attributes for a specified ground area, which have been stored on separate thematic layers.

pixel Like a raster grid cell, a pixel (picture element) is the basic element of a digital image (e.g., satellite image or scanned image)—whose resolution represents a defined amount of geographic space. See Chapter 11 ("Satellite Imagery").

point A zero-dimensional entity that specifies geometric location (e.g., by coordinates).

relational database A database which stores attribute and spatial data in the form of tables which all can be linked through a common identifier (e.g., primary key)—such as a coordinated spatial locator.

routing Networks are analyzed to determine optimal travel paths or flow from one point to another.

rubber sheeting This is a process of arbitrarily reconciling (stretching) data from different sources which may not fit perfectly together because of problems occasioned by differing scales, orientations, and projections.

scale In GIS and mapping, scale refers to the size ratio between a ground distance or area and the depiction of that ground distance or area on a map or raster grid.

scanner This is a raster input device which scans a document one line at a time and records the location of scanned objects in a raster matrix.

segmentation This is the creation of new lines or arcs as changes occur to the original line or arc—such as those that occur when there is a change in attributes (a two-lane road becomes a four-lane road, or a valve is inserted into a water distribution line) or the creation of new segments caused by intersecting lines or polygons.

string A sequence of line segments.

thematic layer Computer storage layers or maps that display the geographic locations of spatial entities which have one or more common attributes.

TIGER Topologically Integrated Geographic Encoding and Referencing System—this is a topological data model designed by the U.S. Bureau of Census. Much data digitized for this model have come from small-scale USGS maps.

vertex Similar to a node, a vertex point entity signals a simple change in direction of a line or arc (but not an intersection).

9.10 INTERNET WEBSITES

The websites listed here include links to various related sites. Although these websites were verified at the time of writing, some changes are inevitable. Corrected site addresses and new sites may be accessed by searching the links and websites provided here, or by conducting a search via an Internet search engine.

American Association of Geographers (AAG) www.aag.org

American Association of State Highway and Transportation Officials (AASHTO) www.aashto.orgwww.transportation1.org/aashtonew

American Congress on Surveying and Mapping (ACSM) www.acsm.net

American Society for Photogrammetry and Remote Sensing (ASPRS) www.asprs.org

Association for Geographic Information www.agi.org.uk/

Blue Marble Geographics www.bluemarblegeo.com

California Technology Agency Image Server for NAIP Photography http://www.cio.ca.gov/wiki/Imagery.ashx

Canadian Geodetic Survey http://www.geod.nrcan.gc.ca

Clark Labs (Idrisi) www.clarklabs.org

CORS information http://www.ngs.noaa.gov/CORS/

Earth Resources Observation and Science (EROS) http://edc.usgs.gov/

Environmental Systems Research Institute (ESRI) www.esri.com

ESRI (free viewer) www.esri.com/software/arcexplorer/index.html

Farm Service Agency, Aerial Photography Field Office http://www.apfo.usda.gov/FSA/apfoapp?area=apfohome&subject=landing&topic=landing

Geographic Coordinate Database http://www.blm.gov/wo/st/en/prog/more/gcdb.html

GeoConnections (Canada) www.geoconnections.ca

Kingston University Centre for Geographical Information System (UK) http://www.kingston.ac.uk/centreforGIS/

Intergraph Corporation http://www.intergraph.com/photo/default.aspx

Land Survey Information (BLM) www.geocommunicator.gov/

National Atlas (public land survey) www.nationalatlas.gov/

National Center for Geographic Information and Analysis www.ncgia.ucsb.edu/

National Geodetic Survey (NGS) homepage http://www.ngs.noaa.gov/

North American Atlas http://geogratis.cgdi.gc.ca/clf/en?action=northAmericanAtlas

Open GIS Consortium (OGC) www.opengis.org/techno/specs

Penn State Geographic Visualization Science Technology www.geovista.psu.edu/

POB (Point of Beginning) Magazine http://www.pobonline.com

Professional Surveyor Magazine http://www.profsurv.com

Solid terrain modeling www.stm-usa.com

Trimble GIS www.trimble.com/mappinggis.html

U.S. Census Bureau www.census.gov

U.S. Geological Survey http://www.usgs.gov/

U.S. Spatial Data Transfer Standard (SDTS) information http://mcmcweb.er.usgs.gov/sdts/

Urban and Regional Information Systems Association (URISA) www.urisa.org

9.11 PUBLICATIONS

Chrisman, N. (1997), *Exploring Geographic Information Systems*, John Wiley and Sons, New York.

DeMers, M. (1997), *Fundamentals of Geographic Information Systems*, John Wiley and Sons, New York.

Easa, S., and Chan, Y. (2000), *Urban Planning and Development Applications of GIS*, American Society of Civil Engineers (ASCE), Reston, VA.

ERDAS (1999), *Field Guide*, 5th edition, Atlanta, Georgia.

Foresman, Timothy (1998), *The History of Geographic Information Systems*, Prentice Hall Inc., Upper Saddle River, NJ.

Heywood, I., Cornelius, S., and Carver, C. (1998), *An Introduction to Geographical Information Systems*, Prentice Hall Inc., Upper Saddle River, NJ.

Korte, G. B. (2001), *The GIS Book*, 5th edition, Onward Press, Australia.

Worboys, Michael F. (1995), *GIS: A Computing Perspective*, Taylor and Francis Inc., Bristol, PA.

Questions

9.1 How do scale and resolution affect the creation and operation of a GIS?

9.2 What are the various ways that scale can be shown on a map?

9.3 When is it advantageous to use cylindrical projections? To use conical projections? (See also Chapter 11.)

9.4 Which features are best modeled using vector techniques? Using raster techniques?

9.5 List and describe as many GIS applications as you can.

9.6 Accepting that GIS is difficult to define, in your own words describe GIS, as you see it.

9.7 What are the similarities and differences between mapping and GIS?

9.8   What are the similarities and differences between GIS and CAD?

9.9   What modern developments enabled the creation of GIS?

9.10   Why is it important to tie spatial data to a recognized reference system?

9.11   Why is it that the conversion of raster data to vector data has the potential for locational errors?

9.12   How do you measure polygon areas in a vector model? In a raster model?

9.13   What would the appropriate positional precision be for a GIS showing the following information?

Watershed limits.

Assessing the change of redwood growths in western United States.

Mapping row crops in the Midwest.

Locating water and sewer features in a mid-size city.

Determining legal street right of ways in a mid-size city.

9.14   For the list in the above question, what might be the appropriate tool for collecting the positional information for each item listed?

9.15   If a digital image covered an area of 2 km by 2 km and the pixel size is 1 m, how many pixels are there in the total image?

9.16   In the question above, if the pixel size is 0.5 m, how many pixels are there?

9.17   What coordinate system would you expect a digital raster graphic (DRG) from USGS of a 7½-minute Quad sheet to be on?

CHAPTER TEN

CONTROL SURVEYS

10.1 GENERAL BACKGROUND

The highest order of control surveys was once thought to be national or continental in scope. With the advent of the global positioning system (GPS) (see Chapter 7), control surveys are now based on frameworks that cover the entire surface of the earth; as such, they must take into account the ellipsoidal shape of the earth.

The early control net of the United States was tied into the control nets of both Canada and Mexico, giving a consistent continental net. The first major adjustment in control data was made in 1927, which resulted in the North American Datum (NAD27). Since that time, a great deal more has been learned about the shape and mass of the earth; these new and expanded data come to us from releveling surveys, precise traverses, very long baseline interferometry (VLBI), satellite laser ranging (SLR) surveys, earth movement studies, GPS observations, gravity surveys, and other measurements. The data thus accumulated have been utilized to update and expand existing control data. The new data have also provided scientists with the means to define more precisely the actual geometric shape of the earth.

The reference solid previously used for this purpose (the Clarke spheroid of 1866) was modified to reflect the knowledge of the earth's dimensions at that time. Thus, a World Geodetic System, first proposed in 1972 (WGS72) and later endorsed in 1979 by the International Association of Geodesy (IAG), included proposals for an earth mass–centered ellipsoid (GRS80 ellipsoid)

that would represent more closely the planet on which we live. The ellipsoid was chosen over the spheroid as a representative solid model due to the slight bulging of the earth near the equator caused by the earth spinning on its polar axis. See Table 10-1 for the parameters of four models.

GRS80 was used to define the North American Datum of 1983 (NAD83), which covers the North American continent, including Greenland and parts of Central America. All individual control nets were included in a weighted simultaneous computation. A good tie to the global system was given by satellite positioning. The geographic coordinates of points in this system are latitude (φ) and longitude (λ). Although this system is widely used in geodetic surveying and mapping, it has been too cumbersome for use in everyday surveying. For example, the latitude and longitude angles must be expressed to three or four decimals of a second (01.0000″) to give position to the closest 0.01 ft. At latitude 44°, 1 second of latitude equals 101 ft and 1 second of longitude equals 73 ft. Conventional field surveying (as opposed to control surveying) is usually referenced to a plane grid (see Section 10.2).

In most cases, the accuracies between NAD83 first-order stations were better than 1:200,000, which would have been unquestioned in the pre-GPS era. However, the increased use of very precise GPS surveys and the tremendous potential for new applications for this technology created a demand for high-precision upgrades, using GPS techniques, to the control net. See Section 2.1 for vertical control specifications.

Table 10-1 Typical Reference Systems

Reference System	a (Semi-Major) m	b (Semi-Minor) m	1/f (Flattening)
NAD83 (GRS80)	6,378,137.0	6,356,752.3	298.257222101
WGS84	6 378,137.0	6,356,752.3	298.257223563
ITRS	6 378,136.49	6,356,751.75	298.25645
NAD27 (Clarke, 1866)	6,378,206.4	6,356,583.8	294.978698214

10.1.1 Modern Considerations

A cooperative network upgrading program under the guidance of the National Geodetic Survey (NGS), including both the federal and state agencies, began in 1986 in Tennessee and was completed in Indiana in 1997. This High Accuracy Reference Network (HARN)—sometimes called the High Precision Geodetic Network (HPGN)—resulted in about 16,000 horizontal control survey stations being upgraded to either AA-order, A-order, or B-order status. Horizontal AA-order stations have a relative accuracy of 3 mm ± 1:100,000,000 relative to other AA-order stations; horizontal A-order stations have a relative accuracy of 5 mm ± 1:10,000,000 relative to other A-order and AA-order stations; horizontal B-order stations have a relative accuracy of

Table 10-2 Positioning Accuracy Standards

Survey Categories	Order	Minimum Geometric Accuracy Standard (95 Percent Confident Level)		
		Base Error	Line-Length Dependent Error	
		e (cm)	p (ppm)	a (1:a)
HARN				
Federal base network (FBN)				
Global-regional geodynamics	**AA**	0.3	0.01	1:100,000,000
National Geodetic Reference				
System, primary network	**A**	0.5	0.1	1:10,000,000
Cooperative base network (CBN)				
National Geodetic Reference	**B**			
System, secondary networks	**C**	0.8	1	1:1,000,000
Terrestrial based				
National Geodetic Reference System	1	1.0	10	1:100,000
	2-I	2.0	20	1:50,000
	2-II	3.0	50	1:20,000
	3	5.0	100	1:10,000

From Geometric Geodetic Accuracy Standards Using GPS Relative Positioning Techniques. [Federal Geodetic Control Subcommittee (FGCS) 1988]. Publications are available through the National Geodetic Survey (NGS), (301) 443-8631.

8 mm ± 1:1,000,000 relative to other AA-order, A-order, and B-order stations. Of the 16,000 survey stations, the NGS has committed to the maintenance of about 1,400 AA- and A-order stations, named the federal base network (FBN); individual states maintain the remainder of the survey stations, the B-order stations, called the cooperative base network (CBN). See Table 10-2 for accuracy standards for the new HARN using GPS techniques and for traditional (pre-GPS) terrestrial techniques.

The Federal Geodetic Control Subcommittee (FGCS) has published guidelines for the GPS field techniques (see Chapter 7) needed to achieve the various orders of surveys shown in Table 10-2. For example, order AA, A, and B surveys require the use of receivers having both L1 and L2 frequencies, whereas the C-order results can be achieved using only a single-frequency (L1) receiver.

Work was also completed on an improved vertical control net with revised values for about 600,000 benchmarks in the United States and Canada. This work was largely completed in 1988, resulting in a new North American Datum (NAVD88). The original adjustment of continental vertical values was performed in 1927.

The surface of the earth has been approximately duplicated by the surface of an oblate ellipsoid, that is, the surface developed by rotating an ellipse on its minor axis. An ellipse (defined by major axis and flattening) was originally chosen that most closely conformed to the geoid of the area of interest, which was usually continental in scope. The reference ellipsoid chosen by the United States and Canada was one recommended by the IAG called the Geodetic Reference System 1980 (GRS80). See Table 10-1 for reference ellipsoid parameters, and see Figures 7-16 through 7-18 for more on this topic.

The origin of a three-dimensional coordinate system was defined to be the center of the mass of the earth (geocentric), which is located in the equatorial plane. The z axis of the ellipsoid is defined as running from the origin through the mean location of the North Pole, more precisely, the International Reference Pole as defined by the International Earth Rotation Service (IERS); z coordinates are measured upward (positive) or downward (negative) from the equatorial plane (see Figure 10-1). The x axis runs from the origin to a point of 0° longitude (Greenwich meridian) in the equatorial plane. The x coordinates are

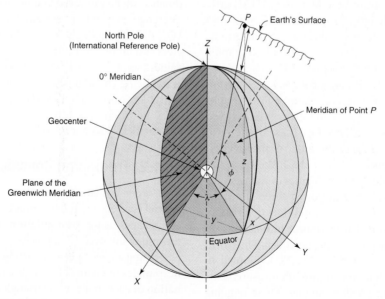

FIGURE 10-1 Ellipsoidal and Geographic Reference Systems

measured from the y-z plane, parallel to the x axis. They are positive for the zero meridian and up to 90° east and west; for the remaining 180° they are negative. The y axis forms a right-handed coordinate frame (easterly from the x axis) 90° to the x axis and z axis. The y coordinates are measured perpendicular to the plane through the zero meridian and are positive in the eastern hemisphere and negative in the western hemisphere. The *semimajor axis* (*a*) runs from the origin to the equator and the *semiminor axis* (*b*) runs from the origin to the earth's North Pole. Another defining parameter used to define ellipsoids is the *flattening* (*f*), which is defined to be $f = (a - b)/a$, or $f = 1 - b/a$ (see Table 10-1).

Figure 10-1 shows the relationships between the ellipsoidal coordinates x, y, and z and the geodetic (or geographic) coordinates of latitude (φ), longitude (λ), and ellipsoidal height (*h*). Note that the x, y, and z dimensions are measured parallel to the x, y, and z axes respectively, and that h is measured vertically up from the ellipsoidal surface to a point on the surface of the earth. For use in surveying, the ellipsoidal coordinates/geodetic coordinates are transformed into plane grid coordinates such as those used for the state plane grid or the universal transverse Mercator (UTM) grid; the ellipsoidal height (*h*) is also transformed into an orthometric height (elevation) by determining the geoid separation at a specified geographic location as described in Section 7.12.

The axes of this coordinate system have not remained static for several reasons; for example, the earth's rotation varies and the vectors between the positions of points on the surface of the earth do not remain constant because of plate tectonics. It is now customary to publish the (x, y, and z) coordinates along with velocities of change (plus or minus), in meters/year, for all three directions (vx, vy, and vz) for each station. For this reason, the axes are defined with respect to positions on the earth's surface at a particular epoch. The North American Datum of 1983 (NAD83), adopted in 1986, was first determined through measurements using VLBI and satellite ranging. These ongoing geodetic measurements together with continuous

GPS observations (e.g., CORS) have discovered discrepancies, resulting in several upgrades to the parameters of NAD83. The IERS continues to monitor the positioning of the coordinates of their global network of geodetic observation stations, which now include GPS observations; this network is known as the International Terrestrial Reference Frame, with the latest reference epoch, at the time of this writing, set at the year 2005 (ITRF2005 or ITRF05). For most purposes, the latest versions of NAD83 and WGS84 are considered identical. Because of the increases in accuracy occasioned by improvements in measurement technology, the NGS is commencing an adjustment to their National Spatial Reference System (NSRS) of all GPS HARN stations (CORS stations' coordinates will be held fixed). This adjustment, when combined with their newest geoid model (GEOID04), should result in horizontal and vertical coordinate accuracies in the 1- to 2-cm range, including orthometric heights. The adjustments will affect all HARN stations in the FBN, specifically in its AA- and A-order stations and all B-order stations in the CBN. This adjustment, scheduled to begin in 2005, and both NAD83 (NSRS) and ITRF05 [or the latest ITRF (e.g., ITRF200x)] positional coordinates will be produced and published. The ITRF reference ellipsoid is very similar to GRS80 and WGS84 with slight changes in the *a* and *b* parameters and more significant changes in the flattening values (see Table 10-1). NGS reports that NAD83 is not being abandoned because many states have legislation specifying that datum. See the latest NGS news on this topic at www.ngs.noaa.gov/NationalReadjustment/.

10.1.2 Traditional Considerations

First-order horizontal control accuracy using terrestrial (pre-electronic) techniques were originally established using **triangulation** methods. This technique involved (1) a precisely measured baseline (using a steel tape) as a starting side for a series of triangles or chains of triangles; (2) the determination of each angle in the triangle using a precise theodolite—which permitted the computation of the lengths of each side; and (3) a check on the work made possible by precisely measuring a side of a subsequent triangle (the spacing of check lines depended on the desired accuracy level). See Figure 10-2.

Triangulation was originally favored because the basic measurement of angles (and only a few sides) could be taken more quickly and precisely than could the measurement of all the distances (the surveying solution technique of measuring only the sides of a triangle is called **trilateration**).

The advent of EDM instruments in the 1960s changed the approach to terrestrial control surveys. It became possible to precisely measure the length of a triangle side in about the same length of time as was required for angle determination.

Figure 10-2 shows some two control survey configurations. Figure 10-2(a) depicts a simple chain of single triangles. In triangulation (angles only) this configuration suffers from the weakness that essentially only one route can be followed to solve for side *KL*. Figure 10-2(b) shows a chain of double triangles or quadrilaterals. This configuration was preferred for triangulation because side *KL* can be solved using different routes (many

more redundant measurements). Modern terrestrial control survey practice favors a combination of triangulation and trilateration (i.e., measuring both the angles and the distances), thus ensuring many redundant measurements even for the simple chain of triangles shown in Figure 10-2(a).

Whereas triangulation control surveys were originally used for basic state or provincial controls, precise traverses and GPS surveys are now used to densify the basic control net. The advent of reliable and precise EDM instruments has elevated the traverse to a valuable role, both in strengthening a triangulation net and in providing its own stand-alone control figure. To provide reliability, traverses must close on themselves or on previously coordinated points. Table 10-3 gives a summary of characteristics and specifications for traverses. Tables 10-4 and 10-5 show summaries of characteristics and specifications for vertical control for the United States and Canada.

More recently, with the advent of the programmed total station, the process called **resection** is used much more often. This process permits the surveyor to set up the total station at any convenient location and then by sighting

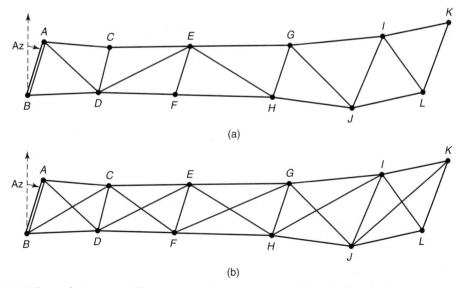

(a)

(b)

FIGURE 10-2 Control survey configurations. AB is the measured baseline, with known (or measured) azimuth. (a) Chain of single triangles. (b) Chain of double triangles (quadrilaterals)

Table 10-3   Traverse Specifications—United States

| Classification | First Order | Second Order | | Third Order | |
		Class I	Class II	Class I	Class II
Recommended spacing of principal stations	Network stations 10–15 km; other surveys seldom less than 3 km	Principal stations seldom less than 4 km, except in metropolitan area surveys, where the limitation is 0.3 km	Principal stations seldom less than 2 km, except in metropolitan area surveys, where the limitation is 0.2 km	Seldom less than 0.1 km in tertiary surveys in metropolitan area surveys: as required for other surveys	
Position closure After azimuth adjustment	0.04 m $\sqrt{K}$ or 1:100,000	0.08 m $\sqrt{K}$ or 1:50,000	0.2 m $\sqrt{K}$ or 1:20,000	0.4 m $\sqrt{K}$ or 1:10,000	0.8 m $\sqrt{K}$ or 1:5,000

Source: From Federal Control Committee, United States, 1974.

(measuring just angles or both angles and distances) to two or more coordinated control stations, the coordinates of the setup station can then be computed. See Section 5.13.4.

In traditional (pre-GPS) surveying, to obtain high accuracy for conventional field control surveys, the surveyor had to use high-precision techniques. Several types of high-precision equipment, used to measure angles, vertical and horizontal/slope distances, are illustrated in Figures 10-3–10-5. Specifications for horizontal high-precision techniques stipulate the least angular count of the theodolite, the number of observations, the rejection of observations exceeding specified limits from the mean, the spacing of major stations, and the angular and positional closures.

The American Congress on Surveying and Mapping (ACSM) and the American Land Title Association (ALTA) collaborated to produce new classifications for cadastral surveys based on present and proposed land use. These 1992 classifications (subject to state regulations) are shown in Table 10-4. Recognizing the impact of GPS techniques on all branches of surveying, in 1997 ACSM and ALTA published positional tolerances for different classes of surveys, but updated those

positional tolerances for all classes of surveys in 2005 to state that the positional tolerance would be 0.07 ft + 50 ppm.

Higher-order specifications are seldom required for engineering or mapping surveys—an extensive interstate highway control survey could be one example where higher-order specifications are used in engineering work. Control for large-scale projects (e.g., interchanges, large housing projects) that are to be laid out using polar ties (angle/distance) by total stations may require accuracies in the range of 1/10,000 to 1/20,000, depending on the project, and would fall between second- and third-order accuracy specifications (Table 10-2). Control stations established using GPS techniques inherently have the potential for higher orders of accuracy. The lowest requirements are reserved for small engineering or mapping projects that are limited in scope—for example, traffic studies, drainage studies, and borrow pit volume surveys. To enable the surveyor to perform reasonably precise surveys and still use plane geometry and trigonometry for related computations, several forms of plane coordinate grids have been introduced.

Table 10-4   American Congress on Surveying and Mapping—Minimum Angle, Distance, and Closure Requirements for Survey Measurements That Control Land Boundaries for ALTA-ACSM Land Title Surveys*

Direct Reading of Instrument†	Instrument Reading, Estimated‡	Number of Observations per Station§	Spread from Mean of D&R Not to Exceed‖	Angle Closure Where N = Number of Stations Not to Exceed	Linear Closure#	Distance Measurement**	Minimum Length of Measurements‡#
20″ <1′> 10″	5″ <0.1′> N.A.	2 D&R	5″ <0.1′> 5″	10″√N	1:15,000	EDM or doubletape with steel tape	81 m††, 153 m‡‡. 20 m§§

*All requirements of each class must be satisfied to qualify for that particular class of survey. The use of a more precise instrument does not change the other requirements, such as number of angles turned, etc.

†Instrument must have a direct reading of at least the amount specified (not an estimated reading), that is, 20″ = micrometer reading theodolite, <1> = scale reading theodolite, 10″ = electronic reading theodolite.

‡Instrument must have the capability of allowing an estimated reading below the direct reading to the specified reading.

§D&R means the direct and reverse positions of the instrument telescope: for example, urban surveys require that two angles in the direct and two angles in the reverse position be measured and meaned.

‖Any angle measured that exceeds the specified amount from the mean must be rejected and the set of angles remeasured.

#Ratio of closure after angles are balanced and closure calculated.

**All distance measurements must be made with a properly calibrated EDM or steel tape, applying atmospheric, temperature, sag, tension, slope, scale factor, and sea level corrections as necessary.

††EDM having an error of 5 mm, independent of distance measured (manufacturer's specifications).

‡‡EDM having an error of 10 mm, independent of distance measured (manufacturer's specifications).

§§Calibrated steel tape.

Table 10-5   State Plane Coordinate Grid Systems

Transverse Mercator System		Lambert System		Both Systems
Alabama	Mississippi	Arkansas	North Dakota	Alaska
Arizona	Missouri	California	Ohio	Florida
Delaware	Nevada	Colorado	Oklahoma	New York
Georgia	New Hampshire	Connecticut	Oregon	
Hawaii	New Jersey	Iowa	Pennsylvania	
Idaho	New Mexico	Kansas	Puerto Rico	
Illinois	Rhode Island	Kentucky	South Carolina	
Indiana	Vermont	Louisiana	South Dakota	
Maine	Wyoming	Maryland	Tennessee	
		Massachusetts	Texas	
		Michigan	Utah	
		Minnesota	Virginia	
		Montana	Virgin Islands	
		Nebraska	Washington	
		North Carolina	West Virginia	
			Wisconsin	

(a)

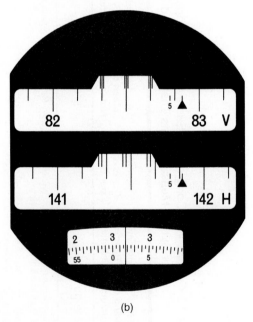

(b)

FIGURE 10-3 (a) Kern DKM 3 precise theodolite; angles directly read to 0.5 seconds; used in first-order surveys. (b) Kern DKM 3 scale reading (vertical angle 82°53′01.8″)

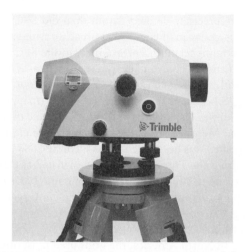

FIGURE 10-4 Precise level. Precise optical levels have accuracies in the range of 1.5 mm to 1.0 mm for 1 km two-way leveling depending on the instrument model and the type of leveling rod used. See also Figure 2-7

(Courtesy of Trimble)

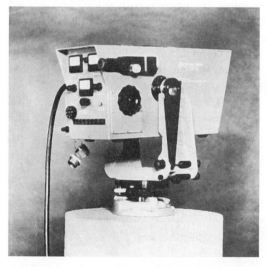

FIGURE 10-5 Kern Mekometer ME 3000, a high-precision EDM [SE (0.2 mm 1 ppm)] with a triple-prism distance range of 2.5 km. Used wherever first-order results are required, for example, deformation studies, network surveys, plant engineering, and baseline calibration

(Courtesy of Kern Instruments, Leica Geosystems)

10.2 PLANE COORDINATE GRIDS

10.2.1 General Background

The earth is ellipsoidal in shape, and portraying a section of the earth on a map or plan makes a certain amount of distortion unavoidable. Also, when creating plane grids and using plane geometry and trigonometry to describe the earth's curved surface, some allowances must be made. Over the years, various grids and projections have been employed. The United States uses the state plane coordinate grid system (SPCS), which utilizes both a transverse Mercator cylindrical projection and the Lambert conformal conic projection.

As already noted, geodetic control surveys are based on the best estimates of the actual shape of the earth. For many years, geodesists used the Clarke 1866 spheroid as a base for their work, including the development of the first North American Datum in 1927 (NAD27). The NGS created the state plane coordinate system (SPCS27) based on the NAD27 datum. In this system map projections are used that best suit the geographic needs of individual states; the Lambert conformal conical projection is used in states with a larger east/west dimensions (see Figures 10-6 and 10-7) and the transverse Mercator cylindrical projection (see Figure 10-8) is used in states with larger north/south dimensions (see Table 10-5).

To minimize the distortion that always occurs when a spherical surface is converted to a plane surface, the Lambert projection grid is limited to a relatively narrow strip of about 158 miles in a north/south direction, and the transverse Mercator projection grid is limited to about 158 miles in an east/west direction. At the maximum distance of 158 miles, or 254 km, a maximum scale factor error of 1:10,000 will exist at the zone boundaries.

As noted earlier, modernization in both instrumentation and technology permitted the establishment of a more representative datum based on the GRS80, which was used to define the new NAD83 datum. A new state plane coordinate

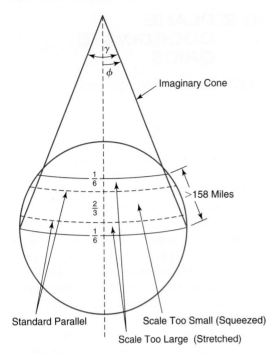

FIGURE 10-6 Lambert secant projection

system of 1983 (SPCS83) was developed based on the NAD83 datum.

Surveyors using both old and new SPCSs are able to compute positions using tables and computer programs made available from the NGS. The SPCS83, which enables the surveyor to work in a

more precisely defined datum than did SPCS27, uses similar mathematical approaches with some new nomenclature. For example, in the SPCS27, the Lambert coordinates were expressed as X and Y with values given in feet (U.S. survey foot, see Table 1-1), and the convergence angle (mapping angle) was displayed as θ; alternately, in the transverse Mercator grid, the convergence angle (in seconds) was designated by $\Delta\lambda''$. The SPCS83 uses metric values for coordinates (designated as eastings and northings) as well as foot units (U.S. survey foot or international foot). The convergence angle is now shown in both the Lambert and transverse Mercator projections as γ.

In North America, the grids used most often are the state plane coordinate grids. These grids are used within each state within the United States; the UTM grid is used in much of Canada. The Federal Communications Commission has mandated that, by 2005, all cell phones must be able to provide the spatial location of all 911 callers. By 2002, about half the telephone carriers opted for network-assisted GPS (A-GPS) for 911 caller location. The other carriers proposed to implement caller location using the enhanced observed time difference (E-OTD) of arrival; this technique utilizes the cellular network itself to pinpoint the caller location.

With such major initiatives in the use of GPS to help provide caller location, some believe that

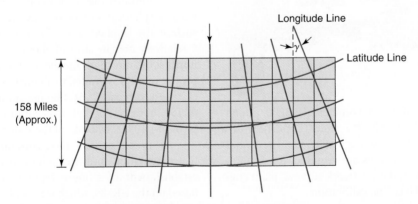

FIGURE 10-7 Lines of latitude (parallels) and lines of longitude (meridians) on the Lambert projection grid

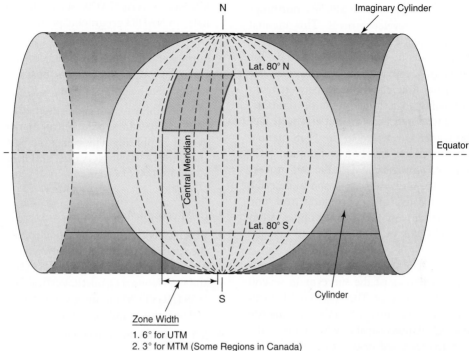

N

Imaginary Cylinder

Lat. 80° N

Central Meridian

Equator

Lat. 80° S

Cylinder

S

Zone Width

1. 6° for UTM
2. 3° for MTM (Some Regions in Canada)
3. About 158 Miles for State Plane Coordinate Grid (United States)

FIGURE 10-8 Transverse Mercator projection cylinder tangent to the earth's surface at the central meridian (CM) (see Figure 10-12 for zone number)

to provide seamless service, proprietary map databases should be referenced to a common grid, that is, a U.S. national grid (USNG) for spatial addressing (see Section 10.5.3). One such grid being considered is the military grid reference system (MGRS), which is based on the UTM grid. In addition to the need for a national grid for emergency (911) purposes, the Federal Geographic Data Committee (FGDC) recognizes the benefits of such a national grid for the many applications now developed for the geographic information system (GIS) field. The ability to share data from one proprietary software program to another depends on a common grid, such as that proposed in the USNG. See www.fgdc.gov/usng.

In addition to supplying tables for computations in SPCS83, the NGS provides both interactive computations on the Internet and PC software available for downloading at www.ngs.noaa.gov. Many surveyors prefer computer-based

computations to working with cumbersome tables. A manual that describes SPCS83 in detail, NOAA Manual NOS NGS 5, *State Plane Coordinate System of 1983*, is available from NGS.*

This manual contains an introduction to SPCS, a map index showing all state plane coordinate zone numbers (zones are tied to state counties), which are required for converting state plane coordinates to geodetic positions; a table showing the SPCS legislative status of all states (1988); a discussion of the (t-T) convergence "second term" correction which is needed for precise surveys of considerable extent (see Figure 10-18); and the methodology required to convert

---

*To obtain NGS publications, contact NOAA, National Geodetic Survey, N/NGS12, 1315 East-West Highway, Station 9202, Silver Springs, MD 20910-3282. Publications can also be ordered by phoning: (301) 713-3242.

NAD83 latitude/longitude to SPCS83 northing/easting; plus the reverse process. This manual also the four equations needed to convert from latitude/longitude to northing/easting (i.e., for northing, easting, convergence, and the grid scale factor) and the four equations to convert from northing/easting to latitude/longitude (latitude, longitude, convergence factor, and grid scale factor). The reader is referred to the NGS manual for these conversion equation techniques. NGS uses the term *conversion* to describe this process and reserves the term *transformation* to describe the process of converting coordinates from one datum or grid to another, for example, from NAD27 to NAD83, or from SPCS27 to SPCS83 to UTM, etc.

The NGS also has a range of software programs designed to assist the surveyor in several areas of geodetic inquiry. The NGS Tool Kit can be found at www.ngs.noaa.gov/TOOLS. This site has online calculations capability for many of the geodetic activities listed below (to download PC software programs, go to www.ngs.noaa.gov and click on the PC software icon):

- DEFLEC99 computes deflections of the vertical at the surface of the earth for the continental United States, Alaska, Puerto Rico, Virgin Islands, and Hawaii.
- G99SSS computes the gravimetric height values for the continental United States.
- GEOID09 computes geoid height values for the continental United States.
- HTDP is a time-dependent horizontal positioning software that allows users to predict horizontal displacements and/or velocities at locations throughout the United States.
- NADCON transforms geographic coordinates between the NAD27, Old Hawaiian, Puerto Rico, or Alaska Island data and NAD83 values.
- OPUS.
- State plane coordinate GPPCGP converts NAD27 state plane coordinates to NAD27 geographic coordinates (latitudes and longitudes) and the converse.

- SPCS83 converts NAD83 state plane coordinates to NAD83 geographic positions and the converse.
- Surface gravity prediction predicts surface gravity at specified geographic position and topographic height.
- The tidal information and orthometric elevations of a specific survey control mark can be viewed graphically. These data can be referenced to NAVD88, NGVD29, and mean lower low water (MLLW) data.
- USNG.
- UTM coordinates.
- VERTCON computes the modeled difference in orthometric height between the North American Vertical Datum of 1988 (NAVD88) and the National Geodetic Vertical Datum of 1929 (NGVD29) for any given location specified by latitude and longitude.
- X Y Z coordinate conversion.

In Canada, software programs designed to assist the surveyor in a variety of geodetic applications are available on the Internet from the Canadian Geodetic Survey at www.geod.nrcan.gc.ca/. Following is a selection of available services, including online applications and programs that can be downloaded:

- Precise GPS satellite ephemerides.
- GPS satellite clock corrections.
- GPS constellation information.
- GPS calendar.
- National gravity program.
- UTM to and from geographic coordinate conversion (UTM is in 6° zones with a scale factor of 0.9996).
- Transverse Mercator (TM) to and from geographic coordinate conversion (TM is in 3° zones with a scale factor of 0.9999, similar to U.S. state plane grids).
- GPS height transformation (based on GSD99; see Section 7.12).

10.2.2 Use of the NGS TOOLS to Convert Coordinates

(a) Convert geodetic positions to state plane coordinates. The user selects http://www.ngs .noaa.gov//TOOLS/spc.html and then selects "Latitude/longitude > SPC"; he or she is asked to choose NAD83 or NAD27, enter the geodetic coordinates, and enter the zone number (the zone number is not really required here as the program automatically generates the zone number directly from the geodetic coordinates of latitude and longitude). The coordinates are entered in degrees, minutes, and seconds (dddmmss format). The longitude degree entry must always be three digits, 079 in this example.

O **NAD 83**
O NAD 27
Latitude   **N421423.0000**
Longitude   **W0792035.0000**
Zone[ ] (This can be left blank)

The program response is:

Latitude INPUT= N421423.0000		Longitude W0792035.0000		Datum NAD83			Zone 3103

| North (Y)
Meters | East (X)
Meters | Area | Convergence | | | Scale |
			DD	MM	SS.ss	
248999.059	287296.971	NY W	− 0	30	38.62	0.99998586

(b) Convert state plane coordinates to geodetic positions. The user selects http://www.ngs.noaa .gov/TOOLS/spc.html and then selects "SPC > Latitude/longitude"; he or she must select either NAD83 or NAD27, enter the state plane coordinates, and enter the SPCS zone number.

O **NAD 83**
O NAD 27
Northing = **248999.059**
Easting = **287296.971**
Zone = **3103**

The program response is:

North (meters) INPUT = 248999.059		East (meters) 287296.971		Datum NAD83			Zone 3103

| Latitude | | | Longitude | | | Area | Convergence | | | Scale Factor |
DD	MM	SS.sssss	DD	MM	SS.sssss		DD	MM	SS.sssss	
42	14	23.00000	079	20	35.00001	NY W	-0E	30	38.62	0.9999859

10.2.3 Use of the Canadian Geodetic Survey Online Sample Programs

The programs can be downloaded free. Use the same geographic position as in Section 10.2.2.

a. Geographic position to universal transverse Mercator. Go to http://www.geod.nrcan.gc.ca/ apps/index_e.php and then select "Geographic to UTM." Enter the geographic coordinates of the point to be converted. For this example, enter the following:

Latitude: **42** degrees **14** minutes **23.0000** seconds **north**

Longitude: **079** degrees **20** minutes **35.0000** seconds **west**

Ellipsoid: **GRS80 (NAD83, WGS84)** Zone Width **6° UTM**

The desired ellipsoid and zone width, 6° or 3°, are selected by highlighting the appropriate entry while scrolling through the list. The program response is

Input Geographic Coordinates

Latitude: 42 degrees 14 minutes 23.0000 seconds North

Longitude: 079 degrees 20 minutes 35.0000 seconds West

Ellipsoid: NAD83 (WGS84)

Zone Width: 6° UTM

Output: UTM Coordinates:

UTM Zone: 17

Northing: 4677721.911 meters North

Easting: 636709.822 meters

b. Universal transverse Mercator (UTM) to geographic. Go to http://www.geod.nrcan.gc.ca/apps/gsrug/index_e.php and then select "UTM to Geographic." Enter the UTM coordinates of the point to be converted. For this example, enter the following:

Zone: 17

Northing: 4677721.911 meters North

Easting: 636709.822 meters

Ellipsoid: GRS80 (NAD83, WGS84)

Zone Width: 6° UTM

The program's response is

Input Geographic Coordinates

UTM Zone 17

Northing: 4677721.911 meters North

Easting: 636709.822 meters

Ellipsoid: NAD83 (WGS84)

Zone Width: 6° UTM

Output geographic coordinates

Latitude: 42 degrees 14 minutes 23.000015 seconds North

Longitude: 79 degrees 20 minutes 34.999989 seconds West

10.3 LAMBERT PROJECTION

The Lambert projection is a conical conformal projection. The imaginary cone is placed around the earth so that the apex of the cone is on the earth's axis of rotation above the North Pole, for northern hemisphere projections, and below the South Pole, for southern hemisphere projections. The location of the apex depends on the area of the ellipsoid that is being projected. Reference to Figures 10-6 and 10-7 confirms that although the east–west direction is relatively distortion-free, the north–south coverage must be restrained (e.g., to 158 miles) to maintain the integrity of the projection; therefore, the Lambert projection is used for states having a greater east–west dimension such as Pennsylvania and Tennessee. Table 10-5 lists all the states, indicating the type of projection being used; New York, Florida, and Alaska utilize both the transverse Mercator and the Lambert projections. The NGS publication *State Plane Coordinate Grid System of 1983* gives a more detailed listing of each state's projection data.

10.4 TRANSVERSE MERCATOR GRID SYSTEM

The transverse Mercator projection is created by placing an imaginary cylinder around the earth with its circumference tangent to the earth along a meridian (central meridian; see Figure 10-8). When the cylinder is flattened, a plane is developed that can be used for grid purposes. At the central meridian the scale is exact [Figures 10-8 and 10-9(a)], and the scale becomes progressively more distorted as the distance east and west of the central meridian increases. This projection is used in states with a more predominant north/south dimension such as Illinois and New Hampshire. The distortion (which is always present when a curved surface is projected onto a plane) can be minimized in two ways. First, the distortion can be minimized by keeping the zone width relatively narrow (158 miles in SPCS); and, second, the distortion can be lessened by reducing the radius of the projection cylinder (secant projection) so that, instead of being tangent to the earth's surface, the cylinder cuts through the earth's surface at an optimal distance on either side of the central meridian [see Figures 10-9(b) and 10-10]. This means that the scale factor at the

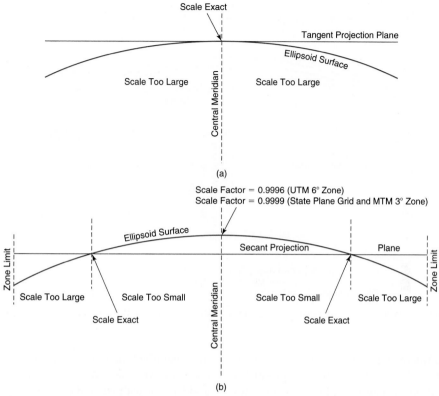

FIGURE 10-9 (a) Section view of the projection plane and earth's surface (tangent projection). (b) Section view of the projection plane and the earth's surface (secant projection)

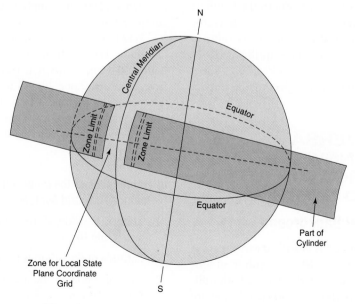

FIGURE 10-10 Transverse Mercato

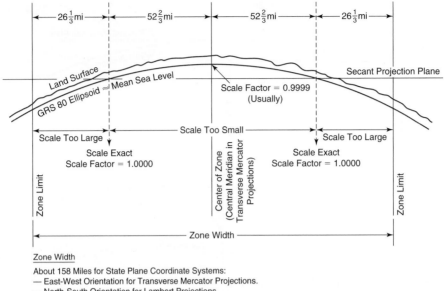

FIGURE 10-11 Section of the projection plane and the earth's surface for state plane grids (secant projection)

central meridian is less than unity (0.9999); it is unity at the line of intersection at the earth's surface and more than unity between the lines of intersection and the zone limit meridians. Figure 10-11 shows a cross section of a SPCS transverse Mercator zone. For both the Lambert and transverse Mercator grids, the scale factor of 0.9999 at the central meridian (this value is much improved for some states in the SPCS83) gives surveyors the ability to work within a specification of 1:10,000 while neglecting the impact of scale distortion.

10.5 UNIVERSAL TRANSVERSE MERCATOR (UTM) GRID SYSTEM

10.5.1 General Background

The UTM grid is much as described earlier except that the zones are wider, set at a width 6° of longitude. This grid is used worldwide for both military and mapping purposes. UTM coordinates are now published (in addition to SPCS and geodetic coordinates) for all NAD83 control stations. With a wider zone width than the SPCS zones, the UTM has a scale factor at the central meridian of only 0.9996. Surveyors working at specifications better than 1:2,500 must apply scale factors in their computations.

UTM zones are numbered, beginning at longitude 180°W, from 1 to 60; reference to Figure 10-12(a) shows that U.S. territories range from zone 1 to zone 20 and that Canada's territory ranges from zone 7 to zone 22. The central meridian of each zone is assigned a false easting of 500,000 m and the northerly is based on a value of zero at the equator.

Characteristics of the Universal Transverse Mercator (UTM) Grid System

1. Zone is 6° wide. Zone overlap of 0°30′ (see also Table 10-6).

2. Latitude of the origin is the equator, 0°.

3. Easting value of each central meridian = 500,000.000 m.

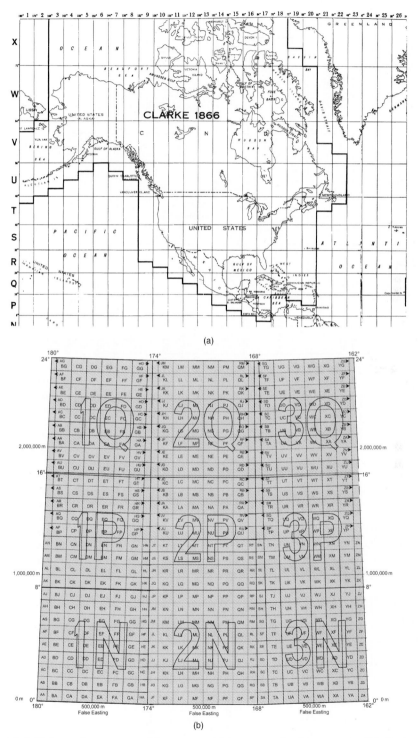

FIGURE 10-12 (a) Universal transverse Mercator grid zone numbering system. (b) Basic plan of the 100,000-m^2 identification of the United States national grid (USNG). (*Continued*)

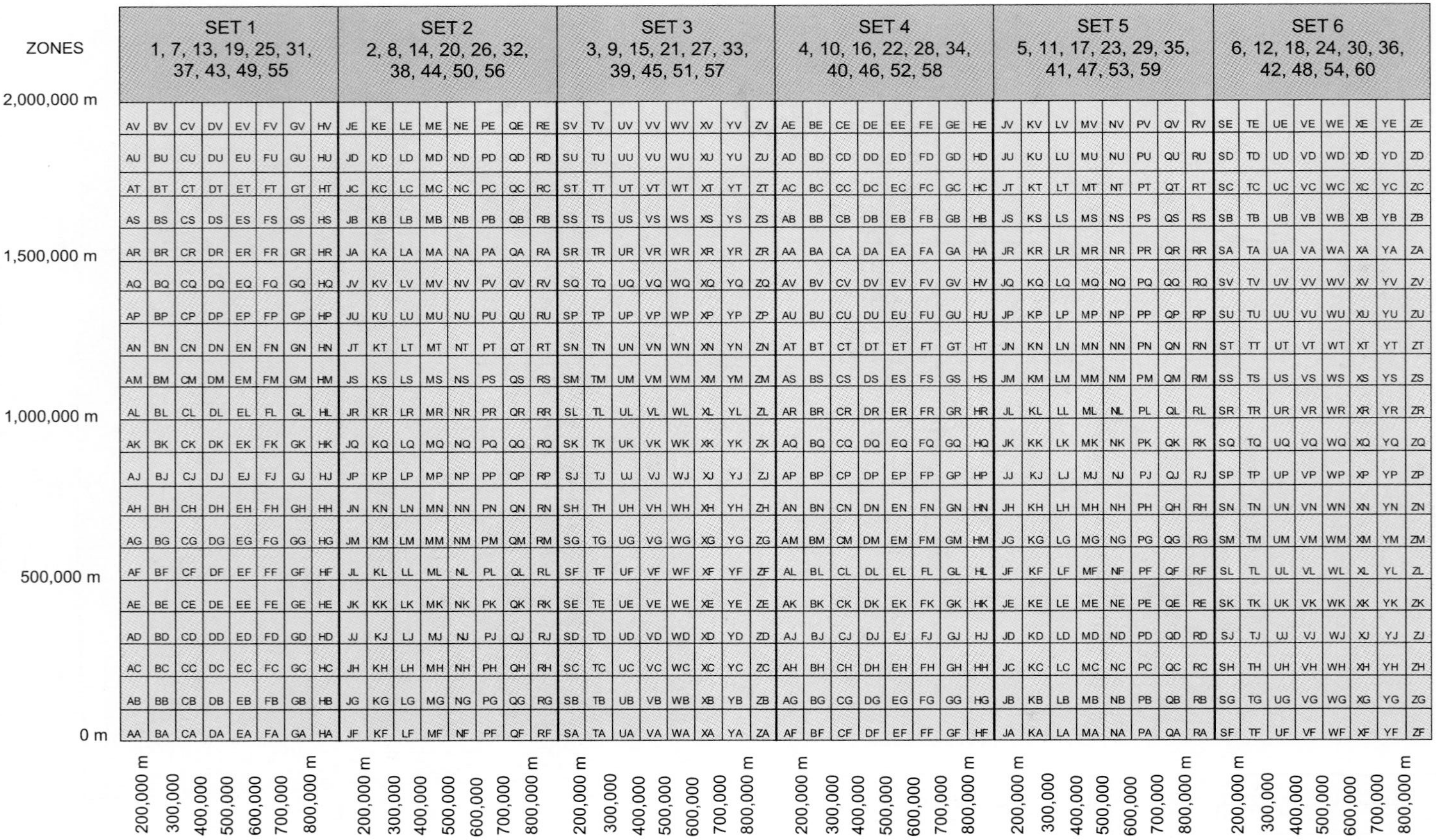

FIGURE 10-12 (c) Organization of the U.S. national grid (USNG) 100,000-m grid squares

ZONES

SET 1 — 1, 7, 13, 19, 25, 31, 37, 43, 49, 55

SET 2 — 2, 8, 14, 20, 26, 32, 38, 44, 50, 56

SET 3 — 3, 9, 15, 21, 27, 33, 39, 45, 51, 57

SET 4 — 4, 10, 16, 22, 28, 34, 40, 46, 52, 58

SET 5 — 5, 11, 17, 23, 29, 35, 41, 47, 53, 59

SET 6 — 6, 12, 18, 24, 30, 36, 42, 48, 54, 60

Table 10-6　UTM Zone Width

North Latitude	Width (km)
42°00′	497.11827
43°00′	489.25961
44°00′	481.25105
45°00′	473.09497
46°00′	464.79382
47°00′	456.35005
48°00′	447.76621
49°00′	439.04485
50°00′	430.18862

Source: Ontario Geographical Referencing Grid, Ministry of Natural Resources, Ontario, Canada.

4. Northing value of the equator = 0.000 m (10,000,000.000 m in the southern hemisphere).

5. Scale factor at the central meridian is 0.9996 (i.e., 1/2,500).

6. Zone numbering commences with 1 in the zone 180°W to 174°W and increases eastward to zone 60 at the zone 174°E to 180°E [see Figure 10-12(a)].

7. Projection limits of latitude 80°S to 80°N.

See Figure 10-13 for a cross section of a 6° zone (UTM).

10.5.2 Modified Transverse Mercator (MTM) Grid System

Some regions and agencies outside the United States have adopted a modified transverse Mercator (MTM) system. The modified projection is based on 3° wide zones instead of 6° wide zones. By narrowing the zone width, the scale factor, at the central meridian, is improved from 0.9996 (1/2,500) to 0.9999 (1/10,000), the same as for the SPCS grids. The improved scale factor permits surveyors to work at moderate levels of accuracy without having to account for projection corrections. The zone width of 3° (about 152 miles wide at latitude 43°) compares very closely with the 158-mile-wide zones used in the United States for transverse Mercator and Lambert projections in the state plane coordinate system.

Characteristics of the 3° Zone

1. Zone is 3° wide.

2. Latitude of origin is the equator, 0°.

3. Easting value of the central meridian is set by the agency, for example, 1,000,000.000 ft or 304,800.000 m.

4. Northing value of the equator is 0.000 ft or m.

5. Scale factor at the central meridian is 0.9999 (i.e., 1/10,000).

Keep in mind that narrow grid zones (1:10,000) permit the surveyor to ignore only corrections for scale, and that other corrections to field measurements, such as those for elevation, temperature, and sag, and the balancing of errors are still routinely required.

See Figures 10-13 and 10-14 for cross sections of the UTM and MTM projection planes.

10.5.3 The United States National Grid (USNG)

As noted in previous sections, points on the surface of the earth can be identified by several types of coordinate systems, for example, the geographic coordinates of latitude/longitude, state plane coordinates, UTM coordinates, and MTM coordinates. Points on the earth's surface in the United States can be identified (georeferenced) by using the USNG; the USNG is an expansion of the long-established MGRS, which is used in many countries. The need for a simpler georeferencing system became apparent with the advent of 911 emergency procedures. A GPS-equipped cell phone (or handheld GPS receiver) can display location information in different coordinate systems such as latitude/longitude, UTM coordinates, and USNG

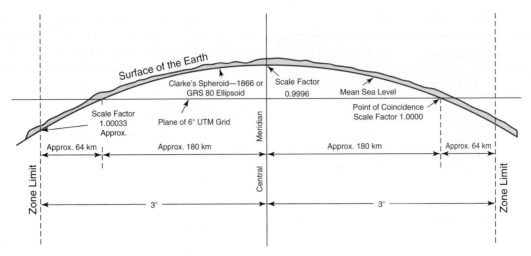

FIGURE 10-13 Cross section of a zone (UTM)

coordinates. Many think that the USNG is easier for the general population to comprehend and that it would be a good choice when a standard grid is selected for nationwide use.

Based on the framework of the world-wide UTM coordinate system, USNG is an alphanumeric reference grid that includes three levels of identity. It covers the earth's surface from 80° south to 84° north. The first level of location precision is denoted by the UTM zone number and the latitude band letter, and is known as the *grid zone designation* (GZD). It usually covers an

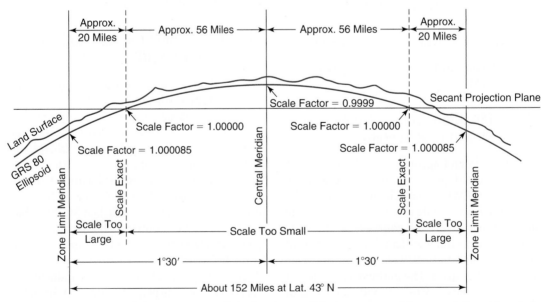

FIGURE 10-14 Section of the projection plane for the modified transverse Mercator (3°) grid and the earth's surface (secant projection)

area 6° in an east/west direction and 8° in a north/south direction. UTM zones, each covering 6° of longitude, are numbered eastward from the 180° meridian [see Figure 10-12(a)]. For example, the state of Florida is in zone 17. In North America, the zones in the conterminous United States run from zone 10 (meridian 126°W) eastward to zone 20 (meridian 60°W), and the zones in Canada run from zone 7 (meridian 144°W) eastwards to zone 22 (meridian 48°W).

Latitude bands, for the most part, cover 8° of latitude (the exception is Band X, which covers from 72° north to 84° north) and are identified by letters. In the northern hemisphere, latitude band coverage is given in Table 10-7.

In the conterminous United States, latitude band letters run northerly from letter R (latitude 24°N) in the southern United States to letter U (latitude 56°N) for the northern states. The latitude bands covering Canada run from the letter T (latitude 40°N) in southern Ontario and Quebec up to letter W (latitude 72°N) for the rest of the country. Refer to Figure 10-12(a) to see that the GZD for the state of Florida is 17R.

The second level of location precision is a 100,000-m^2 designation. This designation is given by two unique letters that repeat every three UTM zones (east/west) and every 2,000,000 m north of the equator; thus there is little opportunity for mistaken identification. These 100,000-m^2 identifiers are defined in the document *United States National Grid*, Standards Working Group, Federal Geographic Data Committee, December 2001 (available at www.fgdc.gov/usng). See Figures 10-12(b) and (c). Note that the northerly progression of letters in the first column (180° meridian) begins at AA, at the equator, and advances northerly for 20 bands (2,000,000 m) to AV (letters I, O, X, W, Y, and Z are not used). Then the letters commence again at AA and continue on northerly. Also note that in the easterly progression of letters, at the equator, the letters progress easterly from AA (at 180° meridian) through 24 squares to AZ (letters I and O are not used). In each zone, the most westerly and easterly columns of "squares" become progressively narrower than (100,000-m) as the meridians converge northerly. Just as with the UTM, the false northing is 0.00 m at the equator and the false eastings are set at 500,000 m at the central meridian of each zone.

The third level of location precision is given by coordinates unique to a specific 100,000-m^2 grid. The coordinates are an even number of digits ranging from 2 to 10. The coordinates are written in a string, with no space between easting and northing values (easting is always listed first). These grid coordinates follow the grid zone designation and the 100,000-m^2 identification letters in the stringed identification. For example, the *United States National Grid* (page 8) shows the following coordinates:

- 18SUJ25 locates a point with 10-km precision. 18S is the grid zone designation (18 is the UTM zone extending from longitude 78°W to 72°W and Band S extends from 32°N to 40°N), UJ is the (100,000-m)^2 identification, and the 25 indicates 20 km east and 50 km north.

Table 10-7 Latitude Bands

Latitude Band	Coverage
N	0° to 8°N
P	8°N to 16°N
Q	16°N to 24°N
R	24°N to 32°N
S	32°N to 40°N
T	40°N to 48°N
U	48°N to 56°N
V	56°N to 64°N
W	64°N to 72°N
X	72°N to 84°N

Note that the letters O, Y, and Z are not used for designating a latitude band.

- 18SUJ2306 locates a point with 1-km precision. The first half of the grid numbers is the grid easting and the second half is the grid northing, thus 23 km east, 6 km north.

- 18SUJ234064 locates a point with 100-m precision.

- 18SUJ23480647 locates a point with 10-m precision.

- 18SUJ2348306479 locates a point with 1-m precision: 23,483 m east and 6,479 m north, measured from the southwest corner of square UJ, in the Band S and in UTM zone 18.

The NGS TOOL kit contains interactive software to convert UTM and latitude/longitude to USNG and the reverse; see www.ngs.noaa.gov/TOOLS/usng.html.

10.6 USE OF GRID COORDINATES

10.6.1 Grid/Ground Distance Relationships: Elevation and Scale Factors

When local surveys (traverse or trilateration) are tied into coordinate grids, corrections must be provided so that

1. grid and ground distances can be reconciled by applying elevation and scale factors, and

2. grid and geodetic directions can be reconciled by applying convergence corrections.

10.6.1.1 Elevation Factor Figures 10-14 and 10-15 show the relationship between ground

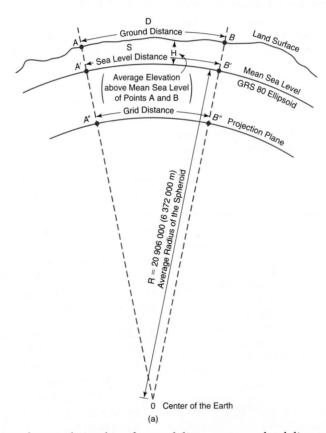

FIGURE 10-15 (a) General case: relationship of ground distances to sea-level distances and grid distances. (*Continued*)

Elevation Factor

$$\frac{S}{D} = \frac{R}{R + h}$$

$$S = D \frac{R}{R + h}$$

$$h = N + H$$

$$S = D \frac{R}{R + N + H}$$

Where
S = Geodetic Distance
D = Horizontal Distance
H = Mean Elevation
N = Mean Geoid Height (negative in figure)
R = Mean Radius of Earth
6,372,000 m or
20,906,000 ft

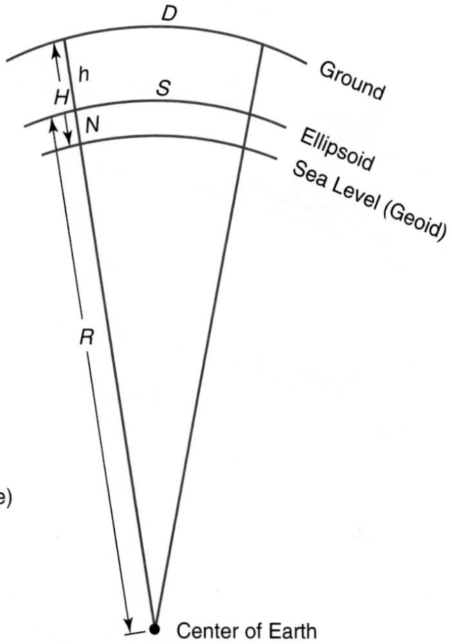

(b)

FIGURE 10-15 (b) Impact of geoid separation on elevation factor determination. Used only on very precise surveys

distances, sea-level distances, and grid distances. It can be seen that a distance measured on the earth's surface must first be reduced for equivalency at sea level, and then it must be further reduced (in this illustration) for equivalency on the projection plane. The first reduction involves multiplication by an **elevation factor** (sea-level factor); the second reduction (adjustment) involves multiplication by the **scale factor**.

The elevation (sea level) factor can be determined by establishing a ratio as is illustrated in Figures 10-15(a) and 10-16:

$$\text{Elevation factor} = \frac{\text{sea-level distance}}{\text{ground distance}} = \frac{R}{R + H}$$

$$(10\text{-}1)$$

where R is the average radius of the earth (average radius of sea-level surface = 20,906,000 ft or 6,372,000 m), and H is the elevation above mean sea level. For example, at 500 ft the elevation factor would be

$$\frac{20,906,000}{20,906,500} = 0.999976$$

and a ground distance of 800.00 ft at an average elevation of 500 ft would become 800 × 0.999976 = 799.98 at sea level.

Figure 10-15(b) shows the case (encountered in very precise surveys) where the geoid separation (N) must also be considered. Geoid separation is the height difference between the sea-level surface and the ellipsoid surface. NOAA Manual NOS NGS 5, *State Plane Coordinate System of 1983*, notes, for example, that a geoid height of −30 m (in the conterminous United States, the ellipsoid is above the geoid) systematically affects reduced distances by −4.8 ppm (1:208,000), which is certainly not a factor in any but the most precise surveys.

10.6.1.2 Scale Factor For state plane projections, the computer solution gives scale factors for positions of latitude difference (Lambert

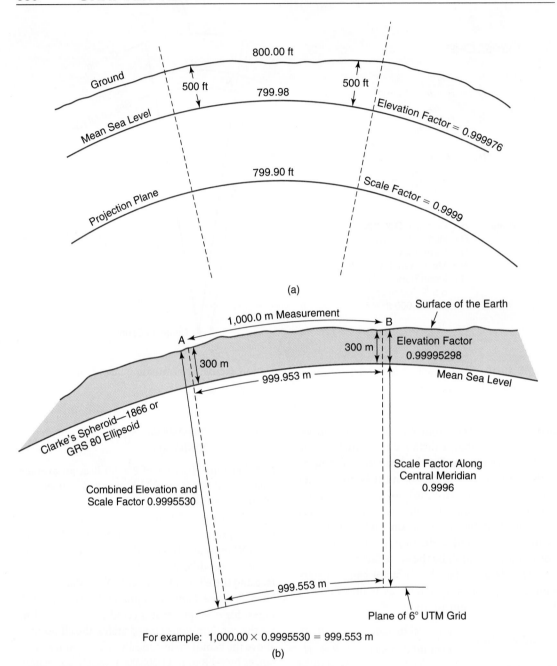

FIGURE 10-16 Conversion of a ground distance to a grid distance using the elevation factor and the scale factor. (a) SPCS 83 grid. (b) Universal transverse Mercator (UTM) grid, 6° zone

projection) or for distances east or west of the central meridian (transverse Mercator projections). By way of illustration, scale factors for the transverse Mercator projections may also be computed by the equation

$$M_p = M_o\left(\frac{1 + x^2}{2R^2}\right) \qquad (10\text{-}2)$$

where M_p = scale factor at the survey station, M_o = scale factor at the central meridian (CM), x = east/west distance of the survey station from the central meridian, and R = average radius of the spheroid. For example, survey stations 12,000 ft from a central meridian having a scale factor of 0.9999 would have a scale factor determined as follows:

$$M_p = 0.9999\left(1 + \frac{12,000^2}{2 \times 20,906,000^2}\right)$$

$$= 0.9999002$$

10.6.1.3 Combined Factor When the *elevation factor* is multiplied by the *scale factor*, the result is known as the *combined factor*.

$$\text{Ground distance} \times \text{combined factor}$$
$$= \text{grid distance} \qquad (10\text{-}3)$$

Stated another way, the equation becomes

$$\text{Grid distance} = \frac{\text{ground distance}}{\text{combined factor}}$$

In practice, it is seldom necessary to use Equations 10-1 and 10-2 because computer programs are now routinely used for computations in all state plane grids, the UTM grid. Previously, computations were based on data from tables and graphs (see Figure 10-17 and Table 10-8).

Example 10-1: Using Table 10-8 (MTM projection), determine the combined scale and elevation factor (grid factor) of a point 125,000 ft from the central meridian (scale factor = 0.9999) and at an elevation of 600 ft above mean sea level. See Table 10-9 for an interpolation technique.

Solution: To determine the combined factor for a point 125,000 ft from the central meridian and at

an elevation of 600 ft, a solution involving double interpolation must be used (see Table 10-9). We must interpolate between combined values for 500 ft and for 750 ft elevation, and between combined values for 100,000 and for 150,000 ft from the central meridian. First, the combined value for 125,000 ft from the central meridian can be interpolated simply by averaging the values for 100,000 and 150,000. Second, the value for 600 ft can be determined as follows. From Table 10-9 for 600 ft:

$$\frac{100}{250} \times 12 = 5$$

$$\text{Combined factor} = 0.999895 - 0.000005$$
$$= 0.999890$$

For important survey lines, combined factors can be determined for both ends and then averaged. For lines longer than 5 miles, intermediate computations are required to maintain high precision.

10.6.2 Grid/Geodetic Azimuth Relationships

10.6.2.1 Convergence In plane grids, the difference between grid north and geodetic north is called convergence (also called the mapping angle). In the SPCS27 transverse Mercator grid convergence was denoted by $\Delta\alpha$, and in the Lambert grid it was denoted as θ; in the SPCS83, convergence (in both projections) is denoted by γ (gamma). On a plane grid, grid north and geodetic north coincide only at the central meridian. Farther east or west of the central meridian, convergence becomes more pronounced.

Using SPCS83 symbols, approximate methods can be determined as follows:

$$\gamma = \Delta\alpha \sin \varphi_P \qquad (10\text{-}4)$$

where γ is the convergence in seconds, $\Delta\alpha$ is the difference in longitude, in seconds, between the central meridian and point P, and φ_P is the latitude of point P. When long sights (>5 miles) are taken, a second term (present in the interactive Internet programs used in Sections 10.2.2 and 10.2.3 and in

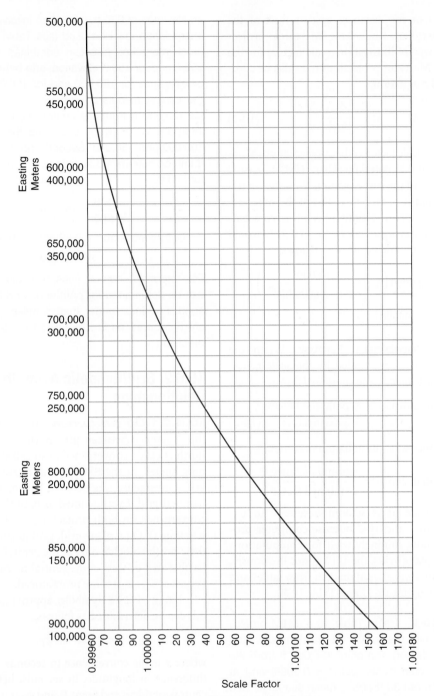

FIGURE 10-17 Universal transverse Mercator grid scale factors

(Courtesy of U.S. Department of the Army, TM5-241-4/1)

Table 10-8 3° MTM Combined Grid Factor Based on Central Scale Factor of 0.9999 for the Modified Transverse Mercator Projection*

Elevation (ft)	Distance from Central Meridian (Thousands of Feet)												
	0	50	100	150	200	250	300	350	400	450	500	550	600
0	0.999900	0.999903	0.999911	0.999926	0.999946	0.999971	1.000003	1.000040	1.000083	1.000131	1.000186	1.000245	1.000311
250	0.999888	0.999891	0.999899	0.999914	0.999934	0.999959	0.999991	1.000028	1.000071	1.000119	1.000174	1.000234	1.000299
500	0.999876	0.999879	0.999888	0.999902	0.999922	0.999947	0.999979	1.000016	1.000059	1.000107	1.000162	1.000222	1.000287
750	0.999864	0.999867	0.999876	0.999890	0.999910	0.999936	0.999967	1.000004	1.000047	1.000095	1.000150	1.000210	1.000275
1,000	0.999852	0.999855	0.999864	0.999878	0.999898	0.999924	0.999955	0.999992	1.000035	1.000083	1.000138	1.000198	1.000263
1,250	0.999840	0.999843	0.999852	0.999864	0.999886	0.999912	0.999943	0.999980	1.000023	1.000072	1.000126	1.000186	1.000251
1,500	0.999828	0.999831	0.999840	0.999854	0.999874	0.999900	0.999931	0.999968	1.000011	1.000060	1.000114	1.000174	1.000239
1,750	0.999816	0.999819	0.999828	0.999842	0.999862	0.999888	0.999919	0.999956	0.999999	1.000048	1.000102	1.000162	1.000227
2,000	0.999804	0.999807	0.999816	0.999830	0.999850	0.999876	0.999907	0.999944	0.999987	1.000036	1.000090	1.000150	1.000216
2,250	0.999792	0.999795	0.999804	0.999818	0.999838	0.999864	0.999895	0.999932	0.999975	1.000024	1.000078	1.000138	1.000204
2,500	0.999781	0.999784	0.999792	0.999806	0.999626	0.999852	0.999883	0.999920	0.999963	1.000012	1.000066	1.000126	1.000192
2,750	0.999796	0.999771	0.999780	0.999794	0.999814	0.999840	0.999871	0.999908	0.999951	1.000000	1.000054	1.000114	1.000180

*Ground distance × combined factor = grid distance.

Source: Adapted from the "Horizontal Control Survey Precis," Ministry of Transportation and Communications, Ottawa, Canada, 1974.

Table 10-9   Distance from Central
Meridian (1,000s of feet)

Elevation	100	150	125 (interpolated)
500	0.999888	0.999902	0.999895
750	0.999876	0.999890	0.999883
		difference = 0.000012	

all precise computations) is required to maintain directional accuracy (see also Section 10.6.2.2).

When the direction of a line from P_1 to P_2 is being considered, the expression becomes

$$\gamma = \Delta\alpha\sin\frac{(\varphi P_1 + \varphi P_2)}{2}$$

Alternatively, if the distance from the central meridian is known, the expression becomes

$$\gamma = 32.370''/\text{km}\, d \tan\varphi \qquad (10\text{-}5)$$

or

$$\gamma = 52.09''/\text{mi}\, d \tan\varphi \qquad (10\text{-}6)$$

where γ = convergence angle, in seconds

d = departure distance from the central meridian

φ = average latitude of the line

See Chapter 14 for further discussion of convergence.

10.6.2.2 Corrections to Convergence

When high precision and/or long distances are involved, there is a *second term* correction required for convergence. This term δ refers to $t - T$ (the grid azimuth minus the projected geodetic azimuth) and results from the fact that the projection of the geodetic azimuth onto the grid is not the grid azimuth but the projected geodetic azimuth symbolized

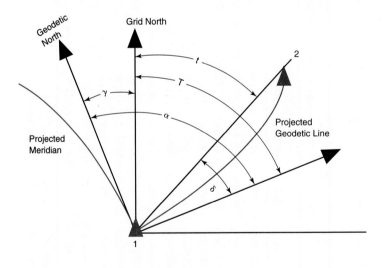

$$t = \alpha - \gamma + \delta$$

α = Geodetic Azimuth Reckoned From North

T = Projected Geodetic Azimuth

t = Grid Azimuth Reckoned From North

γ = Convergence Angle (Mapping Angle)

$\delta = t\text{-}T$ = Second Term Correction = Arc-to-Chord Correction (Negative in Figure)

FIGURE 10-18 Relationships among geodetic and grid azimuths

(From the NOAA Manual NOS NGS 5 State Plane Coordinate System of 1983, Section 2.5)

as *T*. Figure 10-18, *State Plane Coordinate System of 1983*, shows the relationships between geodetic and grid azimuths.

10.7 ILLUSTRATIVE EXAMPLES

These approximate methods are included only to broaden your comprehension in this area. For current precise solution techniques, see NOAA Manual NOS NGS 5 *State Plane Coordinate System of 1983* or the Geodetic Tool Kit discussed in Section 10.2.

Example 10-2: Given the transverse Mercator coordinate grid of two horizontal control monuments and their elevations (see Figure 10-19 for additional given data), compute the ground distance and geodetic direction between them.

Solution: Table 10-10 lists the two monuments and their coordinates. By subtraction, coordinate distances, ΔN = 255.161 ft, ΔE = 919.048 ft. The solution is obtained as follows:

1. Grid distance 870 to 854:

$$\text{Distance} = \sqrt{255.161^2 + 919.048^2} = 953.811\,\text{ft}$$

2. Grid bearing:

$$\tan \text{bearing} = \Delta E/\Delta N = 919.048/255.161$$
$$= 3.6018357$$

$$\text{Grid bearing} = 74.48342°$$
$$= \text{N}\,74°29'00''\,\text{E}$$

3. Convergence: Use method (a) or (b). Average latitude = 43°47′31″; average longitude = 79°20′54″ (Figure 10-19).

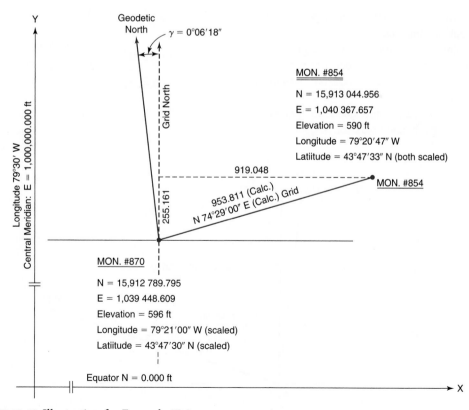

FIGURE 10-19 Illustration for Example 10.2

Table 10-10   Illustration for Example 10.2

Station*	Elevation (ft)	Northing (ft)	Easting (ft)
Monument 870	595	15,912,789.795	1,039,448.609
Monument 854	590	15,913, 044.956	1,040, 367.657

*Scale factor at CM = 0.9999.

$\gamma = 52.09d \tan \varphi$

$= 52.09 \times \dfrac{(40,367.657 + 39,448.609)}{2 \times 5280}$

$\tan 43°47'31''$

a. $= 377.45''$

$\gamma = 0°06'17.5''$

$\gamma = \Delta\lambda \sin \varphi_P$

$= (79°30' - 79°20'54'') \sin 43°47'31''$

$= 546'' \times \sin 43°47'31''$

b. $= 377.85''$

$\gamma = 0°06'17.9''$

Use a convergence of 6°18'. Convergence, in this case, neglects the second-term correction and is computed only to the closest second of arc. (*Note:* The average latitude need not have been computed since the latitude range, 03'', was insignificant in this case.)

For Figure 10-20, the geodetic bearing is equal to the grid bearing plus convergence:

Grid bearing = N 74°29'00'' E + convergence

= 0°06'18''

Geodetic bearing = N 74°35'18'' E

4. Scale factor: The scale factor at CM is $M_o = 0.9999$

$\text{Distance}(x)\text{ from CM} = \dfrac{40,368 + 39,449}{2}$

$= 39,908 \text{ ft}$

The scale factor at the midpoint between 870 and 854 can be found by using Equation 10-2:

$$M_p = M_o\left(1 + \frac{x^2}{2R^2}\right)$$

$$= 0.9999\left(1 + \frac{39,908^2}{2 \times 20,906,000^2}\right)$$

$$= 9999018$$

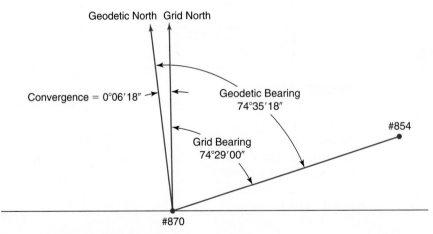

Geodetic North Grid North

Convergence = 0°06'18''

Geodetic Bearing
74°35'18''

#854

Grid Bearing
74°29'00''

#870

FIGURE 10-20 Illustration for Example 10.2

5. The elevation factor can be found by using Equation 10-1:

$$\text{Elevation factor} = \frac{\text{sea-level distance}}{\text{ground distance}} = \frac{R}{R + H}$$

$$= \frac{20,906,000}{20,906,000 + 593} = 0.999716$$

The value of 593 ft is the midpoint (average) elevation in feet.

6. Use method (a) or (b) to find the combined factor.

  a. Combined factor = scale facton
                × elevation factor
            = 0.9999018 × 0.9999716
            = 0.9998734

  b. The combined factor can also be determined through double interpolation of Table 10-8, as follows: The values in Table 10-11 are taken from Table 10-8, required elevation of 593 ft at 39,900 ft from the CM.

After first interpolating the values at 0 and 50 for 39,900 ft from the CM, it is a simple matter to interpolate for the elevation of 593 ft:

$$0.999878 - (0.000012) \times \frac{93}{250} = 0.999873$$

Thus, the combined factor at 39,900 ft from the CM at an elevation of 593 ft is 0.999873.

7. Rearrange Equation 10-3 to determine the ground distance:

It can be seen that ground distance

$$= \frac{\text{grid distance}}{\text{combined factor}}$$

$$\text{Ground distance}(\,\#\,870\text{ to }\#\,854) = \frac{953.811}{0.9998734}$$

$$= 953.93\,\text{ft}$$

Example 10-3: Given the coordinates, on the UTM 6° coordinate grid (based on the Clarke 1866 ellipsoid), of two horizontal control monuments (Mon. 113 and Mon. 115), and their elevations, compute the ground distance and geodetic direction between them [see Figure 10-21(a)].

The following information is also provided (Table 10-12):

Zone 17 UTM; CM at 81° longitude west (see Figure 10-21)

Scale factor at CM = 0.9996

Table 10-11

	Distance from Central Meridian (1,000s of ft)		
Elevation (ft)	0	50	39.9 (interpolated)
500	0.999876	0.999879	0.999878
593			
750	0.999864	0.999867	0.999866

Table 10-12   Data for Example 10.3

Station	Elevation (m)	Northing (m)	Easting (m)
113	181.926	4,849,872.066	632,885.760
115	178.444	4,849,988.216	632,971.593

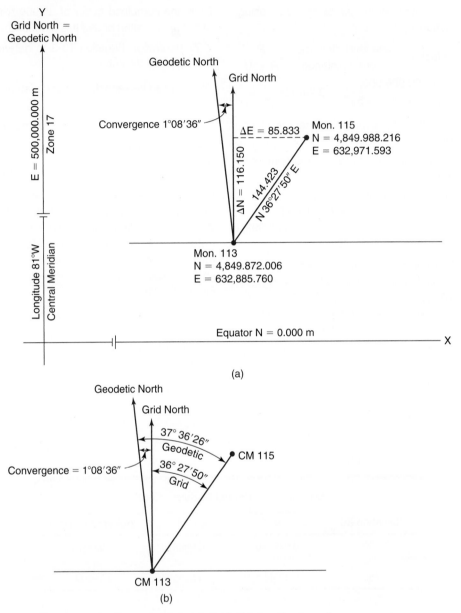

FIGURE 10-21 Illustration for Example 10.3. (a) Grid. (b) Application of convergence

Scaled from topographic map for midpoint of line joining stations 113 and 115:

$$\varphi\,(\text{lat.}) = 43°47'33''$$

$$\lambda\,(\text{long.}) = 79°20'52''$$

Solution: By subtraction, coordinate distances are $\Delta N = 116.150$ m, $\Delta E = 85.833$ m.

1. Grid distance from station 113 to station 115:

$$\text{Distance} = \sqrt{116.150^2 + 85.833^2} = 144.423\,\text{m}$$

Table 10-13

	Distance from CM (km)		
Elevation	130	140	132.9 (interpolated)
100	0.999792	0.999825	0.999802
200	0.999777	0.999810	0.999787

$$0.999802 - (80/100 \times 0.000015) = 0.999790$$

2. Grid bearing:

$$\tan \text{bearing} = \frac{\Delta E}{\Delta N} = \frac{85.833}{116.150}$$

$$\text{Bearing} = 36.463811°$$

$$\text{Grid bearing} = N\,36°27'50''\,E$$

3. Convergence:

$$\gamma = \Delta\lambda \sin \varphi_P$$

$$= (81° - 79°20'52'') \sin 43°47'33''$$

$$= 4,116.3''$$

$$\gamma = 1°08'36''$$

See Figure 10-21(b) for application of convergence. As in the previous example, this technique yields convergence only to the closest second of arc. For more precise techniques, including second-term correction for convergence, see the NGS TOOL kit.

4. Scale factor:

Scale factor at CM = 0.9996

$$\text{Distance from CM} = \frac{132,885.760 + 132,971.593}{2}$$

$$= 132,928.677\,\text{m or}\,132.929\,\text{km}$$

The scale factor at the midpoint on the line between Mon. 113 and Mon. 115 can be found by using Equation 10-2 with 20,906,000 ft or 6,372,000 m as the average radius of the sea-level surface:

$$M_p = M_o\left(1 + \frac{x^2}{2R^2}\right)$$

$$M_p = 0.9996\left(\frac{1 + 132.9292}{2 \times 6372^2}\right) = 0.999818$$

5. The elevation factor can be found by using Equation 10-1:

$$\text{Elevation factor} = \frac{\text{sea-level distance}}{\text{ground distance}} = \frac{R}{R + H}$$

$$= \frac{6,372}{6,372.00 + 0.180} = 0.999972$$

The value of 0.180 is the midpoint elevation divided by 1000.

6. Use method (a) or (b) to find the combined factor.

    a. Combined factor = elevation factor
            × scale factor

$$= 0.999972 \times 0.999818 = 0.999790$$

    b. The values in Table 10-13 come from Table 10-14. All that remains is to interpolate for the elevation of 180 m.

7. To calculate the ground distance, use the following equation:

$$\text{Ground distance} = \frac{\text{grid distance}}{\text{combined factor}}$$

In this example:

$$\text{Ground distance Sta. 113 to Sta. 115} = \frac{144.423}{0.999790}$$

$$= 144.453\,\text{m}$$

10.8 HORIZONTAL CONTROL TECHNIQUES

Typically, the highest-order control is established by federal agencies, the secondary control is established by state or provincial agencies, and the lower-order control is established by municipal agencies or large-scale engineering works' surveyors. Sometimes the federal agency establishes all

Table 10.14 Combined Scale and Elevator Factors (Grid Factors): UTM

Distance from CM (km)	Elevation Above Mean Sea Level (m)										
	0	100	200	300	400	500	600	700	800	900	1,000
0	0.999600	0.999584	0.999569	0.999553	0.999537	0.999522	0.999506	0.999490	0.999475	0.999459	0.999443
10	0.999601	0.999586	0.999570	0.999554	0.999539	0.999523	0.999507	0.999492	0.999476	0.999460	0.999445
20	0.999605	0.999589	0.999574	0.999558	0.999542	0.999527	0.999511	0.999495	0.999480	0.999464	0.999448
30	0.999611	0.999595	0.999580	0.999564	0.999548	0.999533	0.999517	0.999501	0.999486	0.999470	0.999454
40	0.999620	0.999604	0.999588	0.999573	0.999557	0.999541	0.999526	0.999510	0.999494	0.999479	0.999463
50	0.999631	0.999615	0.999599	0.999584	0.999568	0.999552	0.999537	0.999521	0.999505	0.999490	0.999474
60	0.999644	0.999629	0.999613	0.999597	0.999582	0.999566	0.999550	0.999535	0.999519	0.999503	0.999488
70	0.999660	0.999645	0.999629	0.999613	0.999598	0.999582	0.999566	0.999551	0.999535	0.999519	0.999504
80	0.999679	0.999663	0.999647	0.999632	0.999616	0.999600	0.999585	0.999569	0.999553	0.999538	0.999522
90	0.999700	0.999684	0.999668	0.999653	0.999637	0.999621	0.999606	0.999590	0.999574	0.999559	0.999543
100	0.999723	0.999707	0.999692	0.999676	0.999660	0.999645	0.999629	0.999613	0.999598	0.999582	0.999566
110	0.999749	0.999733	0.999717	0.999702	0.999686	0.999670	0.999655	0.999639	0.999623	0.999608	0.999592
120	0.999777	0.999761	0.999746	0.999730	0.999714	0.999699	0.999683	0.999667	0.999652	0.999636	0.999620
130	0.999808	0.999792	0.999777	0.999761	0.999745	0.999730	0.999714	0.999698	0.999683	0.999667	0.999651
140	0.999841	0.999825	0.999810	0.999794	0.999778	0.999763	0.999747	0.999731	0.999716	0.999700	0.999684
150	0.999877	0.999861	0.999845	0.999830	0.999814	0.999798	0.999783	0.999767	0.999751	0.999736	0.999720
160	0.999915	0.999899	0.999884	0.999868	0.999852	0.999837	0.999821	0.999805	0.999790	0.999774	0.999758
170	0.999955	0.999940	0.999924	0.999908	0.999893	0.999877	0.999861	0.999846	0.999830	0.999814	0.999799
180	0.999999	0.999983	0.999967	0.999952	0.999936	0.999920	0.999905	0.999889	0.999873	0.999858	0.999842
190	1.000044	1.000028	1.000013	0.999997	0.999981	0.999966	0.999950	0.999934	0.999919	0.999903	0.999887
200	1.000092	1.000076	1.000061	1.000045	1.000029	1.000014	0.999998	0.999982	0.999967	0.999951	0.999935
210	1.000142	1.000127	1.000111	1.000095	1.000080	1.000064	1.000048	1.000033	1.000017	1.000001	0.999986
220	1.000195	1.000180	1.000164	1.000148	1.000133	1.000117	1.000101	1.000086	1.000070	1.000054	1.000039
230	1.000251	1.000235	1.000219	1.000204	1.000188	1.000172	1.000157	1.000141	1.000125	1.000110	1.000094
240	1.000309	1.000293	1.000277	1.000262	1.000246	1.000230	1.000215	1.000199	1.000183	1.000168	1.000152
250	1.000369	1.000353	1.000338	1.000322	1.000306	1.000290	1.000275	1.000259	1.000243	1.000228	1.000212

three orders of control when requested to do so by the state, province, or municipality.

In triangulation surveys, a great deal of attention was paid to the geometric *strength of figure* of each control configuration. Generally, an equilateral triangle is considered *strong*, whereas triangles with small (less than 10°) angles are considered relatively weak. Formulas involving trigonometric functions can amplify errors as the angle varies in magnitude. The sines of small angles (near zero), the cosines of large angles (near 90°), and the tangents of both small (zero) and large (90°) angles can all amplify errors. That is, relatively large changes in the values of the formulas can result from relatively small changes in angular values. For example, the angular error of 5 seconds in the sine of 10° is 1/7,300, whereas the angular error of 5 seconds in the sine of 20° is 1/15,000, and the angular error of 5 seconds in the sine of 80° is 1/234,000 (see Example 10-4). When sine or cosine functions are used in triangulation to calculate the triangle side distances, it is important to ensure that the formula is not amplifying the solution error more than the specified surveying error limits.

When all angles and distances are measured for each triangle, the redundant measurements ensure an accurate solution, and the configuration strength of figure becomes somewhat less important. However, given the opportunity, most surveyors still prefer to use well-balanced triangles and to avoid using the sine and tangent of small angles and the cosine and tangent of large angles to compute control distances. This concept of strength of figure helps to explain why GPS measurements are more precise when the observed satellites are spread right across the visible sky instead of being bunched together in just one portion of the sky.

Example 10-4: Effect of the Angle Magnitude on the Precision and Accuracy of Computed Distances

a. Find the opposite side X in a right-angle triangle with a hypotenuse 1,000.00 ft long.

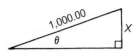

b. Find the opposite side X in a right triangle with an adjacent side 1,000.00 ft long.

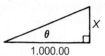

Solution:

a. Use various values for θ to investigate the effect of 05″ angular errors.

  1. $\theta = 10°$ $X = 173.64818$ ft
      $\theta = 10°00'05''$ $X = 173.67205$ ft

  Difference $= 0.02387$ in 173.65 ft,
                 a possible accuracy of 1/7,300

  2. $\theta = 20°$ $X = 342.02014$ ft
      $\theta = 20°00'05''$ $X = 342.04292$ ft

  Difference $= 0.022782$ in 342.02 ft,
                 a possible accuracy of 1/15,000

  3. $\theta = 80°$ $X = 984.80775$ ft
      $\theta = 80°00'05''$ $X = 984.81196$ ft

  Difference $= 0.00421$ in 984.81 ft,
                 a possible accuracy of 1/234,000

b. Use various values for θ to investigate the effect of 05″ angular errors.

  1. $\theta = 10°$ $X = 176.32698$ ft
      $\theta = 10°00?05?$ $X = 176.35198$ ft

  Difference $= 0.025$, a possible accuracy of 1/7,100

  2. $\theta = 45°$ $X = 1,000.00$ ft
      $\theta = 45°00?05?$ $X = 1,000.0485$ ft

  Difference $= 0.0485$, a possible accuracy of 1/20,600

  3. $\theta = 80°$ $X = 5,671.2818$ ft
      $\theta = 80°00'05''$ $X = 5,672.0858$ ft

  Difference $= 0.804$, a possible accuracy of 1/7,100

  4. If the angle can be determined to the closest second, the accuracy would be as follows:

      $\theta = 80°$ $X = 5,671.2818$ ft
      $\theta = 80°00'01''$ $X = 5,671.4426$ ft

  Difference $= 0.1608$, a possible accuracy of 1/35,270

Example 10-4 illustrates that the surveyor should avoid using weak (relatively small) angles in distance computations. If weak angles must be used, they should be measured more precisely than would normally be required. Also illustrated here is the need for the surveyor to pre-analyze the proposed control survey configuration to determine optimal field techniques and attendant precisions.

10.9 PROJECT CONTROL

10.9.1 General Background

Project control begins with either a boundary survey (e.g., large housing projects) or an all-inclusive peripheral survey (e.g., in construction sites). The boundary or site peripheral survey should be tied, if possible, into state or provincial grid control monuments so that references can be made to the state or provincial coordinate grid system. The increasing use of quick-positioning survey-grade (1–2 cm) satellite positioning receivers and techniques has enabled surveyors to establish project control with fewer permanent control monuments. The peripheral survey is densified with judiciously placed control stations over the entire site. The survey data for all control points are entered into the computer for accuracy verification and error adjustment and finally for coordinate determination of all control points. All key layout points (e.g., lot corners, radius points, ₵ stations, curve points, construction points) are also coordinated using coordinate geometry computer programs. Printout sheets can be used by the surveyor to lay out the proposed facility from coordinated control stations. The computer results give the surveyor the azimuth and distance from one, two, or perhaps three different control points to one layout point. Alternatively, the surveyor can upload the coordinates of all control points and layout stations, and then use the total station software to have the instrument oriented and the layout points staked.

Positioning a layout point from more than one control station provides the opportunity for an exceptional check on the accuracy of the work.

Generally, a layout point can be positioned by simultaneous angle sightings from two control points, with the distance being established by EDM from one of those stations, or a layout point can be positioned by simultaneous angle sightings from three control points. Both techniques provide the vital redundancy in measurement that permits positional accuracy determination.

To ensure that the layout points have been accurately located (e.g., with an accuracy level of between 1/5,000 and 1/10,000), the control points themselves must be located to an even higher level of accuracy (i.e., typically better than 1/15,000). These accuracies can be achieved easily using precise GPS techniques and/or total stations with one- or two-second angle precision and having dual-axis compensation.

As noted earlier, in addition to quality instrumentation, the surveyor must use "quality" geometrics in designing the shape of the control net; a series of interconnected equilateral triangles provides the strongest control net.

When positioning control points, the following should be kept in mind:

1. Good visibility to other control points and an optimal number of layout points is important.

2. The visibility factor is considered not only for existing ground conditions but also for potential visibility lines during all stages of construction.

3. A minimum of two (three is preferred) reference ties is required for each control point so that it can be reestablished if destroyed. Consideration must be given to the availability of features suitable for referencing (i.e., features into which nails can be driven or cut crosses chiseled, etc.). Ideally, the three ties should each be about 120° apart.

4. Control points should be placed in locations that will not be affected by primary or secondary construction activity. In addition to keeping clear of the actual construction site positions, the surveyor must anticipate temporary disruptions to the terrain resulting

from access roads, materials stockpiling, and so on. If possible, control points are safely located adjacent to features that will *not* be moved (e.g., electrical or communications towers, concrete walls, large valuable trees).

5. Control points must be established on solid ground (or rock). Swampy area or loose fill areas must be avoided (see Section 10.10 for various types of markers).

Once the control point locations have been tentatively chosen, they are plotted so that the quality of the control net geometrics can be considered. At this stage it may be necessary to go back into the field and locate additional control points to strengthen weak geometric figures. When the locations have been finalized on paper, each station is given a unique identification code number, and then the control points are set in the field. Field notes, showing reference ties to each point, are carefully taken and then filed. Now the actual measurements of the distances and angles, or

positions, of the control net are taken using total station or satellite positioning techniques. When all the field data have been collected, the closures and adjustments are computed. The coordinates of any layout points could then be computed, with polar ties being generated for each layout point, from two or possibly three control stations.

Figure 10-22(a) shows a single layout point being positioned by angle only from three control sights. The three control sights can be referenced simply to the farthest of the control points themselves (e.g., angles *A*, *B*, and *C*), or if a reference azimuth point (RAP) has been identified and coordinated in the locality, it would be preferred because it no doubt would be farther away and thus capable of providing more precise sightings (e.g., angles 1, 2, and 3). RAPs are typically communications towers, church spires, or other identifiable points that can be seen from widely scattered control stations. Coordinates of RAPs are computed by turning angles to the RAP from several project control monuments or preferably

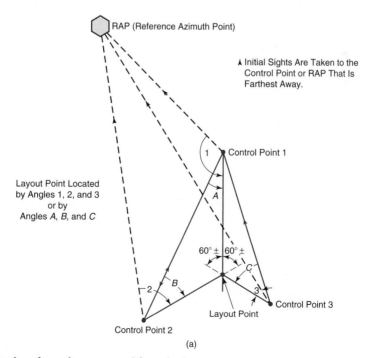

FIGURE 10-22 Examples of coordinate control for polar layout. (a) Single point layout, located by three angles. (*Continued*)

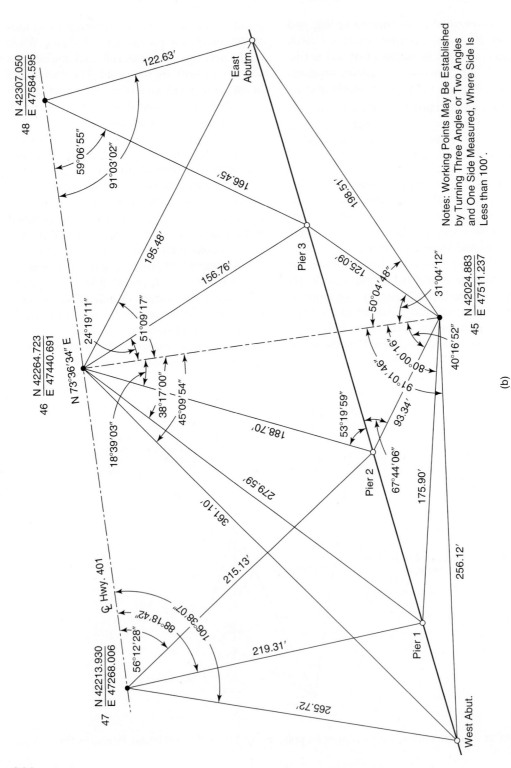

FIGURE 10-22 (b) Bridge layout, located by angle and distance

(Adapted from the Construction Manual, Ministry of Transportation, Ontario)

from state or provincial control grid monuments. Figure 10-22(b) shows a bridge layout involving azimuth and distance ties for abutment and pier locations. It will be noted that although the perfect case of equilateral triangles is not always present, the figures are quite strong, with redundant measurements providing accuracy verification.

Figure 10-23 illustrates a method of recording angle directions and distances to control stations with a listing of derived azimuths. Station 17 can be quickly found by the surveyor from the distance and alignment ties to the hydrant, cut cross on curb, and nail in pole. Had station 17 been destroyed, it could have been reestablished from these and other reference ties. The bottom row marked "check" indicates that the surveyor has "closed the horizon" by continuing to revolve the theodolite, or total station, back to the initial

target point (100 in this example) and then reading the horizontal circle. An angle difference of more than 5″ between the initial reading and the check reading usually means that the series of angles in that column must be repeated.

After the design of a facility has been coordinated, polar layout coordinates can be generated for points to be laid out from selected stations. The surveyor can copy the computer data directly into the field book (see Figure 10-24) for use later in the field. On large projects (expressways, dams, etc.) it is common practice to have bound volumes printed that include polar coordinate data for all control stations and all layout points.

Alternately, modern total station practice permits the direct uploading of the coordinates of control points and layout points to be used in layout surveys (see Chapter 5).

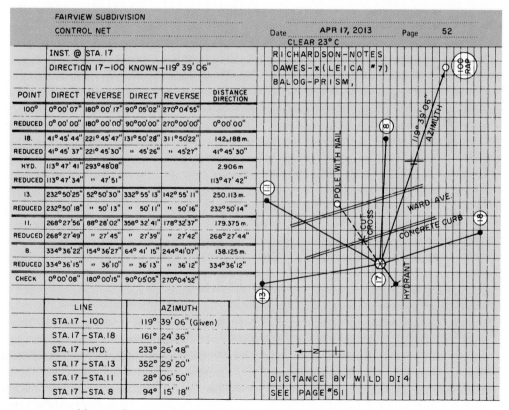

FIGURE 10-23 Field notes for control point directions and distances

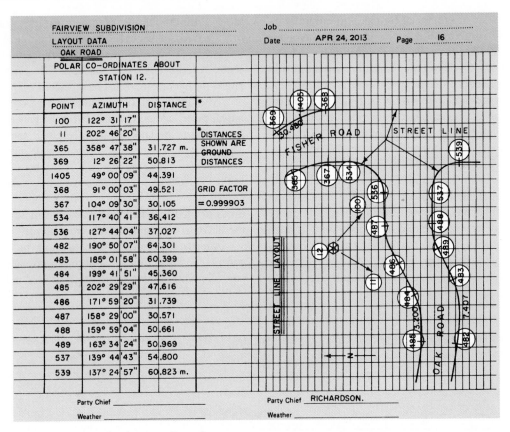

FIGURE 10-24 Prepared polar coordinate layout notes

Figure 10-25 shows a primary control net established to provide control for a construction site. The primary control stations are tied in to a national, state, or provincial coordinate grid by a series of precise traverses or triangular networks. Points on baselines (secondary points) can be tied in to the primary control net by polar ties, intersection, or resection. The actual layout points of the structure (columns, walls, footings, etc.) are established from these secondary points. International standard ISO 4463 (International Standards Organization) points out that the accuracy of key building or structural layout points should not be influenced by possible discrepancies in the state or provincial coordinate grid. For that reason, the primary project control net is analyzed and adjusted independently of the state or provincial coordinate grid. This "free net" is tied to the state or provincial coordinate grid without becoming an integrated adjusted component of that grid. *The relative positional accuracy of project layout points to each other is more important than the positional accuracy of these layout points relative to a state or provincial coordinate grid.*

10.9.2 Positional Accuracies (ISO 4463)

10.9.2.1 Primary System Control Stations

1. Permissible deviations of the distances and angles obtained when measuring the positions of primary points, and those calculated from the adjusted coordinates of these points, shall not exceed

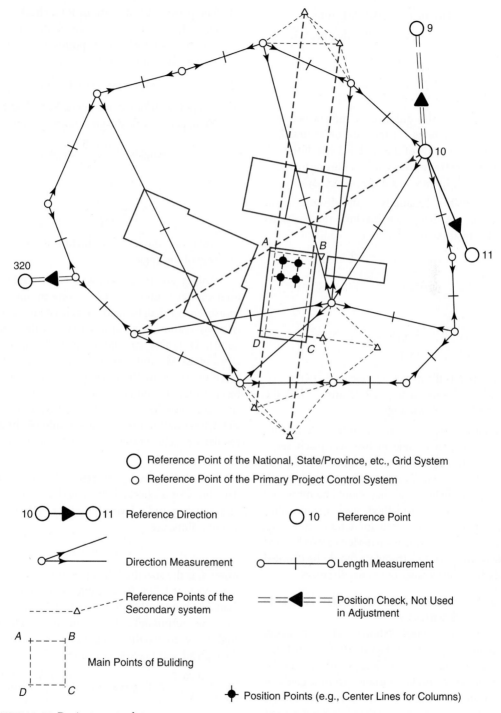

Reference Point of the National, State/Province, etc., Grid System

Reference Point of the Primary Project Control System

10 ◯ ▶ ◯ 11   Reference Direction   ◯ 10   Reference Point

Direction Measurement   ◯—|—◯ Length Measurement

Reference Points of the
Secondary system   = = ◀ = =   Position Check, Not Used
in Adjustment

A +----+ B
| |
| |
D +----+ C   Main Points of Buliding

Position Points (e.g., Center Lines for Columns)

FIGURE 10-25 Project control net

(Adapted from International Organization for Standardization [ISO], Standard 4463)

Distances: $\pm 0.75\sqrt{L}$ mm

Angles: ± 0.045 degrees $\sqrt{L}$

or

$$\pm \frac{0.05}{\sqrt{L}} \text{ gon}$$

where L is the distance in meters between primary stations (in the case of angles, L is the shorter side of the angle); 1 revolution = $360° = 400$ gon (also grad—a European angular unit); 1 gon = 0.9 degrees (exactly).

2. Permissible deviations of the distances and angles obtained when checking the positions of primary points shall not exceed

Distances: $\pm 2\sqrt{L}$ mm

Angles: $\pm \dfrac{0.135}{\sqrt{L}}$ degrees

or

$$\frac{\pm 0.15}{\sqrt{L}} \text{ gon}$$

where L is the distance in meters between primary stations; in the case of angles, L is the shorter side of the angle.

Angles are measured with a 1″ total station, with the measurements made in two sets (each set is formed by two observations, one on each face of the instrument). Distances can also be measured with steel tapes or EDMs, and they should be measured at least twice by either method. Steel tape measurements should be corrected for temperature, sag, slope, and tension: A tension device should be used while taping. EDM instruments should be checked regularly against a range of known distances.

10.9.2.2 Secondary System Control Stations

1. Secondary control stations and main layout points (e.g., *ABCD*, Figure 10-25) constitute the secondary system. The permissible deviations for a checked distance from a given or calculated distance between a primary control station and a secondary point shall not exceed

Distances: $\pm 2\sqrt{L}$ mm

2. The permissible deviations for a checked distance from the given or calculated distance between two secondary points in the same system shall not exceed

Distances: $\pm 2\sqrt{L}$ mm

where L is distance in meters. For L less than 10 m, permissible deviations are ±6 mm.

Angles: $\pm \dfrac{0.135}{\sqrt{L}}$ degrees

or

$$\pm \frac{0.15}{\sqrt{L}} \text{ gon}$$

where L is the length in meters of the shorter side of the angle.

Angles are measured with a theodolite or total station reading to at least one minute. The measurement shall be made in at least one set (i.e., two observations, one on each face of the instrument). Distances can be measured using steel tapes or EDMs and should be measured at least twice by either method. Taped distances should be corrected for temperature, sag, slope, and tension; a tension device is to be used with the tape. EDM/total station instruments should be checked regularly against a range of known distances.

10.9.2.3 Layout Points
The permissible deviations of a checked distance between a secondary point and a layout point, or between two layout points, are

$$\pm K\sqrt{L} \text{ mm}$$

where L is the specified distance in meters, and K is a constant taken from Table 10-15. For L less than 5 m, permissible deviation is $\pm 2K$ mm.

The permissible deviations for a checked angle between two lines, dependent on each other, through adjacent layout points are

$$\pm \frac{0.0675}{\sqrt{L}} K \text{ degrees} \quad \text{or} \quad \pm \frac{0.075}{\sqrt{L}} K \text{ gon}$$

where L is the length in meters of the shorter side of the angle and K is a constant from Table 10-15.

Table 10-15   Accuracy Requirement Constants for Layout Surveys

K	Application
10	Earthwork without any particular accuracy requirement (for example, rough excavation, embankments)
5	Earthwork subject to accuracy requirements (for example, roads, pipelines, structures)
2	Poured concrete structures (for example, curbs, abutments)
1	Precast concrete structures, steel structures (for example, bridges, buildings)

Source: Adapted from Table 8-1, ISO 4463.

Figure 10-26 illustrates the foregoing specifications for the case involving a stakeout for a curved concrete curb. The layout point on the curve has a permissible area of uncertainty generated by ±0.015/m due to angle uncertainties and by ±0.013/m due to distance uncertainties.

10.10 CONTROL SURVEY MARKERS

Generally, **horizontal control survey markers** are used for (1) state or provincial coordinate grids, (2) property and boundary delineation, and (3) project control. The type of marker used varies with the following factors:

- Type of soil or material at the marker site
- Degree of permanence required
- Cost of replacement
- Depth of frost (if applicable)
- Precision requirements

Early North American surveyors marked important points with suitably inscribed 4 in. × 4 in. cedar posts, sometimes embedded in rock cairns. Adjacent trees were used for reference ties, with the initials BT (bearing tree) carved into a blazed portion of the trunk. As time went on, surveyors used a wide assortment of markers (e.g., glass bottles, gun barrels, iron tubes, iron bars, concrete monuments with brass top plates, tablet and bolt markers for embedding into rock and concrete drill holes, and aluminum break-off markers with built-in magnets to facilitate relocation).

Common markers in use today are:

1. For property markers: iron tubes, square iron bars (1″ or ½″ square), and round iron bars.

2. For construction control: reinforcing steel bars (with and without aluminum caps) and concrete monuments with brass caps.

3. For control surveys: bronze tablet markers, bronze bolt markers, post markers, sleeve-type survey markers, and aluminum break-off markers.

The latter type of aluminum marker (see Figure 10-27) has become quite popular for both control and construction monuments. The chief features of these markers are their light weight (compared to concrete or steel) and the break-off feature, which ensures that, if disturbed, the monument will not bend (as iron bars will) and thus give erroneous location. Instead, these monuments will break off cleanly, leaving the lower portion (including the base) in its correct location. The base can be used as a monument, or it can be used in the relocation of its replacement. The base of these monuments is equipped with a magnet (as is the top portion) to facilitate relocation when magnetic locating instruments are used by the surveyor.

Whichever type of monument is considered for use, the key characteristic must be that of horizontal directional stability. Some municipalities establish secondary and tertiary coordinate grid monuments (survey tablet markers, Figure 10-28) in concrete curbs and sidewalks. In areas that experience frost penetration, vertical movement will

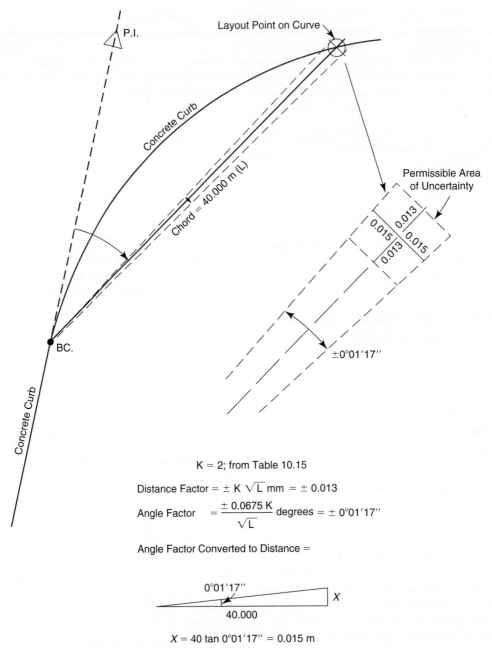

$$K = 2; \text{ from Table 10.15}$$

$$\text{Distance Factor} = \pm\, K \sqrt{L} \text{ mm} = \pm\, 0.013$$

$$\text{Angle Factor} \quad = \frac{\pm\, 0.0675\, K}{\sqrt{L}} \text{ degrees} = \pm\, 0°01'17''$$

$$\text{Angle Factor Converted to Distance} =$$

$$X = 40 \tan 0°01'17'' = 0.015 \text{ m}$$

FIGURE 10-26 Accuracy analysis for a concrete curb layout point (See ISO Standard 4463.)

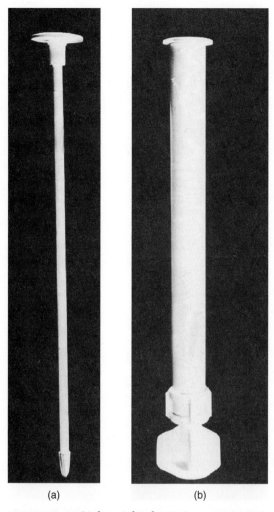

(a)   (b)

FIGURE 10-27 Lightweight aluminium monuments. (a) Rod with aluminum cap. (b) Break-off pipe monument

(Courtesy of Berntsen International, Inc.)

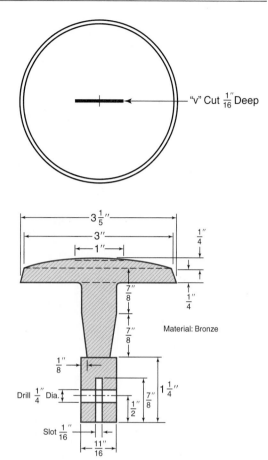

FIGURE 10-28 Survey tablet marker for use in concrete, masonry, or rock

take place as the frost enters and then leaves the soil supporting the curb or walk. Normally, the movement is restricted to the vertical direction, leaving the monument's horizontal coordinates valid.

Vertical control survey markers (benchmarks) are established on structures that restrict vertical movement. It is seldom the case that the same survey marker is used for both horizontal and vertical control. Markers such as those illustrated in Figure 10-28 can be unreliable for

elevation reference unless they have been installed on concrete structures that are not affected by frost or loading movement. The marker shown in Figure 10-29 was designed specifically for use as both a horizontal and vertical marker and could be quite useful in GPS control surveys.

Most agencies prefer to place vertical control markers on vertical structural members (e.g., masonry building walls, concrete piers, abutments, walls), that is, on any structural component whose footing is well below the level of frost penetration. Tablet markers used for benchmarks are sometimes manufactured with a protruding ledge, so that the rod can be easily supported on

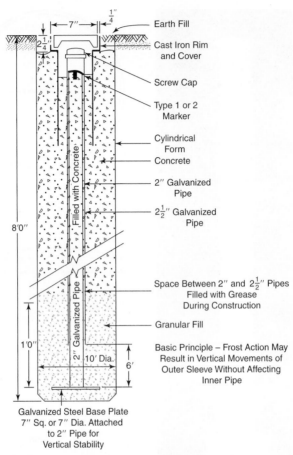

FIGURE 10-29 Sleeve-type survey marker; can be used as a horizontal or a vertical control monument
(Courtesy of Department of Energy, Mines, and Resources, Surveys and Mapping Branch, Ottawa)

the mark. One drawback to using vertical structural members as locations for control monuments is the inability of the marker to be used in GPS surveys, thus combining both horizontal and vertical control. When machine guidance and control techniques are used in construction projects (see Chapter 13), in-ground control monuments, capable of referencing both horizontal and vertical coordinates, must be positioned so that construction machines can occupy their locations and thereby receive confirmation or recalibration of the machine's working coordinates.

Temporary benchmarks (TBMs) can be chiseled marks on concrete, rock, or steel or spikes in utility poles, tree roots, and the like.

The unambiguous description of TBM locations, along with their elevations, is not published (as are regular benchmarks) but is kept on file for same-agency use.

10.11 DIRECTION OF A LINE BY GYROTHEODOLITE

How does a surveyor determine geographic direction while working underground (mining and tunneling surveys)? Several surveying equipment manufacturers produce gyro attachments for use with repeating theodolites (usually 6″ or 20″ theodolites),

FIGURE 10-30 Gyro attachment, mounted on a 20″ theodolite. Shown with battery charger and control unit.

(Courtesy of Sokkia Co. Ltd.)

or total stations. Figure 10-30 shows a gyro attachment mounted on a 20″ theodolite.

The gyro attachment (gyrocompass) consists of a wire-hung pendulum supporting a high-speed, perfectly balanced gyromotor capable of attaining the required speed (for the illustrated instrument) of 12,000 rpm in 1 min. Basically, the rotation of the earth affects the orientation of the spin axis of the gyroscope, such that the gyroscope spin axis orients itself toward the pole in an oscillating motion that is observed and measured in a plane perpendicular to the pendulum. This north-seeking oscillation, which is known as *precession*, is measured on the horizontal circle of the theodolite; extreme left (west) and right (east) readings are averaged to arrive at the meridian direction.

The gyro-equipped theodolite is set up and oriented approximately to north, using a compass; the gyromotor is engaged until the proper angular velocity has been reached (about 12,000 rpm for the instrument shown in Figure 10-30), and then the gyroscope is released. The precession oscillations are observed through the gyro-attachment viewing eyepiece, and the theodolite is adjusted closer to the northerly direction if necessary.

When the theodolite is pointed to within a few minutes of north, the extreme precession positions (west and east) are noted in the viewing eyepiece and then recorded on the horizontal circle; as noted earlier, the position of the meridian is the value of the averaged precession readings. This technique, which takes about a half-hour to complete, is accurate to within 20″ seconds of azimuth.

These instruments can be used well in mining and tunneling surveys for azimuth determination. More precise (3″ to 5″) gyrotheodolites can be used in the extension of surface control surveys.

Questions

10.1 What are the advantages of referencing a survey to a recognized plane grid?

10.2 Describe why the use of a plane grid distorts spatial relationships.

10.3 How can the distortions in spatial relationships (above) be minimized?

10.4 Describe the factors you would consider in establishing a net of control survey monuments to facilitate the design and construction of a large engineering works, for example, an airport.

10.5 Some have said that, with the advent of the global positioning system (GPS), the need for extensive ground control survey monumentation has been much reduced. Explain.

Problems

Problems 10.1 through 10.5 use the following control point data given in Table 10-16.

10.1 Draw a representative sketch of the four control points and then determine the grid distances and grid bearings of sides *AB*, *BC*, *CD*, and *DA*.

Table 10-16   Small-Area Urban Control Monument Data for Problems 10.1–10.5*

Monument	Elevation	Northing	Easting	Latitude	Longitude
A	179.832	4,850,296.103	317,104.062	43°47′33″ N	079°20′50″ W
B	181.356	4,480,218.330	316,823.936	43°47′30″ N	079°21′02″ W
C	188.976	4,850,182.348	316,600.889	43°47′29″ N	079°21′12″ W
D	187.452	4,850,184.986	316,806.910	43°47′29″ N	079°21′03″ W

Average longitude = 079°21′02″ (Monument B)   Easting at CM = 304,800.000 m
Average latitude = 43°47′30″ N (Monument B)   Northing at equator = 0.000 m
Central Meridian (CM) at longitude = 079°30′ W   Scale factor at CM = 0.9999

*Data are consistent with the 3° transverse Mercator projection, related to NAD83.

10.2 From the grid bearings computed in Problem 10.1, compute the interior angles (and their sum) of the traverse *A, B, C, D, A,* thus verifying the correctness of the grid bearing computations.

10.3 Determine the ground distances for the four traverse sides by applying the scale and elevation factors (i.e., grid factors). Use average latitude and longitude.

10.4 Determine the convergence correction for each traverse side and determine the geodetic bearings for each traverse side. Use average latitude and longitude.

10.5 From the geodetic bearings computed in Problem 10.4, compute the interior angles (and their sum) of the traverse *A, B, C, D, A,* thus verifying the correctness of the geodetic bearing computations.

REMOTE SENSING

CHAPTER ELEVEN

SATELLITE IMAGERY

11.1 GENERAL BACKGROUND

Remote sensing is a term used to describe geospatial data collection and interpretive analyses for both airborne and satellite imagery. Satellite platforms collect geospatial data using various digital scanning sensors, for example, multispectral scanners and radar. Digital imagery, now used on both airborne (Chapter 12) and satellite platforms, has some advantages over airborne film-based photography, the traditional form of remotely sensed data acquisition:

- Large digital data-capture areas can be processed more quickly using computer-based analyses.

- Feature identification can be much more effective when combining multispectral scanners with panchromatic imagery. For example, with the analysis of spectral reflectance variations, even tree foliage differentiation is possible (e.g., in near infrared, coniferous trees are distinctly darker in tone than are deciduous trees).

- Ongoing measurements track slowly changing conditions, for example, crop diseases.

- Imaging can be relatively inexpensive. The consumer need purchase only that level of image processing needed for a specific project.

- In digital format, data are ready for computer processing.

As with all measurement techniques, satellite imagery is susceptible to errors and other problems requiring analysis:

- Although individual ground features typically reflect light from unique portions of the spectrum, reflectance can vary for the same type of features.

- The sun's energy, the source of reflected and emitted signals, can vary over time and location.

- Atmospheric scattering and absorption of the sun's radiation can vary unpredictably over time and location.

- Depending on the resolution of the images, some detail may be missed or mistakenly identified.

- Although vast amounts of data can be collected very quickly, processing the data takes considerable time (e.g., the Space Shuttle *Endeavour*'s 10-day radar topography of the earth mission in 2000 collected about 1 trillion images, and it took about 2 years to process the data into map form).

- Scale, or resolution, may be a limiting factor on some projects.

- With satellite-borne instruments, upgrades to reflect the most current advancement in sensors cannot easily be applied.

Because airborne imagery (see Chapter 12) is usually at a much larger scale than is satellite imaging, aerial images are presently more suitable for projects requiring maximum detail and precision, for example, engineering works; however, the functional planning of corridor works, such as in highways, railways, canals, and electrical transmission lines, has been greatly assisted with the advances in satellite imagery.

Remote-sensing data sets do not necessarily have their absolute geographic positions recorded at the time of data acquisition. Each point within an image (pixel or picture element) is located with respect to other pixels within a single image, but the image must be geocorrected before it can be accurately inserted into a GIS or a design document. Data collection and analysis using remote sensing provide much of the information used in GIS, especially those monitoring environmental and land use changes.

11.2 TECHNIQUES OF REMOTE SENSING

Two main categories of remote sensing exist, active and passive. Active remote-sensing instruments (e.g., radar and LiDAR) transmit their own electromagnetic waves and then develop images of the earth's surface as the electromagnetic pulses (known as *backscatter*) are reflected back from the target surface. Passive remote-sensing instruments develop images of the ground surface as they detect the natural energy that is either reflected (if the sun is the signal source) or emitted from the observed target area.

11.3 ELECTROMAGNETIC SPECTRUM

The fundamental unit of electromagnetic radiation is the photon; the photon, which has energy but no mass, moves at the speed of light. The wave theory of light has light traveling in wavelike patterns with its energy characterized by its wavelength or frequency. The speed of light is generally approximately 300,000 km/s or 186,000 miles/s, as we saw in Equation 3.8.

$$c = f\lambda$$

where c is the velocity of light, f is the frequency of the light energy in hertz* (frequency is the number of cycles of a wave passing a fixed point in a given time period), and λ is the wavelength as measured between successive wave crests (see Figure 3-17).

Figure 11-1 shows the electromagnetic spectrum, with wavelengths ranging in size from the very small (cosmic, gamma, and x-rays) to the very large (radio and television waves). In most remote-sensing applications, radiation is described by its wavelength, although those using microwave (radar) sensing have traditionally used frequency instead to describe these much longer wavelength signals.

Visible light is on that very small part of the electromagnetic spectrum ranging from about 0.4 to 0.7 μm (μm is the symbol for micrometer or micron, a unit that is one millionth of a meter; see Table 11-1). Infrared radiation (IR) ranges from about 0.7 to 100 μm and that range can be further subdivided into *reflected* region (ranging from

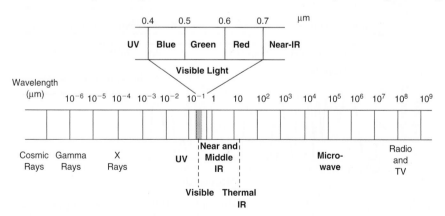

FIGURE 11-1 Electromagnetic spectrum

---

*Hertz (Hz) is a frequency of one cycle per second; kilohertz (kHz) is 10^3 Hz, megahertz (MHz) is 10^6 Hz, and gigahertz (GHz) is 10^9 Hz.

Table 11-1   Visible Spectrum—These Divisions are not Absolute but Generally Accepted Values

Color	Bandwidth
Violet	0.4–0.446 µm
Blue	0.446–0.500 µm
Green	0.500–0.578 µm
Yellow	0.578–0.592 µm
Orange	0.592–0.620 µm
Red	0.620–0.7 µm

about 0.7 to 3.0 µm) and the *thermal* region (ranging from about 3.0 to 100 µm). The other commonly used region of the spectrum is that of microwaves (e.g., radar), with wavelengths ranging from about 1 mm to 1 m.

Only radiation in the visible, infrared (IR), and microwave sections of the spectrum travel through the atmosphere relatively free of blockage; these atmospheric transmission anomalies, known as atmospheric windows, are utilized for most common remote-sensing activities (see Figure 11-2). Note that the relatively short wavelengths of blue light suffer from atmospheric

scattering to the extent that it is often difficult to distinguish features on the resulting blue band image. This scattering is known as the Raleigh effect and is the reason that the sky appears to be blue.

Radiation that is not absorbed or scattered in the atmosphere (*incident* energy) reaches the surface of the earth, where three types of interaction can occur (see Figure 11-3):

- Transmitted energy passes through an object surface with a change in velocity.

- Absorbed energy is transferred through surface features by way of electron or molecular reactions.

- Reflected energy is reflected back to a sensor with the incident angle equal to the angle of reflection (those wavelengths that are reflected determine the color of the surface).

$$\text{Incident energy} = \text{transmitted energy} \\ + \text{absorbed energy} + \text{reflected energy} \tag{11-1}$$

To illustrate, chlorophyll (a constituent of leaves) absorbs radiation in the red and blue wavelengths but reflects the green wavelengths. Thus, at the height of the growing season, when chlorophyll is strongly present, leaves appear greener. As well,

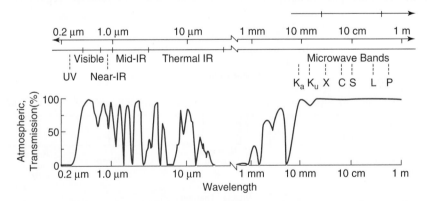

FIGURE 11-2 Electromagnetic spectrum showing atmospheric transmission windows in the visible, near, middle, and thermal infrared and microwave regions. Notice the excellent atmospheric transmission capabilities in the entire range of the microwave

(Courtesy of NASA)

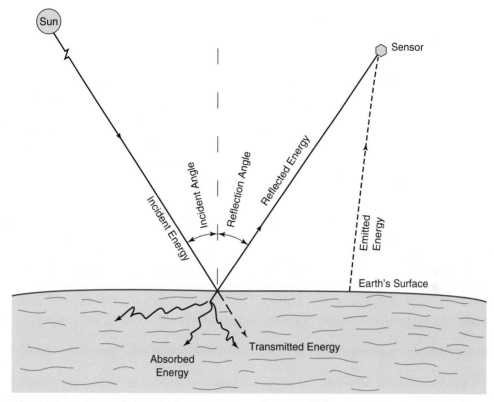

FIGURE 11-3 Interaction of electromagnetic radiation with the earth

water absorbs the longer wavelengths in the visible and near-IR portions of the spectrum and reflects the shorter wavelengths in the blue/green part of the spectrum resulting in the often seen blue/green hues of water.

11.3.1 Reflected Energy

Different types of surfaces reflect radiation differently. Smooth surfaces act like mirrors reflecting most light in a single direction; this is called *specular* reflection. At the other extreme, rough surfaces reflect radiation equally in all directions—this is called *diffuse* reflection. Between these two extremes are an infinite number of reflection possibilities, which are often a characteristic of a specific material or surface condition (see Figure 11-4).

11.4 SELECTION OF RADIATION SENSORS

Satellite sensors are designed to meet various considerations, for example:

- To use those parts of the electromagnetic spectrum that are less affected by scatter and absorption.

- To sense data effectively from different sources (active or passive sensing).

- To provide specific regions within the electromagnetic spectrum suitable for distinguishing specified target surfaces. Multispectral platforms are designed with a specific use in mind, such as crop monitoring, weather forecasting, or ocean temperature measurements.

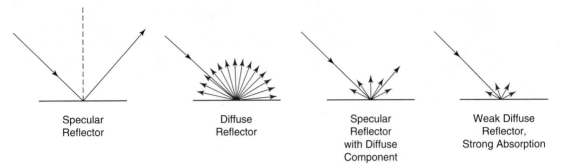

Specular   Diffuse   Specular   Weak Diffuse
Reflector   Reflector   Reflector   Reflector,
                                   with Diffuse   Strong Absorption
                                   Component

FIGURE 11-4 Specular and diffuse reflection

Satellite-based remote-sensing instruments use scanner technology, as opposed to frame or digital camera technology used in aerial photogrammetry (see Chapter 12). The sensors have a spatial resolution ranging from 0.5 m to 1 km. Scanners, as opposed to camera, gather information over a small portion of the surface at any given instant. There are four basic types of scanners used: across-track, along-track, spin, and conical scanners.

Across-track scanners, also called whisk-broom scanners, scan lines perpendicular to the forward motion of the spacecraft using rotating mirrors. As the spacecraft move across the earth, a series of scanned lines are collected producing a two-dimension image of the earth. Each onboard sensor in the vehicle collects reflected or emitted energy in a specific range of wavelengths; these electrical signals are converted to digital data and stored until they are transmitted to earth.

Along-track scanners, also called push-broom scanners, utilize the forward motion of the spacecraft itself to record successive scan lines. Instead of rotating mirrors, these scanners are equipped with linear array charge-coupled devices (CCDs) that build the image as the satellite moves forward.

Spin scanners, common in older meteorological satellites, would spin the satellite relative to the north–south axis of the earth, so the scanner would collect data in an east–west direction, and then after completing a rotation, it would change the angle of view to collect a line north or south of the previous rotation.

Conical scanners, primarily for meteorological purposes, scan in a circular or elliptical pattern. The scanner is oriented to scan about the nadir axis of the system so that the instantaneous field of view (IFOV) and the angle from the nadir remain constant.

Satellites often carry multiple payloads of remote-sensing systems and each system has multiple sensors covering specific ranges within the electromagnetic spectrum. Most systems have four to eight sensors, with the exception of hyperspectral systems, which use imaging spectrometers to sample hundreds of small segments (10 nm) within the electromagnetic spectrum.

Once the data have been captured and downloaded to earth, much more work has to be done to convert the data into useable format. These processes, which include image enhancement and image classification, are beyond the scope of this text. Refer to Section 11.13 at the end of the chapter for further information on these topics, or look for digital imaging processing (DIP) techniques in your library or on the Internet.

11.5 AN INTRODUCTION TO IMAGE ANALYSIS

11.5.1 General Background

Analysis of remotely sensed images is a highly specialized field. Since environmental characteristics include most of the physical sciences, the image interpreter should have an understanding of the basic concepts of geography, climatology, geomorphology, geology, soil science, ecology, hydrology, and civil engineering. For this reason, multidisciplinary teams of scientists, geographers, and engineers, trained in image interpretation, are frequently used to help develop the software utilized in this now largely automated process.

Just as photogrammetry/air photo interpretation helps the analyst to identify objects in an aerial photograph, digital image analysis helps the analyst to identify objects depicted in satellite and airborne images. With the development of computerized image analyses that employ automated digital imagery processing, the relative numbers of trained analysts needed to process data have sharply declined. Satellite imagery has always been processed in digital format and aerial photos (films) can be easily scanned to convert from analog to digital format. Just as the interpretation of aerial photography has now been largely automated using "soft-copy" or "digital" photogrammetry techniques on the digitized photos (see Section 12.10), satellite and airborne imagery analysis is also largely automated. As with aerial photography, satellite images must have ground control and must be corrected for geometric distortion and relief displacement, and also for distortions caused by the atmosphere, by radiometric errors, and by the earth's rotation. Satellite imagery can be purchased with some or many of these distortions removed.

Unlike aerial photographic images, different satellite images of the same spatial area taken by two or more different sensors can be fused together to produce a new image with distinct characteristics and thus provide even more data. A digital image target may be a point, line, or area (polygon) feature that is distinguishable from other surrounding features. Digital images are composed of many pixels each of which is assigned a digital number (DN); the DN represents the brightness level for that specific pixel. For example, in eight-bit imagery, the 256 ($= 2^8$) brightness levels range from 0 to 255, indicating the intensity of the brightness within the range

of the electromagnetic spectrum the sensor is sensitive to. The DN can then be used to display a brightness level on a computer monitor for a specific color gun (red, green, or blue). The combination of these colors gives you the image you see on the monitor. The color shown on the monitor does not have to be the same color that was collected by the sensor, so you can "see" infrared (a nonvisible range within the electromagnetic spectrum) as red by having it displayed by the red gun of the monitor. Each sensor is considered a layer or band within the image.

Each satellite has the capability of capturing and transmitting the data to ground receiving stations that are located around the globe. Modern remote-sensing satellites are equipped with sensors collecting data from selected bands of the electromagnetic spectrum; for example, Landsat 7's sensors collect data from eight bands (channels). Since radiation from various portions of the electromagnetic spectrum reacts in a unique fashion with different materials found on the surface of the earth, it is possible to develop a signature of the reflected and emitted signal responses and, from those signatures, identify the object or class of objects.

When the interactions of reflected, emitted, scattered, and absorbed energy (at the various wavelengths collected by onboard sensors) with the ground surface are plotted, the resultant unique curve, called a spectral signature, can be used to help identify that material. Figure 11-5 shows four signatures of different ground surfaces, with the reflection as a function of wavelength used for plotting purposes. The curves were plotted from data received from the eight sensors onboard Landsat 7. When the results are plotted as percentage reflectance for two or more bands (see Figure 11-6) in multidimensional space, the ability to precisely identify the surface materials increases markedly. This spectral separation permits the effective analysis of satellite imagery, most of which can be performed through computer applications. Image analysis can be productively applied to a broad range of scientific inquiry (see Table 11-2).

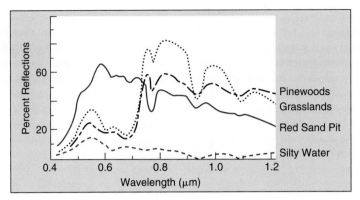

FIGURE 11-5 Spectral reflectance curves of four different targets. Adapted from NASA's *Landsat 7 Image Assessment System Handbook*, 2000

(Courtesy of NASA)

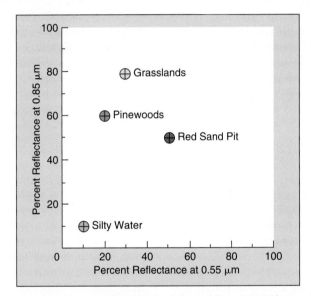

FIGURE 11-6 Spectral separability using just two bands. Adapted from NASA's *Landsat 7 Image Assessment System Handbook*, 2000

(Courtesy of NASA)

11.5.2 Data Resolution

Data resolution defines the detail of information that can be gathered from a system. Higher resolution provides more detail, but with some associated cost. Resolution can change due to sensor design, detector size, focal length, satellite altitude, and changes that occur over time.

11.5.2.1 Spatial Resolution

Generally, the farther away (the greater the altitude) is the sensor, the larger the area that can be seen, but the ability to distinguish some detail may be lost. Some sensors have greater ability to "see detail"—that is, they have greater spatial resolution; spatial resolution refers to the smallest possible feature that can be detected. If a sensor has a spatial resolution of, say, 30 m, and if the

Table 11-2   Applications of Remotely Sensed Images

Field	Applications
Forestry/agriculture	Inventory of crop and timber acreage, estimating crop yields, crop disease tracking, determination of soil conditions, assessment of fire damage, precision farming, and global food analysis.
Land use/mapping	Classification of land use, mapping and map updating, categorization of land use capability, monitoring urban growth, local and regional planning, functional planning for transportation corridors, mapping land/water boundaries, and flood plain management.
Geology	Mapping of major geologic units, revising geologic maps, recognition of rock types, mapping recent volcanic surface deposits, mapping landforms, mineral and gas/oil exploration, and estimating slope failures.
Natural resources	Natural resources exploration and management.
Water resources	Mapping of floods and flood plains, determining the extent of area snow and ice, measurement of glacial features, measurement of sediment and turbidity patterns, delineation of irrigated fields, inventory of lakes, and estimating snow-melt runoff.
Coastal resources	Mapping shoreline changes, mapping shoals and shallow areas, and tracing oil spills and pollutants and erosion.
Environment	Monitoring effect of human activity (lake eutrophication, defoliation), measuring effects of natural disasters, monitoring surface mining and reclamation, assessing drought impact, siting for solid-waste disposal, siting for power plants and other industries, regulatory compliance studies, and development impact analysis.

image is displayed at full resolution, each pixel (picture element) will represent an area of 30 m × 30 m.

Two terms are presently used to describe the spatial resolution of satellite imagery: ground sample distance (GSD), expressed in meters or kilometers, and instantaneous field of view (IFOV), expressed in milliradians (mrad). The IFOV defines the pixel angular size and, together with the altitude, determines the area of terrain or ocean covered by the field of view of a single detector. The general equation for calculating spatial resolution is

$$\text{GSD (meters)} = \text{IFOV (radians)} \times \text{altitude (meters)} \quad (11\text{-}2)$$

For example, a 2.5 mrad (0.0025 rad) IFOV at 1,000 m of altitude results in a GSD of 2.5 m.

The Landsat 7 satellite senses the earth's surface at three different resolutions (see Table 11-3); 30 m for bands 1–5 and 7; 60 m for band 6; and 15 m for band 8. Commercial satellite systems generally have higher spatial resolution, such as GeoEye's IKONOS, which has 0.82 m panchromatic (gray-scale) resolution and 4 m multispectral resolution. The French SPOT panchromatic and color sensor has a resolution of 2 m, whereas the SPOT multispectral (XS) sensor has a resolution of 8 m. Indian remote-sensing satellite CARTOSAT has a panchromatic resolution of about 0.8 m. The 2008 GeoEye-1 has resolutions of 0.41 m for panchromatic and 1.65 m for multispectral. DigitalGlobe's World View 2 has about a 0.5 m panchromatic resolution and a 1.84 m resolution on the multispectral platform.

As a rule of thumb, you need a pixel size of about half the object size in order to distinguish that object. If you were trying to identify utility boxes that were 2 ft × 2 ft, ideally you would have a pixel size or spatial resolution of 1 ft.

Table 11-3   Bandwidth Characteristics—Landsat 5 and 7*

TM and ETM + Spectral bandwidths (μm)

Sensor	Band 1	Band 2	Band 3	Band 4	Band 5	Band 6	Band 7	Band 8
TM	0.45–0.52	0.52–0.60	0.63–0.69	0.76–0.90	1.55–1.75	10.4–12.5	2.08–2.35	N/A
ETM+	0.45–0.52	0.53–0.61	0.63–0.69	0.78–0.90	1.55–1.75	10.4–12.5	2.09–2.35	0.52–0.90
	Blue	Green	Red	Near IR	Short wave IR	Thermal IR	Shortwave IR	Panchromatic
Resolution (pixel size)	30 m	30 m	30 m	30 m	30 m	60 m	30 m	15 m

Band	Use
1	Soil/vegetation discrimination; bathymetry/coastal mapping; cultural/urban feature identification.
2	Green vegetation mapping; cultural/urban feature identification.
3	Vegetated versus nonvegetated and plant species discrimination.
4	Identification of plant/vegetation types, health and biomass content, water body delineation, soil moisture.
5	Sensitive to moisture in soil and vegetation, discriminating snow- and cloud-covered areas.
6	Vegetation stress and soil moisture discrimination related to thermal radiation; thermal mapping (urban, water).
7	Discrimination of mineral and rock types; sensitive to vegetation moisture content.
8	Panchromatic: mapping, planning, design.

*Landsat 5 and 7 are being operated under the direction of NOAA. Data can be obtained from USGS-EROS Data Center, Sioux Falls, SD 57198.

(Courtesy of NOAA)

11.5.2.2 Spectral Resolution Spectral resolution is a measure of the detail of the range of the electromagnetic spectrum sensed by a single sensor. The smaller the range, the greater the detail and the higher the spectral resolution. Surface features can be identified by analyzing the spectral responses over distinct wavelength changes. Commonly seen surfaces such as vegetation and water can be identified using a broad range of the electromagnetic spectrum, whereas a more detailed study, for example, of different tree types and different rock types requires analysis over much finer wavelength ranges. Thus, the higher the spectral resolution of the sensor, the more distinctions can be made of surface materials. Newer satellites record spectral responses over several wavelength ranges at different spectral resolutions; these are called multispectral sensors. More recent advances in multispectral sensing (e.g., hyperspectral sensing) can involve the application of hundreds of very narrow spectral bands, which greatly extends the ability to distinguish between very similar surface features under a variety of seasonal conditions. Hyperspectral remote sensing is becoming popular in exploration geology where certain minerals cannot be detected using broadband scanners. Spectral responses from identical ground surface features may vary for a variety of reasons:

- Atmospheric constituents.
- Angle of sun (for reflected emissions).
- Shadow.
- Smoke.
- Soil moisture.
- Temporal aspects (when comparing images collected on different dates).
- Altitude.
- Topographic surface (whether flat or steeply graded).
- Feature height (heights can be determined using LiDAR, radar, or stereoimage analysis).

11.5.2.3 Radiometric Resolution Radiometric resolution refers to the sensors' ability to detect small changes in energy and reflects the number of bits available for each pixel. Imagery data are represented by positive binary numbers that range from 0 to a selected power of 2. Each bit records an exponent of 2; that is, a 2 bit is $2^2 = 4$, a 4 bit is $2^4 = 16$, and so on. The number of bits used to represent the recorded data determines the number of brightness levels available. Images of the same ground area at both 2-bit and 8-bit resolution would reveal finer detail in the 8-bit image. Many images are now recorded in 8- to 11-bit (and higher) resolutions. This permits a range of 256 to 2,048 DNs of gray tone ranging from black, at a value of zero, to white at the maximum value. Whereas the human eye may only be able to differentiate between 20 and 30 steps of gray tone, some sensors can detect all levels of gray tone. A further advantage to using digital images is that analysis can be largely automated using computer algorithms, thus greatly reducing the need for continual human intervention, with its attendant potential drawbacks in the areas of human error, inconsistent analysis, and higher costs.

11.5.2.4 Temporal Resolution The surface of the earth is always changing. Changes may occur very slowly, as with geologic processes, or changes may occur somewhat more rapidly, as with urban development or shoreline erosion. Changes may occur at a catastrophic rate, as with crop diseases, large fires, flooding, and even troop movements during wartime. As satellites orbit the earth, the earth is also revolving on its axis so that each subsequent swath covers a slightly different portion of the earth's surface. Eventually, the satellites' orbits begin to repeat the surface coverage. The time period required to achieve repeat coverage of the same surface is called the revisit time, which varies from a few days to a month. Revisit times can be greatly shortened for those satellites that have motorized direction-controlled detectors, which can be sent coded instructions from ground stations to "look" toward specified areas as the orbit proceeds.

Temporal variations in surface features can be used to positively identify some features and to systematically track the changes in other features. Additionally, repeat revisiting of all surface features helps to overcome the significant problems that cloud cover poses for passive optical remote sensors.

11.6 CLASSIFICATION

Spectral classification is the process of grouping similar neighboring pixels into classes. Information classification is the process of identifying those groupings as features. Satellite imagery contains a wealth of information, all of which may not be relevant for specific project purposes. A decision must be made as to which feature classifications are to be of most relevance. Once the types of features to be classified have been determined, an image can be created where the pixel value is a class number or a code for the class. This thematic image can then be color-coded based on the class.

There are a number of methods for classification of data. The most common are supervised and unsupervised methods. Supervised classification is where areas with known features are sampled, and then remote-sensing software analyzes all the pixels within an area of interest to determine which pixels fit the samples or signatures of the features. Unsupervised classification is where the number of classifications is defined and the software then determines which class each pixel falls within. In addition, there are hybrid classifications, which use both supervised and unsupervised processes to determine the classifications, and fuzzy classifications, which compensates for the fact that certain features such as vegetation do not have a finite line between types of vegetation, but have a transition area.

When performing any of the foregoing classification techniques, only those bands that will assist in the classification are used. As mentioned before, specific bands are designed to sense specific

types of features; for example, isolating vegetation is best done using the red and near infrared ranges. The classification process is an iterative process where many variables must be considered and reprocessed. Although the software is capable of doing the analysis in a short period of time, creating an acceptable classification often takes hundreds of modifications and runs.

11.7 FEATURE EXTRACTION

A specialized type of classification is feature extraction. Feature extraction is the process of creating closed polygons around specific feature types, such as buildings or trees. The development of feature extraction software is relatively new, with much advancement in the process. Feature extraction software generally uses enhancement techniques to distinguish the features as opposed to standard classification techniques. Enhancement is the process of making the data more interpretable, or isolating out important features. Enhancement does not increase the spatial resolution, but "cleans" up the pixels by correcting data anomalies, radiometric enhancement, and combining like pixels.

11.8 GROUND-TRUTH OR ACCURACY ASSESSMENT

How do we know if the sophisticated image analysis techniques in use are giving us accurate identifications? We have to establish a level of reliability in order to give credibility to the interpretation process. With aerial photo interpretation, we begin with a somewhat intuitive visual model of the ground feature, whereas with satellite imagery we begin with pixels identified by their gray-scale DN, characteristics that are not intuitive.

We need a process whereby we can determine the accuracy and reliability of interpretation results. In addition to providing reliability, such a process enables the operator to correct errors and to compare the successes of the various types of feature extraction that may have been used in the project, additionally. This process is also invaluable in the calibration of imaging sensors.

Ground-truth or accuracy assessment can be accomplished by comparing the automated feature extraction identifications with feature identifications given by other methods for a given sample size. Alternative interpretation or extraction techniques may include air photo interpretation, analyses of existing thematic maps of the same area, and fieldwork involving same-area ground sampling.

Since it is not practical to check the accuracy of each pixel identification, a representative sample is chosen for accuracy assessment. The sample is representative with respect to both the size and locale of the sampled geographic area. Factors to be considered when defining the sample include the following:

- Is the land privately owned and/or restricted to access?

- How will the data be collected?

- How much money can be dedicated to this process?

- Which geographic areas should be sampled: areas that consist of homogeneous features together with areas that consist of a wide mix of features, even overlapping features?

- How can the randomness of the sample points be assured, given practical constraints (e.g., restricted access to specified ground areas, lack of current air photos and thematic maps)?

- If verification data are to be taken from existing maps and plans, what effect will their dates of data collection have on the analysis (the more current the data, the better the correlation)?

The answers to many of the foregoing questions come only with experience. Much of this type of work is based on trial-and-error decisions. For example, when the results of small samples compare favorably with the results of large samples,

it may be assumed that, given similar conditions, such small samples may be appropriate in subsequent investigations. On the other hand, if large samples give significantly different (better) results than do small sample sizes, it may be assumed that such small sample sizes should not be used again under similar circumstances.

Often image analysis has accuracy in the range of 60–70 percent. Is accuracy of 60–70 percent acceptable? That depends on the intended use of the data and, once again, experience is the determining factor. It may be decided that this method of data collection (e.g., satellite imagery) is inappropriate because it is not accurate enough or it is too expensive for the intended purposes of the project. On the other hand, 60–70 percent accuracy may be more than enough for the types of analyses that are to be made using the presented data.

In general, remote-sensing processes have low-accuracy statistics compared to photogrammetry and field survey procedures. However, remote sensing allows for quick measurements over large areas as well as providing information beyond just dimensional information.

11.9 U.S. NATIONAL LAND-COVER DATA (NLCD) 2006

The U.S. Geological Survey (USGS), at the EROS Data Center, completed a land-cover data set of the conterminous United States in 2006—based on data collected by Landsat 5 and Landsat 7. NLCD 2006 was a joint project by the Multi-Resolution Land Characteristics Consortium (MRLC), consisting of a group of federal agencies who coordinate and generate consistent and relevant land-cover information at the national scale. The classification system for NLCD was defined based on the anticipated use of the data and resulted in a listing of 20 classes. The classes are as shown in Table 11-4.

Data sets showing land cover, land-cover changes, and percent developed imperviousness

Table 11-4   NLCD Land-Cover Class Descriptions

Number	Description	Number	Description
11	Open Water	73	Lichens*
12	Perennial Ice/Snow	74	Moss*
21	Developed, Open Space	81	Pasture/Hay
22	Developed, Low Intensity	82	Cultivated Crops
23	Developed, Medium Intensity	90	Woody Wetlands
24	Developed, High Intensity		
31	Barren Land (Rock/Sand/Clay)		
41	Deciduous Forest		
42	Evergreen Forest	95	Emergent Herbaceous Wetlands
43	Mixed Forest		
51	Dwarf Scrub*		
52	Shrub/Scrub		
71	Grassland/Herbaceous		
72	Sedge/Herbaceous*		

*Alaska only.

can be downloaded from the USGS EROS Data Center at http://edc.usgs.gov/ for private or government GIS projects to supplement existing project data or to act as a base for a GIS project. The data can be directly loaded into remote-sensing or GIS software, for example, ERDAS Imagine and ArcGIS. Similar data are available in Canada at Natural Resources Canada, Geogratis (http://geogratis.cgdi.gc.ca).

11.10 REMOTE-SENSING SATELLITES

11.10.1 Landsat Program

The Landsat program was conceived in 1966, with the first satellite Landsat 1, also known as ERTS, launched in 1972. This program has continually monitored the earth's surface since Landsat 1 first started collecting and sending back data. The Landsat program was designed to monitor small-scale processes seasonally on a global scale, for example, cycles of vegetation growth, deforestation, agricultural land use, and erosion. When the spectral signatures of imaged areas have been collected, they are compared to spectral signatures stored in a library to determine the exact nature of the surface feature. Currently, there remain only two operational Landsat satellites. One is Landsat 5, which was launched in 1984, carrying the Multispectral Scanner System (MSS) and the Thematic Mapper (TM). In 1995 the MSS was turned off. In November 2005, Landsat 5 had a solar array problem which took the satellite offline until 2006. Landsat 5 with a design life of 3 years has far exceeded even the most optimistic expectations. The other operational satellite, Landsat 7 was launched in 1999, carrying the enhanced Thematic Mapper Plus (ETM+) system. This system had a panchromatic band with 15 m spatial resolution, and a thermal IR channel with 60 m spatial resolution (see Table 11-3 for the specifics on the bands). Landsat 7 worked without problems until May 2003, when it had a scan line corrector failure. Although this only stopped operations for a short period of time, it has had a continual impact on Landsat 7 in that about 22 percent of any given scene is missing.

Although the Landsat program is only limping along and the spatial resolution of 30 m is not as good as many of the private satellites, the program still has significant achievements:

- The Landsat program has continually imaged the earth's surface for 33 years, with Landsat 7 covering all the earth every 16 days. Most other satellite systems are project based, meaning they are not continually collecting data.

- The data from the Landsat 7 are affordably priced, with much of the data available for free.

- Landsat 7 data are highly calibrated proving extremely accurate data.

- Much of the indexing and classification algorithms used in remote sensing are based on the Landsat 7 sensors.

The Landsat Data Continuity Mission (LDCM) is intended to provide the continual monitoring of the earth started by Landsat in 1972. The next satellite is intended to carry two instruments, an Operation Land Imager (OLI) and Thermal Infrared Sensor (TIRS), which will be similar but an improvement over the Landsat 7 instruments. The scheduled launch date is December 2012.

11.10.2 NASA's Satellites

NASA created a new millennium program to develop new technologies for monitoring our solar system, the universe, and earth. The earth-based satellites are known as Earth Observing 1 (EO-1), which is used to image the earth, and Earth Observing 3 (EO-3), which is used for weather monitoring.

EO-1 was launched on November 21, 2000. It employs a push-broom spectrometer/radiometer—the Advanced Land Imager (ALI)—with many more spectral ranges (hyperspectral scanning) and flies the same orbits at the same altitude as does Landsat 7. It is designed to sample similar surface features at roughly the same time (only minutes apart) for comparative analyses. ALI's panchromatic band has a resolution of 10 m. This experimental satellite was originally

designed as a 1-year project but it was extended. It is also designed to obtain much more data at reduced costs and will influence the design of the next stage (2000–2015) of U.S. exploration satellites. The EO-1 is a testing technology to be used for the LDCM mission.

In December 1999, NASA launched the Terra (EOS AM-1) satellite platform containing five instruments to monitor environmental events and changes on earth. The instruments ASTER, CERES, MISR, MODIS, and MOPITT all have specific objectives they are designed for. Only Advanced Spaceborne Thermal Emission and Reflection Radiometer (ASTER) provides high-resolution spatial data used for common remote-sensing applications. It provides spatial resolutions of 15–90 m and has 14 bands ranging from visible to thermal infrared. ASTER does not collect data continually, but only collects about 8 min of data per orbit.

11.10.3 Other Multispectral Scanning Satellites

Since the 1960s, there have been more than a hundred satellites launched by the scientific and intelligence agencies of many of the earth's leading countries including the United States, France, Russia, China, India, Japan, Europe, Canada, Australia, Israel, and Brazil. By 2000 there were 12 satellites similar to Landsat 7, which include France's SPOT (SPOT 5, 2001) satellites; India's IRS satellites; China and Brazil satellites; U.S./Japan satellite, EOS AM-1, 1999; and the U.S. commercial resource 21 satellites (R-21, A-D). Dozens more orbiting satellites with ever-improving spatial and radiometric resolutions are expected by 2010. See Table 11-5 for a selected summary of current satellites.

11.10.4 High-Resolution Satellites

In the late 1990s several high-resolution satellites were launched by commercial and intelligence agencies. These satellites, which have resolutions as fine as 0.5 m, include IKONOS-2, QuickBird,

Table 11-5   Details about Various Satellite Systems

Satellite	Launch Date	Sensor Data	Ground Resolution (meters)	Orbit Period (minutes)	Revisit Period (days)
Landsat 5	March 1, 1984	Pan, Visible, NIR,TIR	30	99	16
Landsat 7	May 15, 1999	Pan, Visible, NIR,TIR	15–30	99	16
IKONOS	September 24, 1999	Pan, Visible, NIR	0.82 × 3.2		3
Terra	December 18, 1999	Pan, Visible, NIR,TIR	15–30	96	16
QuickBird	October 18, 2001	Pan, Visible, NIR	0.6–2.4	93.4	3–7
SPOT 5	May 1, 2002	Pan, Visible	2.5–20	96.5	
World View 1	September 18, 2007	Panchomatic	0.59	94.6	4.6
Rapid Eye	August 29, 2008	Visible, NIR	6.5	N/A	N/A
GeoEye-1	September 6, 2008	Pan, Visible, NIR	0.41–1.65		Under 3
World View 2	October 8, 2009	Pan, Visible, NIR	0.52–2.08	100	3.7

and OrbImage (all from the United States); EROS A and B from Israel and the United States; and IRS-P6 from India. The IKONOS satellite provides 1 m resolution in panchromatic (black and white) and 4 m resolution for multispectral (true color and infrared) images.

11.10.5 Hyperspectral Satellites

Hyperspectral scanners can have sensors that detect over a spectral range of 15 to more than 300 relatively narrow sequential bands (multispectral sensing is considered by many to be limited to 2–15 bands). Hyperspectral scanning is focused on that range of the spectrum between 0.4 and 2.5 μm. With the use of very narrow bands, the sensors can provide precise differentiation of the composite materials in soil, canopy, and water surface features that have similar reflective properties (a differentiation that multispectral scanners cannot achieve). The first hyperspectral satellite was the U.S. EO-1 (Hyperion scanner), which was launched in 1999. Additional advancements in this field will result in ultraspectral (utilizing thousands of very narrow spectral bands) scanning sensors that are expected to give

analytical results available now only in laboratory settings.

11.11 IMAGING RADAR SATELLITES

Radar (radio detection and ranging) instruments are active scanners that operate in the 1 cm to 1 m microwave range of the electromagnetic spectrum. Because microwaves have a longer wavelength, they are not as prone to scattering as they travel through the atmosphere. Radar waves can penetrate clouds, haze, dust, and all but the heaviest of rainfalls, depending on the channels used. Unlike the passive spectral scanners that record reflected and emitted signals, imaging radar records the signals that are sent from the satellite and then returned (bounced or echoed) from the surface of the earth back to the satellite. As satellites orbit the earth, half the mission is flown in sunlight and the other half in darkness. Passive sensors rely on sunlight for their reflected signals and thus collect only those data on the bright side of the orbit. Radar satellites can record reflected data throughout the entire orbit (ascending or descending).

Unlike aerial photography and most other remote-sensing techniques, synthetic aperture radar (SAR) scans an area below and off to the side of the aircraft or space vehicle. The scanned swath is some distance off to the side of the nadir track, which itself is directly below the vehicle flight track. SAR is a high-resolution, ground-mapping technique that effectively synthesizes a large receiving antenna by processing the phase of the reflected radar return. Radar beamwidth is inversely proportional to the antenna width (also referred to as the aperture), which means that a wider antenna (or aperture) will produce a narrower beam and a finer resolution. Since there is a limit to the size of the antenna that can be carried on aircraft or spacecraft, the forward motion of the craft along with sophisticated processing of backscatter echoes is used to simulate or synthesize a very large antenna and thus increase azimuth resolution. Radar imaging resolution depends on the effective length of the pulse in the slant (across-track) direction and on the width of the beam in the azimuth (along-track) direction.

Interest in radar imaging grew as it became clear that even with the successes of the early Landsat satellites, many regions of the earth remained almost continuously under cloud cover and thus hidden from view. Polar regions too were virtually beyond the capability of existing MSS-type satellites because of the extended periods of darkness. Radar provided a way to obtain constant data for polar regions for climate studies and navigation. Radar also provides a unique look at the earth's surface that complements reflected data and adds a great deal to the store of knowledge about the earth's surface.

Satellite radar imaging began in 1978 with the launch of the U.S. SEASAT (L band). By the 1990s Japan, the USSR (as it was called then), Europe, and Canada had entered this area of exploration. Several enterprises are also involved in aircraft SAR systems that operate in the C, L, P, and X bands and cover much smaller swaths, for example, 3–60 km (slant distance). Resolution of radar imaging ranges from submeter for defense satellites to 10–30 m for most civilian satellites.

11.11.1 Space Shuttle Endeavor

In February 2000, NASA launched the Endeavor, whose mission (Shuttle Radar Topography Mission—SRTM) was to map the earth using the C band (225-km swath) and the X band (56-km swath) of imaging radar. In just 11 days, the Endeavor captured images of almost the entire earth (56°S to 60°N latitudes) in 1 trillion measurements. Traveling at 7.7 km per second, it mapped an area equivalent in size to the state of Florida every 97.5 seconds. The mission lasted 10 days using a deployed mast of 200 ft in length with two separated antennae mounted on the mast. Resolution of the imaging is 20 m. Many of the SRTM images are available free over the Internet.

11.12 SATELLITE IMAGERY VERSUS AIRBORNE IMAGERY

Prior to 2000, the chief differences between aerial imagery and satellite imagery were characterized by scale and by the additional data that could be gathered using spectral scanning. Now that spectral scanning is available for both airborne and satellite platforms, the chief difference (all other things being equal) is now one of scale.

Figures 11-7 and 11-8 show the types of images generally available using either method of remote sensing. Both images show a part of the Niagara frontier. The satellite image (Figure 11-7) shows the Niagara peninsula and northwestern New York state, whereas the aerial image (Figure 11-8, an aerial photograph) shows, in much greater detail, a small part of the area shown in the satellite image: the mouth of the Niagara River and the towns of Youngstown, New York, and Niagara-on-the-Lake, Ontario. See Figure F-1 for a comparison of the level of detail provided by ground techniques versus aerial and satellite techniques. As the resolution of satellite images improves, the major differences between airborne imagery and satellite imagery will become less significant.

FIGURE 11-7 Landsat image showing western New York and the Niagara region

(Courtesy of U.S. Geological Survey, Sioux Falls, SD)

11.13 REMOTE SENSING INTERNET WEBSITES AND FURTHER READING

11.13.1 Satellite Websites

GeoEye-1, http://www.geoeye.com/CorpSite/

Landsat 7, http://landsat7.usgs.gov/

IKONOS, http://www.geoeye.com/CorpSite/

QuickBird, www.digitalglobe.com/

Radarsat, www.radarsat2.info

Space Shuttle Endeavor, http://www.jpl.nasa.gov/srtm

SPOT, www.spot.com/

Terra, http://terra.nasa.gov

World View 1 & 2, www.digitalglobe.com/

11.13.2 General Reference Websites

American Society for Photogrammetry and Remote Sensing (ASPRS), http://www.asprs.org/

Australian Surveying and Land Information Group, http://www.auslig.gov.au/acres

FIGURE 11-8 Aerial photograph, flown at 20,000 ft, showing the mouth of the Niagara River
(Courtesy of U.S. Geological Survey, Sioux Falls, SD)

Earth Observing System (EOS), http://eos.gsfc.nasa.gov

EROS Data Center (source for photos, imagery, elevations, maps), http://edc.usgs.gov/

USGS Land Cover Institute (USA), http://landcover.usgs.gov

Land Cover (Canada), http://geogratis.cgdi.gc.ca

Canada Centre for Remote Sensing, www.ccrs.nrcan.gc.ca

Remote-sensing tutorial (NASA), http://rst.gsfc.nasa.gov

11.14 FURTHER READING

Baker, John C., O'Connell, Kevin, M., and Williamson, Ray, A., *Commercial Observation Satellites* (Santa Monica, CA: RAND, 2001).

Henderson, Floyd M., and Lewis, Anthony J. (eds), *Principles and Applications of Imaging Radar, Manual of Remote Sensing*, Third Edition, Volume 2 (New York: Wiley, 1998).

Jensen, John R., *Remote Sensing of the Environment: An Earth Resource Perspective* (Upper Saddle River, NJ: Prentice Hall Series in Geographic Information Science, 2000).

Lillesand, Thomas M., and Kiefer, Ralph, W., *Remote Sensing and Image Interpretation* (New York, Wiley, 2000).

Questions

11.1 What are the chief differences between maps based on satellite imagery and maps based on airborne imagery?

11.2 What are the advantages and disadvantages inherent in the use of hyperspectral sensors over multispectral sensors?

11.3 Describe all the types of data collection that can be used to plan the location of a highway/utility corridor spanning several counties.

11.4 Why is radar imaging preferred over passive sensors for Arctic and Antarctic data collection?

11.5 Compare and contrast the two techniques of remote sensing, aerial and satellite imagery, by listing the possible uses for which each technique would be appropriate. Use the comparative examples shown in Figures 11-7 and 11-8 as a basis for your response.

11.6 What is the difference between classification and feature extraction?

11.7 What type of remote-sensing projects would Landsat 7 be appropriate for, and which ones would it not be appropriate for?

CHAPTER **TWELVE**

AIRBORNE IMAGERY

12.1 GENERAL BACKGROUND

Airborne imagery introduces the topics of aerial photography and the more recent (2000) topic of airborne digital imagery. Aerial photography has a history dating back to the mid-1800s, when balloons and even kites were used as camera platforms. About 50 years later, in 1908, photographs were taken from early aircraft. During World Wars I and II, the use of aerial photography mushroomed in the support of military reconnaissance. Aerial photography became (from the 1930s to the present) an accepted technique for collecting mapping and other ground data in North America.

This chapter deals with the utilization of aerial imagery in the acquisition of planimetric and elevation ground data. Much airborne imagery is still collected using aerial photographs, although it has been predicted that in the future, most of data capture and analysis will be accomplished using the digital imagery techniques described later in this chapter.

Under the proper conditions, the cost savings for survey projects using aerial surveys instead of ground surveys can be enormous. Consequently, it is critical that the surveyor can identify the situations in which the use of aerial imagery may be beneficial. First, we discuss the basic principles required to use aerial photographs intelligently, the terminology involved, the limitations of their use, and specific applications to various projects.

Photogrammetry is the science of making measurements from aerial photographs. Measurements of horizontal distances and elevations form the backbone of this science. These capabilities result in the compilation of planimetric maps or orthophoto maps showing the horizontal locations of both natural and built features, and topographic maps showing spot elevations and contour lines. Both black and white panchromatic and color film are used in aerial photography. Color film has three emulsions, blue, green, and red light sensitive. In color infrared (IR), the three emulsion layers are sensitive to green, red, and the photographic portion of near IR, which are processed, in false color, to appear as blue, green, and red, respectively.

12.2 AERIAL CAMERA SYSTEMS

12.2.1 General Background

The recent introduction of digital cameras has revolutionized photography. Digital and film-based cameras both use optical lenses. Whereas film-based cameras use photographic film to record an image, digital cameras record image data with electronic sensors: charge-coupled devices (CCDs) or complementary metal-oxide semiconductor (CMOS) devices. One chief advantage to digital cameras is that the image data can be stored, transmitted, and analyzed electronically. Cameras onboard satellites can capture photographic data and then have these data, along with other sensed data, transmitted back to earth for further electronic processing.

Although airborne digital imagery is already making a significant impact in the remote-sensing field, we will first discuss film-based photography,

375

a technology that still accounts for much of aerial imaging. The 9″ × 9″ format used for most film-based photographic cameras captures a wealth of topographic detail, and the photos, or the film itself, can be scanned efficiently, thus preparing the digital image data for electronic processing.

Two main types of camera systems are commonly used for acquiring aerial photographs. The first is the single camera illustrated in Figure 12-1, which uses a fixed focal length, large-format negative, usually 9″ × 9″ (230 by 230 mm). This camera is used strictly for aerial photography and is equipped with a highly corrected lens and vacuum pressure against the film to minimize distortion. This sophisticated system is required to obtain aerial photographs for photogrammetric purposes due to the stringent accuracy requirements. The second camera system consists of one or more cameras using a smaller photographic format (negative size), such as the common 35 mm used for ground photographs. These smaller-format systems do not have the high-quality lens that is required to meet the normal measurement accuracies for the production of standard maps using photogrammetry. Smaller-format camera systems

are very useful and inexpensive for updating land-use changes and for the acquisition of special types of photography to enhance particular terrain aspects, such as vegetative health and algae blooms on lakes.

12.2.2 Large-Format Aerial Camera

The aerial camera illustrated in Figure 12-1 is distinguished by the complexity and accuracy of its lens assembly. Although the lens is shown in simplified form in Figure 12-1, it is actually composed of several elements involving different types of glass with different optical characteristics. This setup is necessary to satisfy the requirements of high resolution and minimal distortion of the image created on the film by the passage of the light rays through the lens. The quality of the lens is the most important consideration and hence represents the greatest cost factor for this type of camera.

The drive mechanism is housed in the camera body, as shown in Figure 12-1. This mechanism is motor driven, and the time between exposures to achieve the required overlap is set based on the

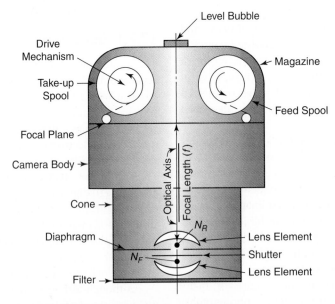

FIGURE 12-1 Components of aerial survey camera (large format)

photographic scale and the ground speed of the aircraft. The film is thus advanced from the feed spool to the take-up spool at automatic intervals. The focal plane is equipped with a vacuum device to hold the film flat at the instant of exposure (the lower air pressures outside the aircraft tend to pull the film away from the focal plane toward the lens, resulting in incorrect focusing). The camera also has four fiducial marks built in, so that each exposure can be oriented properly to the camera calibration. Most cameras also record the frame number, time of exposure, and height of the aircraft on each exposure.

The camera mount permits flexible movement of the camera for leveling purposes. The operator uses the level bubble mounted on the top of the camera body as the indicator, and every attempt is made to have the camera as level as possible at the instant of exposure. This requires constant attention by the operator because the aircraft is subject to pitching and rolling, resulting in a tilt when the photographs are taken. The viewfinder is mounted vertically to show the area being photographed at any time.

Two points, N_F and N_R, are shown on the optical axis in Figure 12-1. These are the front and rear nodal points of the lens system, respectively. When light rays are directed at the front nodal point (N_F), they are refracted by the lens, so that they emerge from the rear nodal point (N_R) parallel with their original direction. The focal length (f) of the lens is the distance between the rear nodal point and the focal plane along the optical axis, as shown in Figure 12-1. The value of the focal length is determined accurately through calibration for each camera. The most common focal length for aerial cameras is 6″ (152 mm).

Because atmospheric haze contains an excessive amount of blue light, a filter is used in front of the lens to absorb some of the blue light, thus reducing the haze on the actual photograph. A yellow, orange, or red filter is used, depending on atmospheric conditions and the flying height of the aircraft above mean ground level. The shutter of a modern aerial camera is capable of speeds ranging from 1/50 s to 1/2,000 s. The range is commonly between 1/100 s and 1/1,000 s. A fast shutter speed minimizes blurring of the photograph, known as image motion, which is caused by the ground speed of the aircraft.

12.3 PHOTOGRAPHIC SCALE

The scale of a photograph is the ratio between a distance measured on the photograph and the ground distance between the same two points. The features, both natural and cultural, shown on a photograph are similar to those on a planimetric map produced from aerial photographs, but with one important difference. The planimetric map has been rectified through ground control, so that the horizontal scale is consistent among any points on the map. The air photo will contain scale variations unless the camera was perfectly level at the instant of exposure and the terrain being photographed was also level. Because the aircraft is subject to tip, tilt, and changes in altitude due to updrafts and downdrafts, the focal plane is unlikely to be level at the instant of exposure. In addition, the terrain is seldom flat. As illustrated in Figure 12-2, any change in elevation causes scale variations. The basic problem is transferring an uneven surface like the ground to the flat focal plane of the camera.

In Figure 12-2(a), points $A, O,$ and B are at the same elevation and the focal plane ab is level. Therefore, all scales are true on the photograph because the distance $AO = A'O'$ and $OB = O'B'$. A, O, and B are points on a level reference datum that would be comparable to the surface $A'O'B'$ of a planimetric map. Therefore, under these ideal circumstances, the scale of the photograph is uniform because the ratio $ao:AO$ is the same as the ratio $ob:OB$.

Figure 12-2(b) illustrates a more realistic situation: The focal plane is tilted and the topographic relief is variable. Points $A, B, O, D, E,$ and F are at different elevations and B'', D'' are the true locations on the datum plane. Although $A'B'$ equals $B'O'$, the ratio $ab:bo$ is far from equal. Therefore, the photographic scale for points between a and b will be significantly different than that for points

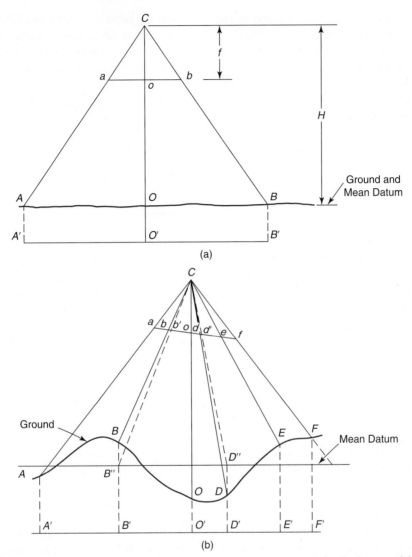

FIGURE 12-2 Scale differences caused by tilt and topography. (a) Level focal plane and level ground. (b) Tilted focal plane and hilly topography

between b and o. The same variations in scale can also be seen for the points between d and e and between e and f.

The overall average scale of the photograph is based partially on the elevations of the mean datum shown in Figure 12-2(b). The mean datum elevation is intended to be the average ground elevation. This is determined by examining the most accurate available contour maps of the area and selecting the apparent average elevation. Distances between points on the photograph that are situated at the elevations of the mean datum will be at the intended scale. Distances between points having elevations above or below the mean datum will be at a different photographic scale, depending on the magnitude of the local relief.

The scale of a vertical photograph can be calculated from the focal length of the camera and the flying height above the mean datum. Note that the flying height above the mean datum and the altitude are different elevations having the following relationship:

Altitude = flying height + mean datum

By similar triangles, as shown in Figure 12-2(a):

$$\frac{ao}{AO} = \frac{Co}{CO} = \frac{f}{H}$$

where AO/ao = the scale ratio between the ground and the photograph
f = the focal length
H = the flying height above the mean datum

Therefore, the scale ratio is

$$SR = \frac{H}{f} \qquad (12\text{-}1)$$

For example, if $H = 1,500$ m and $f = 150$ mm:

$$SR = \frac{1,500}{0.150} = 10,000$$

Therefore, the average scale of the photograph is 1:10,000. In the foot system, the scale would be stated as 1 in. = 10,000/12, or 1 in. = 833 ft. The conversion factor of 12 is required to convert both dimensions to the same unit.

12.4 FLYING HEIGHTS AND ALTITUDE

When planning an air photo acquisition mission, the flying height and altitude must be determined. This is particularly true if the surveyor is acquiring supplementary aerial photography using a small-format camera.

The flying height is determined using the same relationships discussed in Section 12.3 and illustrated in Figure 12-2, and using the relationship in Equation 12.1, $H = SR \times f$.

Example 12-1: What is the flying height if the desired scale ratio (SR) is 1:10,000 and the focal length of the lens (f) is 150 mm?

Solution: Use the relationship in Equation 12.1:

$$SR = \frac{H}{f}$$
$$H = SR \times f$$
$$H = 10,000 \times 0.150$$
$$H = 1,500 \text{ m } (4,920 \text{ ft})$$

Example 12-2: What is the flying height if the desired scale ratio (SR) is 1:5,000 and the focal length of the lens (f) is 50 mm?

Solution: Use the relationship in Equation 12.1:

$$SR = \frac{H}{f}$$
$$H = SR \times f$$
$$H = 5,000 \times 0.050 \text{ mm}$$
$$H = 250 \text{ m } (820 \text{ ft})$$

The flying heights calculated in Examples 12-1 and 12-2 are the vertical distances that the aircraft must fly above the mean datum, illustrated in Figure 12-2. Therefore, the altitude at which the plane must fly is calculated by adding the elevation of the mean datum to the flying height. If the elevation of the mean datum had been 330 ft (100 m), the altitudes for Examples 12-1 and 12-2 would be 1,600 m (5,250 ft) and 350 m (1,150 ft), respectively. These are the readings for the aircraft altimeter throughout the flight to achieve the desired average photographic scale.

If the scale of existing photographs is unknown, it can be determined by comparing a distance measured on the photograph with the corresponding distance measured on the ground or on a map of known scale. The points used for this comparison must be easily identifiable on both the photograph and the map, such as road intersections, building corners, and river or stream intersections. The photographic scale is found using the following relationship:

$$\frac{\text{Photo scale}}{\text{Map scale}} = \frac{\text{photo distance}}{\text{map distance}}$$

Because this relationship is based on ratios, the scales on the left side must be expressed in the same units. The same applies to the measured distances on the right side of the equation. For example, if the distance between two identifiable points on the photograph is 5.75 in. (14.38 cm) and on the map it is 1.42 in. (3.55 cm), and the map scale is 1:50,000, the photo scale is:

$$\frac{\text{Photo scale}}{1/50,000} = \frac{5.75 \text{ in.}}{1.42 \text{ in.}}$$

$$\text{Photo scale} = \frac{5.75}{50,000 \times 1.42}$$

$$= 1/12.349$$

If the scale is required in inches and feet, it is determined by dividing the 12,349 by 12 (number of inches per foot), which yields 1,029. Therefore, the photo scale is 1 in. = 1,029 ft between these points only. The scale will be different in areas that have different ground elevations.

12.5 RELIEF (RADIAL) DISPLACEMENT

Relief displacement occurs when the point being photographed is not at the elevation of the mean datum. As previously explained, and as illustrated in Figure 12-2(a), when all ground points are at the same elevation, no relief displacement occurs. However, the displacement of point b on the focal plane (photograph) in Figure 12-2(b) is illustrated. Because point B is above the mean datum, it appears at point b on the photograph rather than at point b', the location on the photograph for point B', which is on the mean datum.

Fiducial marks are placed precisely on the camera back plate, so that they reproduce in exactly the same position on each air photo negative. These marks are located either in the corners, as illustrated in Figure 12-3, or in the middle of each side, as illustrated in Figure 12-7. Their primary function is the location of the principal point, which is located at the intersection of straight lines drawn between each set of opposite fiducial marks, as illustrated in Figure 12-7.

Relief displacement depends on the position of the point on the photograph and the elevation

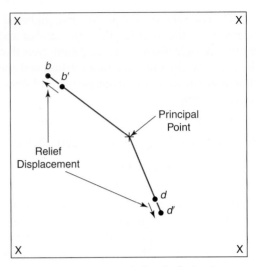

FIGURE 12-3 Direction of relief displacement [compare with Figure 12-2(b)]; X denotes fiducial marks

of the ground point above or below the mean datum. Note the following in Figure 12-2(b):

- The displacement at the center, or principal point, represented by O on the photograph, is zero.

- The farther that the ground point is located from the principal point, the greater is the relief displacement. The displacement dd' is less than bb', even though the ground point D is farther below the mean datum than point B is above it.

- The greater the elevation of the ground point above or below the mean datum, the greater is the displacement. If the ground elevation of point B were increased, so that it was farther above the mean datum, the displacement bb' would increase correspondingly.

Relief displacement is radial to the principal point of the photograph, as illustrated in Figure 12-3.

The practical aspects of relief displacement relate primarily to the proper horizontal location of all points and secondarily to the assemblage of mosaics. A mosaic is a series of overlapping aerial photographs that form one continuous picture (see Section 12.7). This technique involves matching the terrain features on adjacent photographs as closely as possible. Because the relief displacement of the same ground point will vary substantially both in direction and magnitude, for the reasons discussed previously, this presents a real difficulty in matching identical features on adjacent photographs when assembling a mosaic. A partial solution to this problem involves using only the central portion of each photograph in assembling the mosaic because relief displacement is smaller near the center.

12.6 FLIGHT LINES AND PHOTOGRAPH OVERLAP

It is important to understand the techniques by which aerial photographs are taken. Once the photograph scale, flying height, and altitude have been calculated, the details of implementing the mission are carefully planned. Although the planning process is beyond the scope of this text, the most significant factors include the following:

- A suitable aircraft and technical personnel must be arranged, including their availability if the time period for acquiring the photography is critical to the project. The costs of having the aircraft and personnel available, known as mobilization, are extremely high.

- The study area must be outlined carefully and the means of navigating the aircraft along each flight line, using either ground features or magnetic bearings, must be determined. Global positioning system (GPS) is used more often now in aircraft to maintain proper flight line alignment.

- The photographs must be taken under cloudless skies. The presence of high clouds above the aircraft altitude is unacceptable because of the shadows they cast on the ground. Therefore, the aircraft personnel are often required to wait for suitable weather conditions. This downtime can be very expensive. Most aerial photographs are taken between 10 A.M. and 2 P.M. to minimize the effect

of long shadows obscuring terrain features. Consequently, the weather has to be suitable at the right time.

To achieve photogrammetric mapping and to examine the terrain for air photo interpretation purposes, it is essential that each point on the ground appear in two adjacent photographs along a flight line, so that all points can be viewed stereoscopically. Figure 12-4 illustrates the relative locations of flight lines and photograph overlaps, both along the flight line and between adjacent flight lines. An area over which it has been decided to acquire air photo coverage is called a block. The block is outlined on the most accurate available topographic map. The locations of the flight lines required to cover the area properly are then plotted, such as flight lines A and B in Figure 12-4(a). The aircraft proceeds along flight line A, and air photos are taken at time intervals calculated to provide 60 percent forward overlap between adjacent photographs. As illustrated in Figure 12-4(a), the format of each photograph is square. Therefore, the hatched area represents the forward overlap between air photos 2 and 3 in flight line A. The minimum overlap to ensure that all ground points show on two adjacent photographs is 50 percent. However, at least 60 percent forward overlap is standard because the aircraft is subject to altitude variations, tip, and tilt as the flight proceeds. The extra 10 percent forward overlap allows for these contingencies. The air photo coverage of flight line B overlaps that of A by 25 percent, as illustrated in Figure 12-4(a). This ensures that the photographs will not have gaps of unphotographed ground, and it also extends control between flight lines for photogrammetric methods, often called sidelap.

The flight line is illustrated in Figure 12-4(b) in profile view. The single-hatched areas represent the forward overlap between air photos 1 and 2 and photos 2 and 3. Due to the forward overlap of 60 percent, the ground points in the double-hatched area will appear on each of the three photographs, thus permitting full photogrammetric treatment. (See also Section 12.10, where the

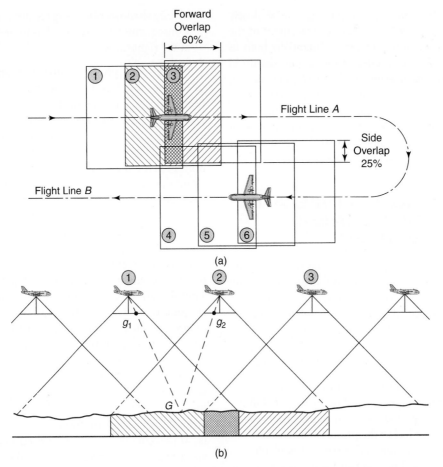

FIGURE 12-4 Flight lines and photograph overlap. (a) Photographic overlap. (b) Overlap along flight line

airborne digital sensor records ground points in three separate images.)

The problems of mosaic assemblage caused by relief displacement (discussed in Section 12.8) are solved partially by increasing the forward overlap to as high as 80 percent. Although this results in roughly twice the total number of air photos at a similar scale, a smaller portion of the central portion of each photo, which is least affected by relief displacement, can be used to assemble the mosaic.

The number of air photos required to cover a block or study area is a very important consideration. Each air photo has to be cataloged and stored, or scanned into a digital file. Most

important, these photographs have to be used individually and collectively and/or examined for photogrammetric mapping and/or air photo interpretation purposes. All other factors being equal, such as focal length and format size, the photographic scale is the controlling factor regarding the number of air photos required. The approximate number of air photos required to cover a given area stereoscopically (every ground point is shown on at least two adjacent photos along the flight line) can be calculated readily. The basic relationships required for this computation are set out next for a forward overlap of 60 percent and a side overlap of 25 percent.

For a photographic scale of 1:10,000, the area covered by one photograph, accounting for loss of effective area through overlaps, is $0.4\,\text{mi}^2$, or $1\,\text{km}^2$. Therefore, the number of air photos required to cover a 200-mi^2 (500-km^2) area is $200/0.4 = 500$, or $500/1 = 500$.

The approximate number of photographs varies as the *square* of the photographic scale. For example, if the scale is 1:5,000 versus 1:10,000 in Figure 12-4, the aircraft would be flying at one-half the altitude. Consequently, the ground area covered by each air photo would be reduced in half in *both* directions. Twice the number of air photos would be required along each flight line, and the number of flight lines required to cover the same area would be doubled. The following examples illustrate the effect on the total number of air photos required, based on the 500 photos required at the scale 1:10,000:

1. For a scale of 1:5,000, the number of photographs is $500 \times (10,000/5,000)^2 = 2,000$.

2. For a scale of 1:2,000, the number of photographs is $500 \times (10,000/2,000)^2 = 12,500$.

3. For a scale of 1:20,000, the number of photographs is $500 \times (10,000/20,000)^2 = 125$.

Thus, you can see that the proper selection of scale for mapping or the air photo interpretation purposes intended is critical. The scale requirements for photogrammetric mapping depend on the accuracies of the analytical equipment to be used in producing the planimetric maps. For general air photo interpretation purposes, including survey boundary line evidence (cut lines, land-use changes), photographic scales of between 1:10,000 and 1:20,000 are optimal.

12.7 GROUND CONTROL FOR MAPPING

As stated previously, the aerial photograph is not perfectly level at the instant of exposure and the ground surface is seldom flat. As a result, ground control points are required to manipulate the air photos physically or mathematically before

mapping can be done. Two situations require ground control. One involves the establishment of control points where *existing* photography is to be used for the mapping. The other requires the establishment of ground control points *prior* to the acquisition of the air photos. Although the principles for both are similar, each technique is described below.

Ground control is required for each data point positioning. The accuracy with which the measurements must be made varies in each case, depending on the following final product requirements:

- Measurements of distances and elevations, such as building dimensions, highway or road locations, and cross-sectional information for quantity payments for cut and fill in construction projects.

- Preparation of topographic maps, usually including contour lines at a fixed interval.

- Construction of controlled mosaics.

- Construction of orthophotos and rectified photographs (see Section 12.12).

Acquisition of ground control data can be a costly aspect of map preparation using air photos. The surveyor should therefore give considerable thought and planning to every detail of ground control requirements.

Recent advances in kinematic GPS techniques have resulted in the development of GPS receivers, which are designed both to compute position and to control the rate of photographic exposures. Two or more receivers, both on the ground (base station) and on the aircraft, can locate each photo precisely with respect to the ground, and onboard inertial measuring units (IMUs) can assist in the determination of altitude variations, tip, and tilt for each exposure, thus greatly reducing processing needs. See also Section 12.12.

12.7.1 Existing Photographs

When mapping of an area is required and air photos with a suitable scale are available, by far the most economical procedure involves using the existing photographs. The minimum ground control for one pair of overlapping air photos is three points. Because leveling of the stereoscopic model (area within the overlap) is the objective, requirements are analogous to the minimum number of legs required by an ordinary table to stand by itself. Vertical control (elevations) is required for all three points, and at least one horizontal distance between two points is also required. Because all three control points must be accessed on the ground, the normal procedure involves the acquisition of north and east coordinates as well as the elevation of each point.

The selection of ground control points must be based on the following criteria:

1. The ground control points must be separated in the overlap area. If the points are clearly separated, the model will be more stable and therefore the results will be more accurate. Using the table analogy, the table will be supported better if its legs are near the corners rather than grouped in the center.

2. The control points must be easily identifiable on both adjacent air photos. If the ground control points are not easy to identify, the control point is useless. Because the photographs are taken at an angle that varies directly with the distance from the principal point or air photo center, features such as trees or shadows from adjacent buildings can obscure the control point on one of the air photos. This difficulty can be identified by closing one eye and then the other while viewing the overlaps stereoscopically. If the potential control point disappears on either photograph, it is not acceptable.

3. Permanent significant changes may have taken place in the overlap area since the existing photographs were taken. The ground points should be selected on the assumption that they will still exist when the ground survey is carried out. Points such as building corners, main road intersections, angles or corners of year-round docking facilities, and fence intersections clear of overhanging

trees are suitable. Natural features generally do not provide good control points because they are subject to erosion, landslides, and cultural activities such as timbering. A clearly defined intersection of rock fracture lines is satisfactory because permanency is almost guaranteed.

4. The surveyor should consider ease of access to all control points to minimize open-ended traverse lines, particularly over heavily wooded terrain with high topographic relief. GPS stations should be in relatively accessible areas, free from obstructions that may block satellite signals.

The selected identifiable points on the photograph are termed either *photo points* or *picture points*. The horizontal control between these points is usually obtained by electronic distance measurement (EDM) between the control points if clear lines of sight exist, by a closed traverse connecting the points, or by GPS positioning. Vertical control is obtained using GPS positioning, trigonometric leveling (total station), or differential leveling. The field method chosen depends on field costs and the required accuracy, which depends on the scale of the map or mosaic and the contour interval to be mapped. While a minimum of three control points is required for controlling a single overlap or "model" between two adjacent photographs, each model along a flight line or within a block does not require three ground control points. This saving in ground surveying is achieved through a process termed *bridging*, which is discussed in Section 12.10.

12.7.2 New Photography

While several situations can arise in which new photography is required, the three listed here are common, and assume that a high degree of accuracy is required for the photogrammetric mapping:

1. Areas containing few identifiable ground control points. Because natural features do

not make good ground control points (for reasons previously discussed), premarked points, or targets, must be used.

2. Legal surveys of densely developed areas. If the property lines in a municipality require resurveying as a group, considerable savings can be achieved by placing targets over or close to all known boundary corners and subsequently obtaining the horizontal coordinates of each, using photogrammetric methods.

3. Municipal surveys of roads and services. As discussed in Chapter 13, municipalities often require accurate maps showing the location of both aboveground and underground services. Targets are easily set by painting the appropriate symbol on existing roads or sidewalks. This ground control network can be surveyed at convenient times of low-traffic volumes. Aerial photography is obtained under low-traffic conditions, and features such as manhole covers, hydro lines, roads, and sidewalks can be recognized on the photographs and mapped accurately (field verification is often required to properly identify storm/sanitary/water supply manhole covers).

The targets for each of these surveys must be placed prior to the acquisition of the air photos. This should be done as close to the anticipated flight time as possible, particularly in populated areas. The targets sometimes attract attention and get removed by people. For example, when a property boundary survey for a small town used photogrammetric methods, a total of 650 targets were placed during the early morning of the day of the flight, which was to be carried out that afternoon. By the time of the flight, however, about 50 targets did not show on the air photos.

The targets come in several shape configurations. For the preceding example, small (1 ft or 0.3 m) square plywood targets were centered over the monuments. In other situations, such as higher-altitude photography, a target in the form of a cross is common. Painted targets on roads and sidewalks for municipal surveys are more flexible

with regard to shape. In addition, they cannot be easily removed and are often used for more than one flight.

The photographic tone of the terrain on which the target is placed is a critical consideration. The camera records only *reflected* light. For example, a black or dark gray asphalt highway will commonly appear light gray to white on the photograph because of its smooth reflective surface. Therefore, a white target on what appears to be a dark background will disappear on the air photos. Too much contrast between the target and the background will result in lateral image spread, which is caused by a gradient in the film density from the light object to the dark. This condition is caused primarily by scattering of light in the film emulsion at the edge of the target. The effect of this phenomenon can render the target unidentifiable, at worst, or cause difficulties in locating the center point, at best. The surveyor is well advised to consider the following:

1. Examine previous air photos of the area, usually available from the appropriate government agency or a private aerial survey company, to determine the relative gray tones of various backgrounds, such as asphalt roads, gravel roads, grass, and cultivated fields.

2. Determine the best gray tone of the targets for each type of background to achieve the optimum balance of contrast. For example, a medium to dark gray target on a white background (gravel road) is logical. Also, a white target on a medium gray background (grass) is suitable.

3. Bare or recently disturbed soils, a situation created by digging to uncover an existing monument, usually photograph as a light tone, even though they may appear medium to dark gray from ground observation. Therefore, a medium to dark gray target is required for identification on the air photo.

4. Ground control points are normally targeted using a configuration that is different from the property corners for ease of recognition during the photogrammetric mapping process.

Differential GPS techniques are becoming more common; they use GPS ground base stations and GPS receivers mounted in the aircraft. The need for the traditional ground control techniques described here is becoming less significant.

12.8 MOSAICS

A mosaic is an assembly of two or more air photos to form one continuous picture of the terrain. Mosaics are extremely useful for one or more of the following applications because of the wealth of detail portrayed:

1. Plotting of ground control points at the optimum locations to ensure the required distribution and strength of figure (see Section 10.8).

2. A map substitute for field checkpoint locations and approximate locations of natural and cultural features. A mosaic is not an accurate map because of relief displacement (see Section 12.5) and minor variations in scale due to flying height differences during the flight.

3. A medium for presenting ground data. Using standard photographic procedures, a copy negative is produced. It is then common practice to produce a transparency with an air photo background. Economical whiteprints can be produced easily from this transparency, using standard blue or white printing equipment. If the contrast in the air photo background is reduced through a photographic process called screening, sufficient terrain detail will still show on the print. The advantage of screening is that valuable information can be drafted onto the transparency and still be clearly read because it is not obscured by the darker-toned areas of the air photo background.

After photograph film has been processed, each negative of a flight line is numbered consecutively. The flight line number is also shown. Other information that may also be shown is the roll number and the year that the photographs were taken. Because these numbers are always shown in one corner of the photograph prints, it is useful to construct a mosaic by arranging the photographs in order and matching the terrain features shown on each so that all numbered information is visible. These mosaics, termed *index mosaics*, are useful to determine the photograph numbers required to cover a particular area. The mosaics are often reproduced photographically in a smaller size for ease of storage.

For projects in which having photo numbers on the finished product is not a concern, every second photograph is set out. For example, in Figure 12-4, photos 1 and 3 from flight line *A*, and photos 4 and 6 from flight line *B* would form the mosaic. Photo 2 from flight line *A* and photo 5 from flight line *B* are then available to permit stereoscopic viewing of the mosaic by simply placing the single photos properly on top of the mosaic. Information can then be transferred directly onto the mosaic during the stereoscopic viewing process.

This type of mosaic and the index mosaic are uncontrolled. The only practical way to adjust the overall mosaic scale involves photographing it and producing a positive print to the desired scale. If the mosaic is constructed using alternative prints and is to be used for stereoscopic viewing (see Section 12.9), the scale *cannot* be adjusted because it renders stereoscopic viewing impossible using the single photos that are not part of the mosaic.

It is often illogical to take the original mosaic to the field for on-site investigations because of possible damage or loss. Instead, a positive print at the same scale can be made on photographic paper at a nominal cost. The single photos not used in the mosaic, but necessary for stereoscopic viewing, can be taken to the field with the positive print of the mosaic, thus permitting stereo viewing in the field.

If an uncontrolled mosaic is to be used for graphical presentation purposes or as a base for mapping terrain information, it is necessary to feather the photograph to avoid shadows along the edges of the overlapping air photos, as well as to improve the appearance of the mosaic.

Feathering is accomplished by cutting through the emulsion with a razor-edge knife and pulling the outside of the photograph toward the photo center, thus leaving only the thin emulsion where the photographs join. The overlapping photograph edges are matched to the terrain features on both adjoining photos as accurately as possible. The best means of permanently attaching the adjacent photos is using a hot roller to apply a special adhesive wax to the underside of the overlapping photograph. Forms of rubber cement are satisfactory for this work, but the photo edges tend to curl. The joins between the photos are then taped securely on the back of the mosaic using masking tape. The mosaics may be constructed by pasting the photographs to a mounting board. Two facts should be kept in mind when using mounting boards: (1) Portability for field use is limited and (2) requirements for storage space increase substantially.

The advantages and disadvantages of mosaics versus maps prepared by ground survey methods are as follows:

Advantages

1. The mosaic can be produced more rapidly because the time requirements to carry out the ground surveys and to plot the related information on a map are extensive.

2. The mosaic is less expensive, even if the cost of acquiring the air photos is included.

3. The mosaic shows more terrain detail because all natural and cultural features on the ground surface show clearly on the air photo. Ground surveys are carried out to locate only the features specified by the contract or features that can be shown by standard symbols in the legend.

4. For air photo interpretation purposes, subtle terrain characteristics such as tone, texture, and vegetation must be visible. Therefore, the use of mosaics for these purposes is essential.

Disadvantages

1. Horizontal scale measurements between any two points on a mosaic, regardless of the degree of ground and photo control employed, are limited in accuracy primarily due to relief displacement.

2. Mosaics are not topographic maps and therefore do not show elevations.

12.9 STEREOSCOPIC VIEWING AND PARALLAX

Stereoscopic viewing is defined as observing an object in three dimensions (see also Section 12.15). To achieve stereoscopic vision, it is necessary to have two images of the same object taken from different points in space. The eyes thus meet this requirement. A person with vision in only one eye cannot see stereoscopically.

Figure 12-5(a) shows a simple pocket stereoscope being used on stereo-pair air photos. Use of the stereoscope on the stereo-pair photos in this section will assist you in learning the basics about this type of image analysis. The following list offers some suggestions on the use of a stereoscope:

- Adjust the width between the lenses to accommodate each individual.

- Adjust the two photos so that identical features are directly below each of the lenses. Features will then appear to be superimposed.

- Relax and let your eyes find their own focus. After a few seconds, the 3D (stereo) image will appear.

- If the 3D image does not readily appear, try shifting the photos slightly, adjusting the lens width, or rotating the stereoscope slightly.

On two adjacent air photos, over half the ground points on both photos are imaged from two different points in space because of forward overlap. For example, the image of ground point G in Figure 12-4(b) is located at g_1 and g_2 on photos 1 and 2, respectively. If the observer looks at point g_1 on photo 1 with the left eye and at point g_2 on photo 2 with the right eye, point G can be seen stereoscopically.

The eyes are accustomed to converging on an object and therefore resist diverging. Divergence is necessary if each eye is to focus on the images of the same point on two adjacent photographs because the air photos have to be separated by 2 in. (5 cm) to 3 in. (7.5 cm). This is illustrated in Figure 12-5, where a pocket or lens stereoscope with two-power magnification is placed over two adjacent air photos. The stereoscope assists in allowing the eyes to diverge. Pocket stereoscopes

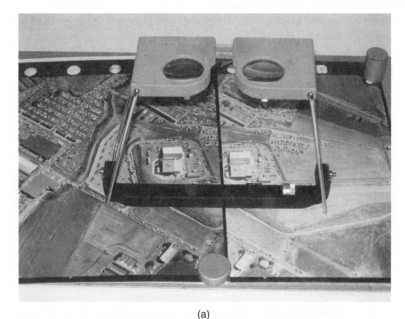

(a)

(b)

FIGURE 12-5 Lens and mirror stereoscopes. (a) Lens (pocket) stereoscope. (b) Mirror stereoscope

are easily portable for fieldwork and are also inexpensive. A mirror stereoscope is often used for office work because the internal mirror system allows the adjacent air photos to be placed a greater distance apart, as illustrated in Figure 12-5(b). As a result, the observer can view the total area covered by the overlap of the two photos. This permits the stereoscopic examination to take place more rapidly. Also, the degree of magnification can be varied easily by substituting the removable binoculars shown in Figure 12-5(b). The eye base for both types of stereoscopes is adjustable to suit the observer. Also, the adjacent air photos must be adjusted slightly until the images coincide to provide stereoscopic viewing.

Figure 12-6 provides an excellent example of an area of high topographic relief. The mountain labeled d_1 and d_2 is approximately 800 ft (270 m) high. The two images (d_1, d_2) of the same area on these adjacent air photos are set at an average

spacing between identical points of 2.4 in. (6 cm). A pocket or lens stereoscope can be placed over this stereo pair, and the high relief can be seen clearly. The distance between the lenses, known as the interpupillary range, is adjustable from 2.2 in. (55 mm) to 3.0 in. (75 mm). This adjustment assists the observer in viewing any stereo pair stereoscopically without eye strain.

The ratio between the distance between the principal points, $PP1$ and $PP2$ on Figure 12-7, to the flying height is much greater than the ratio of the interpupillary distance to the distance from the stereoscope lens to the photo. This fact, combined with long-range focusing when using a stereoscope, thus creates the phenomenon of vertical exaggeration. The height of a feature appears higher through stereoscopic viewing of air photos than its actual height. This discrepancy is usually exaggerated by a factor of between 2:1 and 2.5:1, depending on the interpupillary distance of the

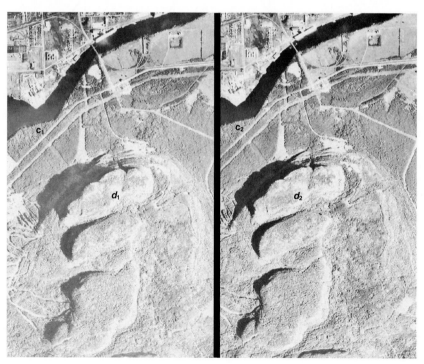

FIGURE 12-6 Stereo pair illustrating high topographic relief
(Courtesy of Bird & Hale Ltd., Toronto)

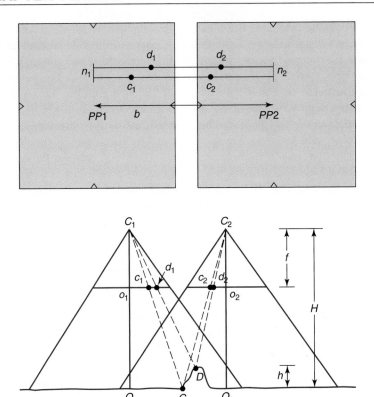

FIGURE 12-7 X parallax along the flight line

individual. This also applies to the examination and estimation of terrain slopes, an important factor in air photo interpretation. Vertical exaggeration is useful for both photogrammetry and air photo interpretation because it emphasizes ground elevation differences, thus rendering them easier to observe and measure. However, this exaggeration must be kept in mind when estimating slopes for terrain analysis purposes.

In Figure 12-7, the mountain shown on the stereo pair in Figure 12-6 has been photographed from two consecutive camera stations, $PP1$ and $PP2$. The images for both the bottom and top of the mountain are designated c_1 and d_1 for the left photo (photo 1) and c_2 and d_2 for the right photo (photo 2). The same designations are shown on both Figures 12-6 and 12-7.

Parallax is the displacement along the flight line of the same point on adjacent aerial photographs. For example, the difference between the distances c_1c_2 and d_1d_2 in Figure 12-7 is the displacement in the X direction (the direction along the flight line), or the difference in parallax between the image points c and d. This difference in parallax is a direct indication of the elevation difference of the height of an object. If points c and d were at the same elevation (not true in this example), the difference in parallax would be zero.

$$h = \frac{\delta p \times H}{b - d_1 d_2}$$

where h = difference in elevation

δp = difference in parallax = $c_1 c_2 - d_1 d_2$

H = flying height

b = photo base (distance between the two adjacent principal points, $PP1$ to $PP2$, in this example)

Examples 12-3 and 12-4 illustrate the calculation of h using both measurement systems. All pertinent data are illustrated in Figures 12-6 and 12-7, and the calculations determine the h between points C and D (the height of the mountain). The data supplied or measured are as follows:

$$H = 18,000 \text{ ft } (5,486 \text{ m})$$
$$b = 3.00 \text{ in. } (76.2 \text{ mm})$$
$$\delta p = 0.17 \text{ in. } (4.32 \text{ mm})$$

Note that δp is determined in this example by measuring $c_1 c_2$ and $d_1 d_2$ and taking the difference: $2.41 - 2.24 = 0.17$ in. (4.32 mm). See Figure 12-7.

Example 12-3: Calculate the difference in elevation using foot units.

Solution:

$$h = \frac{0.17 \, (18,000)}{3.00 - 2.24} = 4,030 \text{ ft}$$

Example 12-4: Calculate the difference in elevation using metric units.

Solution:

$$h = \frac{4.32 (5,486)}{76.2 - 56.9} = 1,230 \text{ m}$$

The following procedure is used to determine b and δp:

1. The air photos are aligned so that (a) the principal points shown in Figure 12-8 form a straight line, as illustrated, and (b) the air photos are the proper distance apart for stereoscopic viewing by a mirror stereoscope, as shown in Figure 12-5(b).

2. The air photos are taped in position, and an instrument known as a parallax bar is used to measure the distances along the flight line between the identical image points for which differences in elevation are required. As illustrated in Examples 12-3 and 12-4, δp is obtained by calculating the difference between these two measurements.

3. The principal points of each air photo are determined by the intersection of lines between opposite fiducial marks (see Figure 12-7). These marks show the midpoint of each side of the air photo and are set permanently in the focal plane of each aerial camera. As shown in Figure 12-8, the principal points of each adjacent air photo are plotted on the other photograph by transferring the ground point through stereoscopic viewing. The value of b is determined by averaging the distances $PP1$ to $PP2'$ and $PP1'$ to $PP2$, thus canceling the effects of relief displacement.

The parallax bar is designed so that a small dot is placed on the bottom side of the plastic plates at each end, as illustrated in Figure 12-5(b). The adjustment dial at the left end of the bar on this figure allows the distances between the dots to be adjusted. When viewed stereoscopically with two adjacent air photos, the dots at each

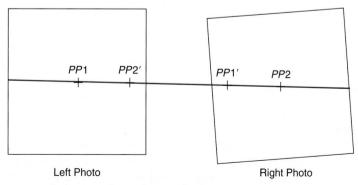

Left Photo   Right Photo

FIGURE 12-8 Positioning of adjacent photos for parallax measurements

end of the bar appear as one floating mark. The bar is adjusted until the floating mark appears to be at the same elevation as the point being examined, and the distance is read from the bar scale. This process is repeated for the other point involved in the difference of elevation calculation. The two measurements are then subtracted to calculate δp.

The parallax bar is a useful piece of equipment, and the theory of parallax is one of the keys of photogrammetric measurement. The use of the parallax bar to measure difference of parallax is reliable only when carried out by an experienced person. Initially, it is difficult to determine when the floating mark is directly on top of a terrain feature.

12.10 PHOTOGRAM-METRIC STEREOSCOPIC PLOTTING TECHNIQUES

Stereoplotters have traditionally been used for image rectification, that is, to extract planimetric and elevation data from stereo-paired aerial photographs for the preparation of topographic maps. The photogrammetric process includes the following steps:

1. Establish ground control for aerial photos.
2. Obtain aerial photographs.
3. Orient adjacent photos so that the ground control matches.
4. Use aerotriangulation to reduce the number of ground points needed.
5. Generate a digital elevation model (DEM).
6. Produce an orthophoto.
7. Collect data using photogrammetric techniques.

Steps 3–7 are accomplished using stereoplotting equipment and techniques. Essentially, stereoplotters incorporate two adjustable projectors that are used to duplicate the attitude of the camera at the time the film was exposed.

Camera tilt and differences in flying height and flying speed can be noted and rectified. A floating mark can be made to rest on the ground surface when the operator is viewing stereoscopically, thus enabling the skilled operator to trace planimetric detail and deduce both elevations and contours.

In the past 50 years, aerial photo stereoplotting has undergone four distinct evolutions. The original stereoplotter (Kelsh Plotter) was a heavy and delicate mechanical device. Then came the analog stereoplotter, and after that the analytical stereoplotter, an efficient technique that utilizes computer-driven mathematical models of the terrain. The latest technique (developed in the 1990s) is soft-copy (digital) stereoplotting. See Figure 12-9. Each new generation of stereoplotting reflects revolutionary improvements in the mechanical, optical, and computer components of the system. Common features of the first three techniques were size, complexity, high capital costs, high operating costs of the equipment, and the degree of skill required by the operator. Soft-copy photogrammetry utilizes (1) high-resolution scanners to digitize the aerial photos and film and (2) sophisticated algorithms to process the digital images on workstations (e.g., Sun, Unix) and on personal computers with the latest Microsoft operating systems.

Plotters that use photographic prints rather than stable-base diapositives as the basis for photogrammetric measurements are available. These measurements are not as accurate because the photo prints do not have the dimensional stability of the diapositives. Several instruments that have the capabilities of measuring from photo prints are available. Some provide for mechanical adjustment of the prints for tip and tilt corrections for orientation, although the solutions are approximate. Other instruments, such as the Zeiss Stereocord (shown in Figure 12-10), use digital computers to carry out the orientation. These computers correct for model deformation, relief displacements, and tip and tilt. Elevations are measured using the parallax bar concept, and the resulting data are provided to very close

FIGURE 12-9 SOCET SET® Windows NT workstation. This shot shows a human operator using the passive polarized viewing system, in this case the StereoGraphics® Corporation ZScreen® system: A polarizing bezel is placed and secured in front of an off-the-shelf, high-performance monitor. The operator wears passive, polarized spectacles
(Courtesy of Leica Geosystems Inc.)

approximations. The stage on which the photographs are mounted is moved manually under the stereoscope and parallax bar. The digitizer coordinates the points along the travel path, and these locations are plotted digitally.

This latest generation of stereoplotting technique (the successor to analytical stereoplotting) uses digital raster images (soft-copy medium) rather than aerial photographs (hardcopy medium) to perform the photogrammetric

FIGURE 12-10 Computer-assisted plotter for use with photo prints. From left to right: plotter, computer, digitizer, and binocular system for viewing stereoscopic model

process. Aerial photographs, in the form of 23-cm (9″) photographs or continuous-roll films, are processed through high-resolution scanners to provide the digital images used in the process. The photo-scanner converts light transmitted through the photographic image into picture elements (called pixels) of fixed size, shape, spacing, and brightness. The size of the pixel is important, with manufacturers claiming that a size of 7–10 μm (μm is a micrometer—a millionth of a meter) is needed for this type of image processing. This scanning process can be bypassed if aerial digital cameras (introduced in 2000) are used (see Section 12.11).

Once the digital image files have been created, stereo-paired images can be observed in three dimensions on a computer monitor (21″ is recommended) by an operator wearing stereo glasses. The operator can, at this stage, perform the same functions available with the highly efficient analytical plotters. But because of the digital nature of the image files, much more of the process can be accomplished automatically. The five steps in aerial photography digital processing are as follows:

1. Scanning of aerial photos or film (if film-based cameras are used).

2. Aerotriangulation.

3. Elevation mapping (DEM/DTM).

4. Orthophoto (and mosaics) production.

5. Planimetric features mapping.

DEM refers to digital elevation model, and DTM refers to digital terrain model. A DTM can be thought of as a DEM with the break lines included, which are needed to define the elevation surface properly (as in contouring).

Software can divide the huge data file into subfiles (called tiles) that can be manipulated more easily by the computers at the appropriate time. Computer processors should have large storage and high speed. High refresh-rate monitors and various graphics accelerator boards permit efficient processing of the huge data files. With the development of software compatible with Windows NT systems, users no longer have to invest huge

sums in stereoplotters or workstation computers; instead, the principal part of the hardware process is a readily available, off-the-shelf computer that is also capable of performing a host of other functions, ranging from the use of computer-aided design (CAD) and geographic information system (GIS) programs to the use of business software. The photogrammetry-specific hardware components now needed are the scanner, stereo glasses, floating mark controller, 3D mouse or pointer, and appropriate hard-copy plotters.

Soft-copy photogrammetry has several advantages over earlier generations of stereoplotting:

1. It produces perspective views.

2. It handles all types of remote-sensing data, not just digitized photos.

3. It prepares data for use in GIS and CAD software.

4. It can import external files such as GIS, CAD, TIGER.

5. The smaller size and stability of computers that are unaffected by vibration (versus conventional stereoplotters) means that the system can be placed in any desired location. Some report that required rental space has been reduced by half. (Stereoplotters require a large floor area that is well supported to take the substantial weight.)

6. The computer base provides a direct link among photogrammetry, remote sensing, GIS, CAD, and GPS control point surveys (see Figure 1-7).

7. It provides a large advancement in automation (with some operator editing), with potential for even more automation.

8. It reduces the number of highly trained operators needed.

9. Equipment calibration time and costs (mostly limited to the scanner) are much reduced.

10. All data file formats can be in TIFF (raster files) and PostScript (raster and vector files), which are used industry-wide in most related processing operations.

11. Hard-copy output can be achieved by any of the available printers and plotters, or high-resolution data files can be contracted to various agencies in the printing industry.

12.11 AIRBORNE DIGITAL IMAGERY

In Section 12.10, we noted that photogrammetric analysis had become almost completely automated through the techniques of soft-copy (digital) photogrammetry. In that process, the only step not digitized was that of image capture. The traditional $9'' \times 9''$ aerial photograph held such a wealth of detail that the incentive to change to digital techniques was somewhat weaker than it was for the other steps in the process. Once the traditional (film-based) photos or the film itself was digitized using high-speed scanners, the remainder of the digital processing and analysis could proceed mostly using automated techniques.

12.12 LIDAR MAPPING

Light detection and ranging (LiDAR) is a laser-mapping technique that has recently become popular in both topographic and hydrographic surveying. Over land, laser pulses can be transmitted and then returned from ground surfaces. The time required to send and then receive the laser pulses is used to create a DEM of the earth surface. Processing software can separate rooftops from ground surfaces and also treetops and other vegetation from the bare ground surface beneath the trees. Although the laser pulses cannot penetrate very heavily foliaged trees, they can penetrate tree cover and other lower-growth vegetation at a much more efficient rate than does either aerial photography or digital imaging because of the huge number of measurements, thousands of terrain measurements every second. "Bare-earth" DEMs are particularly useful for design and estimating purposes.

One of the important advantages to using this technique is the rapid processing time. One supplier claims that 1,000 km² of hilly, forested terrain can be surveyed by laser in less than 12 hours, and that the DEM data are available within 24 hours of the flight. Thus, data processing takes little longer than data collection. Because each laser pulse in the point cloud is individually georeferenced, there is no need for the orthorectification steps needed in aerial photo processing. Also, the direct georeferencing of laser pulses in three dimensions permits 3D viewing (on the computer monitor) from any selected elevation and any selected direction. Additional advantages include the following:

- Laser mapping can be flown during the day or at night when there are fewer clouds and calmer air.
- Vertical accuracies of ± 3 cm (or better) can be achieved.
- There are no shadow or parallax problems, as with aerial photos.
- Laser data are digital and are thus ready for digital processing.
- It is less expensive than aerial imaging for the creation of DEMs.

LiDAR can be mounted in a helicopter or fixed-wing aircraft, and the lower the altitude, the better the resulting ground resolution. Typically, LiDAR can be combined with a digital imaging sensor (panchromatic and multispectral); an IMU to provide data to correct pitch, yaw, and roll; and GPS receivers to provide precise positioning. When the data-gathering package also includes a digital camera to collect panchromatic imagery, then it is possible, with appropriate processing software, to produce high-quality orthorectified aerial imagery so that each pixel can be assigned X, Y, and Z values.

Ground LiDAR operates in the near-IR portion of the spectrum; this part of the spectrum produces wavelengths that tend to be absorbed by water (including rain and fog) and asphalt surfaces (such as roofing and highways). These surfaces produce "holes" in the coverage that can be recognized as such and then edited during

FIGURE 12-11 Highway design features such as signs and ramps can be developed using visualization software based on data extracted from LiDAR imaging
(Courtesy of Texas Department of Transportation)

data processing. Even rolling traffic on a highway at the time of data capture can be removed through editing. See Figures 12-11–12-13 for examples of LiDAR instruments and use. The importance of the 3D aspects of this science are clearly shown in Figure F-4, where the elevation differentials along the Niagara River and Niagara Falls are dramatically illustrated through color-coding.

The growing number of applications for LiDAR include the following:

- Highway design and redesign.
- Flood plain mapping.
- Forest inventory, including canopy coverage, tree density and heights, and timber output.
- Line-of-sight modeling using LiDAR-generated 3D building renderings, used in telecommunications and airport facilities design.

- Generation of lower-cost DEMs. See Section 8.14.10 for airborne laser bathymetry (ALB).

12.13 AERIAL SURVEYING AND PHOTOGRAM-METRIC MAPPING

12.13.1 Advantages

The advantages of using aerial surveying and photogrammetric mapping over traditional ground surveying methods are as follows:

1. Cost saving, which is related to the size of the area to be mapped, assuming that the ground is not obscured from view by certain dense vegetation types such as coniferous trees. The survey manager is advised to cost the mapping using all methods and

FIGURE 12-12 Complete ALTM 3100 system, including control rack, sensor, laptop computer (operator interface), and pilot display
(Courtesy of Optech Incorporated, Toronto)

to select the most economical. As the size of the mapping area increases, aerial and photogrammetric methods become rapidly cheaper on a per-acre or per-hectare basis. *Note:* The development and use of total stations, including ground-based LiDAR scanners, with their wide variety of programmed functions and error-free transfer of data to the computer, have greatly speeded up the fieldwork portion of a survey project (see Chapter 5). The creation of sophisticated data processing and plotting software programs has also provided additional efficiencies to this aspect of the survey project.

2. Reduction in the amount of fieldwork required. It is generally recognized that fieldwork is a very high cost component of any surveying project. It also depends on access and weather. The relatively few control points required for photogrammetric mapping thus reduce the field problems and reduce the time of data acquisition substantially.

3. Speed of compilation. The time required to prepare maps using digital methods is small compared to the time required to carry out the ground survey and process the data.

4. Freedom from inaccessible terrain conditions.

5. Provision of a temporal record. The air photos and images provide an accurate record of the terrain features at one instant in time. This is useful for direct comparisons with imagery taken at other times to record either subtle or major changes in the landscape.

6. Flexibility. Photogrammetric mapping can be designed for any map scale, provided that the proper flying heights, focal lengths, ground control point placement, and plotting instruments are selected. Scales vary from 1:200 upward to 1:250,000. Contour intervals as small as 0.5 ft (0.15 m) can be determined.

American Falls   Horseshoe Falls

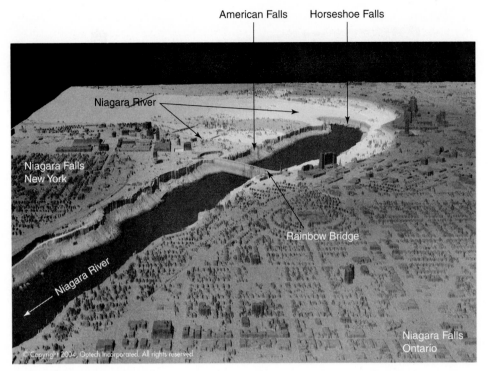

Niagara River

Niagara Falls
New York

Rainbow Bridge

Niagara River

Niagara Falls
Ontario

FIGURE 12-13 Color-coded elevation data of Niagara Falls, surveyed by ALTM 3100 (see also Figure F-4)
(Courtesy of Optech Incorporated, Toronto)

7. New technology. The recent development of new technology in GPS airborne control, soft-copy photogrammetry, airborne digital imagery, and LiDAR mapping make the first six advantages all the more relevant.

12.13.2 Disadvantages

Aerial surveying and photogrammetric mapping also have some disadvantages, which are explained below:

1. Viewing terrain through dense vegetation cover. If the aerial images have been acquired under full-leaf cover or if the vegetative cover is coniferous, the ground may not be visible in the stereo model. This problem can be overcome by obtaining leaf-free air photos in the spring or fall if the tree cover is deciduous, or by supplementing the mapping with field measurements if the vegetation is coniferous, or possibly by using LiDAR measurements.

2. Contour line locations in flat terrain. It is more difficult to place the floating mark accurately in the plotter on areas of flat terrain. Consequently, it is sometimes necessary to carry out additional field measurements in such areas.

3. Going to the site. A visit to the site is necessary to determine the type of roads and surfacing, locations of certain utility lines not easily visible on the air photos, and roads and place names. This can usually be achieved effectively and is usually combined with spot checks of the mapping to ensure that the relief is represented properly.

12.14 AERIAL PHOTOGRAPHY INTERPRETATION

Image interpretation is achieved by a combination of direct human analysis and by automated soft-copy processes. With the introduction in 2000 of airborne multispectral sensors, the ability to interpret the images has greatly increased. Image interpretation techniques are based on three fundamental assumptions:

1. The remotely sensed images are records of the results of long- and short-term natural and human processes.

2. The surface features seen on the image can be grouped together to form patterns that are characteristic of particular environmental conditions.

3. The environmental conditions and their reflected image patterns are repeated within major climatic zones; that is, similar environments produce similar image patterns, whereas different environments usually produce different image patterns.

The terrain elements that collectively produce patterns on remotely sensed images include the topography (geometry of the surface), regional drainage, local erosion, vegetation, and cultural features.

The 3D viewing of stereo-pair images was discussed earlier in this chapter. Use of this technique greatly assists the photo analyst in interpreting photo detail. When using soft-copy techniques, the operator wears stereo glasses while viewing the computer monitor (see Figure 12-9), whereas when using aerial photos, the operator must use a stereoscope (see Figures 12-5 and 12-6).

Before attempting to interpret terrain, the interpreter must know something of the properties inherent in the photographs themselves. The conventional black-and-white aerial photo, used as a record of surface conditions, is created by the reflection of certain spectra of electromagnetic energy from the surface of objects. A panchromatic photo records the surface only as tones of gray, and not, as does the human eye, in variations of color. Thus, two widely different colors that would be readily differentiated by the human eye may have the same gray tone on a black-and-white photograph. In addition, the photo is a much-reduced image of the actual situation; that is, it is a *scaled* representation. Familiar pattern elements may become so reduced in size that they present an unfamiliar appearance on an aerial photograph; for example, the familiar linear pattern of a plowed field becomes an even gray tone without perceptible pattern when the scale of the photo is very small. In general, photographs have two basic properties: (1) photo tone, which is the variation in tones of gray on a photograph and (2) photo texture, which is the combination of tones of gray on an exceedingly small scale, so that individual features are close to the limit of resolution but in which an overall pattern of variation can still be detected.

Air photo interpretation has, in the past, required professionals highly skilled in the interpretive process. With the advent of combining photographs supplemented by multispectral scannings, however, the computerized interpretive process has become much more automated.

12.15 APPLICATIONS OF AIR PHOTO INTERPRETATION FOR THE ENGINEER AND THE SURVEYOR

Air photo interpretation of physical terrain characteristics is used for a wide variety of projects. The main advantages of this technique are as follows:

1. The identification of land forms and consequently site conditions, such as soil type, soil depth, average topographic slopes, and soil and site drainage characteristics, is made before going to the field to carry out either engineering or surveying fieldwork. When

the fieldwork begins, the surveyor should find the site familiar because of careful examination of the air photos.

2. The surveyor can examine the topographic slopes, areas of wet or unstable ground, and the density and type of vegetation cover. Therefore, he or she can become familiar with the ease or difficulty to be expected in carrying out the field survey. From the air photos, the surveyor can also determine the location of property or section boundaries, some of which are extremely difficult to see on the ground.

3. Air photos provide an excellent overview of the site and the surrounding area, which cannot be achieved through groundwork alone. For example, if evidence of soil movement such as landslides is indicated on the air photos, the surveyor can avoid placing monuments in such areas because they would be subject to movement with the soils. The engineer would avoid using this area for any heavy structures because of the potential for gradual or sudden soil failure.

4. Soil test holes should always be used to verify the results of the air photo interpretation, and these can be preselected carefully on the air photo prior to doing the fieldwork. Existing road or stream bank cuts can be preidentified for use as field checkpoints, thus saving on drilling costs.

12.16 AIRBORNE IMAGING WEBSITES

Optech [airborne laser terrain mapper (ALTM)], http://www.optech.ca

Scanning Hydrographic Operational Airborne LiDAR Survey (SHOALS) system, U.S. Army Corps of Engineers (USACE), http://shoals.sam.usace.army.mil/

12.17 FURTHER READING

American Society for Photogrammetry and Remote Sensing, *Digital Photogrammetry: An Addendum to the Manual of Photogrammetry* (Bethesda, MD: ASPRS, 1996).

Falkner, Edgar, *Aerial Mapping: Methods and Applications* (Boca Raton, FL: CRC Press, 1995).

Lillesand, Thomas M., and Kiefer, Ralph W., *Remote Sensing and Image Interpretation*, 4th edition (New York: Wiley, 2000).

Wolf, Paul R., and Dewitt, Bon A., *Elements of Photogrammetry*, 3rd edition (New York: McGraw-Hill, 2000).

Questions

12.1 Why can't aerial photographs be used for scaled measurements?

12.2 What is the chief advantage of digital cameras over film-based cameras?

12.3 What effects does aircraft altitude have on aerial imagery?

12.4 Why are overlaps and sidelaps designed into aerial photography acquisition?

12.5 Describe the adjustments that can be made when the use of a stereoscope fails to produce 3D images.

12.6 Discuss the fundamental components of a landscape with respect to aerial image interpretation.

12.7 What advantages does LiDAR imaging have over aerial photography?

Problems

12.1 Calculate the flying heights and altitudes, given the following information.

   a. Photographic scale = 1:20,000, lens focal length = 153 mm, and elevation of mean datum = 180 m.

   b. Photographic scale is 1 in. = 20,000 ft, lens focal length = 6.022 in., and elevation of mean datum = 520 ft.

12.2 Calculate the photo scales, given the following data.

   a. Distance between points A and B on a topographic map (scale 1:50,000) = 4.75 cm; distance between the same points on an air photo = 23.07 cm.

   b. Distance between points C and D on a topographic map (scale 1:100,000) = 1.85 in.; distance between the same points on an air photo = 6.20 in.

12.3 Calculate the approximate numbers of photographs required for stereoscopic coverage (60 percent forward overlap and 25 percent sidelap) for each of the following conditions.

   a. Photographic scale = 1:30,000; ground area to be covered is 30 km by 45 km.

   b. Photographic scale = 1:15,000; ground area to be covered is 15 miles by 33 miles.

   c. Photographic scale is 1 in. = 500 ft; ground area to be covered is 10 miles by 47 miles.

12.4 Calculate the dimensions of the area covered on the ground in a single stereo model having a 60 percent forward overlap if the scale of the photograph (9 by 9 in. format) is:

   a. 1:10,000.

   b. 1 in. = 400 ft.

12.5 a. Calculate how far the camera would move during the exposure time for each of the following conditions:

    1. Ground speed of aircraft = 350 km/h; exposure time = 1/100 s.

    2. Ground speed of aircraft = 350 km/h; exposure time = 1/1,000 s.

    3. Ground speed of aircraft = 200 miles/h; exposure time = 1/500 s.

   b. Compare the effects on the photograph of all three situations in part (a).

12.6 Does a longer focal-length camera increase or decrease the relief displacement? Briefly explain the reason for your answer.

12.7 Would a wide-angle lens increase or decrease relief displacement? Briefly explain the reason for your answer.

12.8 Choose the best alternative from the following list for the assemblage of an uncontrolled mosaic, using alternative photographs, over rolling terrain having vertical relief differences of up to 50 ft (15 m). Give the main reasons for your choice.

   a. Focal length = 150 mm; scale = 1:20,000.

   b. Focal length = 150 mm; scale = 1:5,000.

   c. Focal length = 12 in. (300 mm); scale 1 in. = 1,000 ft.

   d. Focal length = 12 in. (300 mm); scale 1 in. = 2,000 ft.

12.9 Calculate the time interval between air photo exposures if the ground speed of the aircraft is 100 miles/h (160 km/h), the focal length is 50 mm (2 in.), the format is 70 mm (2.8 in.), the photographic scale is 1:500 (approximately 1 in = 40 ft), and the forward overlap = 60 percent.

SURVEYING APPLICATIONS

CHAPTER THIRTEEN

ENGINEERING SURVEYS

13.1 ROUTE SURVEYS AND HIGHWAY CURVES

Highway and railroad routes are chosen only after a complete and detailed study of all possible locations has been completed. Functional planning and route selection usually involve the use of aerial imagery, satellite imagery, and ground surveys as well as the analysis of existing plans and maps. One route is selected because it satisfies all design requirements with minimal social, environmental, and financial impact.

The proposed centerline ($\mathcal{C}$) is laid out in a series of straight lines (tangents) beginning at $0 + 00$ ($0 + 000$, metric) and continuing to the route terminal point. Each time the route changes direction, the deflection angle between the back tangent and forward tangent is measured and recorded. Existing detail that could have an effect on the highway design is tied in either by conventional ground surveys, by aerial surveys, or by a combination of the two methods; typical detail would include lakes, streams, trees, structures, existing roads and railroads, and so on.

In addition to the detail location, the surveyor determines elevations along the proposed route (profile), as well as across the route width at right angles to the $\mathcal{C}$ at regular intervals (full stations, half stations, etc.) and at locations dictated by changes in the topography (cross sections). The elevations thus determined are used to aid in the design of horizontal and vertical alignments; in addition, these elevations (cross sections) also form the basis for the calculation of construction

cut-and-fill quantities (see Chapter 8). Advances in aerial imaging, including LiDAR and radar imaging (Chapter 12), have resulted in ground surface measuring techniques that can eliminate much of the time-consuming field surveying techniques traditionally employed.

The location of detail and the determination of elevations are normally confined to that relatively narrow strip of land representing the highway right of way (ROW). Exceptions would include potential river, highway, and railroad crossings, where approach profiles and sight lines (railroads) may have to be established.

13.2 CIRCULAR CURVES: GENERAL BACKGROUND

It was noted in the previous section that a highway route survey was initially laid out as a series of straight lines (tangents). Once the $\mathcal{C}$ location alignment has been confirmed, the tangents are joined by circular curves that allow for smooth vehicle operation at the speeds for which the highway was designed. Figure 13-1 illustrates how two tangents are joined by a circular curve and shows some related circular curve terminology. The point at which the alignment changes from straight to circular is known as the BC (beginning of curve). The BC is located distance T (subtangent) from the PI (point of tangent intersection). The length of circular curve (L) is dependent on the central angle and the value of R (radius). The point

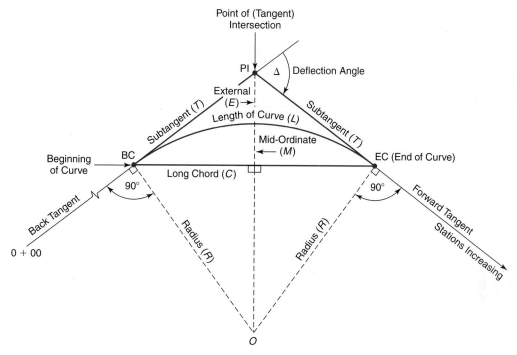

FIGURE 13-1 Circular curve terminology

at which the alignment changes from circular back to tangent is known as the EC (end of curve). Since the curve is symmetrical about the PI, the EC is also located distance T from the PI. From geometry we recall that the radius of a circle is perpendicular to the tangent at the point of tangency. Therefore, the radius is perpendicular to the back tangent at the BC and to the forward tangent at the EC. The terms "BC" and "EC" are also referred to by some agencies as PC (point of curve) and PT (point of tangency), and by others as TC (tangent to curve) and CT (curve to tangent).

13.3 CIRCULAR CURVE GEOMETRY

Most curve problems are calculated from field measurements (Δ and the chainage or stationing of the PI) and from design parameters (R). Given

R (which is dependent on the design speed) and Δ, all other curve components can be computed.

Analysis of Figure 13-2 will show that the curve deflection angle at the BC (PI-BC-EC) is $\Delta/2$, and that the central angle at O is equal to Δ, the tangent deflection angle.

The line (O-PI), joining the center of the curve to the PI, effectively bisects all related lines and angles.

Tangent in triangle BC-O-PI:

$$\frac{T}{R} = \tan \frac{\Delta}{2}$$

$$T = R \tan \frac{\Delta}{2} \qquad (13\text{-}1)$$

Chord in triangle BC-O-B:

$$\frac{\frac{1}{2}C}{R} = \sin \frac{\Delta}{2}$$

$$C = 2R \sin \frac{\Delta}{2} \qquad (13\text{-}2)$$

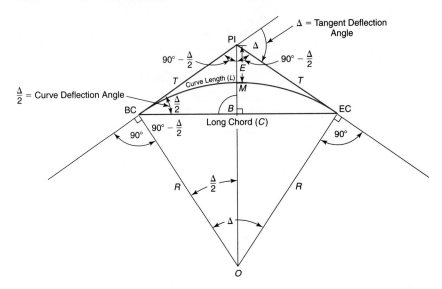

FIGURE 13-2 Geometry of the circle

Midordinate:

$$\frac{OB}{R} = \cos\frac{\Delta}{2}$$

$$OB = R\cos\frac{\Delta}{2}$$

But

$$OB = R - M$$

$$R - M = R\cos\frac{\Delta}{2}$$

$$M = R\left(1 - \frac{\cos\Delta}{2}\right) \quad (13\text{-}3)$$

External in triangle BC-O-PI:

$$O \text{ to } PI = R + E$$

$$\frac{R}{R + E} = \cos\frac{\Delta}{2}$$

$$E = R\left(\frac{1}{\cos\Delta/2} - 1\right) \quad (13\text{-}4)$$

$$= R\left(\sec\frac{\Delta}{2} - 1\right) \quad (\text{alternate})$$

From Figure 13-3, we can develop the following relationships:

$$\text{Arc:} \frac{L}{2\pi R} = \frac{\Delta}{360°}, \quad L = \frac{2\pi R\Delta}{360°} \quad (13\text{-}5)$$

where Δ is expressed in degrees and decimals of a degree.

The sharpness of the curve is determined by the choice of the radius (R); large radius curves are relatively flat, whereas small radius curves are relatively sharp. Some highway agencies use the concept of degree of curve (D) to define the sharpness of the curve. Degree of curve D is defined to be that central angle subtended by 100 ft of arc. (In railway design, D is defined to be the central angle subtended by 100 ft of chord.) From Figure 13-3 D and R:

$$\frac{D}{360°} = \frac{100}{2\pi R} \quad (13\text{-}6)$$

$$D = \frac{5{,}729.58°}{R}$$

Arc:

$$\frac{L}{100} = \frac{\Delta}{D} \quad (13\text{-}7)$$

$$L = 100\left(\frac{\Delta}{D}\right)\text{ft}$$

Example 13-1: Refer to Figure 13-4. Given the following relationships:

$$\Delta = 16°38'$$

$$R = 1{,}000\,\text{ft}$$

$$\text{PI at } 6 + 26.57$$

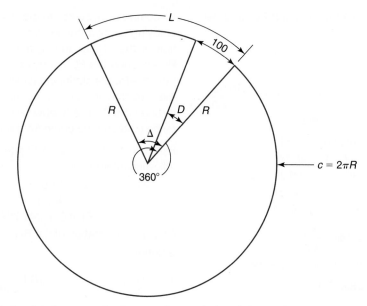

FIGURE 13-3 Relationship between the degree of curve (*D*) and the circle

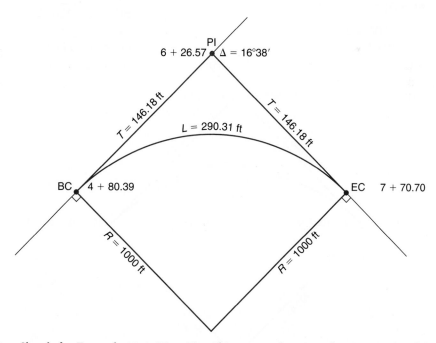

FIGURE 13-4 Sketch for Example 13-1. *Note:* To aid in comprehension, the magnitude of the angle has been exaggerated in this section

Calculate the station of the BC and EC; also calculate lengths C, M, and E.

$$T = R \tan\left(\frac{\Delta}{2}\right) \qquad L = 2\pi R\left(\frac{\Delta}{360°}\right)$$

$$= 1{,}000 \tan 8°19' \qquad = 2\pi \times 1{,}000 \times \frac{16.6333}{360}$$

$$= 146.18 \text{ ft} \qquad\qquad = 290.31 \text{ ft}$$

$$
\begin{array}{ll}
\text{PI at} & 6 + 26.57 \\
-T & 1 \quad 46.18 \\
\hline
\text{BC} = & 4 + 80.39 \\
+L & 2 \quad 90.31 \\
\hline
\text{EC} = & 7 + 70.70
\end{array}
$$

$$C = 2R \sin\left(\frac{\Delta}{2}\right)$$

$$= 2 \times 1{,}000 \times \sin 8°19'$$

$$= 289.29 \text{ ft}$$

$$M = R\left(1 - \cos\frac{\Delta}{2}\right)$$

$$= 1{,}000(1 - \cos 8°19')$$

$$= 10.52 \text{ ft}$$

$$M = R\left(\sec\frac{\Delta}{2} - 1\right)$$

$$= 1{,}000\,(\sec 8°19' - 1)$$

$$= 10.63 \text{ ft}$$

A common mistake made by students first studying circular curves is to determine the station of the EC by adding the T distance to the PI. Although the EC is physically a distance of T from the PI, the stationing (chainage) must reflect the fact that the centerline ($\mathcal{C}$) no longer goes through the PI. The $\mathcal{C}$ now takes the shorter distance (L) along the curve from the BC to the EC.

Example 13-2: Refer to Figure 13-5. Given the following information:

$$\Delta = 12°51'$$
$$R = 400\,\text{m}$$
$$\text{PI at } 0 + 241.782$$

Calculate the station of the BC and EC.

Solution:

$$T = R \tan\frac{\Delta}{2} \qquad L = 2\pi R\left(\frac{\Delta}{360°}\right)$$

$$= 400 \tan 6°25'30'' \qquad = 2\pi \times 400 \times \frac{12.850}{360}$$

$$= 45.044 \text{ m} \qquad\qquad = 89.710 \text{ m}$$

$$
\begin{array}{ll}
\text{PI at } 0 + & 241.782 \\
-T & 45.044 \\
\hline
\text{BC} = 0 + & 196.738 \\
+L & 89.710 \\
\hline
\text{EC} = 0 + & 286.448
\end{array}
$$

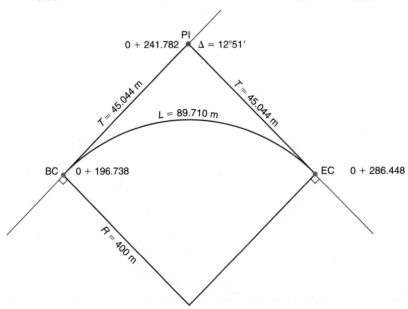

FIGURE 13-5 Sketch for Example 13-2

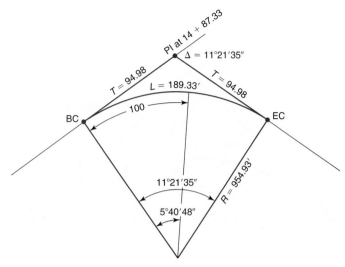

FIGURE 13-6 Sketch for Example 13-3

Example 13-3: Refer to Figure 13-6. Given the following information:

$$\Delta = 11°21'35''$$
$$PI \text{ at } 14 + 87.33$$
$$D = 6°$$

Calculate the station of the BC and EC.

Solution:

$R = 5{,}729.58/D = 954.93\,\text{ft}$

$T = R \tan \Delta/2 = 954.93 \tan 5.679861° = 94.98\,\text{ft}$

$L = 100\Delta/D = 100 \times 11.359722/6 = 189.33\,\text{ft}$

or

$L = 2\pi R\,\Delta/360 = 2\pi \times 954.93 \times 11.359722/360$
$\qquad = 189.33\,\text{ft}$

$$
\begin{array}{lrr}
\text{PI at} & 14 + & 87.33 \\
-T & & 94.98 \\ \hline
BC = & 13 + & 92.35 \\
+L & 1 & 89.33 \\ \hline
EC = & 15 + & 81.68
\end{array}
$$

13.4 CIRCULAR CURVE DEFLECTIONS

A common method of locating a curve in the field is by deflection angles. Typically, the theodolite or total station is set up at the BC, and the deflection angles are turned from the tangent line (see Figure 13-7). (Alternately, the coordinates of the stations on the curve to be laid out are first computed and then uploaded into the total station; using a typical layout command, the point can be laid out while the total station is occupying any control point, sighted on any coordinated back-sight point that has been also coordinated.)

If we use the data from Example 13-2:

$$BC \text{ at } 0 + 196.738$$
$$EC \text{ at } 0 + 286.448$$
$$\frac{\Delta}{2} = 6°25'30'' = 6.4250°$$
$$L = 89.710\,\text{m}$$
$$T = 45.044\,\text{m}$$

and if the layout is to proceed at 20-m intervals, the procedure would be as follows:

1. Compute the deflection angles for the three required arc distances: (deflection angle = arc/L × Δ/2):
   a. BC to first even station (0 + 200):
   $(0 + 200) - (0 + 196.738) = 3.262$

   $3.262/89.710 \times 6.4250 = 0.2336° = 0°14'01''$

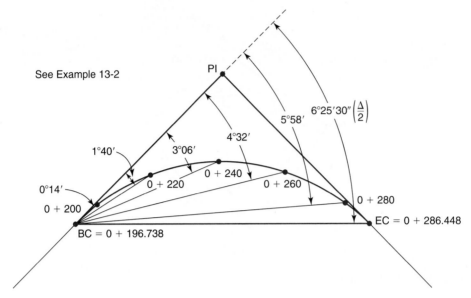

FIGURE 13-7 Field location for deflection angles. See Example 13-2

b. Even station interval:

$$20/89.710 \times 6.4250 = 1.4324° = 1°25'57''$$

c. Last even station (0 + 280) to EC:

$$6.448/89.710 \times 6.4250 = 0.4618° = 0°27'42''$$

2. Prepare a list of appropriate stations together with *cumulative* deflection angles.

BC	0 + 196.738	0°00'00''
	0 + 200	0°14'01''(+1°25'57'')
	0 + 220	1°39'58''(+1°25'57'')
	0 + 240	3°05'55''(+1°25'57'')
	0 + 260	4°31'52''(+1°25'57'')
	0 + 280	5°57'49''(+1°25'57'')
EC	0 + 286.448	6°25'31'' ≈ 6°25'30'' = Δ/2

For many engineering layouts, the deflection angles are rounded to the closest minute or half-minute.

Another common method (see above and Chapter 5) of locating a curve in the field is by using the "setting out" feature of total stations. The coordinates of each station on the curve are first uploaded into the total station, permitting its processor to compute the angle and distance from the instrument to each layout station.

13.5 CHORD CALCULATIONS

In the previous example, it was determined that the deflection angle for station 0 + 200 was 0°14'01''; it follows that 0 + 200 could be located by placing a stake on the theodolite/total station line at 0°14' and at a distance of 3.262 m (200 – 196.738) from the BC. Furthermore, station 0 + 220 could be located by placing a stake on the theodolite/total station line at 1°40' (rounded) and at a distance of 20 m along the arc from the stake locating 0 + 200. The remaining stations could be located in a similar manner. However, it must be noted that this technique contains some error as the distances measured with a steel tape or EDM are not arc distances; they are straight lines known as *subchords*.

To calculate the subchord, Equation 13-2, $C = 2R \sin \Delta/2$, may be used. This equation, derived from Figure 13-2, is the special case of the long chord and the total deflection angle. The general case can be stated as follows:

$$C = 2R \sin \text{ (deflection angle)} \qquad (13\text{-}8)$$

and any subchord length can be calculated if its deflection angle is known.

Relevant chords for the previous example can be calculated as follows (see Figure 13-8):

First chord: $C = 2 \times 400 \sin 0°14'01'' = 3.2618\,\text{m}$
$= 3.262\,\text{m}$ (at three decimals, chord $=$ arc)

Even station chord: $C = 2 \times 400 \sin 1°25'57''$
$= 19.998\,\text{m}$

Last chord: $C = 2 \times 400 \sin 0°27'42'' = 6.448\,\text{m}$

If these chord distances were used, the curve layout could proceed without significant error.

Although the calculation of the first and last subchord shows the chord and arc to be equal (i.e., 3.262 m and 6.448 m), the chords are always marginally shorter than the arcs. In the cases of short distances (above) and in the case of flat (large radius) curves, the arcs and chords can often appear to be equal. If more decimal places are introduced into the calculation, the marginal difference between arc and chord becomes evident.

13.6 METRIC CONSIDERATIONS

Agencies that use metric (SI) units have adopted for highway use a reference station of 1 km (e.g., $1 + 000$), cross sections at 50-m, 20-m, and 10-m intervals, and a curvature design parameter based on a rational (even meter) value radius, as opposed to a rational value degree (even degree) of curve (D). The degree of curve originally found favor with most highway agencies because of the somewhat simpler calculations associated with its use, a factor that was significant in the pre-electronics age when most calculations were performed by using logarithms. A comparison of techniques involving both D and R (radius) shows that the only computation in which the rational aspect of D is carried through is that for the arc length, that is, $L = 100\Delta/D$ (Equation 13-7), and even in that one case, the ease of calculation depends on Δ also being a rational number. In all other formulas, the inclusion of trigonometric functions or π ensures a more complex computation requiring the use of a calculator or computer.

In fieldwork, the use of D (as opposed to R) permits quick determination of the deflection angle for even stations. For example, in foot units, if the degree of curve were 2°, the deflection angle for a full station (100 ft) would be $D/2$ or 1°; for 50 ft, the deflection would be 0°30'; and so on.

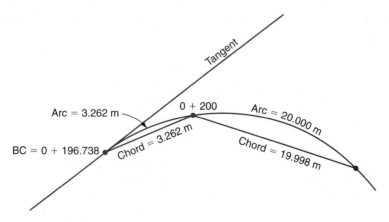

Arc $=$ 3.262 m

0 + 200

Arc $=$ 20.000 m

BC $=$ 0 + 196.738

Chord $=$ 3.262 m

Chord $=$ 19.998 m

Tangent

FIGURE 13-8 Curve arcs and chords

In metric units, the degree of curve would be the central angle subtended by 100 m of arc, and the deflections would be similarly computed. That is, for a metric D of 6°, the deflections would be as follows: 3° for 100 m, 1°30′ for 50 m, 0°36′ for 20 m, and 0°18′ for 10 m. The metric curve deflections here are not quite as simple as in the foot system, but they are still uncomplicated and rational. However, curve stakeouts require more stations than just those on the even stations; for example, the BC and EC, catch basins or culverts, vertical curve stations, and the like, usually occur on odd stations, and the deflection angles for those odd stations involve irrational number calculations requiring the use of a calculator or computer.

The widespread use of handheld calculators and office computers has greatly reduced the importance of techniques that permit only marginal reductions in computations. Surveyors are now routinely solving their problems with calculators and computers rather than using the seemingly endless array of tables that once characterized the back section of survey texts.

An additional reason for the lessening importance of D in computing deflection angles is that many curves (particularly at interchanges) are now being laid out by control point–based polar or intersection techniques (i.e., angle/distance or angle/angle) instead of deflection angles (see Chapter 5).

Those countries using the metric system, almost without exception, use a rational value for the radius (R) as a design parameter.

13.7 FIELD PROCEDURE

With the PI location and Δ angle measured in the field, and with the radius or degree of curve (D) chosen consistent with the design speed, all curve computations can be completed. The surveyor then goes back out to the field and measures off the tangent (T) distance from the PI to locate the BC and EC on the appropriate tangent lines. The theodolite/total station is then set up at the BC and zeroed and sighted in on the PI. The $\Delta/2$

angle (6°25′30″ in Example 13-2) is then turned off in the direction of the EC mark (wood stake, nail, etc.). If the computations for T and the field measurements of T have been performed correctly, the line of sight of the $\Delta/2$ angle will fall over the EC mark. If this does not occur, the T computations and then the field measurements are repeated.

The $\Delta/2$ line of sight over the EC mark will, of necessity, contain some error. In each case, the surveyor must decide if the resultant alignment error is acceptable for the type of survey in question. For example, if the $\Delta/2$ line of sight misses the EC mark by 0.10 ft (30 mm) in a ditched highway ℄ survey, the surveyor would probably find the error acceptable and then proceed with the deflections. However, a similar error in the $\Delta/2$ line of sight in a survey to lay out an elevated portion of urban freeway would not be acceptable; in that case an acceptable error would be roughly one-third of the preceding error (0.03 ft or 10 mm).

After the $\Delta/2$ check has been satisfactorily completed, the curve stakes are set by turning off the deflection angle and measuring the chord distance for the appropriate stations. If possible, the theodolite/total station is left at the BC (see the next section) for the entire curve stakeout, whereas the distance measuring moves continually forward from station to station. If using a steel tape, the rear taping surveyor keeps his or her body to the outside of the curve to avoid blocking the line of sight from the instrument.

A final verification of the work is available after the last even station has been set, as the chord distance from the last even station to the EC stake is measured with a tape and compared to the theoretical value; if the check indicates an unacceptable discrepancy, the work is checked and the discrepancy is removed.

After the curve has been deflected in, the party chief usually walks the curve, looking for any abnormalities. If a mistake has been made (e.g., putting in two stations at the same deflection angle is a common mistake), it will probably be very evident. The circular curve's symmetry is such that even minor mistakes are obvious in a visual check.

Note that many highway agencies use polar layout for interchanges and other complex features. If the coordinates of centerline alignment stations are determined, they can be used to locate the facility. In this application, the total station is placed at a known (or resection) station and aligned with another known station, so that the instrument's processor can compute and display the angle and distance needed for layout (see also Chapter 5).

13.8 MOVING UP ON THE CURVE

The curve deflections shown in Section 13.4 are presented in a form suitable for deflecting in while set up at the BC, with a zero setting at the PI. However, it often occurs that the entire curve cannot be deflected in from the BC, and two or more instrument setups may be required before the entire curve has been located. The reasons for this include a loss of line of sight due to intervening obstacles (i.e., detail or elevation rises).

In Figure 13-9, the data of Example 13-2 are used to illustrate the geometric considerations in moving up on the curve. In this case, station 0 + 260 cannot be established with the instrument at the BC (as were the previous stations). A large tree obscures the line of sight from the BC to 0 + 260. To establish station 0 + 260, the instrument is moved forward to the last station (0 + 240) established from the BC. The horizontal circle is zeroed, and the BC is then sighted with the telescope in its inverted position. When the telescope is transited, the theodolite/total station is once again oriented to the curve; that is, to set off the next (0 + 260) deflection, the surveyor refers to the previously prepared list of deflections and sets the appropriate deflection (4°32' rounded) for the desired station location and then for all subsequent stations.

Figure 13-9 shows the geometry involved in this technique. A tangent to the curve is shown by a dashed line through station 0 + 240 (the proposed setup location). The angle from that

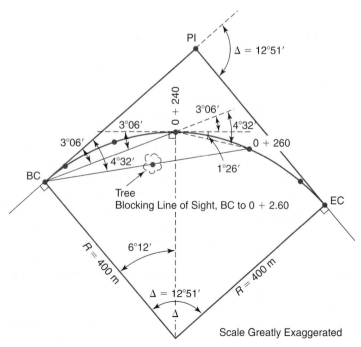

FIGURE 13-9 Moving up on the curve

tangent line to a line joining 0 + 240 to the BC is the deflection angle 3°06′. When the line from the BC is produced through station 0 + 240, the same angle (3°06′) occurs between that line and the tangent line through 0 + 240 (opposite angles). It was determined that the deflection angle for 20 m was 1°26′ (Section 13.4). When 1°26′ is added to 3°06′, the angle of 4°32′ for station 0 + 260 results, the same angle previously calculated for that station.

This discussion has limited the move up to one station; in fact, the move up can be repeated as often as necessary to complete the curve layout. The technique can generally be stated as follows: *When the instrument is moved up on the curve and the instrument is back-sighted with the telescope inverted at any other station, the theodolite will be "oriented to the curve" if the horizontal circle is first set to the value of the deflection angle for the sighted station.* That is, in the case of a BC sight, the deflection angle to be set is obviously zero; if the instrument were set on 0 + 260 and sighting 0 + 240, a deflection angle of 3°06′ would first be set on the scale.

When the inverted telescope is transited to its normal position, all subsequent stations can then be sighted by using the original list of deflections. This is the meaning of "theodolite/total station oriented to the curve," and this is why the list of deflections can be made first, before the instrument setup stations have been determined and (as we shall see in the next section) even before it has been decided whether to run in the curve centerline (℄) or whether it would be more appropriate to run in the curve on some offset line.

13.9 OFFSET CURVES

Curves being laid out for construction purposes must be established on offsets, so that the survey stakes are not disturbed by construction activities. Many highway agencies prefer to lay out the curve on ℄ (centerline) and then offset each ℄ stake a set distance left and right (left and right are oriented by facing to a forward station). The stakes can be offset to one side by using the swung-arm

technique described in Section 8.7.2, with the hands pointing to the two adjacent stations. If this is done with care, the offsets on that one side can be established on radial lines without significant error. After one side has been offset in this manner, the other side is then offset by lining up the established offset stake with the ℄ stake and measuring out the offset distance, ensuring that all three stakes are visually in a straight line. Keeping the three stakes in a straight line ensures that any alignment error existing at the offset stakes will steadily diminish as one moves toward the ℄ of the construction works.

In the construction of most municipal roads, particularly curbed roads, the centerline may not be established; instead, the road alignment is established directly on offset lines that are located a safe distance from the construction works. To illustrate, consider the curve in Example 13-2 used to construct a curbed road, as shown in Figure 13-10. The face of the curb is to be 4.00 m left and right of the centerline. Assume that the curb layout stakes can be offset 2 m (each side) without interfering with construction (generally, the less cut or fill required, the smaller can be the offset distance).

Figure 13-11 shows that if the layout is to be kept on radial lines through the ℄ stations, the station arc distances on the left-side (outside) curve will be longer than the corresponding ℄ arc distances, whereas the station arc distances on the right-side (inside) curve will be shorter than the corresponding ℄ arc distances. (See arc computations in the next section.) *By keeping the offset stations on radial lines, the surveyor is able to use the ℄ deflections previously computed.*

When using "setting-out" programs in total stations to locate offset stations in the field, the surveyor can simply identify the offset value (when prompted by the program), so that the processor can compute the offset stations' coordinates and then inverse to determine (and display) the required angle and distance from the instrument station. Civil coordinate geometry (COGO) software can be used to compute coordinates of all offset stations, and the layout angles and distances from selected proposed instrument stations.

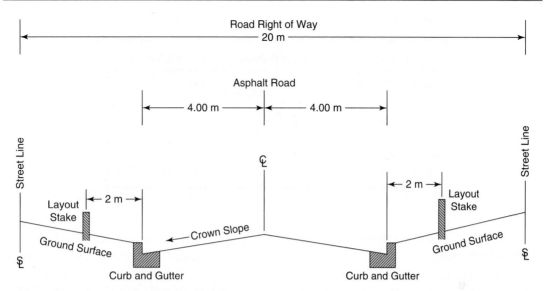

FIGURE 13-10 Municipal road cross section

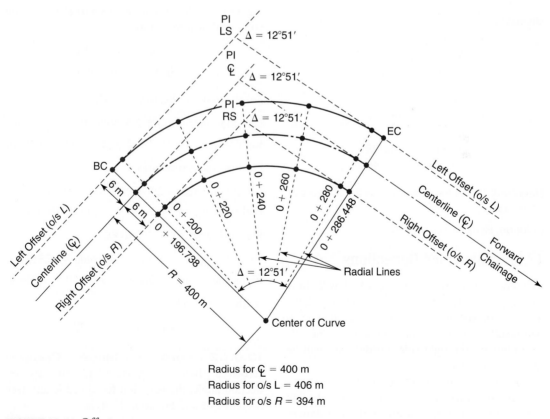

Radius for ℄ = 400 m
Radius for o/s L = 406 m
Radius for o/s R = 394 m

FIGURE 13-11 Offset curves

Table 13-1

Station	Computed Deflection	Field Deflection
BC 0 + 196.738	0°00'00"	0°00'
0 + 200	0°14'01"	0°14'
0 + 220	1°39'58"	1°40'
0 + 240	3°05'55"	3°06'
0 + 260	4°31'52"	4°32'
EC 0 + 286.448	6°25'31"	$6°25'30 = \dfrac{\Delta}{2}$; Check

Example 13-4: Illustrative Problem for Offset Curves (Metric Units)

Consider the problem of a construction offset layout using the data of Example 13-2, the deflections developed in Section 13.4, and the offset of 2 m introduced in the previous section.

Given data:

$$\Delta = 12°51'$$
$$\text{\textcentoldstyle} R = 400\,\text{m}$$
$$\text{PI at } 0 + 241.782$$

Calculated data:

$$T = 45.044\,\text{m}$$
$$L = 89.710\,\text{m}$$
$$\text{BC at } 0 + 196.738$$
$$\text{EC at } 0 + 286.448$$

Required: Curbs to be laid out on 2-m offsets at 20-m stations

Solution: Refer to Table 13-1.

13.9.1 Calculated Deflections

Reference to Figures 13-10 and 13-11 will show that the left-side (outside) curb face will have a radius of 404 m. A 2-m offset for that curb will result in an offset radius of 406 m. The offset radius for the right-side (inside) curb will be $400 - 6 = 394\,\text{m}$.

Because we are going to use the deflections already computed, it only remains to calculate the corresponding left-side arc or chord distances and the corresponding right-side arc or chord distances. Although layout procedure (angle and distance) indicates that chord distances will be required, for illustrative purposes we will compute both the arc and chord distances on offset.

13.9.1.1 Arc Distance Computations
Figure 13-12 shows that the offset (o/s) arcs can be computed by direct ratio:

$$\frac{\text{o/s arc}}{\text{\textcentoldstyle arc}} = \frac{\text{o/s radius}}{\text{\textcentoldstyle radius}}$$

For the first arc (BC to 0 + 200):

Left side: o/s arc $= 3.262 \times 406/400 = 3.311\,\text{m}$

Right side: o/s arc $= 3.262 \times 394/400 = 3.213\,\text{m}$

For the even station arcs:

Left side: o/s arc $= 20 \times 406/400 = 20.300\,\text{m}$

Right side: o/s arc $= 20 \times 394/400 = 19.700\,\text{m}$

For the last arc (0 + 280 to EC):

Left side: o/s arc $= 6.448 \times 406/400 = 6.545\,\text{m}$

Right side: o/s arc $= 6.448 \times 394/400 = 6.351\,\text{m}$

Arithmetic check:

$$\text{LS} - \text{\textcentoldstyle} = \text{\textcentoldstyle} - \text{RS}$$

13.9.1.2 Chord Distance Computations
Refer to Figure 13-13. For any deflection angle, the equation for chord length (see Section 13.5 and Equation 13-8) is

$$C = 2R \sin\,(\text{deflection angle})$$

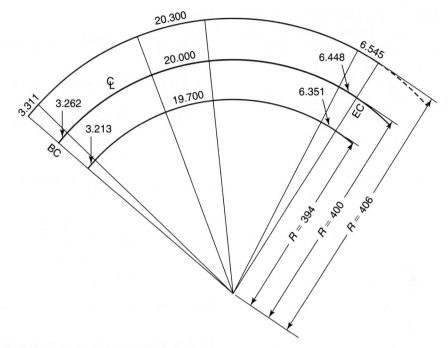

FIGURE 13-12 Offset arc lengths calculated by ratios

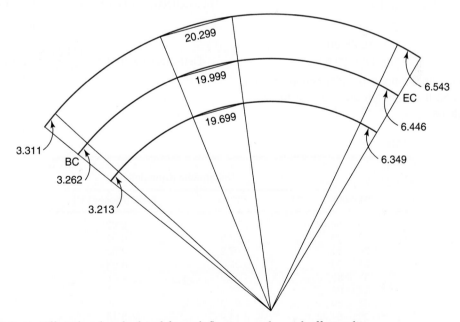

FIGURE 13-13 Offset chords calculated from deflection angles and offset radii

In this problem, the deflection angles have been calculated previously, and the radius (R) is the variable. For the first chord (BC to 0 + 200),

Left side: $C = 2 \times 406 \times \sin 0°14'01''$
$= 3.311\,\mathrm{m}$

Right side: $C = 2 \times 394 \times \sin 0°14'01''$
$= 3.213\,\mathrm{m}$

For the even station chords,

Left side: $C = 2 \times 406 \times \sin 1°25'57''$
$= 20.299\,\mathrm{m}$

Right side: $C = 2 \times 394 \times \sin 1°25'57''$
$= 19.699\,\mathrm{m}$

See Section 13-4. For the last chord,

Left side: $C = 2 \times 406 \times \sin 0°27'42''$
$= 6.543\,\mathrm{m}$

Right side: $C = 2 \times 394 \times \sin 0°27'42''$

$= 6.349\,\mathrm{m}$

Arithmetic check:

LS chord $-$ ℄ chord $=$ ℄ chord $-$ RS chord

Example 13-5:   Curve Problem (Foot Units)

Given the following ℄ data:

$$D = 5°$$
$$\Delta = 16°28'30''$$
$$\text{PI at } 31 + 30.62$$

you must furnish stakeout information for the curve on 50-ft offsets left and right of the ℄ at 50-ft stations.

Solution:

$$℄R = \frac{5{,}729.58}{D} = 1{,}145.92 \text{ ft}$$

$$T = R\tan\frac{\Delta}{2} = 1{,}145.92 \tan 8°\,14'15'' = 165.90 \text{ ft}$$

$$L = 100\frac{\Delta}{D} = \frac{100 \times 16.475}{5} = 329.50 \text{ ft}$$

or

$$L = 2\pi R\frac{\Delta}{360} = 329.50 \text{ ft}$$

$$
\begin{array}{lrr}
\text{PI at} & 31 & + 30.62 \\
-T & 1 & 65.90 \\
\hline
\text{BC} = & 29 & + 64.72 \\
+L & 3 & 29.50 \\
\hline
\text{EC} = & 32 & + 94.22
\end{array}
$$

Refer to Table 13-2 for the station deflections.

13.9.2 Computation of Deflections

$$\text{Total deflection for curve} = \frac{\Delta}{2} = 8°14'15''$$
$$= 494.25'$$

$$\text{Deflection per foot} = \frac{494.25}{329.50} = 1.5'\,\text{per ft}$$

Table 13-2

	Deflections (Cumulative)	
Stations	**Office**	**Field (Closest 1')**
BC 29 + 64.72	0°00.0'	0°00'
30 + 00	0°52.9'	0°53'
30 + 50	2°07.9'	2°08'
31 + 00	3°22.9'	3°23'
31 + 50	4°37.9'	4°38'
32 + 00	5°52.9'	5°53'
32 + 50	7°07.9'	7°08'
EC 32 + 94.22	8°14.2'	8°14'
	$\approx 8°14.25'$	
	$= \dfrac{\Delta}{2}$; Check	

Table 13-3   Chord Calculations for Example 13-5

Interval	Left Side	℄	Right Side
BC to 30 + 00	$C = 2 \times 1{,}195.92$ $\times \sin 0°52.9'$ $= 36.80\,\text{ft}$	$C = 2 \times 1{,}145.92$ $\times \sin 0°52.9'$ $= 35.27\,\text{ft}$	$C = 2 \times 1{,}095.92$ $\times \sin 0°52.9'$ $= 33.73\,\text{ft}$
		↖ Diff. = 1.53 ↗ ↖ Diff. = 1.54 ↗	
50-ft stations	$C = 2 \times 1{,}195.92$ $\times \sin 1°15'$ $= 52.18\,\text{ft}$	$C = 2 \times 1{,}145.92$ $\times \sin 1°15'$ $= 50.00\,(\text{to 2 decimal})$	$C = 2 \times 1{,}095.92$ $\times \sin 1°15'$ $= 47.81\,\text{ft}$
		↖ Diff. = 2.18 ↗ ↖ Diff. = 2.19 ↗	
32 + 50 EC	$C = 2 \times 1{,}195.92$ $\times \sin 1°06.3'$ $= 46.13\,\text{ft}$	$C = 2 \times 1{,}145.92$ $\times \sin 1°06.3'$ $= 44.20\,\text{ft}$	$C = 2 \times 1{,}095.92$ $\times \sin 1°06.3'$ $= 42.27\,\text{ft}$
		↖ Diff. = 1.93 ↗ ↖ Diff. = 1.93 ↗	

Because $D = 5°$, the deflection for 100 ft is $D/2$ or $2°30' = 150'$. The deflection, therefore, for 1 ft is 150/100 or $1.5'$ per ft.

Deflection for first station: $35.28 \times 1.5' = 52.92'$
$\qquad\qquad\qquad\qquad\qquad\qquad\quad = 0°52.9'$

Deflection for even 50-ft stations: $50 \times 1.5' = 75'$
$\qquad\qquad\qquad\qquad\qquad\qquad\qquad\quad = 1°15'$

Deflection for last station: $44.22 \times 1.5' = 66.33'$
$\qquad\qquad\qquad\qquad\qquad\qquad\quad = 1°06.3'$

Chord calculations for left- and right-side curves on 50-ft (from ℄) offsets (see Table 13-3 and Figure 13-14) are as follows:

$$\text{Radius for ℄} = 1{,}145.92\,\text{ft}$$
$$\text{Radius for LS} = 1{,}195.92\,\text{ft}$$
$$\text{Radius for RS} = 1{,}095.92\,\text{ft}$$

The differences shown in Table 13-3 provide a check on the calculations.

13.10 VERTICAL CURVES: GENERAL BACKGROUND

Vertical curves are used in highway and street vertical alignment to provide a gradual change between two adjacent gradelines. Some highway and municipal agencies introduce vertical curves at every change in gradeline slope, whereas other agencies introduce vertical curves into the alignment only when the net change in slope direction exceeds a specific value (e.g., 1.5 percent or 2 percent).

In Figure 13-15, vertical curve terminology is introduced: g_1 is the slope (percent) of the lower chainage gradeline, g_2 is the slope of the higher chainage gradeline, BVC is the beginning of the vertical curve, EVC is the end of the vertical curve, and PVI is the point of intersection of the two adjacent gradelines. The length of vertical curve (L) is the projection of the curve onto a horizontal surface and as such corresponds to plan distance. The algebraic change in slope direction is A, where $A = g_2 - g_1$. For example, if $g_1 = +1.5$ percent and $g_2 = -3.2$ percent, A would be equal to $(-3.2 - 1.5) = -4.7$ percent.

The geometric curve used in vertical alignment design is the vertical axis parabola. The parabola has the desirable characteristics of (1) a constant rate of change of slope, which contributes to smooth alignment transition, and (2) ease of computation of vertical offsets, which permits easily computed curve elevations. The general equation of the parabola is

$$y = ax^2 + bx + c \qquad (13\text{-}9)$$

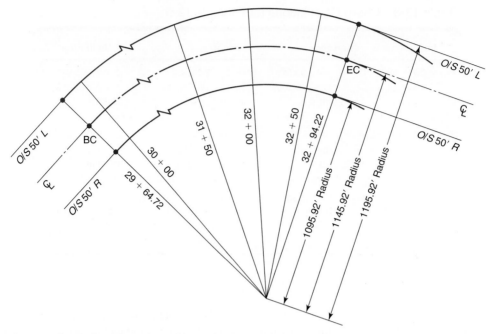

FIGURE 13-14 Sketch for the curve problem of Example 13-5

The slope of this curve at any point is given by the first derivative,

$$\frac{dy}{dx} = 2ax + b \qquad (13\text{-}10)$$

and the rate of change of slope is given by the second derivative,

$$\frac{d^2y}{dx^2} = 2a \qquad (13\text{-}11)$$

which, as was previously noted, is a constant. The rate of change of slope ($2a$) can also be written as A/L.

If, for convenience, the origin of the axes is placed at the BVC (see Figure 13-16), the general equation becomes

$$y = ax^2 + bx$$

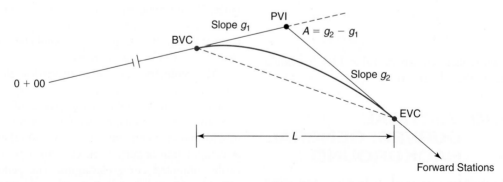

FIGURE 13-15 Vertical curve terminology (profile view shown)

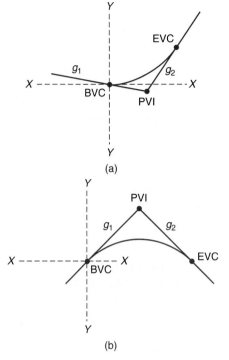

(a)

(b)

FIGURE 13-16 Types of vertical curves. (a) Sag curve. (b) Crest curve

and because the slope at the origin is g_1, the expression for slope of the curve at any point becomes

$$\frac{dy}{dx} = \text{slope} = 2ax + g_1 \quad (13\text{-}12)$$

The general equation can finally be written as

$$y = ax^2 + g_1x \quad (13\text{-}13)$$

or

$$y = \frac{g_2 - g_1}{2L}x^2 + g_1x$$

13.11 GEOMETRIC PROPERTIES OF THE PARABOLA

Figure 13-17 illustrates the following relationships.

- The difference in elevation between the BVC and a point on the g_1 gradeline at a distance x units (feet or meters) is g_1x (g_1 is expressed as a decimal).

- The tangent offset between the gradeline and the curve is given by ax^2, where x is the horizontal distance from the BVC; that is, tangent

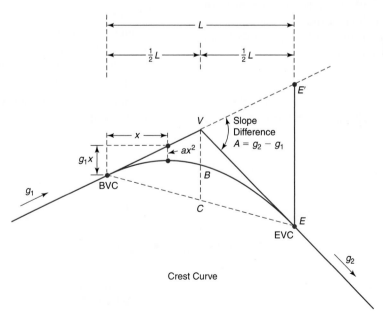

Crest Curve

FIGURE 13-17 Geometric properties of the parabola

offsets are proportional to the squares of the horizontal distances.

- The elevation of the curve at distance x from the BVC is given by

$$BVC + g_1 x + ax^2 = curve\ elevation$$

where a is positive in a sag curve and negative in a crest curve.

- The gradelines (g_1 and g_2) intersect midway between the BVC and the EVC; that is, BVC to $V = \frac{1}{2} L = V$ to EVC.

- Offsets from the two gradelines are symmetrical with respect to the PVI (V).

- The curve lies midway between the PVI and the midpoint of the chord; that is, $CB = BV$.

13.12 COMPUTATION OF THE HIGH OR LOW POINT ON A VERTICAL CURVE

The locations of curve high and low points (if applicable) are important for drainage considerations; for example, on curbed streets, catch basins must be installed precisely at the drainage low point.

It was noted earlier in Equation 13-12 that the slope was given by the equation

$$Slope = 2ax + g_1$$

Figure 13-18 shows a sag vertical curve with a tangent drawn through the low point; it is obvious

that the tangent line is horizontal with a slope of zero; that is,

$$2ax + g_1 = 0 \qquad (13\text{-}14)$$

Had a crest curve been drawn, the tangent through the high point would have exhibited the same characteristics. Because $2a = A/L$, Equation 13-14 can be rewritten as

$$x = -g_1 \left(\frac{L}{A} \right) \qquad (13\text{-}15)$$

where x is the distance from the BVC to the high or low point.

13.13 PROCEDURE FOR COMPUTING A VERTICAL CURVE

1. Compute the algebraic difference in grades: $A = g_2 - g_1$.

2. Compute the chainage of the BVC and EVC. If the chainage of the PVI is known, $\frac{1}{2} L$ is simply subtracted and added to the PVI chainage.

3. Compute the distance from the BVC to the high or low point (if applicable) using Equation 13-15 and determine the station of the high or low point.

4. Compute the tangent gradeline elevation of the BVC and the EVC.

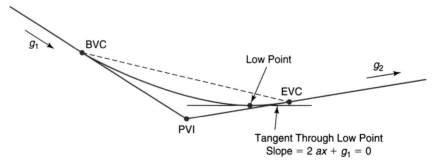

FIGURE 13-18 Tangent at curve low point

5. Compute the tangent gradeline elevation for each required station.

6. Compute the midpoint of chord elevation:

$$\frac{\text{Elevation of BVC} + \text{elevation of EVC}}{2}$$

7. Compute the tangent offset (d) at the PVI (i.e., distance VM in Figure 13-17):

$$d = \frac{\begin{array}{c}\text{difference in elevation of PVI}\\\text{and mid-point of chord}\end{array}}{2}$$

8. Compute the tangent offset for each individual station (see line ax^2 in Figure 13-19):

$$\text{Tangent offset} = \frac{d(x)^2}{(L/2)^2}, \quad \text{or} \qquad (13\text{-}16)$$
$$\frac{(4d)}{L^2}x^2$$

where x is the distance from the BVC or EVC (whichever is closer) to the required station.

9. Compute the elevation on the curve at each required station by combining the tangent offsets with the appropriate tangent gradeline elevations (add for sag curves and subtract for crest curves).

The techniques used for vertical curve computations are illustrated in Example 13-6.

Example 13-6: Given that $L = 300$ ft, $g_1 = -3.2$ percent, $g_2 = +1.8$ percent, PVI at 30 + 30 with elevation = 465.92 ft, determine the location of the low point and elevations on the curve at even stations, as well as at the low point.

Solution:

1. $A = 0.018 - (-0.032) = 0.05$

2. $\text{PVI} - \frac{1}{2}L = \text{BVC};$ BVC at $(30 + 30) - 150$
$$= 28 + 80.00$$

$\text{PVI} + \frac{1}{2}L = \text{EVC};$ EVC at $(30 + 30) + 150$
$$= 31 + 80.00$$

$\text{EVC} - \text{BVC} = L; (31 + 80) - (28 + 80)$
$$= 300; \text{Check}$$

3. Elevation of PVI = 465.92 ft

    150 ft at 3.2 percent = 4.80 (see Figure 13-19)

    Elevation BVC = 470.72 ft

    Elevation PVI = 465.92 ft

    150 ft at 1.8 percent = 2.70

    Elevation EVC = 468.62 ft

4. Location of low point is calculated using Equation 13-15:

$$x = \frac{0.032 \times 300}{0.05} = 192.00 \text{ ft} \quad \text{(from the BVC)}$$

5. Tangent gradeline computations are entered in Table 13-4. For example:

Elevation at 29 + 00 = $470.72 - (0.032 \times 20)$
$$= 470.72 - 0.64 = 470.08 \text{ ft}$$

6. Midchord elevation:

$$\frac{470.72 \text{ (BVC)} + 468.62 \text{ (EVC)}}{2} = 469.67 \text{ ft}$$

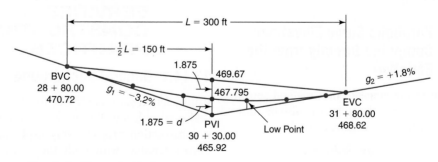

FIGURE 13-19 Sketch for Example 13-6

Table 13-4   Parabolic Curve Elevations by Tangent Offsets

Station	Tangent Elevation	+	Tangent Offset $\left(\dfrac{x}{1/2\,L}\right)^2 d*$	=	Curve Elevation
BVC 28 + 80	470.72		$(0/150)^2 \times 1.875 = 0$		470.72
29 + 00	470.08		$(20/150)^2 \times 1.875 = .03$		470.11
30 + 00	466.88		$(120/150)^2 \times 1.875 = 1.20$		468.08
PVI 30 + 30	465.92		$(150/150)^2 \times 1.875 = 1.875$		467.80
Low Point 30 + 72	466.68		$(108/150)^2 \times 1.875 = .97$		467.65
31 + 00	467.18		$(80/150)^2 \times 1.875 = .53$		467.71
EVC 31 + 80	468.62		$(0/150)^2 \times 1.875 = 0$		468.62

See Section 13.16.3

30 + 62	466.50		$(118/150)^2 \times 1.875 = 1.16$		467.66
30 + 72	466.68		$(108/150)^2 \times 1.875 = 0.97$		467.65
30 + 82	466.86		$(98/150)^2 \times 1.875 = 0.80$		467.66

*Where x is the distance from BVC or EVC, whichever is closer.

7. Tangent offset at PVI (d):

$$d = \frac{\text{difference in elevation of PVI and mid-point of chord}}{2}$$

$$= \frac{469.67 - 465.92}{2} = 1.875 \text{ ft}$$

8. Tangent offsets are computed by multiplying the distance ratio squared $(x/[L/2])^2$, by the maximum tangent offset (d). See Table 13-4.

9. The computed tangent offsets are added (in this example) to the tangent elevation in order to determine the curve elevation (see Table 13-4).

13.13.1 Parabolic Curve Elevations Computed Directly from the Equation

In addition to the tangent offset method shown earlier, vertical curve elevations can also be computed directly from the general equation:

$$y = ax^2 + bx + c$$

where

$a = (g_2 - g_1)/(2L)$

$y =$ elevation on the curve at distance x from the BVC

$L =$ horizontal length of vertical curve

$x =$ horizontal distance from BVC

$b = g_1$

$c =$ elevation at BVC

This technique is illustrated in Table 13-5 using the data from Example 13-6.

13.14 MUNICIPAL SERVICES CONSTRUCTION PRACTICES

13.14.1 General Background

Construction surveys provide for the horizontal and vertical layout for every key component of a construction project. Only experienced surveyors familiar with both the related project

Table 13-5 Parabolic Curve Elevations from the Equation
$$y = ax^2 + bx + c$$

Station	Distance from BVC, x	ax^2	bx	c	y (Elevation on the Curve)
BVC 28 + 80	0				470.72
29 + 00	20	0.03	−0.64	470.72	470.11
30 + 00	120	1.20	−3.84	470.72	468.08
PVI 30 + 30	150	1.88	−4.80	470.72	467.80
Low 30 + 72	192	3.07	−6.14	470.72	467.65
31 + 00	220	4.03	−7.04	470.72	467.71
EVC 31 + 80	300	7.50	−9.60	470.72	468.62

design and the appropriate construction techniques can accomplish this provision of **line and grade**. A knowledge of related design is essential to effectively interpret the design drawings for layout purposes, and knowledge of construction techniques is required to ensure that the layout is optimal for both line-and-grade transfer and construction scheduling. The word *grade* has several different meanings. In construction work alone, it is often used in three distinctly different ways:

1. To refer to a proposed elevation.
2. To refer to the slope of profile line (i.e., gradient).
3. To refer to cuts and fills: vertical distances below or above grade stakes or grade marks.

Engineering surveyors should be aware of these different meanings and always note the context in which the word *grade* is being used.

We have seen that data can be gathered for engineering (*preengineering* surveys) and other works in a variety of ways. Modern practice favors total station surveys for high-density areas of limited size, and LiDAR/photogrammetric techniques for high-density areas covering large tracts. Additionally, global positioning system (GPS) techniques are now being successfully implemented in areas of moderate density. The technique chosen by a survey manager is usually influenced by the costs (per point) and the reliability of the various techniques.

With regard to *construction surveys* (layout or setting-out surveys), modern practice has become dependent on the use of total stations, and more recently, on the use of GPS receivers working in "real" time (RTK techniques). In theory, GPS techniques seem to be ideal, as roving receiver-equipped surveyors quickly move to establish precise locations for layout points, in which both line and grade are promptly determined and marked. Attained accuracies can be observed by simply remaining at the layout point as the position location is continuously updated. Points located through the use of GPS techniques must be verified, if possible, through independent surveys (e.g., check GPS surveys, tape measurements taken from point to point where feasible, total station sightings, etc.). In the real world of construction works, the problems surrounding layout verification are compounded by the fact that the surveyor often doesn't have unlimited time to perform measurement checks; the contractor may actually be waiting on site to commence construction. A high level of planning and a rigid and systematic performance (proven successful in past projects) are necessary for performing the GPS survey.

Unlike other forms of surveying, construction surveying is often associated with speed of operation. Once contracts have been awarded, contractors may wish to commence construction immediately, because they probably have definite commitments for their employees and equipment and cannot accept delay. However, a hurried surveyor is more likely to make mistakes in measurements and calculations, and thus even more vigilance than normal is required. Construction surveying is therefore not an occupation for the faint of heart; the responsibilities are great and the working conditions often less than ideal. However, the sense of achievement when viewing the completed facility can be very rewarding.

The elimination of mistakes and the achievement of required accuracy have been stressed in this text. In no area of surveying are these qualities more important than in construction surveying. All field measurements and calculations are suspect until they have been verified by independent means or by repeated checks. Mistakes have been known to escape detection in as many as three independent, conscientious checks by experienced personnel. These comments apply to all tape, total station, and GPS layouts.

13.14.1.1 Construction Control   Depending on the size and complexity of the project, the survey crew should arrive on site one day or several days prior to the commencement of construction. The first on-site job for the construction surveyor is to relocate the horizontal and vertical control used in the preliminary survey (see Chapter 10). Usually, a number of months, and sometimes even years, have passed between the preliminary survey, the project design based on the preliminary survey, and the budget decision to award a contract for construction.

It may be necessary to reestablish the horizontal and vertical control in the area of proposed construction. If this is the case, extreme caution is advised, as the design plans are based on the original survey fabric, and any deviation from the original control could well lead to serious problems in construction. If the original control (or most of it)

still exists in the field, it is customary to check and verify all linear and angular dimensions that could directly affect the project.

The same rigorous approach is required for vertical control. If local benchmarks have been destroyed (as is often the case), the benchmarks must be reestablished accurately. Key existing elevations shown on design drawings (e.g., connecting invert elevations for gravity-flow sewers, or connecting beam seat elevations on concrete structures) must be resurveyed to ensure that (1) the original elevation shown on the plan was correct and (2) the new and original vertical control are, in fact, both referenced to the same vertical datum. In these areas, absolutely nothing is taken for granted.

Once the original control has been reestablished or verified, the control must be extended over the construction site to suit the purposes of each specific project. This operation and the specification of line and grade are discussed in detail for most types of projects in subsequent sections.

13.14.2 Measurement for Interim and Final Payments

On most construction projects, partial payments are made to the contractor at regular intervals and final payment is made upon completion and acceptance of the project. The payments are based on data supplied by the project inspector and the construction surveyor. The project inspector records items, such as daily progress, staff and equipment in use, and materials used, whereas the construction surveyor records items that require a surveying function (e.g., excavation quantities, concrete placed in structures, placement of sod).

The discussion in Chapter 1 covered field-book layout and stressed the importance of diaries and thorough note-taking. These functions-Maintaining diaries and thorough note-taking are important in all survey work but especially so in construction surveying. Since a great deal of money can depend on the integrity of daily notes and records, it is essential that the construction surveyor's notes and diaries be complete and accurate with respect to dates, times, locations, quantities, method of measurement, personnel, design changes, and so on.

13.14.3 Final Measurements for As-Built Drawings

Upon completion of a project, it is essential that a final plan be drawn showing the actual details of construction. The final plan, known as the *as-built drawing*, is usually quite similar to the design plan, with the exception that revisions are made to reflect changes in design that invariably occur during the construction process. Design changes result from problems that become apparent only after construction is under way. It is difficult, especially on complex projects, to plan for every eventuality that may be encountered; however, if the preliminary surveyor and the designer have both done their jobs well, the design plan and the as-built plan are usually quite similar.

13.14.4 Road Geometrics

13.14.4.1 Classification of Roads

The plan shown in Figure 13-20 depicts a typical municipal road pattern. The **local** roads shown have the primary purpose of providing access to individual residential lots. The **collector** roads [both major (e.g., S. Ridge Avenue.) and minor (e.g., E. Fairview Street)] provide the dual service of lot access and traffic movement. The collector roads connect the local roads to **arterial** roads (e.g., E. Central road and Highway #14); the main purpose of the arterial roads is to provide a relatively high level of traffic movement service.

Municipal works engineers base their road design on the level of service to be provided. The

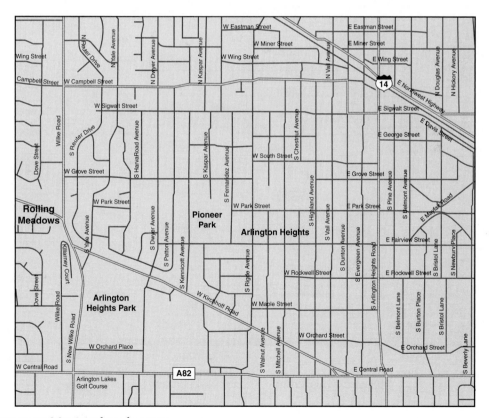

FIGURE 13-20 Municipal road pattern

proposed cross sections and geometric alignments vary in complexity and cost from the fundamental local roads to the more complex arterials. The highest level of service is given by the **freeways**, which provide high-velocity, high-volume routes with limited access (interchanges only), ensuring continuous traffic flow when design conditions prevail.

13.14.4.2 Road Allowances/Rights of Way

The road allowance varies in width from 40 ft (12 m) for small locals to 120 ft (35 m) for major arterials. In parts of North America, the local road allowances originally were 66 ft wide (one Gunter's chain), and when widening was required due to increased traffic volumes, it was common to take 10-ft widenings on each side, initially resulting in an 86-ft road allowance for major collectors and minor arterials. Further widenings left major arterials at 100- and 120-ft widths.

13.14.4.3 Road Cross Sections

A full-service municipal road allowance will usually have asphalt pavement, curbs, storm and sanitary sewers, water distribution pipes, hydrants, catch basins, and sidewalks. Additional utilities such as natural gas pipelines, electrical supply cables, and cable TV are also often located on the road allowance. The essential differences between local cross sections and arterial cross sections are the widths of pavement and the quality and depths of pavement materials. The construction layout of sewers and pipelines is covered in subsequent sections. See Figure 13-21 for a typical municipal road cross section.

The cross fall (height of crown) used on the pavement varies from one municipality to

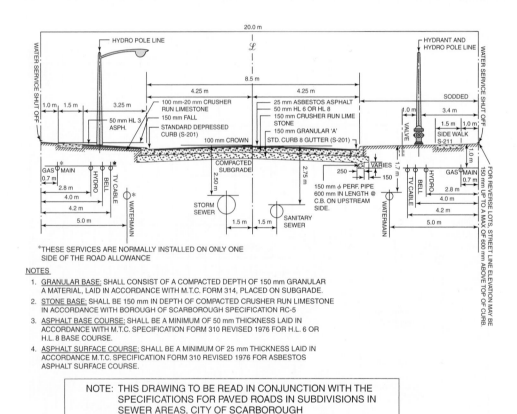

FIGURE 13-21 Typical cross section of a local residential road

another, but is usually close to a 2 percent slope (down to each curb face from ℄) to provide adequate drainage. The curb face is often 6 in. (150 mm) high, except at driveways and crosswalks, where the height is restricted to about 2 in. (50 mm) for vehicle and pedestrian access. The slope on the boulevard from the curb to the street line usually rises at a 2 percent minimum slope, thus ensuring that roadway storm drainage does not run onto private property.

13.14.4.4 Plan and Profile A typical plan and profile are shown in Figure 13-22. The plan and profile, which usually also show the cross-section details and construction notes, are the "blueprint" from which the construction is accomplished. The plan and profile and cross

section, together with the contract specifications, spell out in detail precisely where and how the road (in this example) is to be built.

The plan portion of the plan and profile gives the horizontal location of the facility, including curve radii, whereas the profile portion shows the key elevations and slopes along the road centerline, including vertical curve information. The cross section shows cross slopes, vertical dimensions, and cross horizontal dimensions for all municipal services. Both the plan and profile relate all data to the project stationing established as horizontal control.

13.14.4.5 Establishing Centerline (℄) Using the example of a ditched residential road being upgraded to a paved and curbed road, the first job for the construction surveyor is

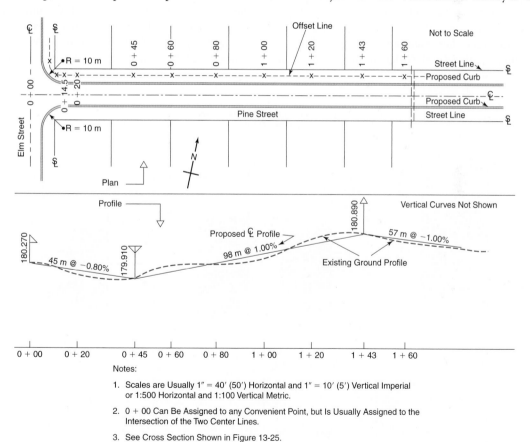

FIGURE 13-22 Plan and profile

to reestablish the centerline (℄) of the roadway. Usually this entails finding several property markers delineating street line (ℒ). Fence and hedge lines can be used initially to guide the surveyor to the approximate location of the property markers. When the surveyor finds one property marker, he or she can measure from that marker frontage distances shown on the property plan (plat) to locate a sufficient number of additional markers. Usually, the construction surveyor has the notes from the preliminary survey showing the location of property markers used in the original survey. If possible, the construction surveyor uses the same evidence used in the preliminary survey, taking the time, of course, to verify the resultant alignment. If the evidence used in the preliminary survey has been destroyed, as is often the case when a year or more elapses between the two surveys, the construction surveyor takes great care to see to it that his or her results are not appreciably different from those of the original survey, unless, of course, a blunder occurred on the original survey. If the evidence used in the original survey was coordinated, a GPS receiver can be used to navigate to the correct location.

The property markers can be square or round iron bars (including rebars), or round iron or aluminum pipes, magnetically capped. The markers can vary from 18″ to 4 ft in length. It is not unusual for the surveyor to have to use a shovel, as the tops of the markers are often buried. The surveyor can use an electronic or magnetic metal detector to aid in locating buried markers.

Sometimes even an exhaustive search of an area does not turn up a sufficient number of markers to establish the ℄. The surveyor must then extend the search to adjacent blocks or backyards in order to reestablish the missing markers. The surveyor can also approach the homeowners in the area and inquire as to the existence of a mortgage survey plan or house survey (see Figure 13-23) for the specific property. Such a

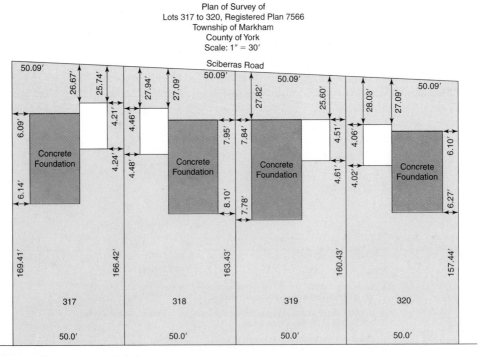

FIGURE 13-23 House survey (plat)

plan is required in most areas before a financial institution will provide mortgage financing. The mortgage survey plan shows dimensions from the building foundation to the street line and to both sidelines. Information thus gained can be used to narrow the search for a missing marker or can be used directly to establish points on the street line.

Once a number of points have been established on both sides of the roadway, the ℄ can be marked from each of these points by measuring (at right angles) half the width of the road allowance. The surveyor then sets up a theodolite, or total station, on a ℄ mark near one of the project extremities and sights in on the ℄ marker nearest the other project extremity. The surveyor can then see if all markers line up in a straight line (assuming tangent alignment) (see Figure 13-24). If the markers do not all line up, the surveyor will check the affected measurements; if discrepancies still occur (as often is the case), the surveyor will make the "best fit" of the available evidence. Depending on the length of the roadway involved and the number of ℄ markers established, the number of markers lining up perfectly varies. Three well-spaced, perfectly aligned markers is the absolute minimum number required for the establishment of construction ℄. The reason that all markers do not line up is that over the years most lots are resurveyed; some lots may be resurveyed several times. The land surveyor's prime area of concern is the area of the plan immediately adjacent to the client's property, and the surveyor must ensure that the property stakeout is consistent for both evidence and plan intentions. Over a number of years, cumulative errors and mistakes can significantly affect the overall alignment of the street lines.

If the ℄ is being marked on an existing road, as in this example, the surveyor uses nails with washers and red plastic flagging to establish marks. The nails can be driven into gravel, asphalt, and, in some cases, concrete surfaces. The washers will keep the nails from sinking below the road surface, and the red flagging helps in relocation.

If the project had involved a new curbed road in a new subdivision, the establishment of the ℄ would have been much simpler. The recently

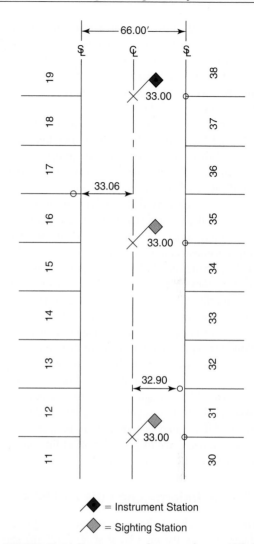

FIGURE 13-24 Property markers used to establish centerline

set property markers would for the most part be intact, and discrepancies between markers would be minimal (as all markers would have been set in the same comprehensive survey operation). The ℄ in this case would be marked by wood stakes $2'' \times 2''$, or $2'' \times 1''$ and $18''$ long.

13.14.4.6 Establishing Offset Lines
In this example, the legal fabric of the road allowance as given by the property markers constitutes

the horizontal control. Construction control consists of offset lines referenced to the proposed curbs with respect to line and grade. In the case of ditched roads and most highways, the offset lines are referenced to the proposed centerline with respect to line and grade.

The offset lines are placed as close to the proposed location of the curbs as possible. It is essential that the offset stakes not interfere with equipment and form-work; it is also essential that the offset stakes be far enough removed, so that they are not destroyed during cut or fill operations. Ideally, the offset stakes, once established, will remain in place for the duration of construction. This ideal can often be realized in municipal road construction, but it is seldom achieved in highway construction due to the significant size of cuts and fills. If cuts and fills are not too large, offset lines for curbs can be 3–5 ft (1–2 m) from the proposed face of the curb. An offset line this close allows for very efficient transfer of line and grade.

In the case of a ditched gravel road being upgraded to a curbed paved road, the offset line has to be placed far enough away on the boulevard to avoid the ditch-filling operation and any additional cut and fill that may be required. In the worst case, it may be necessary to place the offset line on the street line, an 18- to 25-ft offset (6- to 8-m).

13.14.5 Determining Cuts and Fills

The offset stakes (with nails or tacks for precise alignment) are usually placed at 50-ft (20-m) stations and at any critical alignment change points. The elevations of the tops of the stakes are usually determined by rod and level, based on the vertical control established for the project. It is then necessary to determine the proposed elevation for the top of the curb at each offset station. In Figure 13-22 elevations and slopes have been designed for a portion of the project. The plan and profile were simplified for illustrative purposes, and offsets are shown for one curb line only.

Given the ℂ elevation data, the construction surveyor must calculate the proposed curb elevations. The surveyor must calculate the relevant elevations on ℂ and then adjust for crown and curb height differential, or the surveyor can apply the differential first and work out curb elevations directly.

To determine the difference in elevation between ℂ and the top of the curb, the surveyor must analyze the appropriate cross section. In Figure 13-25 you can see that the cross fall is $4.5 \times 0.02 = 0.090$ m (90 mm). The face of the curb is 150 mm; therefore, the top of the curb is 60 mm above the ℂ elevation.

A list of key stations (see Figure 13-22) is prepared and ℂ elevations at each station are calculated (see Table 13-6). The ℂ elevations are then adjusted to produce curb elevations (the controlling grade of curbed-road layouts). Since superelevation is seldom used in municipal design, it is safe to say that the curbs on both sides of the road are normally parallel in both line and grade. A notable exception to this can occur when intersections of collectors and arterials are widened to allow for turn lanes or bus lanes.

The construction surveyor can then prepare a grade sheet (see Table 13-7), copies of which are given to the contractor and project inspector. The top of stake elevations, determined by level and rod or total station, are assumed in this example.

The grade sheet, signed by the construction surveyor, also includes the street name, date, limits of the contract, and, most important, the offset distance to the face of the curb. It should be noted that the construction grades (cut and fill) refer only to the vertical distance to be measured down or up from the top (usually) of the grade stake to locate the proposed elevation. Construction grades do not define with certainty whether the contractor is in a cut or fill situation at any given point. For example (see Figure 13-26), at 0 + 20, a construction grade of cut 0.145 is given, whereas the contractor is actually in a fill situation (i.e., the proposed top of the curb is *above* the existing ground at that station). This lack of correlation between construction grades and the construction process can become more pronounced as the offset distance lengthens. For example, if the grade stake at station 1 + 40 had been located

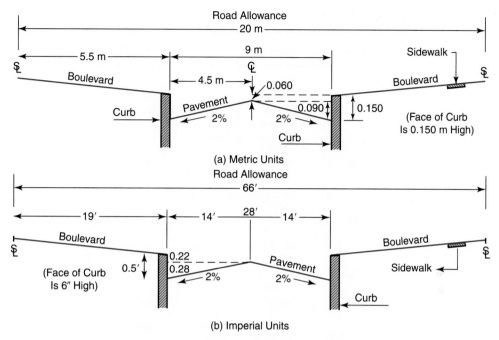

FIGURE 13-25 Cross section showing the relationship between the centerline and the top of the curb elevations. (a) Typical Metric units. (b) Typical Imperial units

at the street line, the construction grade would have been cut, whereas the construction process is almost entirely in a fill operation.

When the layout is performed using foot units, the basic station interval is 50 ft. The dimensions are recorded and calculated to the closest one hundredth (0.01) of a foot. Although all survey measurements are in feet and decimals of a foot, for the contractors' purposes the final cuts and fills are often expressed in feet and inches. The decimal foot–inch relationships are soon committed to memory by surveyors working in the construction field (see Table 13-8). Cuts and fills are usually expressed to the closest 1/8″ for concrete, steel, and pipelines, and to the closest 1/4″ for highway, granular surfaces, and ditch lines.

The grade sheet in Table 13-9 illustrates the foot–inch relationships. The first column and the last two columns are all that are required by the contractor, in addition to the offset distance, to construct the facility properly.

In some cases the cuts and fills (grades) are written directly on the appropriate grade stakes. This information, written with lumber crayon (keel) or felt markers, is always written on the side of the stake facing the construction. The station is also written on each stake and is placed on that side of each stake facing the lower chainage or stationing.

13.14.6 Treatment of Intersection Curb Construction

Intersection curb radii usually range from 30 ft (10 m) for two local roads intersecting to 60 ft (20 m) for two arterial roads intersecting. The angle of intersection is ideally 90°; however, the range from 70° to 110° is allowed for visibility design purposes.

Following is an example of curb intersection analysis based on the plan and profile data shown in Figure 13-22 and the intersection geometry shown in Figure 13-27. As you can see from

Table 13-6   Grade Computation

Station	₵ Elevation		Curb Elevation
0 + 00	180.270		
	−0.116		
BC 0 + 14.5	180.154	+0.060	180.214
	−0.044		
0 + 20	180.110	+0.060	180.170
	−0.160		
0 + 40	179.950	+0.060	180.010
	−0.040		
0 + 45	<u>179.910</u>	+0.060	179.970
	+0.150		
0 + 60	180.060	+0.060	180.120
	+0.200		
0 + 80	180.260	+0.060	180.320
	+0.200		
1 + 00	180.460	+0.060	180.520
	+0.200		
1 + 20	180.660	+0.060	180.720
	+0.200		
1 + 40	180.860	+0.060	180.920
	+0.030		
1 + 43	<u>180.890</u>	+0.060	180.950
etc.			

Table 13-7   Grade Sheet (metric)

Station	Curb Elevation	Stake Elevation	Cut	Fill
0 + 14.5	180.214	180.325	0.111	
0 + 20	180.170	180.315	0.145	
0 + 40	180.010	180.225	0.215	
0 + 45	179.970	180.110	0.140	
0 + 60	180.120	180.185	0.065	
0 + 80	180.320	180.320	On grade	
1 + 00	180.520	180.475		0.045
1 + 20	180.720	180.710		0.010
1 + 40	180.920	180.865		0.055
1 + 43	180.950	180.900		0.050
etc.				

Figure 13-27, the curb radius is 10 m and the curb elevation at the BC (180.214) is determined from the plan and profile of Pine Street (see Figures 13-22 and 13-25, and Table 13-7). The curb elevation at the EC is determined from the plan and profile of Elm Street (assume the EC elevation = 180.100). The length of the curved curb can be calculated from Equation 13.5:

$$L = \frac{\pi R \Delta}{180°}$$
$$= 15.708 \text{ m}$$

The slope from BC to EC can be determined:

$$180.214 - 180.100 = 0.114 \text{ m}$$

The fall is 0.114 over an arc distance of 15.708 m, which is −0.73 percent.

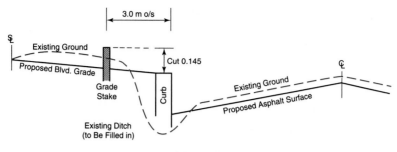

Station 0 + 20 Cut Grade

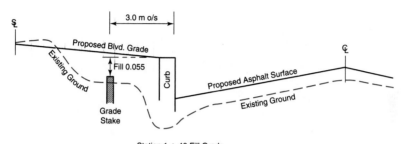

Station 1 + 40 Fill Grade

FIGURE 13-26 Cross sections showing cut-and-fill grades

Table 13-8   Decimal Foot–Inch Conversion

1 ft = 12 in.	1 in. = $\frac{1}{12}$ ft + 0.083 ft	
1 in. = 0.08(3) ft	7 in. = 0.58 ft	$\frac{1}{8}$ in. = 0.01 ft
2 in. = 0.17 ft	8 in. = 0.67 ft	$\frac{1}{4}$ in. = 0.02 ft
3 in. = 0.25 ft	9 in. = 0.75 ft	$\frac{1}{2}$ in. = 0.04 ft
4 in. = 0.33 ft	10 in. = 0.83 ft	$\frac{3}{4}$ in. = 0.06 ft
5 in. = 0.42 ft	11 in. = 0.92 ft	
6 in. = 0.50 ft	12 in. = 1.00 ft	

Table 13-9   Grade Sheet Showing Foot–Inch Conversion*

Station	Curb Elevation	Stake Elevation	Cut	Fill	Cut	Fill
0 + 30	470.20	471.30	1.10		1 ft 1$\frac{1}{4}$ in.	
0 + 50	470.40	470.95	0.55		0 ft 6$\frac{5}{8}$ in.	
1 + 00	470.90	470.90	On grade		On grade	
1 + 50	471.40	471.23		0.17		0 ft 2 in.
2 + 00	471.90	471.46		0.44		0 ft 5$\frac{1}{4}$ in.
2 + 50	472.40	472.06		0.34		0 ft 4$\frac{5}{8}$ in.

*Refer to Table 13-8 for foot–inch conversion.

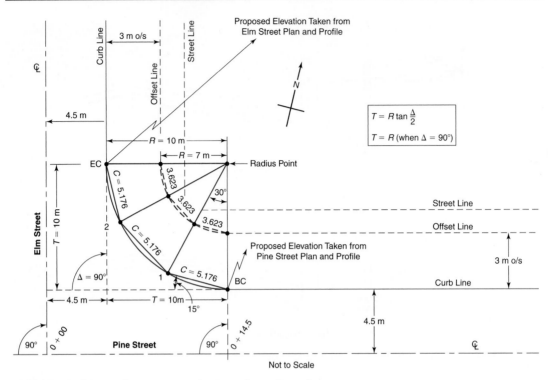

FIGURE 13-27 Intersection geometry (one quadrant shown)

These calculations indicate that a satisfactory slope (0.5 percent is the usual minimum for curb construction) joins the two points. The intersection curve is located by four offset stakes, BC, EC, and two intermediate points. In this case, 15.708/3 or 5.236 m is the distance measured from the BC to locate the first intermediate point, the distance measured from the first intermediate point to the second intermediate point, and the distance used as a check from the second intermediate point to the EC.

In actual practice, the chord distance is used rather than the arc distance. Since the curve deflection angle $\Delta/2 = 45°$, and we are using a factor of $\frac{1}{3}$, the corresponding deflection angle for one-third of the arc would be 15°.

$$C = 2R \sin \text{(deflection angle)}$$
$$= 2 \times 10 \times \sin 15° = 5.176 \text{ m}$$

These intermediate points can be deflected in from the BC or EC, or they can be located by the use of two tapes: one surveyor at the radius point (holding 10 m, in this case) and the other surveyor at the BC or intermediate point (holding 5.176), while the third surveyor holds the zero point of both tapes. The latter technique is the most often used on these small-radius problems. The only occasion when these curves are deflected in by theodolite occurs when the radius point (curve center) is inaccessible (due to fuel pump islands, front porches, etc.).

The proposed curb elevations on the arc are as follows:

BC	0 + 14.5 =	180.214
		−0.038
Stake 1	=	180.176
		−0.038
Stake 2	=	180.138
		−0.038
EC	=	180.100 Check
Are interval	=	5.236 m

Difference in elevation $= 5.236 \times 0.0073 = 0.038$

Grade information for the curve can be included on the grade sheet. The offset curve can be established in the same manner, after making allowances for the shortened radius (see Figure 13-11).

For an offset (o/s) of 3 m, the radius becomes 7 m. The chords required can be calculated using Equation 13.8:

$$C = 2R \sin (\text{deflection})$$
$$= 2 \times 7 \times \sin 15° = 3.623 \text{ m}$$

13.14.7 Sidewalk Construction

The sidewalk is constructed adjacent to the curb or at some set distance from the street line. If the sidewalk is adjacent to the curb, no additional layout is required as the curb itself gives line and grade for construction. In some cases the concrete for this curb and sidewalk is placed in one operation.

When the sidewalk is to be located at some set distance from the street line (𝕊), a separate layout is required. Sidewalks located near the 𝕊 give the advantages of increased pedestrian safety and boulevard space for the stockpiling of a winter's accumulation of plowed snow in northern regions. Figure 13-28 shows the typical location of a sidewalk on road allowance.

Sidewalk construction usually takes place after the curbs have been built and the boulevard has been brought to sod grade. The offset distance for the grade stakes can be quite short (1–3 ft). If the sidewalk is located within 1–3 ft of the 𝕊, the 𝕊 is an ideal location for the offset line. In many cases only a line layout is required for construction because the grade is already established by boulevard grading and because the permanent elevations at 𝕊 are seldom adjusted in municipal work. The cross slope of the walk (toward the curb) is usually given as being 1/4 in./ft (2 percent).

When the sidewalk is being located as near as 1 ft to the 𝕊, greater care is required by the surveyor to ensure that the sidewalk does not encroach on private property throughout its length. Due to numerous private property surveys performed by different surveyors over the years, the actual location of 𝕊 may no longer conform to plan location.

13.15 HIGHWAY CONSTRUCTION

Highways, like the municipal roads, are classified as locals, collectors, arterials, and freeways. The bulk of the highways, in mileage, are arterials that join towns and cities together in a state or provincial network. Unlike municipal roads, highways do not usually have curbs and storm sewers, relying instead on ditches for removal of storm drainage. Whereas in municipal work the construction layout and offsets are referenced to the curb lines, in highway work the layout and offsets are usually all referenced to the centerline of construction.

The construction surveyor must first locate the right-of-way (ROW) legal markers and set up the construction ℄ in a manner similar to that

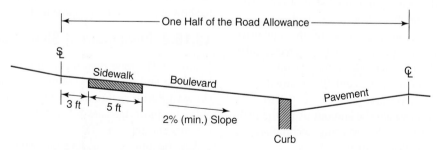

Note: The Sidewalk Is Always Constructed so that It Slopes Toward the Road—Usually @ 1/4 in. per foot (2%)

FIGURE 13-28 Typical location of a sidewalk on the road allowance

described in the preceding section. Mistakes can be eliminated if the surveyor, armed with a complete set of construction and legal plans, takes the time to verify all evidence by checking plan measurements against field measurements. The construction surveyor next reestablishes the stationing used for the project. Stations established in the preliminary survey can be reestablished from reference monuments or crossroad intersections. Stations are reestablished from at least three independent ties to ensure that verification is possible.

Highways are usually laid out at 100-ft (30- or 40-m) stations, and additional stations are required at all changes in horizontal alignment (including BC, EC, and spiral stations) and at all changes in vertical direction (including BVC, EVC, and low points). The horizontal and vertical curve sections of highways are often staked out at 50-ft (15- to 20-m) intervals to ensure that the finished product closely conforms to the design. When using foot units, the full stations are at 100-ft intervals (e.g., 0 + 00, 1 + 00). In metric units, municipalities use 100-m full station intervals (0 + 00, 1 + 00), whereas most highway agencies use 1,000-m (kilometer) intervals for full stations (e.g., 0 + 000, 0 + 100, ..., 1 + 000).

The ₵ of construction is staked out using a steel tape and plumb bobs, or total stations, using specifications designed for 1/3,000 accuracy (minimum). The accuracy of ₵ layout can be verified at reference monuments and road intersections, and by using GPS positioning.

Highways can also be laid out from random coordinated stations using total station polar layouts. Most interchanges are now laid out using polar methods, whereas most of the highways between interchanges are laid out using rectangular layout offsets. The methods of polar layout, which are covered in Chapters 5 and 10, require higher-order control survey precision (e.g., 1:10,000).

Highways can also be laid out using the real-time differential GPS surveying techniques, as described in Sections 7.10 and 7.11.

The profile grade, shown on the contract drawing, can refer to the top of granular elevation or it can refer to the top of asphalt elevation; the surveyor must ensure that the proper reference is used before calculating subgrade elevations for the required cuts and fills. As well, resultant cuts and fills must be clearly identified with respect to the reference feature—that is, top of subgrade, top of granular, or top of asphalt or concrete.

13.15.1 Clearing, Grubbing, and the Stripping of Topsoil

Clearing and **grubbing** are the terms used to describe the cutting down of trees and the removal of all stumps and rubbish. The full highway width is staked out, approximating the limits of cut and fill, so that the clearing and grubbing can be accomplished. The first construction operation after clearing and grubbing is the stripping of topsoil. The topsoil is usually stockpiled for later use. In cut sections, the topsoil is stripped for full width, which extends to the points at which the far-side ditch slopes intersect the original ground (OG) surface (see Figure 13-29). In fill sections the topsoil is usually stripped for the width of highway embankment (see Figure 13-30). Most highway agencies do not strip the topsoil where heights of fill exceed 4 ft (1.2 m), believing that this water-bearing material cannot damage (frost) the road base below that depth. The bottom of fills (toe of slope) and the top of cuts (top of slope) are marked by slope stakes. These stakes, which are driven in angled away from ₵, delineate not only the limits of stripping but also indicate the limits for cut and fill, such operations taking place immediately after the stripping operation. Lumber crayon (keel) or felt markers are used to show station and s/s (slope stake).

13.15.2 Placement of Slope Stakes

Figure 13-31 shows typical cut and fill sections in both foot and metric dimensions. The side slopes shown are 3:1 (i.e., three horizontal to one vertical), although most agencies use a steeper slope (2:1) for cuts and fills over 4 ft (1.2 m). To locate slope stakes, the difference in elevation between the profile grade at ₵ and the invert of ditch (cut section) or the toe of embankment (fill section) must first be determined. In Figure 13-31(a), the 1.5 ft depth of granular fill tapers to zero over a distance of

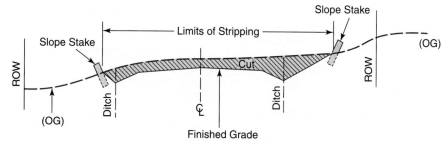

FIGURE 13-29 Highway cut section

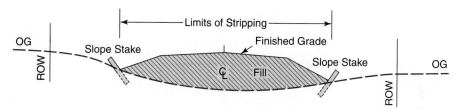

FIGURE 13-30 Highway fill section

1.5/(1/3) = 4.50 ft. The 1.5 ft ditch depth falls over a distance of 1.5/(1/3) = 4.5 ft, so the center of the ditch is 20 + 4.50 + 4.50 = 29.00 ft from the ℄.

The difference in elevation consists of:

$$\text{Depth of granular} = 1.50\,\text{ft}$$

$$\begin{array}{r} \text{Subgrade cross fall at} \\ \text{3 percent over 24.95 ft} = 0.75\,\text{ft} \end{array}$$

$$\underline{\text{Depth of ditch (assumed)} = 1.50\,\text{ft}}$$

$$\text{Total difference in elevation} = 3.75\,\text{ft}$$

If we can further assume (for this illustration) that the proposed ℄ elevation and the elevation across to the ROW limit are the same, the back slope of 3:1 will rise this elevation over a distance of 3.75/(1/3) = 11.25 ft. The slope stake should be placed 29.45 + 11.25 = 40.69 ft.

In Figure 13-31(b), the 0.45 m depth of granular fill tapers to zero over 0.45/(1/3) = 1.48 m, and the (minimum depth of ditch assumed here = 0.5 m) ditch falls 0.5 m over 0.5/(1/30) = 1.5 m, so the center of the ditch is 9.08 m from the ℄.

The difference in elevation between ℄ elevation and the invert of the ditch consists of:

$$\text{Depth of granular} = 0.45\,\text{m}$$

$$\text{Fall at 3 percent over 7.58 m} = 0.23\,\text{m}$$

$$\underline{\text{Minimum depth of ditch} = 0.50\,\text{m}}$$

$$\text{Total difference in elevation} = 1.18\,\text{m}$$

In Figure 13-31(c), the difference in elevation consists of:

$$\text{Depth of granular} = 1.50\,\text{ft}$$

$$\underline{\text{Fall at 3 percent over 24.95 ft} = 0.75\,\text{ft}}$$

$$\text{Total difference in elevation} = 2.25\,\text{ft}$$

The difference from ℄ elevation of construction to this point, where the subgrade intersects the side slope, is 24.95 ft.*

In Figure 13-31(d), the difference in elevation consists of:

$$\text{Depth of granular} = 0.45\,\text{m}$$

$$\underline{\text{Fall at 3 percent over 7.58 m} = 0.23\,\text{m}}$$

$$\text{Total difference in elevation} = 0.68\,\text{m}$$

The distance from the ℄ of construction to this point, where the subgrade intersects the side slope, is 7.58 m*.

---

*In the last two examples, the original ground elevation would be needed to compute the additional distances (at 3:1) required to place the slope stakes.

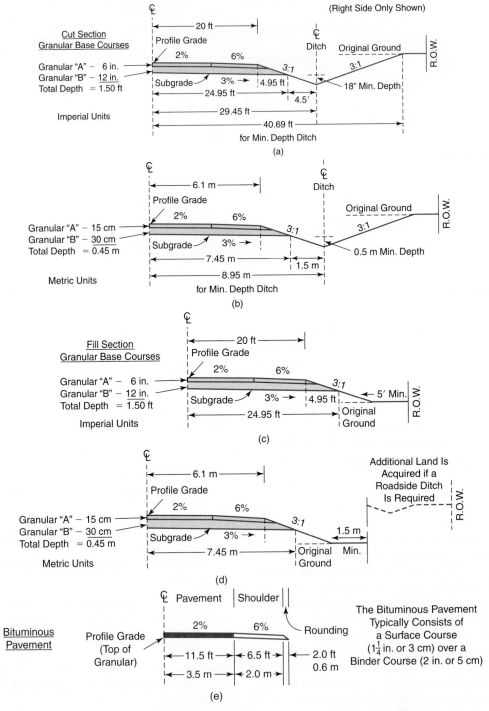

FIGURE 13-31 Typical two-lane highway cross section. (a) Cut section, foot units. (b) Cut section, metric units. (c) Fill section, foot units. (d) Fill section, metric units. (e) Bituminous pavement section

Figures 13-28 and 13-29 illustrate the techniques used when surveying in slope stakes in fill sections. Alternatively, the slope stake distance from centerline can be scaled from cross sections or topographic plans. In most cases, cross sections (see Chapter 8) are drawn at even stations (100 ft or 30–40 m). The cross sections are necessary to calculate the volume estimates used in contract tendering. The location of the slope stakes can be scaled from the cross-section plan. In addition, highway contract plans are now usually developed from aerial photos or other aerial images; these plans show contours that are precise enough for most slope staking purposes. The surveyor can usually scale off the required distances from the ℄ by using either the cross-section plan or the contour plan to the closest 1.0 ft or 0.3 m. The cost savings gained from determining this information in the office should usually outweigh any resultant loss of accuracy. It is now possible to increase the precision through advances in computers and photogrammetric equipment. Occasional field checks can be used to check on these "scale" methods. LiDAR imaging also holds great potential for these types of ground surface measurement applications.

If scale methods are employed, total stations can be used to establish the horizontal distance from the ℄ to the slope stake. These methods will be more accurate than using a cloth tape on deep cuts or high fills when "breaking tape" may be required several times.

13.15.3 Layout for Line and Grade

In municipal work, it is often possible to put the grade stakes on offset, issue a grade sheet, and then go on to other work. The surveyor may be called back to replace the occasional stake knocked over by construction equipment, but usually the layout is thought to be a one-time occurrence.

In highway work, the surveyor must accept the fact that the grade stakes will be laid out several times (unless machine guidance and/or control

techniques are used). The repetitive nature of this work makes it a natural choice for the use of machine guidance techniques. The chief difference between the municipal and highway work is the large values for cut and fill encountered in highway construction. For the grade stakes to be in a "safe" location, they must be located beyond the slope stakes. Although this location is used for the initial layout, as the work progresses this distance back to the ℄ becomes too cumbersome to allow for accurate transfer of alignment and grade. As a result, as the work progresses, the offset lines are moved ever closer to the ℄ of construction, until the final location for the offsets is 2–3 ft (1 m) from each edge of the proposed pavement. The number of times that the layout must be repeated is a direct function of the height of fill or depth of cut involved.

In highway work, the centerline is laid out at the appropriate stations. The centerline points are then individually offset (o/s) at convenient distances on both sides of the ℄. For the initial layout, the ℄ stakes, offset stakes, and slope stakes are all put in at the same time. The cuts and fills are written on the o/s grade stakes, referenced either to the top of the stake or to a mark on the side of the stake that gives even foot (even decimeter) values. The cuts and fills are written on that side of the stake facing the ℄, whereas the stations are written on that side of the stake facing back to the 0 + 00 location, as noted previously.

As the work progresses and the cuts and fills become more pronounced, care should be taken so that the horizontal distances are maintained. The centerline stakes are offset by turning 90°, either with a right-angle prism or, more usually, by the swung-arm method. Fiberglass tapes can be used to lay out the slope stakes and offset stakes. Once a ℄ station has been offset on one side, care is taken when offsetting to the other side to ensure that the two offsets and the ℄ stake are all in a straight line.

When the cut and/or fill operations have brought the work to the proposed subgrade (bottom of granular elevations), the subgrade must be verified by cross sections before the contractor is permitted to place the granular material. Usually, a tolerance of 0.10 ft (30 mm) is allowed. Once the top of granular profile has been reached, layout for pavement (sometimes a separate contract) can commence. The final layout for pavement is usually on a very close offset (3 ft or 1 m). If the pavement is to be concrete, more precise alignment is provided by nails driven into the tops of the stakes.

When the highway construction has been completed, a final survey is performed. The *final survey*, also called an *as-built survey*, includes cross sections and locations that are used for final payments to the contractor and for the completion of an as-built drawing. The final cross sections are taken at the same stations used in the preliminary survey.

The description here has referred to two-lane highways. The procedure for layout of a four-lane divided highway is very similar. The same control is used for both sections; grade stakes can be offset to the center of the median and used for both sections. When the lane separation becomes large and the vertical alignment is different for each direction, the project can be approached as being two independent highways. The layout for elevated highways, often found in downtown urban areas, follows the procedures used for structures layout, entailing the use of more precise methods and instrumentation that provides the higher accuracy required by these types of surveys.

13.15.4 Grade Transfer

The grade stakes can be set so that the tops of the stakes are at "grade." Stakes set to grade are colored red or blue on the top to differentiate them from all other stakes. This procedure is time-consuming and often impractical except for final pavement layout. Generally, the larger the offset distance, the more difficult it is to drive the tops of the stakes to grade.

As noted earlier, the cut and fill can refer to the top of the grade stake or to a mark on the side

of the grade stake referring to an even number of feet (decimeters) of cut or fill. The mark on the side of the stake is located by sliding the rod up and down the side of the stake until a value is read on the rod that gives the cut or fill to an even foot (decimeter). This procedure of marking the side of the stake is best performed by two people, one to hold the rod and one other to steady the bottom of the rod and then to make the mark on the stake. The cut or fill can be written on the stake or entered on a grade sheet, one copy of which is given to the contractor.

To transfer the grade (cut or fill) from the grade stake to the area of construction, a means of transferring the stake elevation in a horizontal manner is required. When the grade stake is close (within 6 ft or 2 m), the grade transfer can be accomplished using a carpenter's level set on a piece of sturdy lumber [see Figure 13-32(a) and 13-32(b)]. The recently developed laser torpedo level permits the horizontal reference (visible laser beam) to extend beyond the actual location of the level itself (that is right across the grade). When the grade stake is far from the area of construction, a string line level can be used to transfer the grade (cut or fill) (see Figure 13-33). In this case, a fill of 1 ft 0 in. is marked on the grade stake (as well as the offset distance). A guard stake has been placed adjacent to the grade stake, and the grade mark is transferred to the guard stake. A 1-ft distance is measured up the guard stake, and the grade elevation is marked.

A string line is then attached to the guard stake at the grade elevation mark; a line level is hung from the string near the halfway mark, and the string is pulled taut so as to eliminate most of the sag (it is not possible to eliminate all the sag). The string line is adjusted up and down until the bubble in the line level is centered; with the bubble centered, one can see quickly at ℂ how much more fill may be required to bring the highway, at that point, to grade. Use of rotating construction lasers allows one person, working alone to transfer and check grades much more quickly and efficiently.

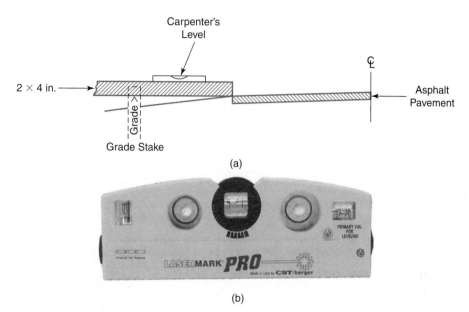

(a)

(b)

FIGURE 13-32 (a) Grade transfer using a conventional carpenter's level. (b) 8″ laser torpedo level, featuring accuracy up to $\frac{1}{4}$″ (0°02′) at 100 ft (6 mm at 30 m); three precision glass vials, horizontal, vertical, and adjustable; and three AA cell batteries providing approximately 40 hours of intermittent use
(Courtesy of CST/Berger, IL)

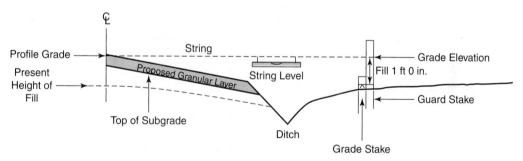

FIGURE 13-33 Grade transfer using a string level

Figure 13-34 shows that more fill is required to bring the total fill to the top of subgrade elevation. The surveyor can convey this information to the grade inspector, so that the fill can be properly increased. As the height of fill approaches the proper elevation (top of subgrade), the grade checks become more frequent.

In the preceding example, the grade fill was 1 ft 0 in.; had the grade been cut 1 ft, the procedure with respect to the guard stake would have been the same; that is, measure up the guard stake 1 ft, so that the mark now on the guard stake would be 2 ft above the ℄ grade. The surveyor or grade inspector at ℄ would simply measure down 2 ft using a tape measure from the level string at ℄. If the measurement down to the present height of fill exceeded 2 ft, it would indicate that more fill was required; if the measurement down to the present height of fill were less than 2 ft, it would indicate that too much fill has been placed and that an appropriate depth of fill must be removed.

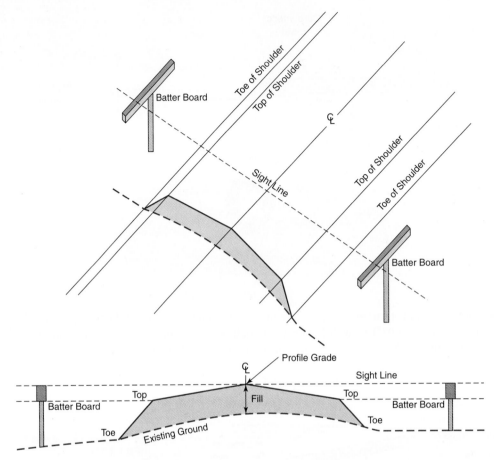

FIGURE 13-34 Grade transfer using batter boards

Another method of grade transfer used when the offset is large and the cuts or fills are significant is boning rods (*batter boards*; see Figure 13-34). In the preceding example, a fill grade is transferred from the grade stake to the guard stake; the fill grade is measured up the guard stake and the grade elevation is marked (any even foot/decimeter cut or fill mark can be used so long as the relationship to profile grade is marked clearly). A crosspiece is nailed on the guard stake at the grade mark and parallel to the ℄. A similar guard stake and crosspiece are established on the opposite side of the ℄. The surveyor or grade inspector can then sight over the two crosspieces to establish a profile grade datum at that point. Another worker can move across the section with a rod, and the progress of

the fill or cut operation can be visually checked. In some cases, two crosspieces are used on each guard stake, one indicating the ℄ profile grade and the lower one indicating the shoulder elevation.

13.15.5 Ditch Construction

The ditch profile often parallels the ℄ profile, especially in cut sections. When the ditch profile does parallel the ℄ profile, no additional grades are required to assist the contractor in construction. However, it is quite possible to have the ℄ profile at one slope (even 0 percent) and the ditch profile at another slope (0.3 percent is often taken as a minimum slope to give adequate drainage). If the ditch grades are independent of ℄ profile, the contractor must be given these cut or fill grades, either from the existing grade stakes or from grade stakes specifically referencing the ditch line.

In the extreme case (e.g., a spiraled highway going over the brow of a hill), the contractor may require five separate grades at one station [i.e., ₵, two edges of pavement (superelevated), and two different ditch grades]; it is even possible in this extreme case to have the two ditches flowing in opposite directions for a short distance.

13.16 SEWER CONSTRUCTION

Sewers are usually described as being in one of two categories. Sanitary sewers collect residential and industrial liquid waste and convey these wastes (sewage) to a treatment plant. Storm sewers are designed to collect runoff from rainfall and to transport this water (sewage) to the nearest natural receiving body (e.g., creek, river, and lake). The rainwater enters the storm sewer system through ditch inlets or through catch basins located at the curb line on paved roads. The design and construction of sanitary and storm sewers are similar in the respect that the flow of sewage is usually governed by gravity. Since the sewer gradelines (flow lines) depend on gravity, it is essential that the construction grades be given precisely.

Figure 13-35 shows a typical cross section of a municipal roadway. The two sewers are typically located 5 ft (1.5 m) either side of the ₵. The sanitary sewer is usually deeper than the storm sewer, as it must be deep enough to allow for all house connections. The sanitary house connection is usually at a 2 percent (minimum) slope. If sanitary sewers are being added to an existing residential road, the preliminary survey must include the basement floor elevations. The floor elevations are determined by taking a rod reading on the windowsill and then, after getting permission to enter the house, measuring from the windowsill down to the basement floor. As a result of deep basements and long setbacks from the ₵, sanitary sewers often have to be at least 9 ft (2.75 m) below the ₵ grade.

In the southern United States, the depth of storm sewers below the ₵ grade depends on the traffic loading. In the northern United States and most of Canada, it depends on traffic loading or on the depth of frost penetration, whichever is larger. The minimum depth of storm sewers ranges from 3 ft (1 m) in some areas of the south to 8 ft (2.5 m) in the north. The design of the inlets and catch basins depends on the depth of sewer and the quality of the effluent. The minimum slope for storm sewers is usually 0.50 percent, whereas the minimum slope for sanitary

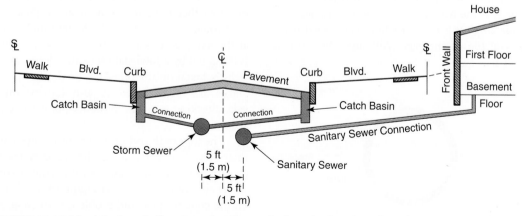

FIGURE 13-35 Municipal road allowance showing typical service locations (see also Figure 13-21)

sewers is often set at 0.67 percent. In either case, the designers try to achieve self-cleaning velocity 2.5 to 3 ft/s (0.8–0.9 m/s) to avoid excessive sewer maintenance costs.

Manholes (maintenance holes, MHs) are located at each change in direction, slope, or pipe size. In addition, manholes are located at 300- to 450-ft (100-m to 140-m) intervals maximum. Catch basins are located at 300-ft (100-m) maximum intervals; they are also located at the high side of intersections and at all low points. The 300-ft (100-m) maximum interval is reduced as the slope on the road increases.

For construction purposes, sewer layout is considered only from one manhole to the next. The stationing (0 + 00) commences at the first (existing) manhole (or outlet) and proceeds only to the next manhole. If a second leg is also to be constructed, that station of 0 + 00 is assigned to the downstream manhole and proceeds upstream only to the next manhole. Each manhole is described by a unique manhole number to avoid confusion with the stations for extensive sewer projects. Figure 13-36 shows a section of sewer pipe. The **invert** is the inside bottom of the pipe. The invert grade is the controlling grade for construction and design. The sewer pipes may consist of vitrified clay, metal, some of the newer plastics, or (as usually is the case) concrete. The pipe wall thickness depends on the diameter of the pipe. For storm sewers, 12 in. (300 mm) is usually taken as the minimum diameter. The **springline** of the pipe is at the halfway mark, and connections are made above this reference line. The **crown** is the outside top of the pipe. Although this term is

relatively unimportant (sewer cover is measured to the crown) for sewer construction, it is important for pipeline (pressurized pipes) construction as it gives the controlling grade for construction.

13.16.1 Layout for Line and Grade

As in all other construction work, offset stakes can be used to provide line and grade for the construction of sewers. Before deciding on the offset location, it is wise to discuss the matter with the contractor. The contractor will be excavating the trench and casting the material to one side or loading it into trucks for removal from the site. Additionally, the sewer pipe will be delivered to the site and positioned conveniently alongside its future location. The position of the offset stakes should not interfere with either of these operations.

The surveyor positions the offset line as close to the pipe centerline as possible, but seldom is it possible to locate the offset line closer than 15 ft (5 m) away. 0 + 00 is assigned to the downstream manhole or outlet, the chainage proceeding upstream to the next manhole. The centerline of construction is laid out with stakes marking the location of the two terminal points of the sewer leg. The surveyor uses survey techniques giving 1/3,000 accuracy as a minimum for most sewer projects. Large-diameter (6 ft or 2 m) sewers require increased precision and accuracy.

The two terminal points on the ℄ are occupied by a theodolite/total station, and right angles are turned to locate precisely the terminal points at the assigned offset distance. 0 + 00 on offset is occupied by a total station or theodolite, and a sight is taken on the other terminal offset point. Stakes are then located at 50-ft (20 m) intervals, checking in at the terminal point to verify accuracy.

The tops of the offset stakes are surveyed and their elevations determined, the surveyor taking care to see that the leveling is accurate; the existing invert elevation of MH 3 (see Figure 13-37) shown on the contract plan and profile is verified at the same time. The surveyor next calculates the sewer

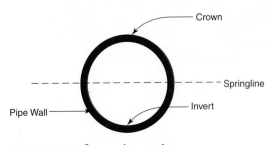

FIGURE 13-36 Sewer pipe section

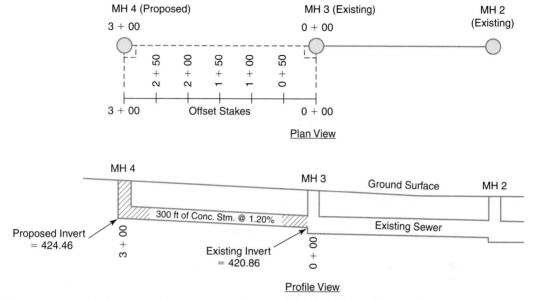

FIGURE 13-37 Plan view (a) and profile view (b) of a proposed sewer (foot units)

invert elevations for the 50-ft (20 m) stations. He or she then prepares a grade sheet showing the stations, stake elevations, invert grades, and cuts. The following examples will illustrate the techniques used.

Assume that an existing sewer (see Figure 13-37) is to be extended from existing MH 3 to proposed MH 4. The horizontal alignment will be a straight-line production of the sewer leg from MH 2 to MH 3. The vertical alignment is taken from the contract plan and profile (see Figure 13-37). The straight line is produced by setting up the total station or theodolite at MH 3, sighting MH 2, and double-centering to the location of MH 4. The layout then proceeds as described previously.

The stake elevations, usually determined by differential leveling, are shown in Table 13-10. At station 1 + 50, the cut is $8'3\frac{7}{8}''$ (see Figure 13-38). To set the cross-trench batter board at the next even foot, measure up $0'8\frac{1}{8}''$ to top of the batter board. The offset distance of 15 ft can be measured

Table 13-10   Sewer Grade Sheet: Foot Units*

Station	Invert Elevation	Stake Elevation	Cut
MH 3 0 + 00	420.86	429.27	$8.41 = 8\text{ ft }4\frac{7}{8}$ in.
0 + 50	421.46	429.90	$8.44 = 8\text{ ft }5\frac{1}{4}$ in.
1 + 00	422.06	430.41	$8.35 = 8\text{ ft }4\frac{1}{4}$ in.
1 + 50	422.66	430.98	$8.32 = 8\text{ ft }3\frac{7}{8}$ in.
2 + 00	423.26	431.72	$8.46 = 8\text{ ft }5\frac{1}{2}$ in.
2 + 50	423.86	431.82	$7.96 = 7\text{ ft }11\frac{1}{2}$ in.
MH 4 3 + 00	424.46	432.56	$8.10 = 8\text{ ft }1\frac{1}{4}$ in.

*Refer to Table 13-8 for foot–inch conversion.

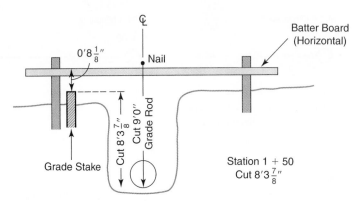

FIGURE 13-38 Use of cross-trench batter boards

and marked at the top of the batter board over the pipe ℄, and a distance of 9 ft measured down to establish the invert elevation. This even foot measurement from the top of the batter board to the invert is known as the grade rod distance. A value can be picked for the grade rod so that it is the same value at each station. In this example, 9 ft appears to be suitable for each station as it is larger than the largest cut. The arithmetic shown for station 1 + 50 is performed at each station so that the batter boards can be set at 50-ft intervals. The grade rod (a 2 in. × 2 in. length of lumber held by a worker in the trench) has a foot piece attached to the bottom at a right angle to the rod so that the foot piece can be inserted into the pipe and allow measurement precisely from the invert.

This method of line-and-grade transfer has been shown first because of its simplicity; it has *not*, however, been widely used in the field in recent years. With the introduction of larger and faster trenching equipment, which can dig deeper and wider trenches, this method would only slow the work down as it involves batter boards spanning the trench at 50-ft intervals. Most grade transfers are now accomplished by freestanding offset batter boards or laser alignment devices, with laser alignment progressively taking over. Using the data from the previous example, it can be illustrated how the technique of freestanding batter boards is used (see Figure 13-39).

These batter boards (3 ft or 1 m wide) are erected at each grade stake. As in the previous

example, the batter boards are set at a height that results in a grade rod that is an even number of feet (decimeters) long. However, with this technique, the grade rod distance will be longer, as the top of the batter board should be at a comfortable eye height for the inspector. The works inspector usually checks the work while standing at the lower chainage stakes and sighting forward to the higher chainage stakes. The line of sight over the batter boards is a straight line parallel to the invert profile, and in this example (see Figure 13-39) the line of sight over the batter boards is rising at 1.20 percent.

As the inspector sights over the batter boards, he or she can include in the field of view the top of the grade rod, which is being held on the most recently installed pipe length. The top of the grade rod has a horizontal board attached to it in a fashion similar to the batter boards. The inspector can visually determine whether the line over the batter boards and the line over the grade rod are at the same elevation. If an adjustment up or down is required, the worker in the trench makes the necessary adjustment and has the work rechecked. Grades can be checked to the closest $\frac{1}{4}$ in. (6 mm) in this manner. The preceding example is now worked out using a grade rod of 14 ft (see Table 13-11).

The grade rod of 14 ft requires an eye height of $5'7\frac{1}{8}''$ at 0 + 00; if this is considered too high, a grade rod of 13 ft could have been used, which would have resulted in an eye height of $4'7\frac{1}{8}''$ at the first batter board. The grade rod height is chosen to suit the needs of the inspector.

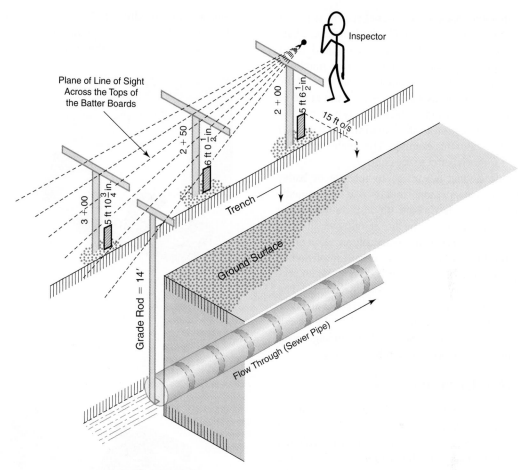

FIGURE 13-39 Freestanding batter boards

Table 13-11   Sewer Grade Sheet for Batter Boards

Station	Invert Elevation	Stake Elevation	Cut	Stake to Batter Board
MH 3 0 + 0	420.86	429.27	8.41	$5.59 = 5$ ft $7\frac{1}{8}$ in.
0 + 50	421.46	429.90	8.44	$5.56 = 5$ ft $6\frac{3}{4}$ in.
1 + 00	422.06	430.41	8.35	$5.65 = 5$ ft $7\frac{3}{4}$ in.
1 + 50	422.66	430.98	8.32	$5.68 = 5$ ft $8\frac{1}{8}$ in.
2 + 00	423.26	431.72	8.46	$5.54 = 5$ ft $6\frac{1}{2}$ in.
2 + 50	423.86	431.82	7.96	$6.04 = 6$ ft $0\frac{1}{2}$ in.
MH 4 3 + 00	424.46	432.56	8.10	$5.90 = 5$ ft $10\frac{3}{4}$ in.

In some cases, an additional station is put in before 0 + 00 (i.e., 0 – 50). The grade stake and batter board refer to the theoretical pipe ₵ and invert profile produced back through the first manhole. This batter board will, of course, line up with all the others and will be useful in checking the grade of the first few pipe lengths placed. Many agencies use 25-ft (10-m) stations, rather than the 50-ft (20-m) stations used in this example. The smaller intervals allow for much better grade control.

One distinct advantage to the use of batter boards in construction work is that an immediate visual check is available on *all* the survey work involved in the layout. The line of sight over the tops of the batter boards (which is actually a vertical offset line) must be a straight line. If, upon completion of the batter boards, all the boards do not line up precisely, it is obvious that a mistake has been made. The surveyor checks the work by first verifying all grade computations and, second, by releveling the tops of the grade stakes. Once the boards are in alignment, the surveyor can move on to other projects.

13.16.2 Laser Alignment

Laser devices are now in use in most forms of construction work. Lasers are normally used in a fixed direction and slope mode or in a revolving, horizontal or sloped pattern. One such device (see Figure 13-40) can be mounted in a sewer manhole, aligned for direction and slope, and used with a target for the laying of sewer pipe. Since the laser beam can be deflected by dust or high humidity, care must be taken to overcome these factors.

Lasers used in sewer trenches allow for setting slope within the limits of −10° to 30°. Some devices have automatic shutoff capabilities when the device is disturbed from its desired setting. Figure 13-41 illustrates additional laser applications.

13.16.3 Catch Basin (CB) Construction Layout

CBs are constructed along with the storm sewer or at a later date, just prior to curb construction. Usually, the CB is located by two grade stakes, one on each side of the CB. The two stakes are on

curb line and are usually 5 ft (2 m) from the center of the CB. The cut or fill grade is referenced to the CB grate elevation at the curb face. The ₵

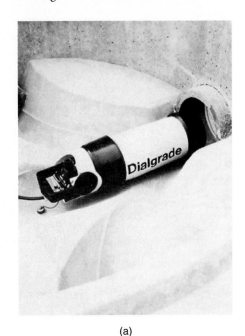

(a)

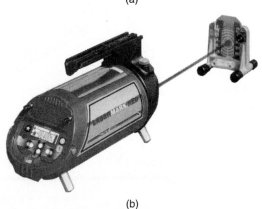

(b)

FIGURE 13-40 (a) Pipeline laser-mounted in a stormpipe manhole (Courtesy of Trimble, CA). (b) Pipe-laying laser featuring red or green visible laser beam, working range of 650 ft, grade range settings of −15 percent to 50 percent vertical, self-leveling range of 10 percent vertical, and accompanying pipe targets sized to fit various pipe diameters

(Courtesy of CST/Berger, IL)

FIGURE 13-41 Spectra precision rotating laser level and an electronic receiver attached to a grade rod, shown being used to set concrete forms to the correct elevation
(Courtesy of Trimble)

pavement elevation is calculated, and from it the crown height is subtracted to arrive at the top of grate elevation (see Figures 13-42 and 13-43).

At low points, particularly at vertical curve low points, it is usual practice for the surveyor to lower the catch-basin grate elevation (below design grade) to ensure that ponding does not occur on either side of the completed CB (please

see Table 13.4). It has been noted that the longitudinal slope at vertical curve low points is virtually flat for a significant distance. The CB grate elevation can be arbitrarily lowered as much as 1 in. (25 mm) to ensure that the gutter drainage goes directly into the CB without ponding. The CB, which can be of concrete poured in place, is now more often prefabricated concrete which that

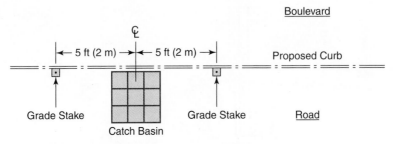

FIGURE 13-42 Catch basin layout (plan view)

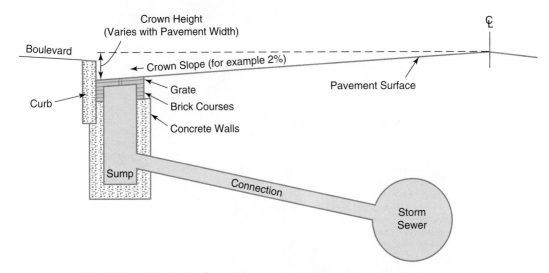

FIGURE 13-43 Typical catch basin (with sump)

is delivered to the job site and set below finished grade until the curbs are constructed. At the time of curb construction, the finished grade for the CB grate is achieved by installing one or more courses of brick on top of the concrete walls.

13.17 PIPELINE CONSTRUCTION

Pipelines are designed to carry water, oil, or natural gas while under pressure. Because pressure systems do not require close attention to gradelines, the layout for pipelines can proceed at a much lower order of precision than is required for gravity pipes. Pipelines are usually designed so that the cover over the crown is adequate for the

loading conditions expected and also adequate to prevent damage due to frost penetration, erosion, and the like.

The pipeline location can be determined from the contract drawings; the grade stakes are offset an optimal distance from the pipe ℄ and are placed at 50- to 100-ft (15-m to 30-m) intervals. When existing ground elevations are not being altered, the standard cuts required can be simply measured down from the ground surface. Required cuts in this case would equal the specified cover over the crown plus the pipe diameter plus the bedding, if applicable (see Figure 13-44).

In the case of proposed general cuts and fills, grades must be given to establish suitable crown elevation so that *final* cover is as specified.

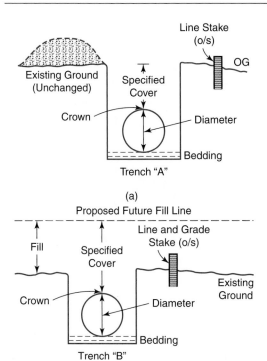

FIGURE 13-44 Pipeline construction. (a) Existing ground to be unchanged. (b) Existing ground to be altered

Additional considerations and higher precisions are required at major crossings (e.g., rivers, highways, utilities). Final surveys show the actual location of the pipe and appurtenances (valves and the like). As-built drawings, produced from final surveys, are especially important in urban areas where underground construction is extensive.

13.18 CULVERT CONSTRUCTION

The plan location and invert grade of culverts are shown on the construction plan and profile. The intersection of the culvert ℄ and the highway ℄ will be shown on the plan and will be identified by its highway station. In addition, when the proposed culvert is not perpendicular to the highway ℄, the skew number will be shown (see Figure 13-45).

The construction plan will show the culvert location (℄ station), skew number, and length of culvert; the construction profile will show the inverts for each end of the culvert. One grade stake will be placed on the ℄ of the culvert, offset a safe distance from each end (see Figure 13-46).

The grade stake will reference the culvert ℄ and will give the cut or fill to the top of footing for open footing culverts, to the top of slab for concrete box culverts, or to the invert of pipe for pipe culverts. If the culvert is long, intermediate grade stakes may be required. The stakes may be offset 6 ft (2 m) or longer distances if site conditions warrant.

In addition to the grade stake at either end of the culvert, it is customary when laying out concrete culverts to place two offset line stakes to define the end of the culvert. These end lines are normally parallel to the ℄ of construction, or perpendicular to the culvert ℄. See Figure 13-47 for types of culverts.

13.19 BUILDING CONSTRUCTION

All buildings must be located with reference to the property limits. Accordingly, the initial stage of the building construction survey involves the careful retracing and verification of the property lines. Once the property lines are established, the building is located according to plan, with all corners marked in the field. At the same time, original cross sections will be taken, perhaps using one of the longer wall lines (or its offset) as a baseline.

The corners already established will be offset an optimum distance, and batter boards will be erected (see Figure 13-48). The crosspieces will either be set to first-floor elevation (finished) or to a set number of feet (decimeters) above or below the first-floor elevation (four-foot marks are commonly used). The contractor is always informed of the precise reference datum. The batter boards for each wall line will be joined by string or wire

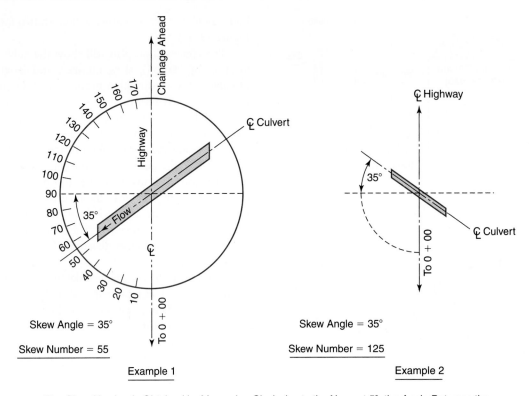

Skew Angle = 35°

Skew Number = 55

Example 1

Skew Angle = 35°

Skew Number = 125

Example 2

The <u>Skew Number</u> Is Obtained by Measuring Clockwise to the Nearest 5°, the Angle Between the Back Tangent ℄ of the Highway and the ℄ of the Culvert.

FIGURE 13-45 Culvert skew numbers, showing the relationship between the skew angle and the skew number

running from nails or saw cuts at the top of each batter board. It is the string (wire) that is at (or referenced to) the finished first-floor elevation (four-foot marks are commonly used). After the excavation for footings and basement has been completed, final cross sections can be taken to determine excavation quantities and costs. In addition to layout for walls, it is usual to stake out all columns or other critical features, given the location and proposed grade.

Once the foundations are complete, it is only necessary to check the contractor's work in a few key dimension areas on each floor. Optical plummets, lasers, and theodolites and total stations can all be used for high-rise construction [see Figures 13-49(a) and (b)].

13.20 OTHER CONSTRUCTION SURVEYS

The techniques for survey layout of heavy construction (i.e., dams, port facilities, and other large-scale projects) are similar to those already described. Key lines of the project will be located according to plan and referenced for the life of the project. All key points will be located precisely, and in many cases these points will be coordinated and tied to a state or province coordinate grid. The two key features of construction control on large-scale projects are (1) high precision and (2) durability of the control monuments and the overall control net.

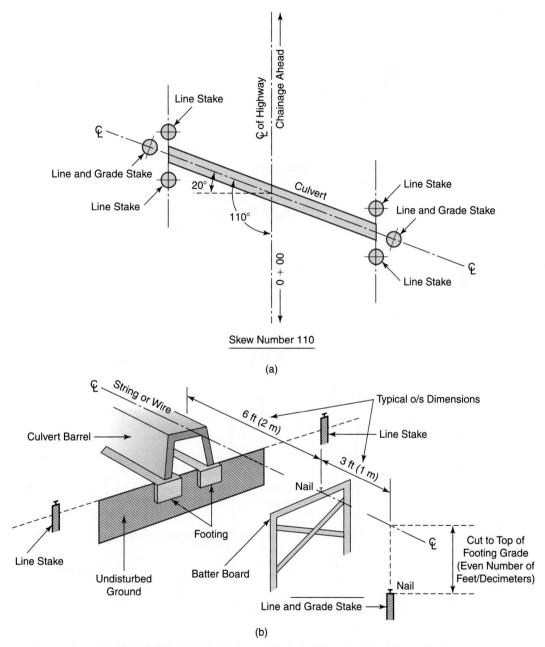

FIGURE 13-46 Line and grade for culvert construction. (a) Plan view. (b) Perspective view
(Courtesy of the Ministry of Transportation, Ontario)

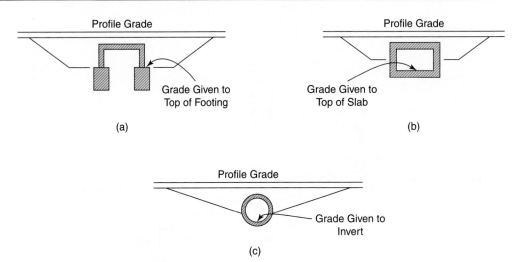

Grade Given to
Top of Footing

(a)

Grade Given to
Top of Slab

(b)

Grade Given to
Invert

(c)

FIGURE 13-47 Types of culverts. (a) Open footing culvert. (b) Concrete box culvert. (c) Circular, arch, etc., culvert

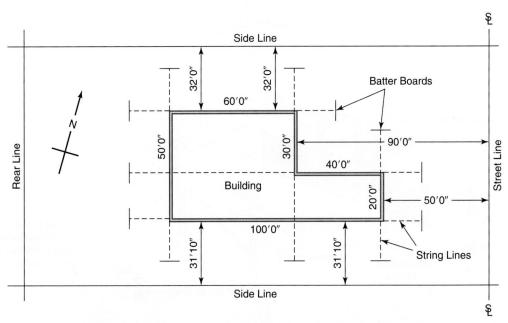

Property Plan (Plat) Showing Location of Proposed Building with Respect to the Property Lines.

This Plan also Shows the Location of the Batter Boards and the String Lines for Each Building Wall Footing.

FIGURE 13-48 Building layout

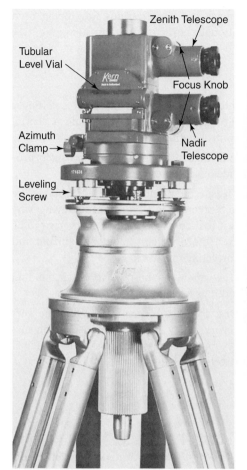

FIGURE 13-49 (a) Kern OL precise optical plummet. SE in 100 m for a single measurement (zenith or nadir) = ± 1 mm (using a coincidence level). Used in high-rise construction, towers, shafts, and the like (Courtesy of Kern Instruments–Leica Geosystems). (b) Rotating laser positioned to check plumb orientation of construction wall

(Courtesy of Laser Alignment, Inc.–Leica Geosystems)

Prior to construction, horizontal and vertical controls are established over the project site. Most of the discussion in this chapter has dealt with rectangular layout techniques, but as noted earlier, some projects lend themselves in whole or part to polar layout techniques using coordinated control and coordinated construction points. Heavy construction is one project area usually well suited for polar layouts and/or GPS layouts.

13.21 CONSTRUCTION SURVEY SPECIFICATIONS

By far the bulk of all construction work is laid out using 1/3,000 (e.g., sewers, highways) or 1/5,000 (e.g., curbed roads, bridges) specifications. The foregoing is true for rectangular layouts, but as noted in Chapter 10, the control for layout by polar techniques will be accomplished using survey techniques allowing for 1/10,000 to 1/20,000 accuracy ratios (depending on the construction materials and techniques). Section

10.9.2 gives the specifications now being used internationally in construction surveying (see Table 10-13).

Questions

13.1 Why are curves used in roadway horizontal and vertical alignments?

13.2 Curves can be established by occupying ℄ or offset stations and then turning off appropriate deflection angles, or curves can be established by occupying a central control station and then turning off angles and measuring out distances as determined through coordinates analyses; what are the advantages and disadvantages of each technique?

13.3 Why is it that sometimes chord and arc lengths for the same curve interval appear to be equal in value?

13.4 Describe techniques that could be used to check the accuracy of the layout of a set of

interchange curves using polar techniques (i.e., angle/distance layout from a central control station).

13.5 Why do construction surveyors usually provide offset stakes?

13.6 Measurement and calculation mistakes are to be avoided in all forms of surveying; why is it that construction surveyors need to be even more vigilant about recognizing and correcting mistakes?

13.7 How can water be used to check building footings for level?

13.8 Describe factors that may affect the surveyors' safety in highway construction surveying, pipeline surveying, and municipal streets surveying?

Problems

13.1 Given PI at 9 + 27.26, $\Delta = 29°42'$, and $R = 700$ ft, compute tangent (T) and length of arc (L).

13.2 Given PI at 15 + 88.10, $\Delta = 7°10'$, and $D = 8°$, compute tangent (T) and length of arc (L).

13.3 From the data in Problem 13.1, compute the stationing of the BC and EC.

13.4 From the data in Problem 13.2, compute the stationing of the BC and EC.

13.5 A straight-line route survey, which had PIs at 3 + 81.27 ($\Delta = 12°30'$) and 5 + 42.30 ($\Delta = 10°56'$), later had 600-ft-radius circular curves inserted at each PI. Compute the BC and EC stationing (chainage) for each curve.

13.6 Given PI at 5 + 862.789, $\Delta = 12°47'$, and $R = 300$ m, compute the deflections for even 20-m stations.

13.7 Given PI at 8 + 272.311, $\Delta = 24°24'20''$, and $R = 500$ m, compute E (external), M (midordinate), and the stations of the BC and EC.

13.8 Given PI at 10 + 71.78, $\Delta = 36°10'30''$ RT, and $R = 1,150$ ft, compute the deflections for even 100-ft stations.

13.9 From the distances and deflections computed in Problem 13.6, compute the three key ℄ chord layout lengths, that is, (1) BC to first 20-m station, (2) chord distance for 20-m (arc) stations, and (3) from the last even 20-m station to the EC.

13.10 From the distances and deflections computed in Problem 13.8, compute the three key ℄ chord layouts as in the previous problem.

13.11 From the distances and deflections computed in Problem 13.8, compute the chords (6) required for layout directly on offsets 50-ft right and 50-ft left of ℄.

13.12 Two highway ℄ tangents must be joined with a circular curve of radius 1,000 ft. See Figure 13-50. The PI is inaccessible, because its location falls in a river. Point A is established near the river on the back tangent and point B is established near the river on the forward tangent. Distance AB is measured to be 615.27 ft. Angle $\alpha = 51°31'20''$, and angle $\beta = 32°02'45''$. Perform the calculations required to locate the BC and the EC in the field.

13.13 Two street curb lines intersect with $\Delta = 71°36'$. See Figure 13-51. A curb radius must be selected so that an existing catch basin (CB) abuts the future curb. The curb-side of the CB ℄ is located from point V: V to CB = 8.713 m and angle E-V-CB = 21°41'. Compute the radius

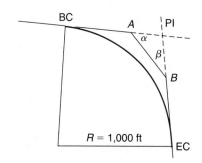

FIGURE 13-50

that will permit the curb to abut the existing CB.

13.14 Given the following vertical curve data: PVI at 7 + 25.712; $L = 100$ m; $g_1 = -3.2$ percent; $g_2 = +1.8$ percent elevation of PVI = 210.440, compute the elevations of the curve low point and even 20-m stations.

13.15 Given the following vertical curve data: PVI at 19 + 00; $L = 500$ ft; $g_1 = +2.5$ percent; $g_2 = -1$ percent; elevation at PVI = 723.86 ft, compute the elevations of the curve summit and even full stations (i.e., 100-ft even stations).

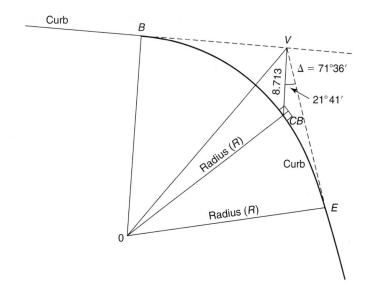

FIGURE 13-51

13.16 A new road is to be constructed beginning at an existing road (C elevation = 472.70 ft) for a distance of 600 ft. The C gradient is to rise at 1.32 percent. The elevations of the offset grade stakes are as follows: 0 + 00 = 472.60; 1 + 00 = 472.36; 2 + 00 = 473.92; 3 + 00 = 475.58; 4 + 00 = 478.33; 5 + 00 = 479.77; 6 + 00 = 480.82. Prepare a grade sheet (similar to Table 13-9) showing the cuts and fills in feet and inches.

13.17 A new road is to be constructed to connect two existing roads. The C intersection with the east road (0 + 00) is at an elevation of 210.500 m, and the C intersection at the west road (1 + 32.562) is at an elevation of 209.603 m. The elevations of the offset grade stakes are as follows: 0 + 00 = 210.831; 0 + 20 = 210.600; 0 + 40 = 211.307; 0 + 60 = 210.114; 0 + 80 = 209.772; 1 + 00 = 209.621; 1 + 20 = 209.308; and 1 + 32.562 = 209.400. Prepare a grade sheet (similar to Table 13-7) showing cuts and fills in meters.

13.18 In Figure 13-31(a), if the ditch invert is to be 3 ft 6 in. deep (below subgrade), how far from the highway C would the C of the ditch be?

13.19 In Figure 13-31(b),

a. If the ditch invert is to be 2.0 m deep (below subgrade), how far from the highway C would the C of the ditch be?

b. If the original ground is level with the C of the highway right across the width of the highway, how far from the C would be the slope stake marking the far side of the ditch at original ground?

13.20 A storm sewer is to be constructed from existing MH 8 (invert elevation = 360.44) at +1.20 percent for a distance of 240 ft to proposed MH 9. The elevations of the offset grade stakes are as follows: 0 + 00 = 368.75; 0 + 50 = 368.81; 1 + 00 = 369.00; 1 + 50 = 369.77; 2 + 00 = 370.22; and 2 + 40 = 371.91. Prepare a grade sheet (similar to Table 13-11) showing stake-to-batter-board distance in feet and inches. Use a 14-ft grade rod.

13.21 A sanitary sewer is to be constructed from existing MH 4 (invert elevation = 150.666) at +0.68 percent for a distance of 115 m to proposed MH 5. The elevations of the offset grade stakes are as follows: 0 + 00 = 152.933; 0 + 20 = 152.991; 0 + 40 = 153.626; 0 + 60 = 153.725; 0 + 80 = 153.888; 1 + 00 = 153.710; and 1 + 15 = 153.600. Prepare a grade sheet (similar to Table 13-11) showing stake-to-batter-board distances in meters. Use a 4-m grade rod.

CHAPTER FOURTEEN

LAND SURVEYS

14.1 GENERAL BACKGROUND

Land surveying involves the establishment of boundaries for public and private properties. It includes both the measurement of existing boundaries and the laying out of new boundaries. Land surveys (performed by land surveyors, also known as professional surveyors) are made for one or more of the following reasons:

1. To subdivide the public lands (United States) or crown lands (Canada) into townships, sections or concessions, and quarter-sections or lots, thus creating the basic fabric to which all land ownership and subsequent surveys will be directly related.

2. To attain the necessary information for writing a legal description and for determining the area of a particular tract of land.

3. To reestablish the boundaries of a parcel of land that has been previously surveyed and legally described.

4. To subdivide a parcel of land into two or more smaller units in agreement with a plan that dictates the size, shape, and dimensions of the smaller units.

5. To establish the position of particular features such as buildings on the parcel with respect to the boundaries.

Land surveys are recommended whenever real property or real estate is transferred from one owner to another. The location of the boundaries should be established to ensure that the parcel of land being transferred is properly located; acceptably close to the size and dimensions indicated by the owner; and free of encroachment by adjacent buildings, roadways, and the like. It is often necessary to determine the location of buildings on the property relative to the boundaries for the purpose of arranging mortgages and ensuring that building bylaws concerning location restrictions have been satisfied.

This section describes the techniques employed and the modes of presentation used for various types of land surveys. Familiarity with the methods employed is essential for anyone involved in any aspect of surveying. For example, construction surveying (see Chapter 13) and layout for new road construction within an allowance for road on the original township survey in Canada or existing road rights of way in the United States necessitate the reestablishment of the legal boundaries of the road allowance or rights of way. Bends and/or jogs in the road allowance alignment, attributable to the methods employed in the original township survey, occur frequently. An understanding of the survey system and boundary retracement used not only explains why these irregularities exist but also warns the surveyor in advance where they may occur.

A recently surveyed subdivision provides ample evidence of property boundary locations. However, as soon as any construction commences, the lot markers may be removed, bent, misplaced, or covered with earth fill. Therefore, reestablishment of the street and lot pattern becomes an essential yet demanding task. It must be carried out before further work may proceed.

14.1.1 Surveyor's Duties

The professional land surveyor must be knowledgeable in both the technical and legal aspects of property boundaries. She or he must have passed state or provincial exams and be properly licensed. Considerable experience is also required, particularly with regard to deciding on the "best evidence" of a boundary location. It is not uncommon for the surveyor to be exposed to conflicting physical as well as legal evidence of a boundary line. Consequently, a form of apprenticeship is often required, in addition to academic qualifications, before the surveyor is licensed to practice by the state, province, or territory.

Regulations are usually required, by law, to assist in standardizing procedures and requirements. Most states and provinces have agencies or associations, which operate under conditions set out in legislative acts. The objectives of a typical association are as follows:

- To regulate the practice of professional land surveying and to govern the profession in accordance with this act, the regulations, and the bylaws.

- To establish and maintain standards of knowledge and skill among its members.

- To establish and maintain standards of professional ethics among its members, in order that the public interest may be served and protected.

14.1.2 Historical Summary

The first North American public land surveys were performed in Ontario in 1783 and north of the Ohio River in 1785. These public land surveys divided the land for settlers in rectangular patterns, with most of the lines following one of the cardinal directions—as determined by compass. Prior to the establishment of public land surveys, land in the northern and eastern United States (e.g., the original 13 colonies) and parts of Canada were obtained by purchase or as a gift from the British Crown. The western United States has many Spanish and Mexican Land Grants that were never in public domain; therefore, they were never a part of the public land surveys. Texas has no public domain land.

Early titles were very vaguely described, but as time went on, description of these holdings was accomplished by metes and bounds. The term *metes* referred to the distances (originally measured in chains or rods) and bearings, referenced to magnetic north. The term *bounds* meant the boundary references which helped define the property. Many of these early parcels followed natural features (e.g., river banks, trails, and roads) rather than the cardinal directions later adopted in the public lands system.

Bounds also included the names of adjacent property owners when that information was available.

A metes and bounds survey, or property description, started at a key point on the property boundary, known as the point of beginning (POB), and then proceeded around each side of the property in turn, giving the distance and bearing, together with boundary references (including survey markers such as blazes, rock cairns, posts, metal bars, and pipes).

Prior to 1910, when the contract system of surveying the public lands was abolished, most of the land surveying had been done by private surveyors under government contract. These early surveys were made with crude instruments such as a compass and a chain, often under adverse field conditions. Consequently, some were incomplete and others showed field notes for lines that were never run in the field. Therefore, the corners and lines are often found in other than their theoretical locations. Regardless, the original corners as established during the original survey, however inaccurate or incomplete, stand as the true corners, except when the government determines the original survey to be fraudulent. An original monument's position is considered the true position of the boundary; therefore, all subsequent surveys must attempt to determine the original monument's position.

In 1946, the Bureau of Land Management in the Department of the Interior was made responsible for public land surveys within the United States, which had been administered by the General Land Office in the Treasury Department since 1812.

In 1789 the surveyor general of Canada was directed to prepare plans of each district using townships 9 miles wide by 12 miles long. Prior to 1835, a variety of nonsectional or "special" township survey systems were authorized. Each presented particular techniques for the establishment of individual lines and generated difficulties for resurveys and access along public road allowances. This confusion was partially ended in 1859, when sectional townships 6 miles square divided into 36 sections, each 1 mile square without road allowances, were authorized through federal legislation.

In 1906, the 1,800-acre sectional system, having townships 9 miles square with 12 concessions and lots of 150 acres each, was authorized by the Ontario government. The methods of surveys in sectional townships and aliquot part divisions were changed. The techniques for establishing the boundaries of aliquot parts were altered by an amendment to the Surveys Act in 1944. The long and somewhat turbulent history of land surveying in both the United States and Canada is evident from the preceding historical summary.

The complexities of survey systems based on French common law, used in states such as Louisiana and provinces like Quebec (or on Mexican law influence in California and Texas), are beyond the scope of this section. Some of these systems are based on the importance of frontage ownership along water bodies, such as rivers, for purposes of water access and transportation. Consequently, the irregularities resulting from following natural boundaries present a totally different resurvey concept from the systems based on English common law.

Because various regions of the United States and Canada have been surveyed under different sets of instructions from 1783 to the present, important changes in detail have occurred. Therefore, the local surveyor should become familiar with the exact methods in use at the time of the original survey, before commencing to retrace land boundaries.

14.2 PUBLIC LAND SURVEYS

14.2.1 General Background

The most common methods for public land surveys in the United States and Canada provided for townships that are 6 miles square, each containing 36 sections. In the United States, each section was thus 1 mile square and was numbered from 1 to 36, as illustrated in Figure 14-1. Section 1 is in the northeast corner of the township. The sections are then numbered consecutively from east to west and from west to east alternately, ending with section 36 in the southeast corner, as illustrated. In Canada, the sections are bounded and numbered as shown in Figure 14-2. Each section was 1 mile square, as in the United States.

6	5	4	3	2	1
7	8	9	10	11	12
18	17	16	15	14	13
19	20	21	22	23	24
30	29	28	27	26	25
31	32	33	34	35	36

FIGURE 14-1 Numbering of sections in the United States

31	32	33	34	35	36
30	29	28	27	26	25
19	20	21	22	23	24
18	17	16	15	14	13
7	8	9	10	11	12
6	5	4	3	2	1

FIGURE 14-2 Numbering of sections in Canada

Initially, only the exterior boundaries of the township were surveyed and mile corners were established on the township lines. However, the plats or original township maps showed subdivisions into sections 1 mile square, numbered as just described. Under these conditions, subsequent surveys had to be carried out to establish the corners and boundaries of sections in the interior of the townships. Changes in the procedures for township surveys led successively to each section corner being marked, then to each quarter-section corner being established. Because the methods of resurvey differ depending on the corners that were originally established, it is essential that the surveyor or engineer understand these techniques.

This section discusses both the American system and the principles of the various sectional systems used in Canada. The Canadian systems were based on the general principles of the American system.

14.2.2 Standard Lines

14.2.2.1 Initial Points   The point at which a survey commences in any area is known as the initial point. A meridian, called the Principal Meridian, and a parallel of latitude, called the baseline, are run through the initial points, as illustrated in Figure 14-3. After the initial point has been established, the latitude and longitude of the point are determined using field astronomical observations.

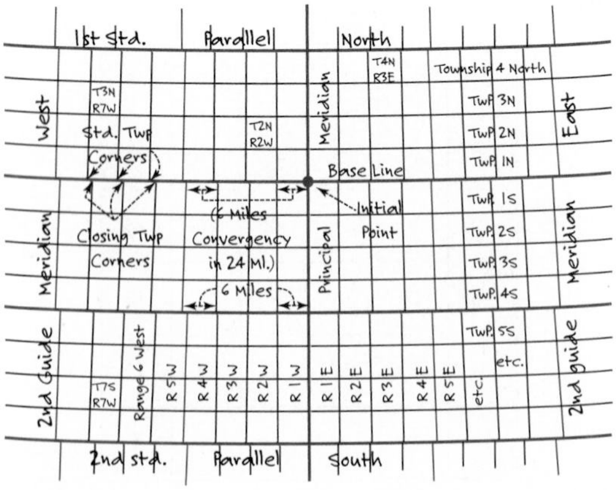

Many of the original township surveys were carried out simultaneously. Therefore, several initial points have been established. The Principal Meridian through an initial point is given a name, and the original surveys governed by each initial point are recorded. For example, the initial point for the Mount Diablo Principal Meridian, governing surveys in the states of California and Nevada, has a longitude of 121°54′47″ W and a latitude of 37°52′54″ N. The original surveys referred to a particular initial point shown on a map entitled "United States, Showing Principal Meridians, Base Lines, and Areas Governed Thereby," published by the Bureau of Land Management (U.S. Government Printing Office, Washington, D.C.).

In the Canadian system, the International Boundary (latitude 49°00′ N) is used as the baseline, where appropriate, such as the western provinces. The Principal Meridian has an approximate longitude of 97°27′09″ W and is located about 12 miles west of the city of Winnipeg, Manitoba. The Second Meridian is located close to longitude 102° W; the Third near 106° W; and so on. Therefore, each initial meridian after the Second is located four degrees west of the preceding one. The Coast Meridian of British Columbia is in a special location due to the configuration of the Pacific Ocean's shoreline.

14.2.2.2 Meridians   The Principal Meridian is a true north/south line that is extended in either direction from the initial point to the limits of the area covered by the township surveys. This line is monumented at 40-chain ($\frac{1}{2}$ mile) intervals in both the United States and Canada.

In Canada, no accuracy tolerances were given in the early instructions and no penalties were levied if the work was incorrect. There was no suggestion that the work should be redone. A common source of error was miscounting the number of chains (66 ft) across the front of a section or lot. Consequently, it is not uncommon to find the older township surveys differing from the intended dimensions by approximate multiples of

FIGURE 14-3 Initial point, parallels, and meridians

66 ft. As the only equipment available to carry out the early surveys was a compass and chain, differences much greater than those stated previously will be found in doing resurveys.

Guide meridians are located at intervals of 24 miles east and west of the Principal Meridian. These lines are true meridians and extend north of the baseline to their intersection with the standard parallel, as illustrated in Figure 14-4. These guide meridians are established in the same manner as the Principal Meridians. Due to the convergence of meridians, the distance between these lines will be 24 miles only at the starting points along the baseline. As illustrated in Figure 14-4, the distance between meridians at all other points is less than 24 miles. As a new monument is established where the guide meridian meets the next standard parallel to the north, two sets of monuments are located on the standard parallels. Those established when

the parallel was first located are called standard corners and apply to the lands north of the parallel. Those established by the intersection with the parallel of the guide meridians from the south are called closing corners.

Similar accuracy requirements apply to the guide meridians as to the Principal Meridians. Monuments are placed at half-mile (40-chain) intervals. All measurement discrepancies are placed in the last half-mile. Consequently, the distance from the first monument south of the standard parallel to the closing monument on the parallel may differ from 40 chains ($\frac{1}{2}$ mile).

14.2.2.3 Convergence of Meridians

As all meridians form a great circle through both the North and South poles, these lines converge toward a pole as they proceed northerly or southerly from the equator. Consequently, a line at

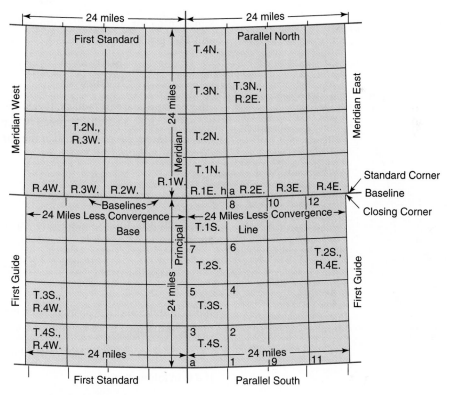

FIGURE 14-4 Standard parallels and guide meridians

right angles (90°) with a meridian will be an east–west line for an infinitely small distance from the meridian. If the line at 90° to the meridian is extended, it will form a great circle around the earth that will not be an east–west line except along the equator. The true east–west line is called a parallel of latitude, which is represented by a small circle, illustrated by line AB in Figure 14-5. The true parallel, due to convergence, will gradually curve to the north (when located north of the equator) of the great circle at right angles to the meridian.

In Figure 14-5, DAP and EBP represent two meridians. P is the earth's North Pole; F is the center of the earth; DE is an arc of the equator; AB is the arc of a parallel of latitude at latitude φ; λ is the difference in longitude between the meridians.

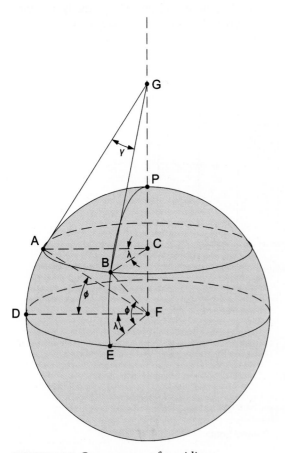

FIGURE 14-5 Convergence of meridians

The angular and linear convergency of the two meridians is to be determined. Convergence is also discussed in Chapter 10.

The difference in longitude between the meridians is

$$\lambda(\text{radians}) = \frac{AB}{BC}, \text{ therefore } AB = BC \times \lambda \text{ (radians)} \quad (14\text{-}1)$$

The latitude of the circle, of which AB is an arc, is

$$\varphi = \text{angle } BFE = \text{angle } BGC \quad (14\text{-}2)$$

Therefore,

$$\sin \varphi = \frac{BC}{BG} \quad \text{and} \quad BG = \frac{BC}{\sin \varphi}$$

The angle of convergency, for all practical purposes, is

$$\gamma(\text{radians}) = \frac{AB}{BG}$$

Substituting for AB and BG from Equations 14-1 and 14-2 gives

$$\gamma = \frac{BC \times \lambda}{BC/\sin \varphi} = \lambda \sin \varphi \quad (14\text{-}3)$$

where γ and λ are in radians. If AB, the distance measured along a parallel between two meridians, is d, and the radius of the earth at the parallel is R, then from Equation 14-1 we have

$$\lambda = \frac{AB}{BC} = \frac{d}{R \cos \varphi}$$

Substitution of this equation into Equation 14-3 gives

$$\gamma(\text{radians}) = \frac{d \sin \varphi}{R \cos \varphi} = \frac{d \tan \varphi}{R} \quad (14\text{-}4)$$

If d is in miles and $R = 3{,}960$ mi (the mean radius of the earth), then from Equation 14-4

$$\gamma(\text{seconds}) = (52.09''/\text{mi})\, d \tan \varphi \quad (14\text{-}5)$$

If d is in kilometers, then

$$\gamma(\text{seconds}) = (32.370''/\text{km})\, d \tan \varphi \quad (14\text{-}6)$$

If the distance between parallels along the meridians, represented by AD and BE in Figure 14-5, is y, and if the difference in arc length along the two parallels, that is, $DE - AB$ in Figure 14-5, is p, then for all practical purposes

$$\gamma = \frac{p}{y}$$

where y is the mean angle of convergency of the two meridians. If the mean latitude is φ, substituting in Equation 14-4 gives

$$p = \frac{d\,y\tan\varphi}{R} \qquad (14\text{-}7)$$

Therefore, the reduction in arc distance along the northerly parallel due to convergence can be calculated using Equation 14-7.

If d and y are in miles and R is the mean radius of the earth, then

$$p\,(\text{feet}) = 1.33\,\text{ft/mi}^2\,dy\,\tan\varphi \qquad (14\text{-}8)$$

$$p\,(\text{chains}) = 0.0202\,\text{chain/mi}^2\,dy\,\tan\varphi \quad (14\text{-}9)$$

$$p\,(\text{meters}) = 0.405564\,\text{m/mi}^2 dy\,\tan\varphi \quad (14\text{-}10)$$

If d and y are in kilometers, then

$$p\,(\text{meters}) = 0.15659\,\text{m/km}^2\,dy\,\tan\varphi \quad (14\text{-}11)$$

Example 14-1: Find the angular convergency between two guide meridians 24 miles (38.63 km) apart at latitude 47°30′.

Solution:

1. Using Equation 14-5:

$$\gamma = 52.09 \times 24 \times \tan 47°30'$$
$$= 1,364'' = 0°22'44''$$

2. Using Equation 14-6:

$$\gamma = 32.37 \times 38.63 \times \tan 47°30'$$
$$= 1.365'' = 0°22'45''$$

Example 14-2: Find the convergency in feet, chains, and meters of two guide meridians 24 miles apart and 24 miles long if the *mean* latitude is 47°30′.

Solution:

1. Using Equation 14-8:

$$p = 1.33 \times 24 \times 24 \times \tan 47°30' = 836\,\text{ft}$$

2. Using Equation 14-9:

$$p = 0.0202 \times 24 \times 24 \times \tan 47°30'$$
$$= 12.70\,\text{chains}$$

3. Using Equation 14-10:

$$p = 0.4655 \times 24 \times 24 \times \tan 47°30' = 255\,\text{m}$$

4. Using Equation 14-11:

$$p = 0.1569 \times 38.63 \times 38.63 \times \tan 47°30'$$
$$= 256\,\text{m}$$

As discussed in the preceding section, two sets of monuments are established on the standard parallels due to convergence of the meridians. Using Example 14-2, the distance between these monuments would be 12.70 chains.

14.2.2.4 Parallel of Latitude

Due to this convergence, the baseline, being a true parallel, must be run as a curve having chords 40 chains (½ mile) long. There are three methods of establishing a parallel of latitude: (1) the solar method, (2) the tangent method, and (3) the secant method, which is the most commonly used.

14.2.2.4.1 Solar Method

The sun is used to determine the true meridian every 40 chains. The true parallel is then established by turning 90° from the meridian. A slight error is incurred using this method because observations are taken at 40-chain intervals rather than at small increments along the parallel. However, the line defined using this method will be well within acceptable accuracies for original surveys.

14.2.2.4.2 Tangent Method

The direction of the tangent is determined by turning a horizontal angle of 90° to the east or west with the meridian. Points on the parallel are established at 40-chain ($\frac{1}{2}$ mile) intervals by offsets to the north from the tangent. The establishment of a baseline for a latitude of 47°30′ is illustrated in Figure 14-6. The illustration is exaggerated to illustrate the offset lengths properly. In fact, the small magnitude of the offset distances compared with the distances along the tangent leads to the

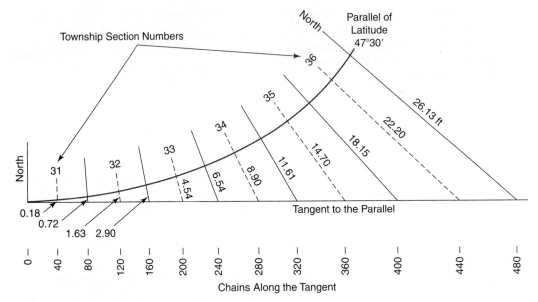

FIGURE 14-6 Baseline using tangent offsets for latitude of 47°30′ (offsets from the tangent o the parallel of latitude are shown in feet)

conclusion that distances measured along the tangent are essentially equal to those measured along the parallel within the accuracy requirements for baseline surveys.

The values of the offsets are dependent on the latitude and the distance from the starting meridian. These values are proportional to the square of the distances from the starting meridian, as illustrated in Table 14-1.

The values of the tangent offsets significantly exceed the width of normal cut lines, as illustrated

in Figure 14-6. Consequently, as the true parallel must be blazed and the landmarks noted, it may be necessary to clear two lines: the tangent and the true parallel. Therefore, this method is more costly and time-consuming than the secant method.

14.2.2.4.3 Secant Method The secant used for laying out a parallel of latitude passes through the 1- and 5-mile points on the parallel, as illustrated in Figure 14-7. The offset is measured southerly from the initial point to the secant

Table 14-1   Offsets from Tangent to Parallel (Feet)
for Latitudes from 30° to 50°

Latitude	1 Mile	2 Miles	3 Miles	4 Miles	5 Miles	6 Miles
30°	0.38	1.54	3.46	6.15	9.61	13.83
35°	0.47	1.86	4.19	7.45	11.65	16.77
40°	0.56	2.23	5.02	8.93	13.95	20.09
45°	0.66	2.66	5.98	10.64	16.62	23.94
50°	0.79	3.17	7.13	12.68	19.81	28.52

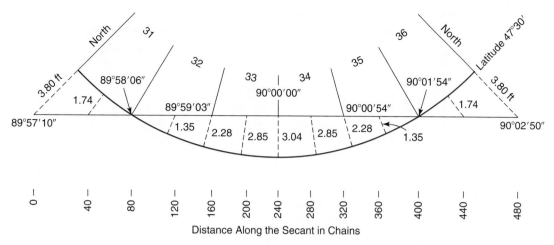

FIGURE 14-7 Baseline using secant offsets for latitude 47°30′(offsets from the secant to the parallel of latitude are shown in feet)

and the angle turned to determine the direction of the secant easterly. The secant is the line actually run, the points on the parallel being located at 40-chain ($\frac{1}{2}$ mile) intervals by offsets from the secant. As illustrated in Figure 14-7, all offsets to the true parallel between $1\frac{1}{2}$ and $4\frac{1}{2}$ miles are southerly and those at the $\frac{1}{2}$-, $5\frac{1}{2}$-, and 6-mile points are to the north.

The secant method is advantageous because the offsets are much smaller than those required for the tangent method, as can be seen by comparing Figures 14-6 and 14-7. Therefore, a cleared line of reasonable width will contain both the secant and the parallel. The measurements made to landmarks along both lines will be essentially the same, thus requiring no modifications in the field notes.

Both the tangent and secant appear as straight lines on the plan view. Due to the convergency of meridians, however, the azimuths of these lines vary along the line. The changes in azimuths along these lines are determined using Equation 14-5, where d is the distance along the parallel from the starting meridian.

The tangent commences at an azimuth of 90° from the meridian and bends gradually southerly, as shown on Figure 14-6. The azimuth of the tangent 6 miles from the starting meridian

is 90°05′41″, the additional 5′41″ being the convergence for a d of 6 miles. Using the tangent method, the 90° azimuth would be reestablished every 6 miles and the process of laying out the parallel repeated as illustrated in Figure 14-6.

The azimuths of the secant are shown in Table 14-2. Assuming that the secant is laid out toward the east, as illustrated in Figure 14-7, the direction of the secant from the starting meridian to the end of the third mile is north of true east. From the 3-mile to the 6-mile points, the azimuth is south of true east. At the 3-mile point, the azimuth of the secant is 90°. At the starting meridian, the secant has an azimuth of 89°57′10″ at a latitude of 47°30′, as illustrated in Figure 14-7 and interpolated from Table 14-2. The difference between the starting azimuth and that at the 3-mile point is 02′50″, the angular convergency for meridians 3 miles apart.

By the same reasoning, the secant azimuth at the 6-mile point is 90°02′50″. The theodolite is set up at the 0-mile offset point, and azimuth of the secant line is laid off using the azimuths given in Table 14-2. For our example at a latitude of 47°30′, the azimuth is 89°57′10″. The secant is then directed in a straight line for 6 miles, and the points are established at ½-mile (40-chain) intervals using the offsets as described previously.

Table 14-2   Azimuths of the Secant

Latitude	0 Mile	1 Mile	2 Miles	Deflection angle 6 Miles
30°	89°58.5′	89°59.0′	89°59.5′	3′00″
35°	89°58.2′	89°58.8′	89°59.4′	3′38″
40°	89°57.8′	89°58.5′	89°59.3′	4′22″
45°	89°57.4′	89°58.3′	89°59.1′	5′12″
50°	89°56.9′	89°57.9′	89°59.0′	6′12″

At the end of 6 miles, the succeeding secant line may be established by either (1) establishing the true meridian and laying off the same azimuth angle as before or (2) turning off a deflection angle, from the preceding secant to the succeeding secant, equal to the convergency of meridians 6 miles apart. These deflection angles are given in the last column of Table 14-2.

14.2.3 Township Boundaries

Most township boundaries in the United States and Canada were established using the sectional systems, wherein the boundary lines were oriented north–south and east–west. Because this was the most common method, it is emphasized in this section.

14.2.3.1 Sectional Systems   The normal method of establishing township boundaries is best understood by reference to Figure 14-4. The procedure is set out as follows:

1. Commencing at point 1, located on the first standard parallel south (a baseline) 6 miles to the east of the Principal Meridian (or guide meridian), a line is run due north for 6 miles to point 2. Monuments are established at intervals of 40 chains ($\frac{1}{2}$ mile).

2. From point 2, a line is run westerly to intersect the Principal Meridian (or guide meridian) at or near point 3. Because this is a trial line, only temporary monuments are set at 40-chain ($\frac{1}{2}$-mile) intervals. Since point 3 has previously been set during the running of the

principal (or guide) meridian, the amount by which the trial line fails to intersect point 3 is measured.

3. The line from point 3 to point 2 is run along the correct course, and the temporary monuments are replaced by permanent monuments in the correct pattern. The true line is properly blazed, and the topographic and vegetative features are recorded along the true line. Due primarily to the convergence of meridians, the length of the line between points 2 and 3 is less than 6 miles. This difference resulting from convergence and measurement errors is placed in the most westerly 40 chains ($\frac{1}{2}$ mile). All other distances are therefore 40 chains.

4. The eastern boundary of the next township to the north is run northerly from point 2 to point 4.

5. The procedure in step 2 is repeated to establish the north boundary of the next township between points 4 and 5. Any discrepancies are again left in the most westerly 40 chains.

6. The preceding procedures are continued, establishing easterly and northerly township boundaries in that order, that is, points 6 and 7.

7. From point 6, the meridian is run to its intersection with the baseline (or standard parallel) at point 8. A closing corner is established at point 8. The distance from the closing corner to the nearest standard corner is measured and recorded. Since all errors in measurement up the meridian from point 1 to

point 8 are placed in the line from point 6 to point 8, the last interval on the meridian may be less or more than 40 chains.

8. The two other meridians, points 9–10 and 11–12 in Figure 14-4, in the 24-mile-square block are run in a manner similar to that described for points 1–8. Therefore, any east-west discrepancies are placed in the westerly 40 chains of each township as before, and the northerly 40 chains of each meridian will absorb the measurement errors.

An understanding of this procedure is important, because all 36 sections in a township are not of equal size. Therefore, the legal requirements relating to the rectangular surveys of the public lands should be examined. Due to the procedures employed in carrying out the township boundary surveys, combined with the effect of convergence of meridians, certain sections within each township vary in area. This is illustrated in Figure 14-8.

Of the 36 sections in each standard township, 25 are considered to contain 640 acres, because these sections are not affected by the convergence of meridians or the placing of measurement errors. The sections containing less than 640 acres are located along the westerly township boundary and along the northerly boundary, primarily due to the measurement errors concentrated in the northerly 40 chains ($\frac{1}{2}$ mile) along the meridian.

14.2.3.2 Numbering and Naming of Townships

In a sectional system, the townships of a survey district are numbered into ranges and tiers (townships) in relation to the Principal Meridian and the baseline used for the district. As illustrated in Figure 14-4, the fourth township south of the baseline is in tier four south (T.4S.). The fourth township west of the Principal Meridian is located in range 4 west. Using this method of numbering, any township is located if its range, tier, and Principal Meridian are stated. The example shown in Figure 14-4 and discussed previously illustrates this system: tier (township) 4 south, range 4 west of the sixth Principal Meridian. The abbreviation for this township is T.4S., R.4W., 6th P.M.

14.2.4 Reestablishing Boundaries

The aim of both the American and Canadian governments was to monument the corners using the procedures discussed in Section 14.2.3. However, for a multitude of reasons (some of the more common are listed later), many of these corner marks have been obliterated. Consequently, the modern surveyor is required to relocate a missing corner. The basic requirements for this process are a thorough understanding of the methods used in establishing the original boundaries of the particular township in question, combined with good judgment regarding the best evidence of the corner location—including copies of the original field notes.

A lost corner is a survey point whose position cannot be determined because no reasonable evidence of its location exists. The only means of reestablishment involves a mathematical solution using clearly defined surveying procedures and the closest-related existing corners.

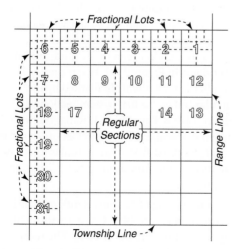

FIGURE 14-8 Relative sizes of sections in sectional townships

The main reasons for lost corners are as follows:

- Obliteration of the marked lines and wooden corner posts by forest fires and logging, often covering large areas.

- Improper marking of the lines and/or monuments during the original township or subdivision surveys.

- Inaccurate placing of lines and monuments.

- Lack of concern for preservation of the original corners and lines, particularly by the early settlers.

The surveyor must first determine if the survey corner is really lost. If the original monuments were wooden, the decaying of the stake will often leave a brownish rust color in the soil. One can look for this telltale color by carefully slicing the earth horizontally using a thin shovel blade. Blazes on trees used as witness points for survey corners, also called **bearing trees**, as well as blazes marking the survey lines, can become totally covered by subsequent tree growth and enlargement of tree diameter.

Much effort is put in to the determination of the status of a survey corner. If all avenues of best evidence are not explored in their entirety, the surveyor will likely be put in the embarrassing and expensive position of defending his or her lack of thoroughness in a court of law. Experience in reestablishing original survey points and boundaries is essential. A young and/ or inexperienced surveyor is well advised, under these circumstances, to carry out such surveys under the direction of an experienced surveyor. This is often arranged by hiring the experienced surveyor in a consultative capacity for the particular survey, searching for evidence under his or her direction, accepting his or her analysis of the results, and requiring his or her explanation leading to these decisions. To understand the reestablishment process, the surveyor must be familiar with the original survey procedures described in this chapter, as the reestablishment of original survey points is based on the intent of the original survey.

14.2.5 Canadian Sectional Systems

A total of six sectional systems were used to subdivide most of Canada. The townships were commonly 6 miles square, containing 36 sections of 640 acres each. Other variations based on section areas, such as 1,000 acres and 800 acres, were used prior to May 1, 1871.

The first, second, and third Canadian sectional systems are illustrated in Figure 14-9. The effects of convergency were offset by running in supplementary baselines used as correction lines. These lines were located every two townships north and south of the main baseline, as shown in Figure 14-9. Guide meridians were run north and south every four townships west of the Principal Meridian, the Second Meridian shown in Figure 14-9. Thus, convergence was "corrected" in this way, resulting in a high irregularity of township configurations and sizes, as illustrated immediately east of the Third Meridian in Figure 14-9.

The main points of difference between the U.S. and Canadian sectional systems are as follows:

- The numbering of townships is different, as illustrated under the "first system" section on the right side of Figure 14-9. The townships number northerly from the International Baseline, for example, from 1 to 24 in Figure 14-9. The ranges number westerly from the nearest Principal Meridian, for example, from I to XXX between the Second and Third Meridians in Figure 14-9. For example, the number for the township marked by the circular dot in the first system section of Figure 14-9 is Township 11, Range 30, west of the First Meridian. The abbreviation would be Tp.11, R.30, 1st P.M. This is similar to but not identical to the American system, discussed in Section 14.2.3.

- The numbering of sections is different, as illustrated in Figure 14-2. Compare this with Figure 14-1 for the U.S. system. The numbering of quarter-quarter sections is also different.

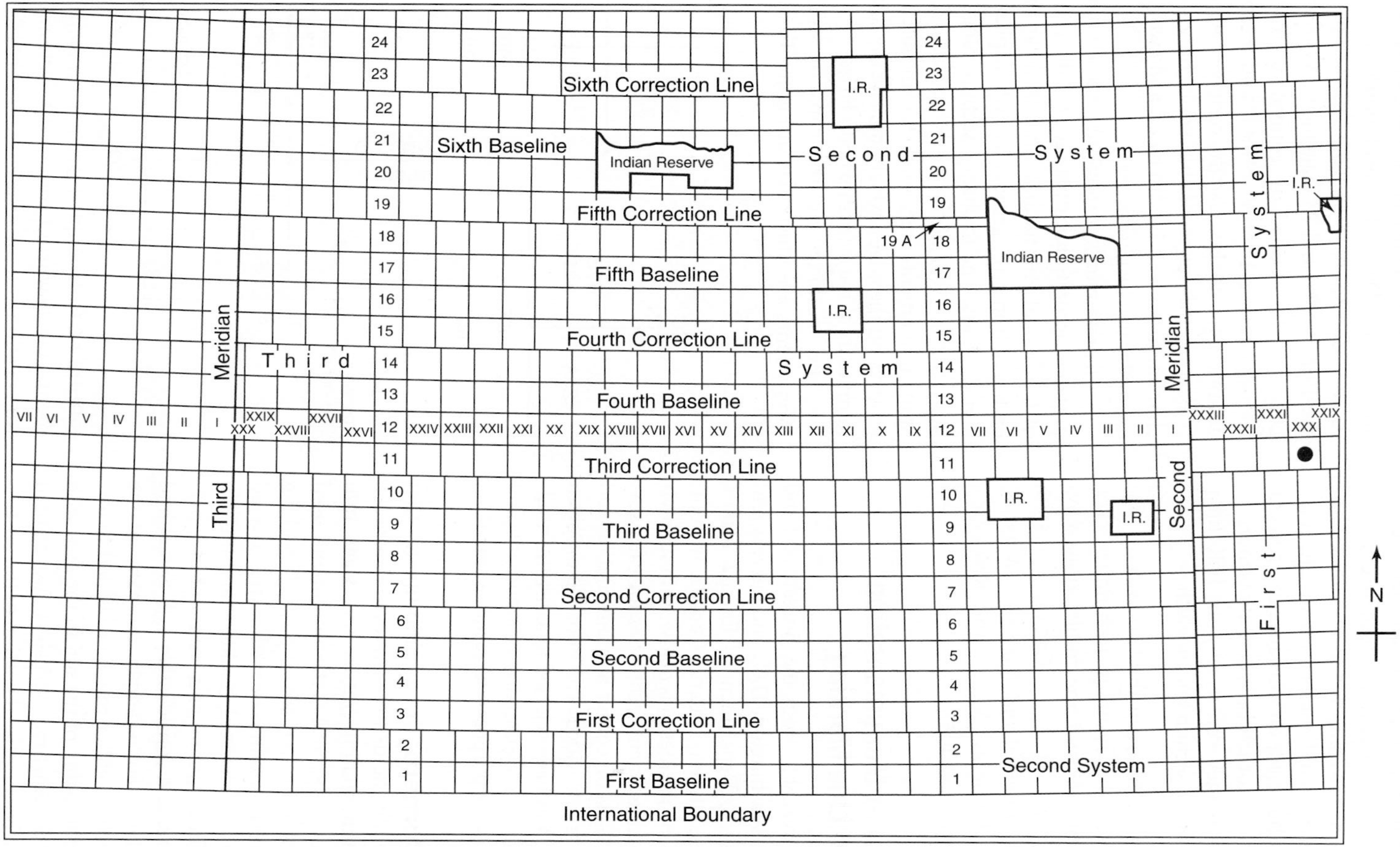

FIGURE 14-9 Subdivision of country into blocks and townships, illustrating the Canadian first system, second system, and third system

(From the *Dominion Land Surveyors Manual*, Department of Energy, Mines, and Resources, Ottawa)

- The Canadian systems provide for road allowances from 1 to 1½ chains in width on either all or some of the township and section lines. The U.S. system often did not set aside specific road allowances, but made an allowance of 5 percent of the total area for roads. Where allowances were specified, they were usually one-half chain either side of the section line in width.

These differences between the two systems significantly affect the resurvey techniques. The surveyor should consult the publications covering the aspects of every Canadian system in detail (*Dominion Land Surveyors Manual*, Department of Energy, Mines, and Resources, Ottawa, Ontario, Canada).

14.2.6 Corner Monumentation and Line Marking

Manuals are available from the Bureau of Land Management, Washington, D.C., and Surveys and Mapping Branch, Department of Energy, Mines, and Resources, Ottawa, Ontario, Canada. In addition to specifying the methods of township subdivision and reestablishment techniques for each survey system, details on corner monumentation and line marking are provided. The information given in this section has been selected to assist the modern surveyor in locating the corner monuments and the survey lines for purposes of reestablishment.

14.2.6.1 Corners It is unfortunate that most of the public lands were surveyed before the present monumentation regulations went into effect. Consequently, most of the monuments used consisted of wood (often cedar) posts. If nothing else was available, mounds of earth, earth-covered charred stakes, or charcoal, boulders, and the like were used. These monuments were not of a permanent nature, and considerable skill is required to relocate them. For example, as many of the wooden posts have either rotted or burned in forest fires, careful slicing of the soil in the area with a sharp shovel sometimes reveals a rust-colored stain from the underground portion of the post.

Corner markings for sectional survey systems are complex. Detailed information can be found in the *Manual of Surveying Instructions* from the Bureau of Land Management, Washington, D.C., or the Canadian equivalent from the Department of Energy, Mines, and Resources, Ottawa, Ontario. Because the variety of markings is beyond the scope of this book, only general characteristics will be discussed and two typical examples will be given.

All monuments are marked using a system that provides accurate information regarding the type of monument and its location. Metal caps placed on top of iron posts are marked with capital letters and numbers. The township, section, range numbers, and the year of the survey are all inscribed. Stone monuments were marked with notches and grooves on the edges, which indicate the distances in miles from the township boundaries.

The markings on the caps of iron post monuments are illustrated in Figure 14-10. The

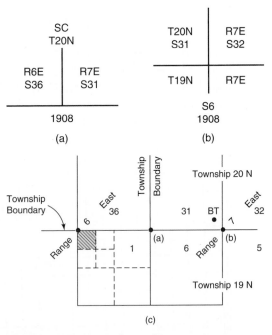

FIGURE 14-10 Example markings on iron monuments. (a) Township corner. (b) Section corner. (c) Location of points from points (a) and (b)

sketch in Figure 14-10(c) illustrates the position of the points in Figure 14-10(a) and (b). Point (a) is a township corner marked by the symbol SC (standard corner). The markings are set to be viewed from the south, so the township to the north is shown: T20N. The ranges on each side of the township line are shown, in this case, range 6 east (R6E) and range 7 east (R7E). The section numbers to the northwest (S36) and to the northeast (S31) are shown below the respective ranges, thus fixing the point. The date of survey (1908) is shown as the bottom number, which assists the modern surveyor in obtaining the original field notes from the appropriate state government authority. Point (b) is a section corner. The township number (T20N) is shown on the upper left of the symbol, and the section to the northwest is indicated below the township number (S31). The range number, in this case, range 7 east (R7E), is shown on the right side, with the northeast section corner (S32) indicated below. The township (T19N) and the range (R7E) to the south of the corner marker are shown as indicated in Figure 14-10(b).

In addition to setting the monuments on the actual section or lot corners in the original survey, additional monuments known as *witness points* were within 5 chains (330 ft) of the monuments. These consist of bearing trees, designated as BT, and *bearing objects*, designated as BO, used where there were no trees.

Trees used as witnesses are blazed (cut to the smooth wood surface), and the smooth wood surface on the blaze is inscribed with letters and figures to aid in the identification of the locations. A tree used as a witness for a section corner, such as point (b) in Figure 14-10, would be inscribed T20N R7E S31 BT, which indicates that the bearing tree (BT) is located as shown in Figure 14-10. The true bearing and horizontal distance are determined, including a description of the tree, and recorded in the field notes. Normally, at least one and usually two bearing trees are established for each corner set in the original survey. Due to changing regulations and/or lack of diligence by the original surveyor, the only markings on a bearing tree are often BT. However,

this does not create a practical problem, because the surveyor doing the retracement usually has a good idea of where the original monument was located. Therefore, the surveyor can usually deduce the location of the found bearing tree referred to in the original field notes. This becomes a problem only when two bearing trees close together have been marked to witness the same monument.

Bearing objects, within 5 chains (330 ft) of the monument, are used where no substantial trees exist. These consist of one or more of the following: (1) significant cliffs or large boulders, (2) stone mounds, and (3) pits dug into the ground. Where the bearing object consists of rock, the point used for the measurements to the monument is marked using a chisel by a cross (x), the letters BO, and the section number, the latter only for sectional survey systems. Stone mounds are used where loose stone is readily available. Where no rocks, stones, or trees are available, the corners may be witnessed by digging pits 18 in. (45 cm) square and 12 in. (30 cm) deep. One side (not corner) of the pit should face the monument. Depending upon the country (United States or Canada) and the regulations in force at the time, up to four pits are sometimes required (placed on all four sides of the monument), and the distance of the pits from the monument varies, although it is normally 3 ft. Because the pit will usually fill in gradually with a different soil, and often revegetates with a different species, the location of the pit(s) may be identified many years or decades later.

14.2.6.2 Lines The marking of the survey lines on the ground was done to preserve the lines between monuments. In addition to setting the monuments and witness points as described in the previous section, natural terrain features along the line were recorded in the field notes, as discussed in Section 14.2.7, and blazes or hack marks were made on living timber at regular intervals along the line. The regulations pertaining to public land surveys in the United States and Crown Land surveys in Canada have always required blazing and hack marks along lines through timber. Trees directly on the line, called line trees, are marked with two horizontal V-shaped notches, called hack marks, on each side of the tree facing along the line in both directions. Trees along the line within one-half chain (50 links, 33 ft) are blazed at chest height, the flat side of the blaze being placed parallel to and facing the line. The frequency of blazes along the lines varies widely, depending on the density of the trees and the thoroughness of the original surveyors.

14.2.7 Field Notes and Survey Records

Field notes for all run survey lines are often of great value in reestablishing lines where no physical ground evidence exists. In many cases, no other evidence exists. The present-day surveyor should bear in mind the conditions under which the original field notes were made. The chain may have been missing one or even two links. The number of chains to a particular feature may have been miscounted, thus creating potential discrepancies of approximately 66 ft or 132 ft. The use of aerial photographs and comparison with the field notes, prior to attending in the field, are usually of great assistance in identifying the discrepancies mentioned (see Chapter 12). This is particularly true for streams, rivers, swamps, and geological features such as ridges or cliffs. Office and subsequent field adjustments of the line on the aerial photographs, bearing the possible discrepancies in mind, will save the modern surveyor considerable time in line reestablishment.

The information to be included in field notes, according to the manuals, is similar in both the United States and Canada. A summary list follows.

1. Course and length of each line run, including offsets, the reason, and the method used.

2. The type, diameter, and bearing and distance from the monument of all bearing trees and bearing objects (see Section 14.2.6), including the material used for the monuments and depth set into the ground.

3. Line tree description: species name, diameter, distance along the line, and markings.

4. Line intersections with either natural or manufactured features, such as the line distance to Native American or Canadian reservations, mining claims, railroads, ditches, canals, electric transmission lines, and changes of soil types, slopes, vegetation, and geological features such as ridges, fractures, outcrops, and cliffs. All pertinent information, such as the bearings of intersecting boundaries, including the margin of heavy timber lines, is recorded.

5. Line intersections with water bodies such as un-meandered rivers, creeks, and intermittent watercourses such as ravines, gullies, and the like. The distance along the line is measured to the center of the smaller watercourses and to each bank for larger rivers.

6. Lakes and ponds on line, describing the type and slopes of banks, water quality (clear or stagnant), and the approximate water depth.

7. Towns and villages; post offices; Native American or Canadian occupancy; houses or cabins; fields; mineral claims; mill sites; and all other official monuments other than survey monuments.

8. Stone quarries and rock ledges, showing the type of stone.

9. All ore bodies, including coal seams, mineral deposits, mining surface improvements, and salt licks.

10. Natural and archeological features, such as fossils, petrifactions, cliff dwellings, mounds, and fortifications.

Two types of field notes were kept. One type listed the chainages (stationing) to pertinent points, down the left side of the page, with each pertinent feature listed shown on the right side opposite the appropriate chainage. The other type was termed *split line*, in which the centerline was widened to accommodate the chainage figures, and the features to the left or right of the line as run were shown on the side of the page where they were located. A typical example of the latter is shown in Figure 14-11.

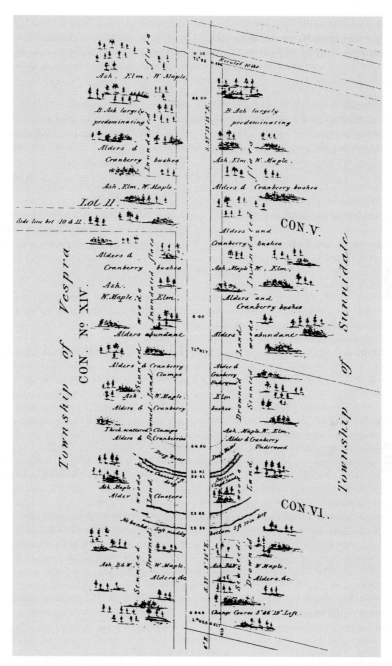

FIGURE 14-11 Split-line field notes for township subdivision

(Courtesy of Ontario Ministry of Natural Resources)

Copies of the field notes and the township plat or original survey maps are kept on record at the Bureau of Land Management, Washington, D.C.; at the Department of Energy, Mines, and Resources, Ottawa, Ontario; and at the majority of the state and provincial government offices. The departments storing these records in the state or provincial governments have a variety of names (Auditor of State; Public Survey Office; Register of State Lands, Natural Resources, Lands, and Forests, etc.). Consequently, the present-day surveyor is advised to determine the appropriate government agency in the state or province where the records are kept.

14.2.7.1 Resurveys and Real Property Boundaries   The filing of township field notes and plats varies considerably depending on the regulations of the federal, state, and provincial jurisdictions in which the property is located. The normal situation is as follows. The original field notes and plans must, by law, be made available to other surveyors or members of the public for a reasonable fee. The maximum fee is usually stipulated by the federal, state, or provincial land surveyors' association. The fee is determined by the time required to locate the proper file and plans in the surveyors' office, combined with the costs of reproduction of the notes and plans.

The same system is generally used for field notes and plans relating to real property boundary surveys. Plans for subdivision are kept on record in the county registry office. These are numbered consecutively as received by the registrar. The degree to which these plans are examined for reliability and accuracy varies greatly between jurisdictions. Some county registry offices accept plans of subdivision without any investigation, while others go to great lengths to verify or check key measurements through field investigation.

14.3 PROPERTY CONVEYANCE

When any piece of land changes ownership, it cannot be by word of mouth, but must be by a written document, usually called a deed, although the original transfer of land from the U.S. government is called a patent. The deed includes a legal description of the property. Knowledge of any rights of way, easements, highway widenings, and the like are also essential. Such title encumbrances are important as the owner's rights on these lands are usually severely curtailed. For example, erection of any structure (on the encumbrance portion) such as even a small building may not allowed, even though the land is still owned by the owner of the lands adjacent to the encumbrance.

Most of the principles of establishing boundaries are well defined. These are based on case law, in the United States, coming from English common law. In addition there are various statute law addressing boundary principles. The topics in this section were selected because they are common to most property boundary surveys.

14.3.1 Definitions of Legal Terms

Some of the most common legal terms relating to conveyance of land are defined next. The definitions are presented as a practical interpretation of the formal definitions presented in legal dictionaries. The real meanings of the terms as they affect the modern surveyor are more important than those couched in formal legal language.

1. Adverse possession. When land is used by a person other than the owner for an extended period of time, the land may be claimed from the owner by the user under certain circumstances. This term is discussed in more detail in Section 14.3.4.

2. Alluvium. When land along the bank of a river or shore of a lake or sea is increased by any form of wave or current action or by dropping of the water level, the additional land is called alluvium. This usually is a gradual process and the rate of addition cannot be accurately or easily determined in relation to small increments of time. As the owner of the lands along the river bank or shoreline may gain or lose land by these processes, the

additional lands become the property of the owner of the shoreline.

3. Avulsion. This involves the sudden removal of the land of one owner to the land of another caused by water forces. When this happens, the transferred property belongs to the original owner. The main difference between avulsion and alluvium is the rate at which the transfer of soil and land occurs. Avulsion can be attributed to a sudden movement caused by an identifiable event, such as a violent storm or flood.

4. Deed description. The deed is the legal document transferring land(s) from one owner to another. The most important documentation in the deed from the surveyor's aspect is the description of the property being transferred. The description is intended to describe the details of the property boundaries, including bearings, distances, and appropriate corner markers. Because the interpretation of the description is of great importance, this topic is covered in detail in Section 14.3.2.

5. Fee simple. The word *fee* indicates that the land and structures thereon belong to their owner and may be transferred to those heirs that the owner chooses or, if necessary, that the law appoints. The word *simple* means that the owner may transfer the property to whomever he or she chooses. Therefore, to hold the title of land in "fee simple" is the most straightforward and simplest form. It places no restrictions on the owner regarding the sale or disposition of the land.

6. High-water mark. The high-water mark occurs where the vegetation changes from aquatic species to terrestrial species. If no vegetation is present, such as a bare rock shoreline, the high-water mark is usually clearly indicated by a distinctive change in the color or tone of the rock along a level line representing the appropriate water level. This line may appear either above or below the existing water level at the time of the survey, depending on local water-level conditions.

7. Metes and bounds. A method of describing property by listing boundary distances and directions, together with a note of adjacent property owners and relevant natural features.

8. Mortgage. The purchaser usually obtains a mortgage, involving financing by a bank or loan corporation, for the remainder of the purchase price of the property and structures thereon, if applicable. Most mortgage-lending institutions require survey documentation to ensure that the buildings are not only entirely on the property but are also within the acceptable clearances from the property lines as established by the local municipality. These surveys are discussed in Section 14.5.

9. Patent. When property is first conveyed by federal, state, or provincial governments to institutions, companies, or individuals, it is called a "patent." This is usually the first entry made in the registry office books kept for sections and lots or concessions.

14.3.2 Deed Descriptions

There are a number of ways land being conveyed can be described in a deed. In general, in the United States, there are three distinct conveyance types. Each type requires different procedures in retracement. One is a public land type conveyance, with the land being described based on the public land survey system as previously discussed in this chapter. A typical description might be the Northwest quarter of the Southwest quarter of Section 14, Township 24 South, Range 15 East, of Mount Diablo Meridian. This would be conveying a nominal 40-acre piece of land. Another type of conveyance is a simultaneous conveyance type. Simultaneous conveyance indicates that land being conveyed was a part of a larger parcel, where all the lots were created at one time, such as a subdivision or tract. A typical description might be Lot 5 of Block B of the Wooded Hills Tract, in the County of Madison, State of Ohio (see Figure 14-12). Generally there will be a filed

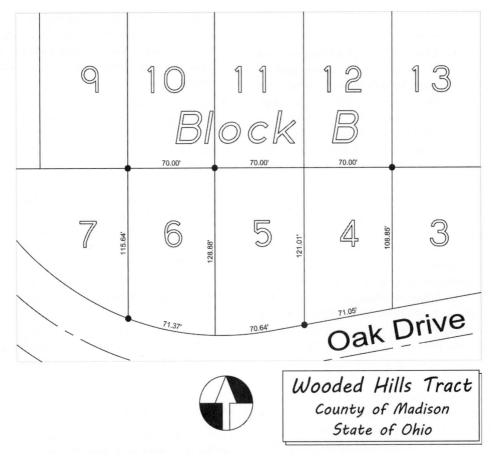

FIGURE 14-12 A typical simultaneous conveyance would be a tract map where all the lots within the tract are created at the time the map is filed

map somewhere showing all the dimensions of the subdivision, which can be used for determining the location of Lot 24. Section 14.4 discusses the general procedures for retracing a simultaneous conveyance.

The final type is a sequential conveyance. Sequential conveyances are where parcels were created at different times. Metes and bounds descriptions, as previously mentioned in this chapter, are one type of sequential conveyance. With a metes and bounds description, there are controlling calls (the bounds portion) and informative calls (often the metes portion). With sequential conveyances,

it is difficult to write the description in such a ways as to correctly convey only the intend land to be transferred. Although it takes much experience to correctly write one, almost anyone used to be allowed to write one, including the land owners, lawyers, or land agents. This has created many poorly written, incorrectly conveyed titles of land over the years. For these reasons, when new parcels are being created these days, it is usually done by simultaneous means; however, there many existing descriptions based on sequential conveyances. Section 14.5 discusses the general procedures for retracing a sequential conveyance.

It is not uncommon for a single description to contain more than one conveyance type, such as a metes and bounds within a quarter-section of land.

14.3.3 Riparian and Littoral Rights

Riparian rights refer to those rights of a property owner of land that borders rivers and streams while littoral rights pertain to rights of owners of land along lakes or tidelands. The rights associated with riparian and littoral ownership are complicated and vary from state to state. These rights may include the use of the water, the ability to access or control the water, and ownership of the land under the water and may cause the limits of the ownership to change over time. Retracing boundaries that involve riparian or littoral rights require an in-depth knowledge of those rights for the area, and often require expertise in determining water marks.

Certain survey systems incorporated a publicly owned strip of land, usually 1 chain (66 ft) wide, parallel with the shoreline or high-water mark. As the owner normally wants to locate buildings as close to the water as possible, the net result is that all or a portion of a private residence is located on public lands.

14.3.4 Adverse Possession

The use of land by other than the registered owner for a sufficiently long and uninterrupted time may lead to a land transfer to the user from the registered owner. The legalities of adverse possession vary considerably among states and provinces. The general principles are discussed herein to warn the surveyor of the responsibilities and the investigative techniques to be employed.

Some common legal aspects are given next. They do not hold in all jurisdictions, and it is the surveyor's responsibility to investigate their validity for the property under investigation.

1. Public property, such as unpatented lands, streets, and highways, cannot be acquired by adverse possession.

2. Usually the land use, whether it involves agriculture, erection of buildings, or fencing, is known or should be known by the owner of the land. In many cases the landowner is not aware that adverse possession exists and, if these other land uses on the land do not interfere with the owner, he or she commonly chooses to ignore them.

3. These land uses under adverse possession must be against the interests of the owner under most jurisdictions. This has to be interpreted carefully. At the beginning of the adverse land uses, the owner could not have been hostile or he or she would have put a stop to the adverse uses of the land. Therefore, the owner becomes hostile when he or she decides to stop the adverse land uses and finds out that, by law, he or she cannot.

4. The adverse possession may consist of public use of a private right of way on private lands. In other words, private land can become public if extended usage is permitted by the owner. Often only an easement is adversely possessed, which is called a prescription or prescriptive easement.

5. If the landowner informs those carrying out the adverse land uses that he or she is the owner of these lands before the adverse land use has continued for the number of years required by law, the time period for adverse possession starts over again. For example, if the legal period for continuous use is 20 years, the adverse land uses started in 1975, and the owner informs those conducting the adverse land uses that he or she is the owner of the lands in 1983, no action can be undertaken for adverse possession until 2003, assuming that the owner does not mention his or her ownership to the user again. The best form of notification is by registered mail, because this means that a record of having received the letter exists. Verbal notification should be carried out in front of witnesses and the date noted.

It is a court, not a land surveyor, that will determine if the conditions of adverse possession have been met. However, land surveyors should note any potential adverse uses on land that they are surveying and advise their clients of such.

14.4 SURVEYS OF SIMULTANEOUS CONVEYANCES

A simultaneous conveyance indicates that a number of lots or parcels were created at one time. This is not always, but most typically be, a subdivision or tract. These lots or parcels are conveyed usually just by referencing the map in the legal description. The legal description might read something like "Lot 5 of Block B of the Wooded Hills Tract, in the County of Madison, State of Ohio."

The original survey would establish the exterior boundary of the tract. The process would be dependent on how the original property was created; if it was a portion of a section, it would be established using public land procedures; it might have been created using metes and bounds or could even be a lot from a previous tract. Once the exterior boundary has been established, the interior lots would be created based on the design of the subdivision or tract. Then, typically, all the corners of the exterior of the tract and interior lots would be monumented. Often the centerlines of streets are also monumented. Generally a map showing all the locations of the monuments as well as physical description of the monuments will be filed with some local agency.

In most areas there are specific regulations and zoning issues that must be met prior to a new subdivision being created. Those regulations may deal with planning, design, infrastructure, and financing issues. However, once all the conditions are met, the monuments are set, and the subdivision is approved by whatever agencies must approve it, the positions of the interior monuments are considered absolute. The map that is filed is intended to represent the field survey, not the other way around.

In retracing a simultaneous conveyance, that is, doing a resurvey of a lot or parcel that has already been created, the land surveyor must determine where that original monument was set. The first step is to go out in the field and look for that monument based on the information provided in the deed and map. This may require measuring from found monuments, searching with a magnetic locator, and carefully digging to find the original monument. Sometimes only traces of the monument are found, which requires the land surveyor to determine if the traces are an indication of the original monument.

If the original monument or evidence thereof is not found, then the land surveyor must determine the position of the original monument through other means. There are a number of factors the land surveyor must consider, but if the monument's location is considered undetermined, then it must be reestablished from other original monuments.

Example 16.3: The northeast corner of Lot 5 in Block B of the Wooded Hills Tract as shown in Figure 14-12 must be reestablished. The record dimensions between the northwest corner of Lot 5 and the northeast corner of Lot 4 is 140.00 ft (70.00 ft along each back line). In the field the original monuments are found at those two corners and the distance between them is measured to be 139.70 ft.

The northeast corner of Lot 5 position would be established along the line from those two found monuments, at a proportional distance between them. The distance from the northwest corner of Lot 5 could be computed as such:

$$\text{Dist}_{\text{NWcor5} \to \text{NEcor5}} = \frac{139.70\,\text{ft}}{140.00\,\text{ft}} \times 70.00\,\text{ft}$$

$$= 69.85\,\text{ft}$$

To establish the southwest corner of Lot 5, use the same concept of proportional distances, except in this case the proportional distance would be along the arc of the curve. Therefore, first the curve would need to be recreated using the found monuments and then the arc distance between those two points would be computed. The computed arc distance would be used for the proportional value.

The concept of distribution of error in a proportional manner comes from the idea that all the lots were created at the same time, so the error should be distributed in an equitable manner to all those affected. In the United States, there is generally one exception to the idea of proportional distribution of error, which is that the public should not share in the error. This has the effect of holding original right-of-way widths when proportioning across a street.

14.5 SURVEYS OF SEQUENTIAL CONVEYANCES

A sequential conveyance indicates there is a time separation between the creation of one parcel and its adjoining parcels. Sequential conveyances are typically the metes and bounds descriptions, but are not always in that format. In writing a sequential conveyance much care must be taken to convey exactly what was intended. Although most lots these days are created using simultaneous conveyance methods (subdivision maps), there are a number of parcels that are still being created sequentially, while many more were originally created sequentially, years ago. Retracement of sequential conveyances can be the most challenging and yet most interesting types of resurveys. It is especially in these types of resurveys where the experience and local knowledge of the surveyor are of utmost importance.

14.5.1 Metes and Bounds Conveyances

As discussed before, a metes and bounds conveyance indicates that there is a course given, as well as some bounding control. Within the description, there will usually be a preamble, giving a general location of the land, such as "lying within the northeast quarter of Section 12, Township 13 South, Range 30 East of Humboldt Meridian," and then the body of the description. In the body of the description, there are

two types of information about the boundary: informative, which gives you some general information, often in the form of a course, such as "south 12°48' east 1320.00 feet"; and controlling, which controls either the end point of a line or the line itself. "To the Northwest corner of Lot 34" would control the end of the line, and "along the northerly line of Section 12" would control the line. Following is a portion of a legal description as shown in Figure 14-13, which is discussed subsequently.

> *Commencing at the Northwest corner of said Section 12; thence along the northerly line of said Section 12, East 2640 feet to the northwest corner of the northeast $\frac{1}{4}$ of the northwest $\frac{1}{4}$ of said section 12, being the True Point of Beginning; thence south 12°48' east 1320.00 feet to the northwest corner of Lot 34, in Tract 345 as shown on the map filed in book 4 of Maps at page 34, in the County Recorder's Office of said County, thence along the northerly line of said Lot 34, north 78°13' east 210.45 feet to the northeast corner of said Lot 34, being marked by a 1" iron Pipe tagged "LS 4819"; thence north 55°13' east 415.57 feet; thence north 82°15' east 400.50 feet; to a 3" × 3" post; thence north 5°00' east 956.78 feet to said northerly line of Section 12; thence along said northerly line, West 1320.00 feet to the True Point of Beginning.*

Looking at the above description, it is important to understand the elements and the effect each element has in locating the land being described in the field.

1. The point of commencement "*Commencing at the Northwest corner of said Section 12*" is used to aid in finding the location of the land being described, but is indicating that it is not a part of the land.

2. The True Point of Beginning "*northwest corner of the northeast $\frac{1}{4}$ of the northwest $\frac{1}{4}$ of said section 12,*" indicates that this is where the description starts to convey the land being described. It is also a controlling call, in that the northwest corner must be found or established in order to determine this point.

3. The course following the true point of beginning is informative, "*thence south 12°48' east*

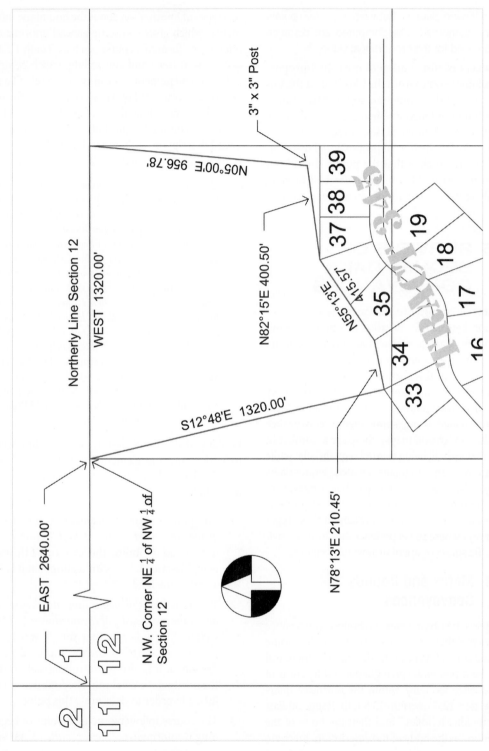

FIGURE 14-13 A typical sequential conveyance would be a metes and bounds description, calling courses as well as bounding control

1320.00 feet," as it is followed by the controlling call, "*the northwest corner of Lot 34, in Tract 345,*" meaning the northwest corner of Lot 34 must be found or established, and the line goes directly from the true point of beginning to this corner of Lot 34. It does not matter if the bearing and distance do not match, as they are informative only. If there is significant variation with the measured course, then the surveyor will need to do some additional review to determine the cause. The information following Tract 345 shows where the map on the tract can be found.

4. "*along the northerly line of said Lot 34*" is controlling the line from the northwest corner of Lot 34, and the easterly terminus is defined by "*the northeast corner of said Lot 34.*" The mention of the iron pipe is an indication of what monument should be found at the northeast corner, but if that monument is found, it must still be verified as the location of the northeast corner.

5. The next course "*thence north 55° 13′ east 415.57 feet*" is controlling because there is no other controlling call. Notice on the plat, it shows that line appearing to run along the north line of some other lots. It may or may not actually run along those lots, as the location would solely be based on the direction and distance given in the deed. The basis for this direction would be the previous line which gave a bearing and controlling line (northerly line of Lot 34).

6. The next course would be controlled by the 3″ × 3″ post, assuming it could be found or reestablished. If the post cannot be found or reestablished, then the bearing and distance would end up controlling the position of this line.

7. The next course "*thence north 5° 00′ east 956.78 feet to said northerly line of Section 12*" in most jurisdictions has a controlling direction of north 5° 00′ east that terminates at the northerly line of Section 12. The distance 956.78 ft is informative only. There are some

jurisdictions where the distance would hold over the direction. The northerly terminus of this line is the northerly line of Section 12.

8. The last course runs along the northerly line to the true point of beginning, which creates a perfectly closed legal description. It is important to understand that just because a description does not mathematically close, it may still legally close. Often the first reaction of a surveyor is to check the mathematical closure on a legal description, which can be important in retracing the description, but does not invalidate the description if it does not mathematically close.

14.5.2 "Of" Conveyances

There are sequential conveyances that are not metes and bounds, one of which is known as "of" descriptions. Samples of these would be "the west $\frac{1}{2}$ 'of' Lot 23" or "the North 100 ft 'of' Lot 23." There are some special considerations when retracing an "of" description.

Proportional "of" descriptions, where a fractional portion of an existing parcel is being conveyed, are considered to be fractional based on area. This is different than public land retracements, where there are specific instructions on how to determine a $\frac{1}{2}$ or $\frac{1}{4}$ of a section. In proportional "of" descriptions the total area of the parcel must be computed and then the fractional portion determined. In addition the direction of the dividing line must be determined. Fixed distance "of" descriptions are generally retraced by measuring the distance perpendicular to the controlling line.

Because there are so many variations on the establishment of the lines in "of" descriptions, it is generally not recommended for new descriptions to use "of" format unless there is wording controlling the establishments of lines.

14.5.3 Senior Rights

With sequential conveyance, as noted before, there is a time sequence in the creation of the parcels of land. This creates potential senior right conflicts, between adjoining parcels. Therefore, once the surveyor has determined the position of the land being described in the description, they must look for senior right conflicts.

A senior right is one where the adjoining parcel was originally created prior to the land being surveyed. It is a senior parcel if it was created out of the same underlying ownership parcel. The surveyor must determine if the parcel being surveyed and the adjoiners were all created in a sequential manner, and if so, the order of the creation of the parcels. If there are senior parcels, those lines abutting the surveyed parcel must also be established.

There is a conflict if the senior parcel lines cross into the description being surveyed, and if there is conflict the senior parcel has ownership of the conflicted area. This is because that land was conveyed first, so the following conveyances had no right to the land.

14.6 TITLE OR MORTGAGE SURVEYS

These surveys involve the detailed positioning of the existing buildings on a parcel of land in relation to the boundaries. The surveys are performed for two primary reasons:

- The finance company or bank granting a mortgage on a property, and usually the buildings thereon, requires proof that the buildings are within the property boundaries.

- Most municipalities have bylaws stating minimum distances to the street line and the sidelines of the lots. The title or mortgage surveys show whether these bylaws have been satisfied or not. Surveys for this purpose are usually carried out as soon as the basement of the house has been constructed up to the first floor level. Hence, if the building location does not satisfy the bylaws and if negotiations for exemptions from these regulations

with the municipality are unsuccessful, the building can be moved before construction is completed.

Figure 14-14 illustrates a typical title or mortgage survey. The steps in conducting such a survey are the following:

1. The lot boundaries as discussed in Section 14.4 are reestablished.

2. The distances from the outside basement walls to the street line and both lot lines are measured. This is usually done by reading the measurement on the tape through the theodolite telescope. The distances required are those at right angles to the lot line, in other words, the shortest distance between the lot line and the building corner.

3. The measurements to the street line, called *setbacks*, and those to the lot lines, called sideyards, are also measured for the buildings on both sides of the lot or parcel being surveyed.

4. Occupation lines such as fences are positioned and shown on the survey plan. If the fence is not on the property line, such as the southerly fence in Figure 14-14, its position is noted.

5. All lot distance measurements are shown. It is common to show both the "plan" and "measured" distances for each boundary, as illustrated in Figure 14-14.

It is necessary to measure the outside dimensions of the buildings in order to plot them to scale. The corner markers may or may not be replaced. Often they are not, as this increases the cost of the survey.

The surveyor should note that title or mortgage survey measurements are very useful for reestablishing property boundaries, as the distances from the buildings are measured to the true property lines. Often, the residents have a copy of the survey among their valuable papers in the house. It is well worth the surveyor's time to ask local residents on nearby properties for a loan of these surveys. The surveyor's name and date of the title or mortgage survey should be recorded and shown in the resurvey field notes.

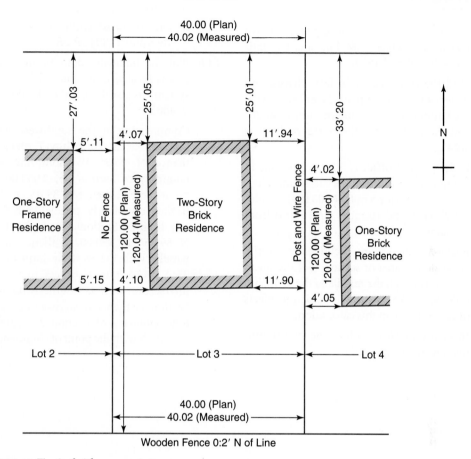

FIGURE 14-14 Typical title or mortgage survey

14.7 CADASTRAL SURVEYING

Cadastral surveying is a general term applied to a number of different types of surveys. It is mentioned here only to make the reader aware of the expression and the broad aspects of its use. A rigid definition of a cadastral survey involves only the information required to define the legal boundaries of a parcel of land, whether rural, urban, or city. Therefore, the monumentation, bearings, distances, and areas would be shown. This definition has now been expanded through common usage to include cultural features, such as building location, drainage features, and topographic information, such as spot elevations or contours.

14.8 LAND SURVEYING WEBSITES

GeoCommunicator, NILS data, http://www .geocommunicator.gov/GeoComm/

LandNet, USA (commercial online mapping/ property data), www.landvoyage.com

U.S. Bureau of Land Management, geographic coordinate database, http://www.blm.gov/wo/ st/en/prog/more/gcdb.html

Problems

14.1 Calculate the angular convergency for two meridians 6 miles (9.66 km) apart at latitudes 36°30′ and 46°30′.

14.2 Find the convergency in feet, chains, and meters of two meridians 24 miles apart if the latitude for the southern limit is 40°20′ and the latitude for the northern limit is 40°41′.

14.3 If the tangent offset is 3.28 ft (1.00 m) at 3 miles (4.83 km) from the meridian, calculate the offsets for 5 miles (8.05 km), 7 miles (11.27 km), and 10 miles (16.10 km) from the same meridian.

14.4 For a 6-mile-square sectional township, calculate the area of section 6, if the mean latitude through the middle of the township is 46°30′. Take into account only the effects of convergence in this calculation.

14.5 In Figure 14-10 show how the iron monument would be marked for

a. The northwest corner of township T.20N., R.7E.

b. The southeast corner of section 32 of township T.20N., R.6E.

14.6 Plot the following description (the property is located in a standard U.S. sectional system) to a scale of either 1:5,000 or 1 in. = 400 ft.

Commencing at the southwest corner of Section 35 in Township T.10N., R.3W., thence N 0°05′ W along the westerly boundary of Section 35, 2,053.00 ft to a point therein; thence N 89°45′ E, 1,050.00 ft; thence southerly, parallel with the westerly limit of Section 35, 670.32 ft; thence N 89°45′ E, 950.00 ft; thence southerly, parallel with the westerly limit of Section 35, 1,381.68 ft, more or less, to the point of intersection with the southerly boundary of Section 35; thence westerly along the southerly boundary of Section 35, 2,000.00 ft, more or less, to the point of commencement.

PART **FOUR**

APPENDIXES

RANDOM ERRORS

A.1 GENERAL BACKGROUND

When a very large number of measurements is taken to establish the value of a specific dimension, the results will be grouped around the true value, much like the case of the range target illustrated in Figure A-1. When all systematic errors and mistakes have been removed from the measurements, the residuals between the true value (dead center of the bull's-eye) and the actual measurements (shot marks) will be due to random errors.

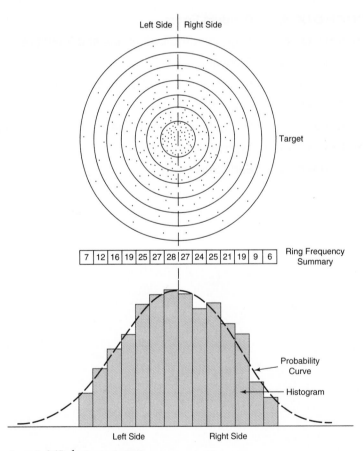

FIGURE A-1 Range target: 265 shots on target

The rifle target shown in Figure A-1 illustrates some of the characteristics of random errors:

- Small random errors (residuals) occur more often than large random errors.
- Random errors have an equal chance of being plus (right) or minus (left).

The number of rifle shots hitting the left and right side of each ring are shown in the ring frequency summary. These results are then plotted directly below in the form of a bar chart called a histogram. You can see that the probability of any target shot hitting a ring (or half-ring) is directly proportional to the area of the histogram rectangles. For example, the probability of one of the target shots hitting the bull's-eye (for a specific rifle and specified conditions) is $(28 + 27)/265 = 0.21$ (or 21 percent).

Table A-1 shows that the total of the ring probabilities is, of course, unity. Geometrically, it can be said that the area under the probability curve is equal to 1, and the probability that an error (residual) falls within certain limits is equal to the area under the curve between those limits. The probabilities for the hits in each ring are shown calculated in Table A-1. For example, given

the same conditions (i.e., the same skill of the shooter, the same precision rifle, and same range conditions) as when the shots in Figure A-1 were fired, one would expect that 40 percent of all future rifle shots would hit the middle two rings $(0.208 + 0.192 = 0.400)$.

Any discussion of probability and probable behavior implies that a very large (infinite) number of observations have been taken. The larger the number of observations, the closer the results will conform to the laws of probability. In the example used, if the number of rifle shots were greatly increased and the widths of the rings greatly narrowed, the resultant plot would take the form of a smooth, symmetrical curve known as the probability curve. This curve is shown superimposed on the histogram in Figure A-1.

In surveying, you cannot take a large number of repetitive measurements. But if you use skilled survey techniques that normally give results that, when plotted, take the form of a probability curve, it is safe to assume that the errors associated with the survey measurements can be treated using random error distribution techniques.

Before leaving the example of the rifle target shot distribution, it is worthwhile to consider the effect of different precisions on the probability curve. If a less precise rifle (technique) is used, the number of shots hitting the target center will be relatively small, and the resultant probability curve [see Figure A-2(a)] will be relatively flat. On the other hand, if a more precise rifle (technique) is used, a larger number of shots will hit the target center and the resultant probability curve will be much higher [see Figure A-2(b)], indicating that all the rifle shots are grouped more closely around the target center.

If the sights of the high-precision rifle were out of adjustment, the target hits would consistently be left or right of the target center. The shape of the resulting probability curve would be similar to that shown in Figure A-2(b), except that the entire curve would be shifted left or right of the target center plot point. This situation illustrates that precise methods can give inaccurate results if the equipment is not adjusted properly.

Table A-1   Target Ring Probabilities

Ring Number	Probability	
1 (bull's-eye)	$\dfrac{28 + 27}{265}$	$= 0.208$
2	$\dfrac{27 + 24}{265}$	$= 0.192$
3	$\dfrac{25 + 25}{265}$	$= 0.189$
4	$\dfrac{19 + 21}{265}$	$= 0.151$
5	$\dfrac{16 + 19}{265}$	$= 0.132$
6	$\dfrac{12 + 9}{265}$	$= 0.079$
7	$\dfrac{7 + 6}{265}$	$= 0.049$
	Total	$\overline{1.000}$

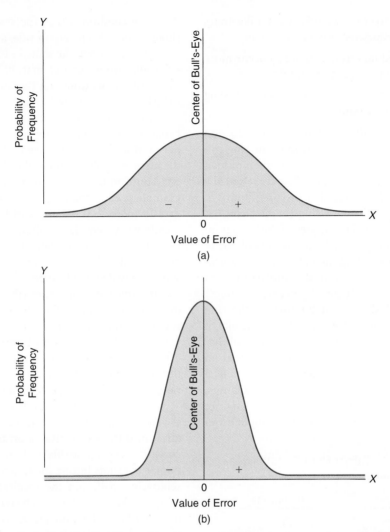

FIGURE A-2 Precision comparisons. (a) Probability curve for target results from a low-precision rifle. (b) Probability curve for target results from a high-precision rifle

A.2 PROBABILITY CURVE

The probability curve shown in Figures A-1 and A-2 has the following equation:

$$y = \frac{1}{\sigma\sqrt{2\pi}}\,e^{-v^2/(2\sigma^2)} \qquad (A\text{-}1)$$

where y is the ordinate value of a point on the curve (probability density of a residual of size v occurring); v is the size of the residual; e is the base of natural logarithms (2.718); and σ is a constant known as the standard deviation or standard error, a measure of precision. Since σ is associated with an infinitely large sample size, the term *standard error* (*SE*) will be used when analyzing finite survey repetitions.

A.3 MOST PROBABLE VALUE

In the preceding section, the concept of a residual (v) was introduced. A residual is the difference between the true value or location (e.g., a bull's-eye center) and the value or location of one occurrence or measurement. When the theory of probability is applied to survey measurements, the residual is in fact the error, that is, the difference between any field measurement and the true value of that dimension. In Section 1.14, the topic of errors was first introduced. We noted that the true value of a measurement is never known, but that for the purpose of calculating errors, the arithmetic mean is taken to be the true, or most probable, value. Since we do not have large (infinite) numbers of repetitive measurements in surveying, the arithmetic mean will itself contain an error (discussed later as the error of the mean):

$$\text{Mean: } \quad \bar{x} = \frac{\Sigma x}{n} \qquad \text{(A-2)}$$

where Σx is the sum of the individual (x) measurements, and n is the number of individual measurements.

A.4 STANDARD ERROR

We saw in Figure A-2 that precision can be depicted graphically by the shape of the probability curve. In statistical theory, precision is measured by the standard deviation. Theoretically,

$$\sigma = \sqrt{\Sigma v^2/n} \qquad \text{(A-3)}$$

where σ is the standard deviation of a very large sample, v is the true residual, and n is the very large sample size. Practically,

$$\text{SE} = \sqrt{\Sigma v^2/(n-1)} \qquad \text{(A-4)}$$

where SE is the standard error of a set of repetitive measurements; v is the error ($x - \bar{x}$), and n is the number of measurements. Since the use of the mean $\bar{x}$ instead of the true value always results in an underestimation of the standard deviation, $n - 1$ is used

in place of n. The term $n - 1$ is known in statistics as degrees of freedom and represents the number of extra measurements taken. That is, if a line were measured ten times, it would have $10 - 1 = 9$ degrees of freedom. Obviously, as the number of repetitions increases, the difference between n and $n - 1$ becomes less significant. The concepts just described are being used increasingly to define and specify the precision of various field techniques.

As we noted earlier, the arithmetic mean contains some uncertainty; this uncertainty can be expressed as the standard error of the mean (SE_m). The laws of probability show that the standard error of a sum of identical measurements be given by the standard error of each measurement multiplied by the square root of the number of measurements: $\text{SE}_{sum} = \text{SE} \sqrt{n}$. The mean is given by the sum divided by the number of occurrences; therefore,

$$\text{SE}_m = \frac{\text{SE}\sqrt{n}}{n} = \frac{\text{SE}}{\sqrt{n}} \qquad \text{(A-5)}$$

This expression shows that the standard error of the mean is inversely proportional to the square root of the number of measurements; that is, if the measurement is repeated by a factor of 4, the standard error of the mean is halved. This relationship demonstrates that, beyond a realistic number, continued repetitions of a measurement do little to reduce uncertainty.

Many instrument manufacturers now specify the precision of their equipment by stating the standard error associated with the equipment use. The terms *standard error*, *standard deviation*, and *mean square error* (*MSE*) are all used to specify the identical concept of precision.

A.5 MEASURES OF PRECISION

Figure A-3 shows the $\pm$ SE, ± 2 SE, and ± 3 SE points plotted under the normal probability curve. It can be shown that the area under the curve between the limits shown is as follows:

$\bar{x} \pm 1$ SE contains 68.27 percent of area under the curve

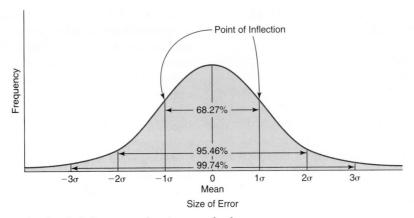

FIGURE A-3 Graph of probability curve showing standard errors

$\bar{x} \pm 2$ SE contains 95.46 percent of area under the curve

$\bar{x} \pm 2$ SE contains 99.74 percent of area under the curve

We saw in Section A.1 that the area under the probability curve equals the probability of the values covered. The preceding relationship can be restated by noting that the probability of measurements deviating from the mean is as follows:

68.27%	[of the measurements will be in range of]	$\bar{x} \pm$ SE
95.46%		$\bar{x} \pm 2$ SE
99.74%		$\bar{x} \pm 3$ SE

A term used in the past, *probable error*, was the 50 percent error (i.e., the limits under the curve representing 50 percent of the total area). Those limits are $\bar{x} \pm 0.6745$ SE. Today's surveyors are more interested in the concept of maximum anticipated error, which varies from 90 percent to the 95 percent probability limits.

A.6 ILLUSTRATIVE PROBLEM

To illustrate the concepts introduced thus far, consider the data in Table A-2. The results of fifteen measurements of a survey baseline are shown

together with the probability computations. It is assumed that all systematic errors have already been removed from the data.

Probability Computations

1. Mean (most probable value), $\bar{x} = \Sigma x/n = 3{,}994.612/15 = 266.3075$ m (Equation A-2)

2. Standard error, $SE = \sqrt{\Sigma v^2/(n-1)} = \sqrt{0.0003565/14} = 0.0050$ m (Equation A-4)

3. Standard error of the mean, $SE_m = SE/\sqrt{n} = 0.0050/\sqrt{15} = 0.0013$ m (Equation A-5)

4. Probable error (50 percent error) $= 0.6745$ SE $= 0.0034$ m (Table A-3)

5. 90 percent error $= 1.6449$ SE $= 0.0082$ m (Equation A-7)

6. 95 percent error $= 1.9599$ SE $= 0.0098$ m (Equation A-8)

Table A-2   Practical Precision Parameters

Name	Value	Probability of Lesser Error (%)
Probable	0.6745 SE	50
Standard	SE	68.27
90%	1.6449 SE	90
95%	1.9599 SE	95

Table A-3   Analysis of Random Distance Errors ($\nu = x - \bar{x}$)

n	Distance x (m)	Residual ν	ν^2
1	266.304	−0.0035	0.0000123
2	266.318	+0.0105	0.0001103
3	266.312	+0.0045	0.0000203
4	266.304	−0.0035	0.0000123
5	266.313	+0.0055	0.0000303
6	266.307	−0.0005	0.0000003
7	266.309	+0.0015	0.0000023
8	266.303	−0.0045	0.0000203
9	266.301	−0.0065	0.0000423
10	266.305	−0.0025	0.0000063
11	266.302	−0.0055	0.0000303
12	266.310	+0.0025	0.0000063
13	266.314	+0.0065	0.0000423
14	266.307	−0.0005	0.0000003
15	266.303	−0.0045	0.0000203
	$\Sigma x = 3{,}994.612$	$\Sigma \nu = -0.0005$	$\Sigma \nu^2 = 0.0003565$

The following observations are taken from the data in Table A-2:

1. The most probable distance is 266.3075 m.

2. The standard error of any one measurement is 0.0050 m.

3. The standard error of the mean is 0.0013 m; that is, there is a 68.27 percent probability that the true length of the line lies within 266.3075 $\pm$ 0.0013 m. There is a 90 percent probability that the true length of the line lies between 266.3075 $\pm$ 0.0082 m (0.0013 $\times$ 1.6449 = 0.0082). There is a 95 percent probability that the true length of the line lies between 266.3075 $\pm$ 0.0098 m (0.0013 $\times$ 1.9599 = 0.0098).

4. With the probable error (50 percent) of 0.0034, it is expected that half of the 15 measurements will lie within 266.3075 $\pm$ 0.0034 (from 266.3041 to 266.3109). In fact, only 5 (33 percent) measurements fall in that range.

5. With the 90 percent error of 0.0082, it is expected that 90 percent of the measurements will lie between 266.3075 $\pm$ 0.0082 (from 266.293 to 266.3157). In fact, 14 out of 15 (93 percent) fall in that range.

6. With the 95 percent error of 0.098, it is expected that 95 percent of the measurements will lie between 266.3075 $\pm$ 0.0098 (from 266.2977 to 266.3173). In fact, 14 out of 15 (93 percent) measurements fall in that range.

Note that expected frequencies are entirely valid only for large samples of normally distributed data. When the number of observations is small, as is the case in surveying measurements, you will often encounter actual data that are marginally inconsistent with predicted frequencies. If the field data differ significantly from probable expectations, however, it is safe to assume that the data are unreliable, due either to untreated systematic error or to undetected mistakes. As a rule of thumb, measurements that fall outside the range of $\bar{x} \pm 3.5$ SE either are rejected from the set of measurements or are repeated in the field.

A.7 PROPAGATION OF ERRORS

This section deals with the arithmetic manipulation of values containing errors (e.g., sums, differences, products).

A.7.1 Sums of Varied Measurements

The sum or difference measurements that have individual mean and SE values are determined as follows. If distance K is the sum of two distances, A and B, then

$$SE_K = \sqrt{SE_A^2 + SE_B^2} \qquad \text{(A-6)}$$

Example A-1: If distance A were found to be 101.318 $\pm$ 0.010 m and distance B were found to be 87.200 $\pm$ 0.008 m, what is the range for distance $K = A + B$?

Solution: Use Equation A-6:

$$SE_K = \sqrt{0.010^2 + 0.008^2} = 0.013 \text{ m}$$
$$K = 188.518 \pm 0.013 \text{ m}$$

Example A-2: If from the preceding data, distance L is the *difference* in the two distances A and B, then find the value of L.

Solution: Use Equation A-6:

$$SE_L = \sqrt{SE_A^2 + SE_B^2}$$
$$= \sqrt{0.010^2 + 0.008^2} = 0.013$$
$$L = (101.318 - 87.200) \pm 0.013$$
$$= 14.118 \pm 0.013 \text{ m}$$

Example A-3: If the difference in elevation between two points is determined by taking two rod readings, each having an SE of 0.005 m, what is the SE of the difference in elevation?

Solution: Use Equation A-6 or Equation A-7 in the following:

SE (diff. of elev.) $= \sqrt{0.005^2 + 0.005^2} = 0.007$ m

or

SE (diff. of elev.) $= 0.005\sqrt{2} = 0.007$ m

A.7.2 Sums of Identical Measurements

The sum of any number of measurements, each one having the same SE, is as follows:

$$SE_{sum} = \sqrt{n \times SE^2}$$
$$SE_{sum} = SE\sqrt{n} \qquad \text{(A-7)}$$

Example A-4: A distance of 700.00 ft is laid out using a 100.00-ft steel tape that has a SE $= 0.02$ ft. Find the standard error of the 700.00-ft distance.

Solution: Use Equation A-7:

$$0.02\sqrt{7} = 0.05 \text{ ft}$$

A.7.3 Products of Measurements

The product of any number of measurements that have individual SE values can be given by the following relationship:

$$SE_{AB} = \sqrt{A^2 SE_B^2 + B^2 SE_A^2} \qquad \text{(A-8)}$$

where A and B are the dimensions to be multiplied.

Example A-5: Consider a rectangular field having $A = 250.00$ ft $\pm$ 0.04 ft and $B = 100.00$ ft $\pm$ 0.02 ft. Find the area of the field and the SE of the area.

Solution: Use Equation A-11:

$$SE_{product} = \sqrt{250^2 \times 0.02^2 + 100^2 \times 0.04^2}$$
$$= 6.40 \text{ ft}$$

Area of field $= 250 \times 100 = 25,000 \pm 6$ ft^2

A.8 WEIGHTED OBSERVATIONS

If the reliability of different sets of measurements varies one to the other, then equal consideration cannot be given to those sets. Some method (weighting) must be used to arrive at a best value. For example, measurements may be made under varying conditions, by people with varying levels of skills, and they may be repeated a varying number of times.

A.8.1 Weight by Number of Repetitions

The simplest concept of weighted values can be illustrated by the following method, where the weighted mean is calculated. A distance was measured six times; the values obtained were 6.012 m, 6.011 m, 6.012 m, 6.012 m, 6.011 m, and 6.013 m. The value of 6.012 was observed three times; 6.011, two times; and 6.013, one time.

Distance, x	Weight, w	x × w
6.012	3	18.036
6.011	2	12.022
6.013	1	6.013
	$\Sigma w = 6$	$\Sigma xw = 36.071$

$$\text{Weighted mean} = \frac{36.6071}{6} = 6.012 \text{ m}$$

That is:

$$\bar{x}_w = \frac{\Sigma xw}{\Sigma w} \qquad \text{(A-9)}$$

where $\bar{x}_w$ is the weighted mean, x is the individual measurement, and w is the weight factor.

If the distance had been measured six times and six different results had occurred, each measurement would have received a weight of 1, and the computation would simply be the same as for the arithmetic mean.

A.8.2 Weight by Standard Error of the Mean (SE$_m$)

The standard error of the mean SE$_m$ was introduced in Section A.4. This measure tells us about the reliability of a measurement set and supplies a weight for the mean of a set of measurements. A set with a small SE$_m$ should receive more weight than a set with a large (less precise) SE$_m$. We saw from Equation A-5:

$$SE_m = \frac{SE}{\sqrt{n}}$$

that the error varies inversely with the square root of the number of measurements; it is also true that the number of measurements varies inversely with the SE$_m^2$.

In the previous section, we saw that weights were proportional to the number (n) of measurements; that is, generally,

$$w_k \alpha \frac{1}{SE_k^2} \qquad \text{(A-10)}$$

A.8.3 Adjustments

When the absolute size of an error is known, and when weights have been assigned to measurements having varying reliabilities, corrections to the field data are made so that the error is eliminated by applying corrections that reflect the various weightings. It is obvious that measurements having large weights are corrected less than measurements having small weights. (The more certain the measurement, the larger the weight.) It follows that correction factors should be in inverse ratio to the corresponding weights.

Example A-6: The angles in a triangle were determined with A measured three times, B measured two times, and C measured once, for a closure error of 20″. What are the correction factors and the adjusted angles?

Solution: The correction factor is simply the inverse of the weight, and the actual correction for each angle is simply the ratio of the correction factor to the total correction factor, all multiplied by the total correction. See Table A-4.

Example A-7: Consider the same angles, except in this case the weights are related to the SE characteristics of three different theodolites. What are the correction factors and the adjusted angles?

Solution: See Table A-5.

Example A-8: *Adjustment of a Level Loop* Errors are introduced into level surveying each time a rod reading is taken. It stands to reason that corrections to elevations in a level loop should be proportional to the number of instrument setups (i.e., the weights should be inversely proportional to the number of instrument setups). Furthermore, since there is normally good correlation between the number of setups and the distance surveyed, corrections can be applied in proportion to the distance surveyed. (If part of a level loop were in unusual terrain, corrections could then be proportional to the number of setups.) What are the correction factors and the adjusted elevations?

Solution: Consider Table A-6, where temporary benchmarks (TBMs) are established in the area of an interchange construction (see also Figure A-4).

Table A-4   Angle Adjustments Using Weight Factors

Angle	Mean Value	Weight	Correction Factor	Correction	Adjusted Angle
A	45°07'32"	3	1/3 = 0.33	0.33/1.83 × 20 = + 4"	45°07'36"
B	71°51'06"	2	1/2 = 0.50	0.50/1.80 × 20 = + 5"	71°51'11"
C	63°01'02"	1	1 = 1.00	1/1.83 × 20 = +11"	63°01'13"
	179°59'40"		1.83	20"	179°59'60"
					= 180°00'00"

Error = −20"   Correction = +20"

Table A-5   Angle Adjustments Using SE*

Angle	Mean Value	SE	(SE²) Correction Factor	Corrected	Correction Angle
A	45°07'34"	±0.50"	0.25	0.25/29.25 × 20 = 0	45°07'40"
B	71°51'06"	±2.00"	4.00	4/29.25 × 20 = 3	71°51'09"
C	63°01'00"	±5.00"	25.00	25/29 × 20 = 17	63°01'17"
	179°59'40"		29.25	20	179°59'60"
					= 180°

Error = −20"   Correction = +20"0

*In this case Equation A-10 was used.

Table A-6   Elevation Adjustments Using Weight Factors

Station	Elevation	Number of Setups Between Station	Corrected Factor	Corrected Elevation
BM 506	172.865 (fixed)			172.865
TBM A-1	168.303	5	5/16 × 0.011 = 0.003	168.306
TBM A-2	168.983	2	7/16 × 0.011 = 0.005	168.988
TBM A-3	170.417	2	9/16 × 0.011 = 0.006	170.423
BM 506	172.854	7	16/16 × 0.011 = 0.011	172.865
		16		

Error = 172.854 − 172.865 = −0.011 m
Correction = +0.011 m

A.9 PRINCIPLE OF LEAST SQUARES

The data in Table A-2 showed that the smaller the sum of the squared errors, $\Sigma(x - \bar{x})^2$, the more precise are the data (i.e., the more closely the data are grouped around the mean $\bar{x}$). If such data were to be assigned weights, it is logical to weight each error $x - \bar{x}$ so that the sum of the squares of the errors is a minimum. This is the principle of least squares: $\Sigma w(x - \bar{x})^2$ is a minimum.

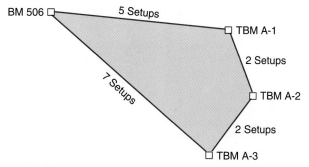

FIGURE A-4 Level loop adjustment

The development of the principle of least squares and the adjustments based on that principle can be found in texts on surveying adjustments. For a least squares adjustment to be valid, a reliable estimate of the SEs of various measuring techniques must be available to identify the individual weights properly.

A.10 TWO-DIMENSIONAL ERRORS

The concept of two-dimensional errors was first introduced in Section 6.7 and Figure 6-12. When the concept of position is considered, two parameters (x and y or r and θ) must be analyzed. In Chapter 6, we saw that traverse closures were rated with respect to relative accuracies (1/5,000, 1/10,000, etc.).

Figure A-5(a) and (b) shows the two-dimensional concept, with an area of uncertainty generated by the uncertainty in distance $\pm E_d$ in combination with the uncertainty (E_a) resulting from the uncertainty in angle $\pm \Delta\theta$. The figure of uncertainty is usually an ellipse with the major and minor axes representing the standard errors in distance and direction. When the distance and the direction have equal standard errors, the major axis equals the minor axis, resulting in a circle as the area of uncertainty, where $E_d = \sigma_x$ and $E_a = \sigma_y$; $(x/\sigma_x)^2 + (y/\sigma_y)^2 = 1$ is the equation of the 1σ ellipse of uncertainty, where x and y denote the errors. See Figure A-6.

In the previous discussion of one-dimensional accuracy, the probability that the true value

was within 1σ (one standard deviation) was 68 percent. In the case of the 1σ ellipse of uncertainty $(x/\sigma_x)^2 + (y/\sigma_y)^2 = 1$, the probability that the true value is within the ellipse is 39 percent. If a larger probability is required, a constant K is introduced so that the $K\sigma$ ellipse

$$(x/\sigma_x)^2 + (y/\sigma_y)^2 = K^2$$

encloses a larger area (see Figure A-6). Values for K are shown in Table A-7.

Example A-9: Assume that a station (A) to be set for construction control is 450.00 ft from a control monument with a position accuracy of ± 0.04 ft. What level of accuracy is indicated for angle and distance?

Solution:

$$\sigma_r = 0.04 = \sqrt{\sigma_x^2 + \sigma_y^2}$$

Since this case represents a circle, $\sigma_x = \sigma_y$:

$$\pm 0.04 = \sqrt{2\sigma_x^2}$$

$$\sigma_x = \frac{0.04}{\sqrt{2}} = 0.028 \text{ ft}$$

Therefore, for an error due to a distance measurement of 0.028, an accuracy of 0.028/450 = 1/16,000 is indicated. And since the error due to angle measurement is also 0.028, the allowable angle error can be determined as follows:

$$\frac{0.028}{450} = \tan(\text{angle error})$$

Therefore, the allowable angle error is $\pm 0°00'13''$. Furthermore, at the 39.4 percent probability level,

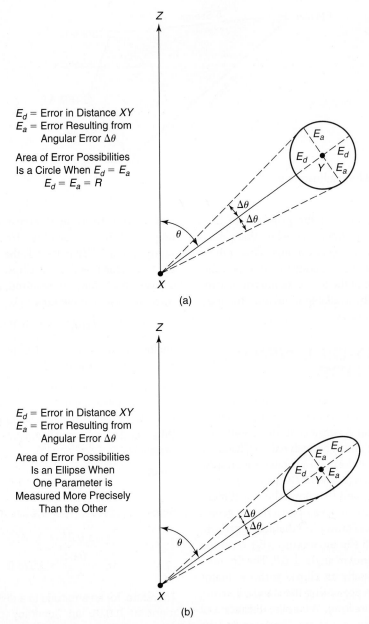

FIGURE A-5 Area of uncertainty. (a) Error circle. (b) Error ellipse

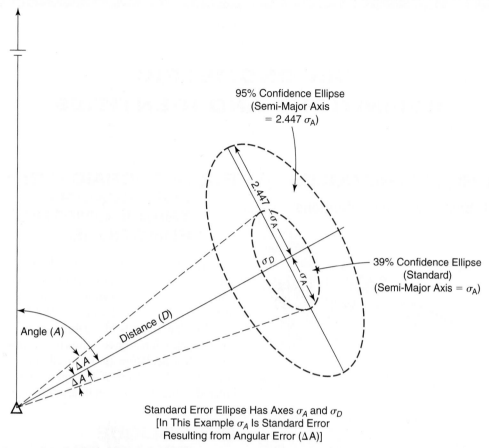

95% Confidence Ellipse
(Semi-Major Axis
= 2.447 σ_A)

2.447 σ_A

σ_D

σ_A

39% Confidence Ellipse
(Standard)
(Semi-Major Axis = σ_A)

Angle (A)

Distance (D)

ΔA

ΔA

Standard Error Ellipse Has Axes σ_A and σ_D
[In This Example σ_A Is Standard Error
Resulting from Angular Error (ΔA)]

FIGURE A-6 Standard error ellipse and the 95 percent confidence ellipse

Table A-7

Probability, P	K
39.4%	1.000
50.0	1.177
90.0	2.146
95.0	2.447
99.0	3.035
99.8	3.500

the distance must be measured to within ±0.028 ft; however, if we wish to speak in terms of the 95 percent level of probability that our point is within ±0.04 ft, we must use the K factor (Table A-7 and Figure A-6) of 2.45. Thus, the limiting error is now

$$\frac{0.028}{2.45} = 0.011 \text{ ft for both distance and angle}$$

The angle limit is given by 0.011/450 = tan (angle error). Therefore, the maximum angle error is now ±0°00'05".

TRIGONOMETRIC DEFINITIONS AND IDENTITIES

B.1 RIGHT TRIANGLES

B.1.1 Basic Function Definitions

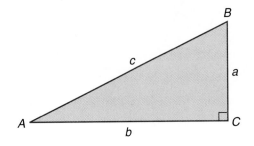

$$\sin A = \frac{a}{c} = \cos B \qquad \text{(B-1)}$$

$$\cos A = \frac{b}{c} = \sin B \qquad \text{(B-2)}$$

$$\tan A = \frac{a}{b} = \cot B \qquad \text{(B-3)}$$

$$\sec A = \frac{c}{b} = \operatorname{cosec} B \qquad \text{(B-4)}$$

$$\operatorname{cosec} A = \frac{c}{a} = \operatorname{cosec} B \qquad \text{(B-5)}$$

$$\cot A = \frac{b}{a} = \tan B \qquad \text{(B-6)}$$

B.1.2 Derived Relationships

$$a = c \sin A = c \cos B = b \tan A$$
$$= b \cot B = \sqrt{c^2 - b^2}$$
$$b = c \cos A = c \sin B = a \cot A$$
$$= a \tan B = \sqrt{c^2 - a^2}$$
$$c = \frac{a}{\sin A} = \frac{a}{\cos B} = \frac{b}{\sin B}$$
$$= \frac{b}{\cos A} = \sqrt{a^2 + b^2}$$

B.2 ALGEBRAIC SIGNS FOR PRIMARY TRIGONOMETRIC FUNCTIONS

The quadrant numbers reflect the traditional geometry approach (counterclockwise) to quadrant analysis. In surveying, the quadrants are numbered 1 (N.E.), 2 (S.E.), 3 (S.W.), and 4 (N.W.). The analysis of algebraic signs for the trigonometric functions (as shown) remains valid. Handheld calculators automatically provide the correct algebraic sign if the angle direction is entered in the calculator in its azimuth form.

B.3 OBLIQUE TRIANGLES

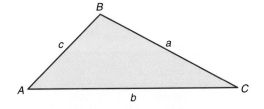

B.3.1 Sine Law

$$\frac{a}{\sin A} = \frac{b}{\sin B} = \frac{c}{\sin C} \qquad \text{(B-7)}$$

B.3.2 Cosine Law

$$a^2 = b^2 + c^2 - 2bc \cos A \qquad \text{(B-8)}$$
$$b^2 = a^2 + c^2 - 2ac \cos B \qquad \text{(B-9)}$$
$$c^2 = a^2 + b^2 - 2ab \cos C \qquad \text{(B-10)}$$

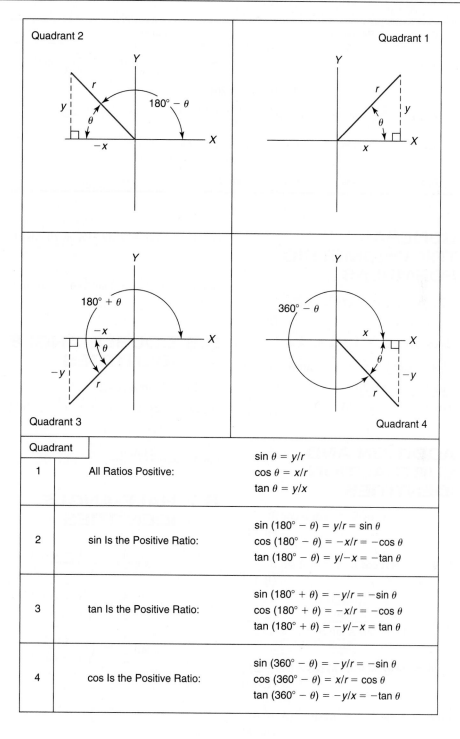

Quadrant		
1	All Ratios Positive:	$\sin \theta = y/r$ $\cos \theta = x/r$ $\tan \theta = y/x$
2	sin Is the Positive Ratio:	$\sin (180° - \theta) = y/r = \sin \theta$ $\cos (180° - \theta) = -x/r = -\cos \theta$ $\tan (180° - \theta) = y/-x = -\tan \theta$
3	tan Is the Positive Ratio:	$\sin (180° + \theta) = -y/r = -\sin \theta$ $\cos (180° + \theta) = -x/r = -\cos \theta$ $\tan (180° + \theta) = -y/-x = \tan \theta$
4	cos Is the Positive Ratio:	$\sin (360° - \theta) = -y/r = -\sin \theta$ $\cos (360° - \theta) = x/r = \cos \theta$ $\tan (360° - \theta) = -y/x = -\tan \theta$

Given	Required	Formulas
A, B, a	C, b, c	$C = 180° - (A + B); b = \dfrac{a}{\sin A} \sin B; c = \dfrac{a}{\sin A} \sin C$
A, b, c	a	$a^2 = b^2 + c^2 - 2bc \cos A$
A, b, c	A	$\cos A = \dfrac{b^2 + c^2 - a^2}{2bc}$
a, b, c	Area	$\text{Area} = \sqrt{s(s - a)(s - b)(s - c)}$ where $s = \frac{1}{2}(a + b + c)$
C, a, b	Area	$\text{Area} = \frac{1}{2} ab \sin C$

B.4 GENERAL TRIGONOMETRIC FORMULAS

$$\sin A = 2 \sin\tfrac{1}{2} A \cos\tfrac{1}{2} A = \sqrt{1 - \cos^2 A}$$
$$= \tan A \cos A \tag{B-11}$$

$$\cos A = 2 \cos^2 \tfrac{1}{2} A - 1 = 1 - 2 \sin^2 \tfrac{1}{2} A$$
$$= \cos^2 \tfrac{1}{2} A - \sin^2 \tfrac{1}{2} A = \sqrt{1 - \sin^2 A} \tag{B-12}$$

$$\tan A = \frac{\sin A}{\cos A} = \frac{\sin 2A}{1 + \cos 2A} = \sqrt{\sec^2 A - 1} \tag{B-13}$$

B.5 ADDITION AND SUBTRACTION IDENTITIES

$$\sin (A \pm B) = \sin A \cos B \pm \sin B \cos A \tag{B-14}$$

$$\cos (A \pm B) = \cos A \cos B \mp \sin A \sin B \tag{B-15}$$

$$\tan (A \pm B) = \frac{\tan A \pm \tan B}{1 \mp \tan A \tan B} \tag{B-16}$$

$$\sin A + \sin B = 2 \sin \tfrac{1}{2}(A + B) \cos \tfrac{1}{2}(A - B) \tag{B-17}$$

$$\sin A - \sin B = 2 \cos \tfrac{1}{2}(A + B) \sin \tfrac{1}{2}(A - B) \tag{B-18}$$

$$\cos A + \cos B = 2 \cos \tfrac{1}{2}(A + B) \cos \tfrac{1}{2}(A - B) \tag{B-19}$$

$$\cos A - \cos B = -2 \sin \tfrac{1}{2}(A + B) \sin \tfrac{1}{2}(A - B) \tag{B-20}$$

B.6 DOUBLE-ANGLE IDENTITIES

$$\sin 2A = 2 \sin A \cos A \tag{B-21}$$
$$\cos 2A = \cos^2 A - \sin^2 A = 1 - 2 \sin^2 A \tag{B-22}$$
$$= 2 \cos^2 A - 1$$

$$\tan 2A = \frac{2 \tan A}{1 - \tan^2 A} \tag{B-23}$$

B.7 HALF-ANGLE IDENTITIES

$$\sin \frac{A}{2} = \sqrt{\frac{1 - \cos A}{2}} \tag{B-24}$$

$$\cos \frac{A}{2} = \sqrt{\frac{1 + \cos A}{2}} \tag{B-25}$$

$$\tan \frac{A}{2} = \sqrt{\frac{\sin A}{1 + \cos A}} \tag{B-26}$$

GLOSSARY

A

absolute positioning   Also called *point positioning*. The direct determination of a station's coordinates by receiving positioning signals from a minimum of four GPS satellites.

absorption   The process by which radiant energy is retained by a substance. The absorbing medium itself may emit energy, but only after an energy conversion has taken place.

accuracy   The conformity of a measurement to the "true" value.

accuracy ratio   The error in a measurement divided by the overall value of the measurement, expressed as a fraction with a numerator of 1 and a denominator rounded to the closest 100 units. For example, an error of 0.01 ft in 30.00 ft would result in an accuracy ratio of 1:3,000.

active control station (ACS)   *See* CORS.

address matching   The process of relating a street address to a geodatabase. This is similar to address geocoding, which determines a horizontal position based on a street address.

aerial survey   Preliminary and final survey using traditional aerial photography and aerial imagery. Aerial imagery includes the use of digital cameras, multispectral scanners, LiDAR, and radar.

alignment   The location of the centerline of a survey or a facility.

ambiguity   The integer number of carrier cycles between the GPS receiver and a satellite.

antenna reference height (ARH)   Height of the measurement center of a GPS receiver above the instrument station.

area   A GIS term describing a surface polygon having two dimensions, and enclosed by a set of lines or chains.

arithmetic check   A check on the reductions of differential leveling involving the sums of the backsights and the foresights.

arterial road (highway)   A road designed primarily for traffic mobility, with some property access consideration.

as-built (final) survey   Postconstruction survey that confirms design execution and records in-progress revisions.

atmospheric window   The ranges of wavelengths at which water vapor, carbon dioxide, and other atmospheric gases only slightly absorb radiation.

automatic level   A surveyor's level in which line of sight is maintained automatically in the horizontal plane, once the instrument is roughly leveled.

azimuth   The direction of a line, measured clockwise (usually) from a north meridian.

B

backsight (BS)   A sight taken with a level to a point of known elevation, thus permitting the surveyor to compute the elevation of the HI. In theodolite/total station work, the backsight is a sighting taken to a point of known position to establish a reference direction.

baseline   A line of reference for survey work; often the legal centerline, the street line, or the centerline of construction is used, although any line can be selected arbitrarily or established. In

GPS, a baseline joins two GPS receivers engaged in relative positioning.

batter boards Horizontal crosspieces on grade stakes or grade rods that refer to proposed elevations.

bearing Direction of a line given by the acute angle from a meridian and accompanied by a cardinal compass direction (N-E, N-W, S-E, or S-W).

bearing tree Tree used as a witness point for a survey corner.

benchmark (BM) A fixed solid reference point with a precisely determined published elevation.

borrow pit A source of fill material that is located off the right of way.

break line A linear series of elevations that define a change in the slope of a surface. Examples are ridge lines, valley lines, tops and bottoms of slope, ditch lines, shorelines.

bucking-in A trial-and-error technique of establishing a theodolite or total station on a line between two points that themselves are not intervisible. Also known as *wiggling in, interlining,* or *balancing-in.*

C

catch basin A structure designed to collect surface water and transfer it to a storm sewer.

central meridian A reference meridian in the center of the zone covered by the plane coordinate grid; at every 6° of longitude in the UTM grid.

circular curve A curve with a constant radius.

clearing The cutting and removal of trees from a construction site.

COGO (coordinate geometry) Software programs that facilitate coordinate geometry computations; used in surveying and civil engineering design.

collector road (highway) A road designed to provide property access with some traffic mobility; it connects local roads to arterials.

compass rule Equation that distributes the errors in latitude and departure for each traverse course in the same proportion as the course distance is to the traverse perimeter: $C_{lat\,AB} \div (\Sigma\,lat) = AB \div P$, or $C_{lat\,AB} = \Sigma\,lat \times (AB \div P)$.

compound curve Two or more circular arcs turning in the same direction that have common tangent points and different radii.

construction survey Surveys that provide line and grade.

contour A line on a map joining points of similar elevation.

control survey Survey taken to establish or reestablish reference points, elevations, and lines for preliminary and construction surveys.

coordinates (plane) A set of numbers (X, Y) defining the two-dimensional position of a point, given by the distances measured north and east of an origin reference point having coordinates of $(0, 0)$.

CORS Continuously operating reference station (GPS). CORS transmitted data can be used by single-receiver surveyors or navigators to permit higher-precision differential positioning.

cross section A profile of the ground and surroundings taken at right angles to a reference line.

crown The high point in a road cross section. Also refers to the vertical change from the edge of the road to the centerline or high point on the road.

culvert A structure designed to provide an opening under a road or other feature, usually for the transportation of storm water.

cut In construction, the excavation of material; also the measurement down from a grade mark.

cycle slip A temporary loss of lock on satellite carrier signals causing a miscount in carrier cycles. Lock must be reestablished to continue positioning solutions.

D

data collector An electronic field book designed to store field data, both measured and descriptive, and in some cases, to control the operation of the instrument.

datum   A fixed reference for horizontal and vertical measurements (e.g., NAD83 and NAVD88).

deflection angle   The angle between the prolongation of the back line measured right (R) or left (L) to the forward line.

departure   The change in easterly displacement of a line (ΔE).

differential leveling   Determining the differences in elevation between points using a surveyor's level.

differential positioning (DGPS)   Obtaining measurements at a known base station to correct simultaneous measurements made at rover receiving stations.

diurnal   Something occurring over 24 hours (i.e., during one rotation of the earth).

Doppler effect   The apparent change in frequency of sound or light waves varying with the relative velocities of both the source and the observer; if the source and the observer come closer together, the emitted frequency appears to be increased.

double centering   A technique of turning angles or producing straight lines involving a minimum of two sightings with a theodolite, once with the telescope direct and once with the telescope inverted.

drainage   The collection and transportation of ground and storm water.

DTM (digital terrain model)   A three-dimensional depiction of a ground surface, usually produced by computer software (sometimes referred to as a DEM, digital elevation model).

dxf (drawing exchange format)   An industry-standard format that permits graphical data to be transferred among various CAD, GIS, and softcopy photogrammetry applications programs.

E

EDM   Electronic distance measurement.

EFB   Electronic field book. *See* Data collector.

elevation   The vertical distance above or below a given datum; also known as *orthometric height*.

elevation factor   The factor used to convert ground distances to sea-level distances.

end area   The area of a cross section. The areas of two adjacent station cross sections can be averaged and then multiplied by the distance between them to determine the volume of cut or fill between those stations.

engineering surveys   Preliminary and layout surveys used for design and construction.

eolian   Surface features or materials created by the wind.

EOS (Earth Observing System)   NASA's study of the earth scheduled to cover the period 2000 to 2015, in which a series of small to intermediate earth observation satellites will be launched to measure global changes. The first satellite (experimental) in the series (TERRA) was launched in 1999.

epoch   An observational event in time that forms part of a series of GPS observations.

error of closure   The difference between the measured location and the theoretically correct location.

ETI⁺ (enhanced thematic mapper)   An eight-band multispectral scanning radiometer, onboard Landsat 7, that is capable of providing relatively high resolution (15 m) imaging information about the earth's surface.

external distance   The distance from the mid-curve to the PI in circular curves.

existing ground (EG)   See original ground.

F

Father Point   A general adjustment of Canadian-Mexican-U.S. leveling observations resulted in the creation of the North American vertical datum of 1988 (NAVD88) in 1991; the adjustment held fixed the height of the primary tidal benchmark located at Father Point, Rimouski, Quebec, on the south shore of the St. Lawrence River.

fiducial marks   Reference marks on the edges of aerial photos, used to locate the principal point on the photo.

fill   Material used to raise the construction level; also, the measurement up from a grade mark.

final survey   *See* as-built survey.

forced centering equipment   The interchanging of theodolites, prisms, and targets into tribrachs, which have been left in position over the station.

foresight (FS)   In leveling, a sight taken to a BM or TP to obtain a check on a leveling operation or to establish a transfer elevation.

free station   A conveniently located instrument station used for construction layout, the position of which is determined after occupation through resection techniques.

freeway   A highway designed for traffic mobility in which access is restricted to interchanges with arterials and other freeways.

G

GDOP (geometric dilution of precision)   A value that indicates the relative uncertainty in position, using GPS observations, caused by satellite vector measurements. A minimum of four widely spaced satellites at high elevations usually produce good results (i.e., lower GDOP values).

dilution of precision (DOP)   An indicator of the quality of the geometric solution in GPS. The better spacing of satellites used in the solution, the lower the DOP and the better the statistical solution. There are five types of DOP; Geometric, Positional, Horizontal, Vertical, and Time.

geocoding   The linking of entity and attribute data to a specific geographic location.

geodetic datum   A precisely established and maintained series of benchmarks referenced to adjusted mean sea level (MSL).

geodetic height (*h*)   The distance from the ellipsoid surface to the ground surface.

geodetic survey   A survey of such high precision or covering such a large geographic area that computations must be based on the ellipsoidal shape of the earth.

geographic information system (GIS)   Analysis and display of selected layers of a spatially and relationally referenced database.

geographic meridian   A line on the surface of the earth joining the poles, in other words, a line of longitude.

geoid   A surface that is approximately represented by mean sea level (MSL) and is, in fact, the equipotential surface of the earth's gravity field.

geoid undulation (*N*)   The distance between the geoid surface and the ellipsoid surface. *N* is negative if the geoid surface is below the ellipsoid surface. Also known as *geoid height.*

geomatics   A term used to describe the science and technology dealing with earth measurement data, including collection, sorting, management, planning and design, storage, and presentation. It has applications in all disciplines and professions that use earth-related spatial data, such as planning, geography, geodesy, infrastructure engineering, agriculture, natural resources, environment, land division and registration, project engineering, and mapping.

geospatial data   Data describing both the geographic location and attributes of features on the earth's surface.

geostationary (geosynchronous) orbit   A satellite orbit in which the satellite appears stationary over a specific location on earth. A formation of geostationary satellites presently provides communication services worldwide.

glacial   Pertaining to surface features and materials produced by the formation and movement of glaciers.

global positioning system (GPS)   A ground positioning (*Y*, *X*, and *Z*) technique based on the reception and analysis of NAVSTAR satellite signals.

GNSS (Global Navigation Satellite System)   The generic term for any of the Satellite Navigation systems, including the two current functional systems GPS and GLONASS, as well as the planned systems such as Galileo and Compass.

gon   A unit of angular measure in which 1 revolution = 400 gon and 100.000 gon = a right angle. Also known as *grad.*

grad *See* gon.

grade sheet A construction report giving line and grade (offsets) and cuts and fills at each station.

grade stake A wood stake (usually) with a cut/fill reference mark to that portion of a proposed facility adjacent to the stake.

grade transfer A technique of transferring cut and fill measurements to the facility using a carpenter's level, stringline level, laser, and batter boards.

gradient The slope of a gradeline.

grid distance A distance on a coordinate grid.

grid factor A factor used to convert ground distances to grid distances.

grid meridian Meridians parallel to a central meridian on a coordinate grid.

ground distance A distance as measured on the ground surface.

grubbing The removal of stumps, roots, and the like from a construction site.

Gunter's chain Early (1800s) measuring device consisting of 100 links, measuring 66 ft long.

haul In highway construction, the distance that each yd^3 (m^3) of cut material is transported to a fill location.

H

hectare $10,000 \text{ m}^2$.

elevation of instrument (HI) Elevation of the line-of-sight of an instrument above a vertical datum; used in leveling.

height of instrument (HI) Height of instrument (optical axis) above the instrument station; used in theodolite/EDM, and total station surveying.

horizontal control survey markers Monuments used for horizontal positions, that are tied to one or more specific datums. These control markers can be maintained by national agencies, regional agencies, local agencies, or private firms.

horizontal line A straight line perpendicular to a vertical line.

hydrographic surveys Surveys designed to define shoreline and underwater features.

I

igneous rock Surface features and materials resulting from the solidification of magma.

interlining (bucking-in) A trial-and-error technique of establishing a theodolite or total station on a line between two points that themselves are not intervisible. Also known as *wiggling in* or *bucking-in*.

intermediate sight (IS) A sight taken with a level and theodolite, or total station, to determine a feature elevation and/or location.

invert The inside bottom of a pipe or culvert.

ionosphere The section of the earth's atmosphere that is about 50–1,000 km above the earth's surface.

ionospheric refraction The variation in the velocity of signals (GPS) as they pass through the ionosphere.

L

lacustrine Surface features or materials resulting from original deposition in lakes.

laser alignment Horizontal or vertical alignment given by a fixed or rotating laser.

latitude (geographic) Angular distance from the earth's center, measured northerly or southerly from the equatorial plane.

latitude (of a course) The change in northerly displacement of a line (ΔN).

layout survey Survey in which the surveyor marks on the ground the features shown on a design plan. Wood stakes, iron bars, aluminum and concrete monuments, nails, spikes, and so on can be used to make these markings.

leveling The procedure for determining differences in elevation between points that are some distance from each other.

level line A line in a level surface.

LiDAR (light detection and ranging) This technique, used in airborne and satellite imagery, utilizes timed laser pulses that are reflected from surface features to obtain DTM mapping detail.

line   A GIS term describing the joining of an ordered set of coordinated points, or by a grid of cells, and having one dimension.

line and grade   The horizontal and vertical position of a facility.

linear error of closure   The line of traverse misclosure representing the resultant of the measuring errors.

littoral   Surface features and materials produced by coastal wave action.

local road (highway)   A road designed for property access, connected to arterials by collectors.

longitude   Angular distance measured in the plane of the equator from the reference meridian through Greenwich, England. Lines of longitude are indicated on globes as meridians.

M

magnetic declination   The horizontal angle between the direction given by a compass needle and geographic north.

magnetic meridians   Lines that are parallel to the directions taken by freely moving magnetized needles, such as a compass.

mask angle   The vertical angle above horizontal below which satellite signals are not recorded or not processed; often a value of 10° or 15° is used. Also known as the *cutoff angle*.

mass diagram   A graphic representation of cumulative highway cuts and fills.

mean sea level (MSL)   A reference datum for leveling.

measured data   Information that is measured and recorded in the field, either in a field book or in a data collector. This does not include data that are computed or reduced from the measured data.

meridian   A north–south reference line; or the line formed by the intersection of the earth's surface with a plane containing the earth's axis of rotation.

metamorphic   Surface features and materials resulting from sedimentary or igneous rock that has been subjected to geoforces of pressure, heat, and/or water.

midordinate distance   The distance from the midchord to the midcurve in circular curves.

mistake   A poor result due to carelessness or a misunderstanding.

monument   A permanent surveying reference marker (usually concrete, steel, or aluminum) for horizontal and vertical positioning.

multispectral scanner   Scanning device, used for satellite and airborne imagery, that records reflected and emitted energy in two or more bands of the electromagnetic spectrum.

N

nadir angle   A vertical angle measured from the nadir direction (straight down) upward to a point.

NAVSTAR   A set of orbiting satellites used in navigation and positioning.

normal tension (P_n)   The amount of tension required in taping to offset the effects of sag.

O

original ground (OG)   The position of the ground surface prior to construction.

orthometric height (H)   The distance from the geoid surface to the ground surface. Also known as *elevation*.

P

page check   Arithmetic check.

parabolic curve   A curve used in vertical alignment to join two adjacent gradelines.

parallax   An error in sighting that occurs when the object and/or the crosshairs of a telescope are focused improperly.

photogrammetry   The science of taking measurements from aerial photographs.

plane survey   A survey of such limited size (most surveys) that computations can be based on plane geometry and trigonometry for all horizontal positioning.

planimeter   A mechanical or electronic device used to measure areas by tracing the outline of the area on the map or plan.

plan view This is a map showing from an overhead view. This shows what features would look like from directly above the features. In plan view elevations are often shown using contours and spot elevations.

plat A plan of survey usually showing property information.

point A GIS term describing a single spatial entity represented by a set of northing/easting coordinates, or by a single pixel location, and having zero dimension.

point of beginning (POB) A referenced point (corner) or a property line from which a description of the property deed begins and at which the description closes.

polar coordinates The location of a feature by angle and distance (r, θ).

polygon A GIS term for a closed chain of points representing an area.

precision The degree of refinement (repeatability) with which a measurement is made.

preengineering survey A preliminary survey that forms the basis for engineering design.

prismoid A solid with parallel ends joined by a plane or warped surface.

preliminary survey The gathering of data (distance, position, elevation, and angles) to locate physical features so that they can be plotted to scale on a map or plan.

profile A series of elevations along the direction of a survey line (e.g., along a road or watercourse $\mathbb{C}$).

property survey A survey to retrace or establish property lines, or to establish the location of buildings within property limits.

pseudorange The uncorrected distance from a GPS satellite to a GPS ground receiver determined by comparing the code transmitted from the satellite to the replica code residing in the GPS receiver. When corrections are made for clock and other errors, the pseudorange becomes the range.

R

random errors Errors associated with the skill and vigilance of the surveyor.

real-time positioning (real-time kinematic, RTK) A technique using a base station to measure the satellites' signals, process the baseline corrections, and then broadcast the corrections (differences) to any number of roving receivers that are simultaneously tracking the same satellites.

rectangular tie-ins The location of a feature by two distances, 90° apart.

rectangular coordinates A plane coordinate system that is created by an x (easting) direction and a perpendicular y (northing) direction. A Cartesian coordinate system.

relative positioning The determination of position through the combined computations of two or more receivers simultaneously tracking the same satellites, resulting in the determination of the baseline vector (X, Y, Z) joining the two receivers.

remote object elevation The determination of the height of an object using total station sightings together with onboard applications software.

remote sensing Geodata collection and interpretive analysis for both airborne and satellite imagery.

resection The solution of the coordinated position of an occupied station by the sighting of angles to three or more coordinated reference stations—two or more stations, if both angles and distances are measured.

right of way (ROW) The legal property limits of a utility or access route.

route survey Preliminary, control, and construction surveys that cover a long but narrow area, as in highway and railroad construction.

S

sag The error introduced when a tape is supported only at its ends.

scale factor The factor used to convert sea-level distances to plane grid distances.

sea-level correction factor The factor used to convert ground distances to sea-level equivalent distances.

sedimentary Surface features and materials first formed by water, wind, or glaciers.

shaft An opening of uniform cross section joining a tunnel to the surface; used for access and ventilation.

shrinkage The decrease in volume of an excavated material.

sidereal day The time taken for one complete revolution of the earth with reference to an infinitely far away object, for example, a star.

skew number A clockwise angle (closest 5°) turned from the back tangent to the centerline of a culvert or bridge.

slope stakes Stakes placed to locate the top or bottom of a slope.

solar day The time taken for one complete revolution of the earth with reference to the sun.

sounding The measurement of water depths in rivers, lakes, and oceans.

spiral curve A transition curve of constantly changing radius placed between a high-speed straight section and a central curve; it permits a safe gradual steering adjustment.

springline The horizontal bisector of a storm or sanitary pipe.

station A point on a baseline that is a specified distance from the point of commencement. The point of commencement is identified as 0 + 00; 100 ft or 100 m are known as full stations (1,000 m in some highway applications); 1 + 45.20 identifies a point distant 145.20 ft/m away from the point of commencement (0 + 145.20 in some highway applications).

superelevation The banking of a curved section of road to help overcome the effects of centrifugal force.

swell The increase in volume of an excavated material.

systematic errors Errors beyond the control of the observer, but whose magnitude and algebraic sign can be determined.

T

tangent A straight line, often referred to with respect to a curve.

temporary benchmark (TBM) A semipermanent point of known elevation.

tension error An error caused by other than standard tension.

three-wire leveling A more precise technique of differential leveling in which rod readings are taken at the stadia hairs in addition to the main crosshair.

toe of slope Bottom of slope.

topographic survey Preliminary survey used to tie in the horizontal and vertical positions of natural and constructed surface features of an area. The features are located relative to one another by tying them all into the same control lines or control grid, or elevation datum.

total station An electronic theodolite combined with an EDM and an electronic data collector capable of measuring and recording horizontal and vertical distances and angles and then computing N, E, and elevation values.

traverse A continuous series of measured (angles and distances) lines.

triangulation A control survey technique involving (1) a precisely measured baseline as a starting side for a series for triangles or chain of triangles; (2) the measurement of each angle in the triangle using a precise theodolite, which permits the computation of the lengths of each side; and (3) a check on the work made possible by precisely measuring a side of a subsequent triangle (the spacing of check lines depends on the desired accuracy level).

trilateration The control surveying solution technique of measuring only the sides of a triangle.

troposphere The part of the earth's atmosphere that stretches from the surface to about 80 km upward (includes the stratosphere as its upper portion).

turning point (TP)   In leveling, a solid point where an elevation is temporarily established so that the level may be relocated.

U

universal time (UT)   Mean solar time at the meridian of Greenwich, England, and kept by atomic clocks.

UTM   Universal transverse Mercator grid system.

V

vertical angle   An angle in the vertical plane measured up (+) or down (–) from horizontal.

vertical curve   A parabolic curve joining two gradelines.

vertical line   A line from the surface of the earth to the earth's center. Also known as a *plumb line* or a *line of gravity*.

W

waving the rod   The slight waving of the leveling rod to and from the instrument that permits the surveyor to take a precise (lowest) rod reading.

Z

zenith angle   A vertical angle measured downward from the zenith (upward plumb line) direction.

ANSWERS TO SELECTED CHAPTER PROBLEMS

CHAPTER 2

2.1 (a) $(c + r) = 0.0206 \times (500/1,000)^2 = 0.005\,\text{ft}$

(b) $(c + r) = 0.0206 \times (4)^2 = 0.033\,\text{ft}$

(c) $(c + r) = 0.0675 \times 0.300^2 = 0.006\,\text{m}$

2.2 (a) i 1.90. ii 1.73

(b) i 1.185 ii 1.150

(c) i 3.06 ii 2.85 (2.84)

(d) i 1.145 ii 1.065

2.4

Station	BS	HI	IS	FS	Elevation
BM #50	1.27	390.34			389.07
TP #1	2.33	387.76		4.91	385.43
TP #2				6.17	381.59

BS = 3.60   FS = 11.08

389.07 + 3.60 = 392.67 − 11.08 = 381.59, Check

2.6

Station	BS	HI	FS	Elevation
BM 100	2.71	317.59		314.88
TP 1	3.62	316.33	4.88	312.71
TP 2	3.51	315.87	3.97	312.36
TP 3	3.17	316.23	2.81	313.06
TP 4	1.47	316.08	1.62	314.61
BM 100			1.21	314.87

BS = 14.48   FS = 14.49

314.88 + 14.48 − 14.49 = 314.87, Check

2.7 Error closure = 0.01 ft; for 1,000 ft, second order (see Table 2-2) permits $0.035\sqrt{1,000/5,280} = 0.015$; therefore results qualify for **second-order** accuracy.

2.8

Station	BS	HI	IS	FS	Elevation
BM 20	8.27	186.04			177.77
TP 1	9.21	192.65		2.60	183.44
0 + 00			11.3		181.4
0 + 50			9.6		183.1
0 + 61.48			8.71		183.94
1 + 00			6.1		186.6
TP 2	7.33	195.32		4.66	187.99
1 + 50			5.8		189.5
2 + 00			4.97		190.35
BM 21				3.88	191.44

BS = 24.81   FS = 11.14

177.7 + 24.81 − 11.14 = 191.44, Check!

2.15 (a) $V = 148.61 \sin 21°26' = 54.30\,\text{ft}$

Elevation of lower station = 318.71

$+ 4.66 − 54.30 − 4.88 = 264.19\,\text{ft}$

(b) $H = 148.61 \cos 21°26' = 138.33'$

Lower station at $110 + 71.25 + 138.33$

$= 112 + 09.58$

2.17

Station	BS	HI	FS	Elevation
BM 130	0.702	189.269		188.567
TP 1	0.970	189.128	1.111	188.158
TP 2	0.559	189.008	0.679	188.449
TP 3	1.744	187.972	2.780	186.228
BM K 110	1.973	188.277	1.668	186.304
TP 4	1.927	188.416	1.788	186.489
BM 132			0.888	187.528

BS = 7.875   FS = 8.914

188.567 + 7.875 − 8.914 = 187.528, Check

2.17 (a) Error = 187.536 − 187.528 = −0.008 m.
Using specifications from Table 2-1,
Third-order accuracy, allowable error
= 0.012√0.780 = 0.011 m. This
error of 0.008 thus qualifies for third-
order accuracy (in both Tables 2-1
and 2-2).

CHAPTER 3

3.1 (c) 129.33 × 66 = 8,535.78 ft × 0.3048
= 2,601.71 m
3.2 tan slope angle = 0.03; slope angle
= 1.718358°
H = 210.23 cos 1.718358° = 210.14 ft
3.8 Error per tape length = +0.004 m; tape used
129.085/30 times,
Error = +0.004 × 129.085/30 = +0.017 m
Corrected distance = 129.085 + 0.017
= 129.102 m
3.9 Error per tape length = 0.02 ft
Tape used
(a) 2.0 times; error = 2.0 × 0.02 = +0.04 ft
(b) 3.50 times; error = 3.50 × 0.02 = +0.07 ft
Corrected layout distance
(a) 200.00 − 0.04 = 199.96 ft
(b) 350.00 − 0.07 = 344.93 ft
(These are the values that, when precor-
rected for errors, result in the required layout
distances.)
3.11 (2 + 33.33) − (0 + 79.23) = 154.10 ft
C_T = 0.00000645(87 − 68)154.10
= +0.019 ft
C_L = 0.02 × 1.5410 = −0.031 ft
C = 0.019 − 0.031 = −0.012 ft
Layout 154.10 + 0.01 = 154.11 ft
3.12 (6 + 11.233) − (5 + 00) = 111.233 m
C_T = 0.0000116(−6 − 20)111.23
= −0.034 m
C_L = .005 × 111.233/30 = −0.019 m
C = −0.053 m
Layout 111.233 + 0.053 = 111.286 m
3.13 C_T = 0.00000645(−14 − 68)198.61 = −0.105 ft
C_L = −0.02 × 1.9861 = −0.040 ft

C = −0.105 − 0.040 = −0.15 ft
Corrected slope distance = 198.61 − 0.15
= 198.46 ft
H = √198.46² − 6.35² = 198.36 ft
3.14 C_T = 0.00000645(46 − 68)219.51 = −0.031 ft
Corrected slope distance = 219.51 − 0.02
= 219.49 ft
H = 219.49 cos 3°18' = 205.54 ft
3.23 C_S = −w²L³/24p² = −0.32² × 48.888³/
(24 × 100²) = −0.050 m
Corrected distance = 48.888 − 0.050
= 48.838 m
3.27 Prism constant = EG − EF − FG
= 426.224 − 277.301 − 148.953
= −0.030 m
3.28 H = 2,556.28 cos 2°45'30" − 2,553.32 ft
V = 2,556.28 sin 2°45'30" = 123.02 ft
Elevation of target station = 322.87 + 123.02
= 445.89 ft
3.29 Inst. @ A, H = 879.209 cos 1°26'50"
= 878.929 m
Elevation difference = 879.209 sin 1°26'50"
= 22.205 m
Inst. @ B, H = 879.230 cos 1°26'38"
= 878.951 m
Elevation difference = 879.230 sin 1°26'38"
= 22.155 m
(a) Horizontol distance = (878.929 + 878.951)/2
= 878.940 m
(b) Elev. B = 163.772 + (22.205 + 22.155)/2
= 185.952 m

CHAPTER 4

4.1 (n − 2)180 = 3 × 180 = 540°00'
A = 121°13'00"
B = 136°44'30"
C = 77°05'30"   539°59'60"
D = 94°20'30"   −427°23'30"
E = ? → → $\overline{E = 112°36'30}$
426°82'90"
= 427°23'30"
4.2 (a) S 30°30' W   (e) N 0° 08' E
4.3 (a) 20°20'   (e) 89°29'
4.4 (a) 30°30'   (e) 180°08'

4.5 (a) S 20°20′ W   (e) S 89°29′ W

4.6 AB N 19°09′ E

  +B   6°25′

  BC N 25°34′ E

  +C   3°54′

  CD N 29°28′ E

  +D   11°47′

  DE N 41°15′ E

4.8 Clockwise Solution for Bearings
  AB N 42°11′10″ E [Given]

DE N 41°15′ E

  −E   20°02′

  EF N 21°13′ E

  −F   7°18′

  FG N 13°55′ E

  +G   1°56′

  GH N 15° 51′ E

  BC N 81°42′20″ E

  CD S 19°48′00″ E

  DE S 87°23′50″ W

  DA N 74°01′50″ W

4.9 Clockwise Solution for Azimuths

  Azimuth AB   42°11′10″   [Given]

           BC   81°42′20″

           CD   160°12′00″

           DE   267°23′50″

           EA   285°58′10″

CHAPTER 6

6.2 (b)

Course	Azimuth	Bearing	Distance	Latitude	Departure
AB	197°17′	S 17°17′ W	636.45	−607.71	−189.09
BC	87°45′	N 87°45′ E	654.45	25.69	653.95
CD	353°54′	N 6°06′ W	382.65	380.48	−40.66
DA	295°24′	N 64°36′ W	469.38	201.33	−424.01
			2,142.93	−0.21	+0.19

$$E = \sqrt{0.21^2 + 0.19^2} = 0.283 \text{ ft}$$

Precision ratio $= E/P = 0.283/2,142.93 = 1/7,572 = 1/7,600$

6.3 (a) Corrections

Course	C_{Last}	C_{Dep}	Adjusted Lat.	Adjusted Dep.
AB	+0.07	−0.06	−607.64	−189.95
BC	+0.07	−0.06	+25.76	+653.89
CD	+0.03	−0.03	+380.51	−40.69
DA	+0.04	−0.04	+201.37	−424.05
	—	—	—	—
	+0.21	−0.19	0.00	0.00

6.3 (b) Coordinates

Station	North	East	
B	1,000.00	1,000.00	
	+25.76	+653.89	
C	1,025.76	1,653.89	
	+380.51	−40.69	
D	1,406.27	1,613.20	
	+201.37	−424.05	
A	1,607.64	1,189.15	
	−607.64	−189.15	
B	1,000.00	1,000.00	Check

6.4 Area by coordinates

$X_B(Y_C − Y_A) = 1,000.00(1,025.76 − 1,607.64) = −581,880$

$X_C(Y_D − Y_B) = 1,653.89(1,406.27 − 1,000.00) = +671,926$

$X_D(Y_A − Y_C) = 1,613.20(1,607.64 − 1,025.76) = +938,689$

$X_A(Y_B − Y_D) = 1189.15(1,000.00 − 1,406.27) = −483,116$

$$2A = 545,619$$

$$A = 272,810 \text{ ft}^2$$

(1 acre $= 43,560$ ft^2)

or, $A = 6.26$ acres

6.5

	Angles
A	101°03′19″
B	101°41′49″
C	102°22′03″
D	115°57′20″
E	118°55′29″

     537°178′120″ = 540°00′00″ Check

6.5 (a, b)

CRS	Azimuth	Bearing	Distance (m)	Latitude	Departure
AE	174°03′19″	S 5°56′41″ E	19.192	−19.089	+1.988
ED	112°58′48″	S 67°01′12″ E	35.292	−13.778	+32.491
DC	48°56′08″	N 48°56′08″ E	37.070	+24.352	+27.950
CB	331°18′11″	N 28°41′49″ W	26.947	+23.637	−12.939
BA	253°00′00″	S 73°00′00″ W	51.766	−15.135	−49.504
			170.267m	−0.013	−0.014

6.5 (c) $E = \sqrt{0.013^2 + 0.014^2} = 0.019\,\text{m}$
Precision ratio $= E/P = 0.019/170.267 =$
$1/8,961 = 1/9,000$

6.6 (a) Balance using Compass Rule

Course	C_{LAT}	C_{DEP}	Corrected Lat.	Corrected Dep.
AE	+0.001	+0.002	−19.088	+1.990
ED	+0.003	+0.003	−13.775	+32.494
DC	+0.003	+0.003	+24.355	+27.953
CB	+0.002	+0.002	+23.639	−12.937
BA	+0.004	+0.004	15.131	−49.500
Check	+0.013	+0.014	0.000	0.000

6.6 (b)

Station	Coordinates North	East
A	1,000.000	1,000.000
	−19.088	+1.990
E	980.912	1,001.990
	−13.775	+ 32.494
D	967.137	1,034.484
	+24.355	+27.953
C	991.492	1,062.437
	+23.639	−12.937
B	1,015.131	1,049.500
	−15.131	−49.500
A	1,000.000	1,000.000 Check

6.7 (a) Area by the Coordinate Method
$X_A(Y_B − Y_E) = 1,000.000(1,015.131 − 980.912)$
$\qquad = +34,219$

$X_E(Y_A − Y_D) = 1,001.990(1,000.000 − 967.137)$
$\qquad = +32,928$

$X_D(Y_E − Y_C) = 1,034.484(980.912 − 991.492)$
$\qquad = −10,945$

$X_C(Y_D − Y_B) = 1,062.437(967.137 − 1,015.131)$
$\qquad = −50,991$
$X_B(Y_C − Y_A) = 1,049.500(991.492 − 1,000.000)$
$\qquad = \underline{−8,929}$
$\qquad 2A = 3,718\,\text{m}^2$
$\qquad$ Area, $A = 1,859\,\text{m}^2$

6.9 $E = 180° − (119°\,57'\,46'' + 30°\,10'\,56'')$
$\qquad = 29°\,51'\,18''$
$CD = \sin 29°\,51'\,18''(1643.655/\sin 119°\,57'\,46'')$
$\qquad = 944.448\,\text{m}$
Bearing $DE = S\ 78°58'45''\ W$

6.17 Coordinates

Station	North	East
K	1,990.000	2,033.000
	+10.000	−33.000
A	2,000.000	2,000.000
K	1,990.000	2,033.000
	+25.271	+19.455
B	2,015.271	2,052.455
K	1,990.000	2,033.000
	−0.311	+38.285
C	1,989.689	2,071.285
K	1,990.000	2,033.000
	−30.055	+7.245
D	1,959.945	2,040.245
K	1,990.000	2,033.000
	−12.481	−30.100
E	1,977.519	2,002.900

6.18
$X_A(Y_E − Y_B) = 2,000.000(1,977.519 − 2,015.271)$
$\qquad = −75,504$
$X_B(Y_A − Y_C) = 2,052.455(2,000.000 − 1,989.689)$
$\qquad = +21,163$
$X_C(Y_B − Y_D) = 2,071.285(2,015.271 − 1,958.945)$
$\qquad = +114,596$

$$X_D(Y_C - Y_E) = 2{,}040.245(1{,}989.689 - 1{,}977.519)$$
$$= +24{,}830$$
$$X_E(Y_D - Y_A) = 2{,}002.900(1{,}959.945 - 2{,}000.000)$$
$$= -80{,}226$$
$$2A = 4{,}859 \text{ m}^2$$
$$\text{Area, } A = 2{,}430 \text{ m}^2$$

6.19

Course	Bearing	Distance (m)
BC	S 36°21′20″E	31.765

CHAPTER 7

Solutions

7.1 *Why is it necessary to observe a minimum of four GPS satellites to solve for position?*
GPS pseudoranging requires the receiver clock to be synchronized with satellite clock. As the receiver clock chip is not synchronized to GPS time, there are basically four unknowns that must be solved for, the three positions (X, Y, Z) and time.

7.2 *How does the United States' GPS constellation compare with the GLONASS constellation and with the proposed Galileo constellation?*
At the time of this writing (June 2011), GPS is the only system that has a full constellation of 24 satellites operational. (A full constellation provides full coverage all the time). GPS currently has 31 operational satellites. GPS primary design is for 60 orbit planes, inclined 55° from the equator, with four satellites in each plane. The Russian GLONASS currently has 22 operational satellites, but should soon have the full constellation of 24 (it had a full constellation when it was operated by the USSR, but went into disrepair during the breakup of the USSR). The system has three orbital planes with eight satellites each and has a 64.8° incline. This system gives better coverage in high northern latitudes than GPS. The European Galileo system currently has only two test satellites launched. It anticipates having its first four operational satellites up by the end of 2012. It has a planned constellation of 30 satellites in three orbital planes with a 56° inclination.

7.4 *What is the difference between range and pseudorange?*
Range in GPS is a term for determining the distance from the satellite to the receiver. Generally GPS determines the distance from the satellite to the receiver by calculating the time shift in the code. This is actually measuring time and then computing distance using the speed of light, so it is known as a "pseudorange." In carrier phase differential solutions, time is not used, but the count of wavelengths and the partial wavelength, so it is considered a direct distance measurement or "range."

7.7 *Describe RTK techniques used for a layout survey.*
For RTK layout or construction surveys, a two-person crew is more efficient than a one-person crew. Once the rover has resolved its position relative to the base and "fixed," the rover must get on the project coordinate system by doing a transformation or localization onto existing control. Verify the solution by staking out to control not used for the transformation.

When setting stakes, have one person using the RTK system and software to walk to the desired point and determine the location, close enough for setting the wood. The second person should drive in the wood (or whatever is used as the marker) and then the final measurement is taken with the RTK. The RTK person calls out the information to be marked on the lath, and the second person marks the lath and then repeats the markings to be verified by the RTK person. Depending on the type of the layout, the RTK person may be able to estimate the location of the next stakeout by pacing.

During the survey if there is any loss of power, or a software issue arises, as soon as the problem is resolved, a check on a previously set point should be made to verify the integrity of the solution.

CHAPTER 8

8.1

Traverse computations

Course	Bearing	Distance	Latitude	Departure
AB	N 3°30′ E	56.05	+55.95	+3.42
BC	N 0°30′ W	61.92	+61.92	−0.54
CD	N 88°40′ E	100.02	+2.33	+99.99
DE	S 23°30′ E	31.78	−29.14	+12.67
EF	S 28°53′ W	69.11	−60.51	+33.38
FG	South	39.73	−39.73	0
GA	N 83°37′ W	82.67	+9.18	−82.16
			0.0	0.00

8.2

Interior angles

A	92°53′	
B	184°00′	Check
C	90°50′	$(n - 2)\,180$
D	112°10′	$5 \times 180 = 900°$
E	127°37′	
F	208°53′	
G	83°37′	
	96°240′	
	$= 900°00′$	

8.5 (d)

Area Computations

First, compute coordinates—assume coordinates of 1,000.00 N and 1,000.00 E for station *A*

Station	Northing	Easting	
A	1,000.00	1,000.00	*Second*, compute the area:
	+ 55.95	+ 3.42	
B	1,055.95	1,003.42	$X_A(Y_B - Y_G) = 1,000.00(1,055.95 - 990.82) = 65,130$
	61.92	−0.54	$X_B(Y_C - Y_A) = 1,003.42(1,117.87 - 1,000.00) = 118,273$
C	1,117.87	1,002.88	$X_C(Y_D - Y_B) = 1,002.88(1,120.20 - 1,055.95) = 64,435$
	+2.33	+99.99	$X_D(Y_E - Y_C) = 1,102.87(1,091.06 - 1,117.87) = -29,568$
D	1,120.20	1,102.87	$X_E(Y_F - Y_D) = 1,115.54(1,030.55 - 1,120.20) = -100,008$
	−29.14	+12.67	$X_F(Y_G - Y_E) = 1,082.16(990.82 - 1,091.06) = -108,476$
E	1,091.06	1,115.54	$X_G(Y_A - Y_F) = 1,082.16(1,000.00 - 1,030.55) = -33,060$
	−60.51	−33.38	Double area $= -23,274$
F	1,030.55	1,082.16	Area $= 11,637$ ft^2 or m^2
	−39.73	0.00	
G	990.82	1,082.16	
	+9.18	−82.16	
A	1,000.00	1,000.00, Check	

CHAPTER 9

Solutions:

9.1 *How do scale and resolution affect the creation and operation of a GIS?*
Scale pertains to GIS design at two levels: First deals with at what display scale should certain information be displayed. This affects the detail of the information shown and the look of the data. The second issue with scale is what resolution can values be stored at. If the scale is going to be such that the whole country or continent needs to be shown, there is a limit as to the precision that information can be shown at. There is a limit as to how many significant figures can be stored. It is not practical to show the whole continent but then to have resolutions to the nearest millimeter. So, when designing a GIS some thought as to the coverage and quality of information must be taken into consideration.

9.2 *What are the various ways that scale can be shown on a map?*
Directly scale can be shown by a scale bar, which shows ground distances along a bar at specific intervals. Scale can be shown by text, such as a representative fraction 1:24,000, which is unit less. The value on the left "1" indicates the measure on the map, and the value on the right "24,000" indicates the measure on the ground. A measure of 1 cm on the map equals 24,000 cm on the ground. Text scales can also have units, such as 1 in. = 2,000 ft. In this case 1 in. on the map equals 2,000 ft on the ground. Note this is the same as 1:24,000 as there are 24,000 in. in 2,000 ft.

Additionally, scale of a map can be determined by having a known distance on the ground and measuring that distance on the map. Say the distance between two fence corners was $\frac{1}{2}$ mile (2,640 ft). The distance on a map measures 33.53 mm. The scale can be calculated as shown below:

$$\text{Scale} = \frac{33.53\,\text{mm}}{2,460\,\text{ft}} \times \frac{1\,\text{ft}}{304.8\,\text{mm}}$$

$$\text{Scale} = \frac{33.53}{804,672} \approx \frac{1}{24,000}$$

Therefore the scale = 1:24,000

9.5 *List and describe as many GIS applications as you can.*
This list is endless these days, but some common ones are as follows:

- Car direction systems
- Google Earth
- Weather predictions
- Survey cadastre
- Census
- Fire management
- Urban and regional planning
- Facilities management
- Military operations
- Precision agriculture
- Forestry management
- Logistics
- Utility mapping
- Environmental management
- Disaster management

CHAPTER 10

10.1

	Average factors	
	Elevation	Scale
A to B; N = –77.773; E = –280.126		
Distance AB = 290.722; Brg = S 74°29'00.4" W	99997166	99990182
B to C; N = –35.982; E = –223.047		
Distance BC = 225.931; Brg = S 80°50'09.5" W	99997094	99990175
C to D; N = +2.638; E = +206.021		
Distance CD = 206.038 m; Brg = N 89°15'59.0" E	99997046	99990156
D to A; N = +111.117; E = +297.152		
Distance DA = 317.248; Brg = N 69°29'50.0" E	99997118	99990163

10.2

A	4°59'10.4"
B	186°21'09.1"
C	8°25'49.5"
D	160°13'51.0"
	358°118'120.0"
	= 360°00'00.0"

10.3 Use Equations 10.1 and 10.2, or Table 10-7—converted to meters.

Monument	Grid factor	Grid dist.	Ground dist.	Course
A	0.999874			
B	0.999873	290.723	290.759	AB
C	0.999872	225.931	225.960	BC
D	0.999872	206.038	206.064	CD
A	0.999874	317.248	317.288	DA

10.4 Use Equation 14.5 ($\gamma = 52.09\ d \tan \varphi$), or Equation 14.6 [$\gamma = 32.370\ d$(km) $\tan \varphi$]

Course	Distance from C.M. (km) (midpoint of course)	Convergence	
		Γ''	Γ
AB	12.164	377.8"	0°06'17.5"
BC	11.912	370.0"	0°06'09.7"
CD	11.904	369.7"	0°06'09.4"
DA	12.155	377.6"	0°06'17.2"

Course	Grid bearing	Geodetic bearing
AB	S 74°29'00.4" W	S 74°35'17.9" W
BC	S 80°50'09.5" W	S 80°56'19.2" W
CD	N 89°15'59.0" E	N 89°22'08.4" E
DA	N 69°29'50.0" E	N 69°36'07.2" E

10.5 A 4°59'11"

CHAPTER 11

Solutions:

11.1 *What are the chief differences between maps based on satellite imagery and maps based on airborne imagery?*
Maps based on satellite imagery generally have a lower resolution (larger pixel size) than do maps based on airborne imagery. Satellite imagery, although usually geo-referenced, does not have high-precision positions, and generally not used for topographic mapping. Airborne imagery often is flown for precise photogrammetric mapping, requiring ground control and overlapping images.

11.2 *What are the advantages and disadvantages inherent in the use of hyperspectral sensors over multispectral sensors?*
The advantage with hyperspectral is the ability to look at a very narrow range on the electromagnetic spectrum. This allows for very detailed spectral signatures and classifications. The disadvantage is the amount of information that needs to be processed. Hyperspectral imagery can contain as much as 60 times more data than multispectral imagery for the same area, requiring much more processing overhead. Determining which set of layers to use to perform classification is more difficult with hyperspectral data, although the results can be far more detailed.

11.7 *What type of remote-sensing projects would Landsat 7 be appropriate for, and which ones would it not be appropriate for?*
The projects that benefit from Landsat can be determined by looking at the resolutions. Landsat has a low spatial resolution (30 m), but a high temporal resolution (16 days) and spectral resolution (eight sensors). This means that worldwide crop monitoring would be ideal with Landsat. Mapping of large features such as forest or desert boundaries would be an appropriate Landsat project. Also mapping of thermal change over large areas would be an appropriate use.

Using Landsat to do urban planning would not be appropriate, nor would the use of Landsat for topographic mapping or route design.

CHAPTER 12

12.1 (a) $H = SR.f = 20,000 \times 0.153 = 3,060$ m

(b) Altitude $= 3,060 + 180 = 3,240$ m

12.2 (a) Photo scale $= 23.07 \times 1{:}50{,}000/4.75$
$= 1{:}10{,}295$

12.3 (a) No. of photos required $= 30 \times 45/1$
$$\frac{(10{,}000)^2}{(30{,}000)^2} = 150$$

12.4 Assume that the negative size is 9 in. (228 mm) square, the normal size.

(a) Dimension across flight line $= 10{,}000 \times 0.228 = 2{,}280$ m or $10{,}000 \times 0.75 = 7{,}500$ ft
Dimension along flight line $= 2{,}280 \times 0.60 = 1{,}368$ m or $7{,}500 \times 0.60 = 4{,}500$ ft

(b) SR is $1{:}400 \times 12$ or $1{:}4{,}800$
Dimension across flight line $= 4{,}800 \times 0.228 = 1{,}094$ m or $4{,}800 \times 0.75 = 3{,}600$ ft
Dimension along flight line $= 1{,}094 \times 0.60 = 656$ m or $3{,}600 \times 0.60 = 2{,}160$ ft

12.5 (a) i Ground speed of aircraft $= 350$ km/hr
$$= \frac{350 \times 1{,}000}{60 \times 60} = 97.2 \text{ m/s}$$
Camera would move $97.2 \times \dfrac{1}{100} = 0.972$ m during exposure.

12.6 A longer focal length decreases the relief displacement. The explanation for this is evident from examining Figure 12-3(b). A longer focal length increases the distance from C to the focal plane. Therefore, the relief displacement, such as *bb'* and *cc'* is reduced.

CHAPTER 13

13.1 $T = R \tan \Delta/2 = 700 \tan 14°51'$
$\quad\quad = 185.60\,\text{ft}$

$$L = \frac{2\pi R\Delta}{360} = 2\pi700 \times \frac{29.7}{360} = 362.85\,\text{ft}$$

13.2 $R = \dfrac{5,729.58}{D} = \dfrac{5,729.58}{8} = 716.20\,\text{ft}$

$\quad\quad T = R \tan \Delta/2 = 716.20 \tan 3°35'$
$\quad\quad\quad = 44.85\,\text{ft}$

$$L = \frac{100\Delta}{D} = \frac{100 \times 7.166}{8} = 89.58\,\text{ft}$$

13.3
PI @	9	+ 27.26
$-T$	1	85.60
BC =	7	+ 41.66
$+L$	3	62.85
EC =	11	+ 04.51

13.4
PI @	15	+ 88.10
$-T$		44.85
BC =	15	+ 43.25
$+L$		89.58
EC =	16	+ 32.83

13.5

Curve #1
$\quad\Delta = 12°30'$
$\quad R = 600\,\text{ft}$
$\quad T_1 = 65.71$
$\quad L_1 = 130.90$

Curve #2
$\quad\Delta = 10°56'$
$\quad R = 600\,\text{ft}$
$\quad T_2 = 57.42$
$\quad L_2 = 114.49$

PI$_1$ at	3 + 81.27		PI$_2$ at 5 + 42.30	
$-T$		65.71	PI$_1$ at 3 + 81.27	
BC$_1$ =	3 + 15.56		Diff. = 161.03	
$+L_1$	1	30.90	$T_1 + T_2 = -123.13$	
			65.71	

EC$_1$ = 4 + 46.46 EC$_1$ to BC$_2$ = 37.90 + 57.42
$\quad\quad\quad\quad\quad\quad\quad\quad\quad\quad\quad\quad\quad = 123.13$

EC$_1$ to BC$_2$ =	37.90
BC$_2$	4 + 84.36
$+L_2$	1 14.49
EC$_2$	5 + 98.85

13.12 Measured values:
$\quad AB = 615.27\,\text{ft}$
$\quad\quad \alpha = 51°31'20''$
$\quad\quad \beta = 32°02'45''$
$\quad\quad \Delta = \alpha + \beta = 83°34'05''$
$\quad\quad T = R \tan \Delta/2$
$\quad\quad\quad = 1,000 \tan 41°47'02''$
$\quad\quad\quad = 893.60\,\text{ft}$

$$\text{Side A, PI} = \sin 32°02'45'' \times \frac{615.27}{\sin 96°25'55''}$$
$$= 328.53\,\text{ft}$$

$$\text{Side PI, B} = \sin 51°31'20'' \times \frac{615.27}{\sin 96°25'55''}$$
$$= 484.71\,\text{ft}$$

To locate *BC*, Layout (893.60 − 328.53); i.e., <u>565.07</u> back from point *A*.

To locate *EC*, Layout (893.60 − 484.71); i.e., <u>408.89</u> forward from point *B*.

13.14

Station	₵ grade	Stake elevation	Cut	Fill	Cut	Fill
0 + 00	472.70	472.60		0.10		$0'1\frac{1}{4}''$
1 + 00	474.02	472.36		1.66		$1'8''$
2 + 00	475.34	473.92		1.42		$1'5''$
3 + 00	476.66	475.58		1.08		$1'1''$
4 + 00	477.98	478.33	0.35		$0'4\frac{1}{4}''$	
5 + 00	479.30	479.77	0.47		$0'5\frac{1}{2}''$	
6 + 00	480.62	480.82	0.20		$0'2\frac{3}{8}''$	

13.15

Station	₡ grade	Stake elevation	Cut	Fill
0 + 00	210.500	210.831	0.331	
0 + 20	210.365	210.600	0.235	
0 + 40	210.230	211.307	1.077	
0 + 60	210.094	210.114	0.020	
0 + 80	209.959	209.772	0.187	
1 + 00	209.823	209.621	0.202	
1 + 20	209.688	209.308	0.380	
1 + 32.562	209.603	209.400	0.203	

13.16 $24.5 + (3.5 \times 3) = 35$ ft from ₡

13.17 (a) $7.45 + (2 \times 3) = 13.45$ m from ₡

    (b) Ditch Invert is 2.67 below original ground; slope stake is $13.45 + (3 \times 2.67) = 21.46$ m from ₡

CHAPTER 14

Solutions:

14.1 *Calculate the angular convergency for two meridians 6 miles (9.66 km) apart at latitude 36°30′ and 46°30′.*

The convergence can be computed using the formula:

γ(seconds) $= 52.09''/$mi $\times$ d(miles) tan φ

Therefore at 36° 30'

γ(seconds) $= 52.09''/$mi $\times$ 6 tan (36.5)
$\gamma'' = 52.09'' \times 6 \times 0.739961$
$= 231'' = 0°3'51''$

At 46°30'

γ(seconds) $= 52.09''/$mi $\times$ 6 tan (46.5)
$\gamma'' = 52.09'' \times 6 \times 1.05378$
$= 3 29'' = 0°5'29''$

14.2 *Find the convergency in feet, chains, and meters of two meridians 24 miles apart if the latitude for the southern limit is 40°20′ and the latitude for the northern limit is 40°41′.*
First the approximate distance from 40°20′ and 40°41′ must be determined. Assume earth's radius as 20,906,000 ft; then the approximate distance would be calculated by

Distance $= 20,906,000$ ft
$\times$ tan $(40°41' - 40°20')$
Distance $= 20,906,000$ ft $\times$ tan $(21')$
$= 127,709$ ft $= 24.19$ mi

Determine the convergency p:

p(feet) $= 1.33$ ft/mi$^2 \times 24$ mi $\times 24.19$ mi
$\times$ tan $(40°30'30'') = 660$ ft
$= 660$ ft

p(chains) $= 0.0202$ ch/mi$^2 \times 24$ mi $\times 24.19$ mi
$\times$ tan $(40°30'30'')$
$= 10$ ch

p(meters) $= 0.4064$ m/mi$^2 \times 24$ mi $\times 24.19$ mi
$\times$ tan $(40°30'30'')$
$= 202$ m

14.4 *For a 6-mile-square sectional township, calculate the area of section 6, if the mean latitude through the middle of the township is 46°30′. Take only the effects of convergence into account in this calculation.*

p(ft) $= 1.33 \times 6 \times 6$ tan $46°30' = 50.4$ft
$= 0.0096$ mi

North boundary of Section 6 $= 1.0000 - 0.0096$
$= 0.9904$ mi

Area of Section 6 $= 1$ mi $\times (1.000 + 0.9904)/2$
$= 0.9952$ sq.mi. $= 636.9$ ac.

14.5 *In Figure 14-10 show how the iron monument would be marked for*

(a) *The northwest corner of township T.20N., R.7E.*

(b) *The southeast corner of section 32 of township T.20N., R.6E.*

Corner (a) is a closing township corner, so it would look something like:

Corner (b) is corner common to four sections and two townships, so it would look something like:

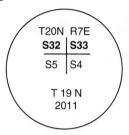

14.6 *Plot the following description (the property is located in a standard U.S. sectional system) to a scale of either 1:5,000 or 1 in. = 400 ft.*

Commencing at the southwest corner of Section 35 in Township T.10N., R.3W., thence N 0°05′ W along the westerly boundary of Section 35, 2,053.00 ft to a point therein; thence N 89°45′ E, 1,050.00 ft; thence southerly, parallel with the westerly limit of Section 35,670.32 ft; thence N 89°45′ E, 950.00 ft; thence southerly, parallel with the westerly limit of Section 35, 1,381.68 ft, more or less, to the point of intersection with the southerly boundary of Section 35; thence westerly along the southerly boundary of Section 35, 2,000.00 ft, more or less, to the point of commencement.

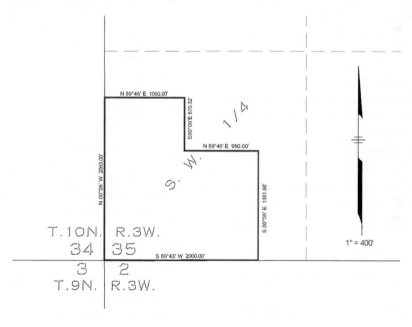

Appendix E

INTERNET WEBSITES

The websites listed here cover surveying, global positioning systems (GPSs), photogrammetry, geographic information systems (GISs), and mapping. Some sites include links to various related sites. Although the websites were verified at the time of writing, some changes are inevitable. Revised Web addresses and new sites may be accessed by searching the links shown at other sites listed here, or by conducting an online search via a search engine.

American Association of State Highway and Transportation Officials (AASHTO), http://www.transportation.org

American Congress on Surveying and Mapping (ACSM), http://www.acsm.net

American Society for Photogrammetry and Remote Sensing, http://www.asprs.org/

American Surveyor magazine, http://www.amerisurv.com/

Bennett (Peter) NMEA-0183 and GPS information, http://vancouver-webpages.com/pub/peter/

Canada Centre for Remote Sensing, http://www.ccrs.nrcan.gc.ca

Canadian Geodetic Survey, http://www.geod.nrcan.gc.ca

Centre for Topographic Information (Canada), http://www.cits.nrcan.gc.ca

CORS information, http://www.ngs.noaa.gov/CORS/

ESRI, http://www.esri.com

Flatirons Surveying, http://www.flatsurv.com

Galileo, http://www.gsa.europa.eu/

GeoCommunicator, NILS data, http://www.geocommunicator.gov/GeoComm

GeoConnections (Canada), http://www.geoconnections.ca

Geographic coordinate database, http://www.blm.gov/wo/st/en/prog/more/gcdb.html

Geospatial One-Stop (map viewer), http://www.geodata.gov

Glonass home page (Russian Federation), http://new.glonass-iac.ru/en/

GPS applications, http://www.gps.gov/applications/

GPS World magazine, http://gpsworld.com

Homeland Security, United States Coast Guard (USCG) Navigation Center, http://www.navcen.uscg.gov/

International Earth Rotation Service (IERS), http://www.iers.org

International GPS Service (IGS), http://igscb.jpl.nasa.gov

Land Surveying and Geomatics, http://surveying.mentabolism.org/

Leica Geosystems, Inc., http://www.leica-geosystems.com/

MAPINFO (mapping), http://www.pbinsight.com/welcome/mapinfo/

MicroSurvey (surveying and design software), http://www.microsurvey.com/

NASA, remote sensing tutorial, http://rst.gsfc.nasa.gov

National Atlas, http://www.nationalatlas.gov/

National Integrated Land System (NILS), http://www.geocommunicator.gov/GeoComm/

National Map, http://nationalmap.usgs.gov

Natural Resources, Canada, http://www.nrcan.gc.ca

NGS home page, http://www.ngs.noaa.gov/

Nikon, http://www.nikonpositioning.com/

Open GIS Consortium, http://www.opengeospatial.org/

Optech (airborne laser terrain mapper), http://www.optech.ca

POB Point of Beginning magazine, http://www.pobonline.com/

Professional Survey magazine, http://www.profsurv.com

Sokkia, http://www.sokkia.com/

Spectra Precision, http://www.spectraprecision.com/

Sun/Polaris tables (Jerry Wahl), http://www.cadastral.com

Topcon Instrument Corporation, http://www.topconpositioning.com/

Trimble, http://www.trimble.com

Trimble (GIS), http://www.trimble.com/mgis.shtml

United States Census Bureau, http://www.census.gov

United States Bureau of Land Management (BLM) (geographic coordinate database), http://www.blm.gov/wo/st/en/prog/more/gcdb.html

United States Bureau of Land Management (BLM) [National Integrated Lands System (NILS)], http://www.geocommunicator.gov/GeoComm/

United States Census Bureau Tiger files, http://www.census.gov/geo/www/tiger/

United States Coast Guard (USCG) Navigation Center, http://www.navcen.uscg.gov

United States Geological Survey (USGS), http://www.usgs.gov/

USGS, EROS Data Center, http://edc.usgs.gov/

University of Maine, http://spatial.umaine.edu/

EXAMPLES OF CURRENT GEOMATICS TECHNOLOGY

Appendix F appears at the end of this textbook in a full-color insert. The figures include examples of traditional mapping and geographic information system (GIS) mapping, aerial imaging, satellite imaging, and soft-copy photogrammetry equipment.

TYPICAL FIELD PROJECTS

The following projects can be performed in either foot units or metric units and can be adjusted in scope to fit the available time.

G.1 FIELD NOTES

Survey field notes can be entered into a bound field book or on loose-leaf field notepaper; if a bound field book is used, be sure to leave room at the front of the book for a title, index, and diary, as required by your instructor. The following list of instructions will guide you in the proper use and layout of a field book:

1. Write your name in ink on the outside cover.
2. Page numbers (e.g., 1–72) are to be on the right side only.
3. All entries are to be in pencil, 2H or 3H.
4. See Figure G-1 for examples of pages 1–7.

5. All entries are to be printed (uppercase, if permitted).
6. Calculations are to be checked and initialed.
7. Sketches are to be used to clarify field notes. Orient the sketch so that the included north arrow is pointing toward the top of the page.
8. All field notes should be updated daily.
9. Show the word *copy* at the top of all copied pages.
10. Field notes are to be entered directly in the field book—not on scraps of paper to be copied later.
11. Mistakes in entered data are to be crossed out, not erased.
12. Spelling mistakes, calculation mistakes, and the like should be erased and reentered correctly.

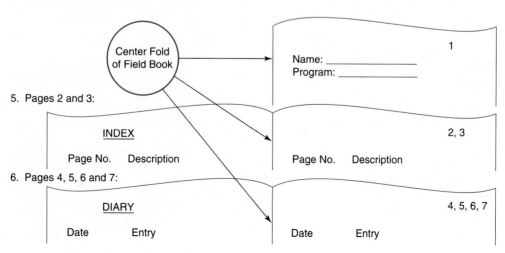

FIGURE G-1 Field book layout

13. Lettering is to be read from the bottom of the page or the right side.

14. The first page of each project should show the date, temperature, project title, crew duties, and so on.

15. The diary should show absentees, weather, description of work, and so on.

G.2 PROJECT 1: BUILDING MEASUREMENTS

Description You will measure the selected walls of an indicated campus building with a cloth or fiberglass tape, and record the measurements on a sketch in the field book, as directed

by the instructor (see Figure G-2 for sample field notes).

Equipment Cloth or fiberglass tape (100 ft or 30 m).

Purpose To introduce you to the fundamentals of note keeping and survey measurement.

Procedure Use the measuring techniques described in class prior to going out.

One crew member will be appointed to take notes for this first project. At the completion of the project (same day), the other crew members will copy the notes (ignoring erroneous data) into their field books, including the diary and index data. (Crew members will take equal turns acting as note keeper over the length of the program.)

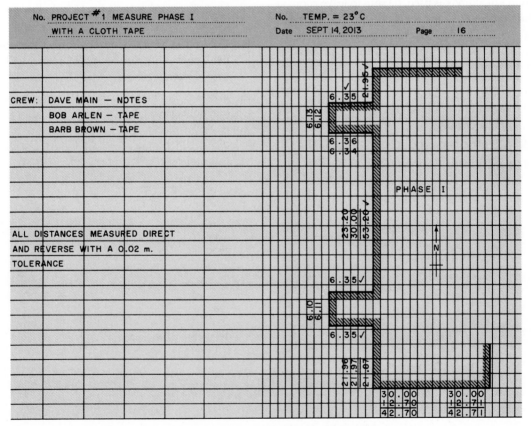

FIGURE G-2 Sample taping field notes for building dimensions—Project 1

Using a straightedge, draw a large sketch of the selected building walls on the right-hand (grid) side of your field book. Show the walls as they would appear in a plan view. For example, ignore overhangs, or show them as dashed lines. Keep the tape taut to remove sag, and try to keep the tape horizontal. If the building wall is longer than one tape length, make an intermediate mark (do not deface the building), and proceed from that point.

After completing all the measurements in one direction, start from the terminal point, and remeasure all the walls. If the second measurement agrees with the first measurement, put a check mark beside the entered data. If the second measurement agrees acceptably (e.g., within ± 0.10 ft or 0.02 m), enter that measurement directly above or below the first entered measurement. If the second measurement disagrees with the first (e.g., by more than 0.10 ft or 0.02 m), enter that value on the sketch, and measure that dimension a third time. Discard the erroneous measurement by drawing a line (using a straightedge) through the erroneous value.

Discussion   If the class results are summarized on the chalkboard, it will be clear that all crews did not obtain the same results for specified building wall lengths. There will be much more agreement among crews on the lengths of the shorter building walls than on the lengths of the longer building walls (particularly the walls that were longer than one tape length). Discuss and enumerate the types of mistakes and errors that can account for measurement discrepancies among survey crews working on this project.

G.3 PROJECT 2: EXPERIMENT TO DETERMINE "NORMAL TENSION"

Description   Project 2 is an experiment in which you will determine the tension required to eliminate errors

due to tension and sag for a 100-ft or 30-m steel tape supported only at the ends. This tension is called normal tension.

Equipment   Steel tape (100.00 ft or 30.000 m), two plumb bobs, and a graduated tension handle.

Purpose   To introduce you to measurement techniques requiring the use of a steel tape and plumb bobs and to demonstrate the "feel" of the proper tension required when using a tape that is supported only at the ends (the usual case).

Procedure

- With a 100-ft or 30-m tape fully supported on the ground and under a tension of 10 lbs or 50 N (standard tension), as determined by use of a supplied tension handle, measure from the initial mark and place a second mark at exactly 100.00 ft or 30.000 m.

- Check this measurement by repeating the procedure (while switching personnel) and correcting if necessary. If this initial measurement is not performed correctly, much time will be wasted.

- Raise the tape off the ground and keep it parallel to the ground to eliminate slope errors.

- Using plumb bobs and the tension handle, determine how many pounds or newtons of tension (see Table 3-1) are required to force the steel tape to agree with the previously measured distance of 100.00 ft or 30.000 m.

- Repeat the process after switching crew personnel. (Acceptable tolerance is ± 2 lbs.)

- Record the normal tension results (at least two) in the field book, as described in the classroom.

- Include the standard conditions for the use of steel tapes in your field notes. See Table G-1.

Discussion   If the class results are summarized on the chalkboard, it will be clear that not all survey crews obtained

Table G-1   Standard Conditions for the Use of Steel Tapes*

English System	or	Metric System
Temperature = 68°F		Temperature = 20°C
Tension = 10 lbs		Tension = 50 N (11.2 lbs)
Tape is fully supported.		Tape is fully supported.

*For this project, you can assume that the temperature is standard, 68°F or 20°C.

the same average value for normal tension. Discuss the reasons for the tension measurement discrepancies and agree on a working value for normal tension for subsequent class projects.

G.4 PROJECT 3: FIELD MEASUREMENTS WITH A STEEL TAPE*

Description   You will measure the sides of a five-sided closed field traverse using techniques designed to permit a precision closure ratio of 1:5,000 (see Table 3-2). The traverse angles will be obtained from Project 5. See Figure G-3 for sample field notes.

Equipment   Steel tape, two plumb bobs, hand level, range pole or plumb bob string target, and chaining pins or other devices to mark the position on the ground.

Purpose   To develop some experience in measuring with surveying equipment (see also Project 5).

	COURSE	DIRECT	REVERSE	MEAN	MEAN (C_T)
	111–112	164.96	164.94	164.95	164.97
	112–113	88.41	88.43	88.42	88.43
	113–114	121.69	121.69	121.69	121.70
	114–115	115.80	115.78	115.79	115.80
	115–111	68.36	68.34	68.35	68.36

No. PROJECT #3   TRAVERSE DISTANCES   No.   Date MARCH 20, 2013   Page 10

BROWN–NOTES
FIELDING–TAPE
SIMPSON–TAPE
TEMP.=83°F

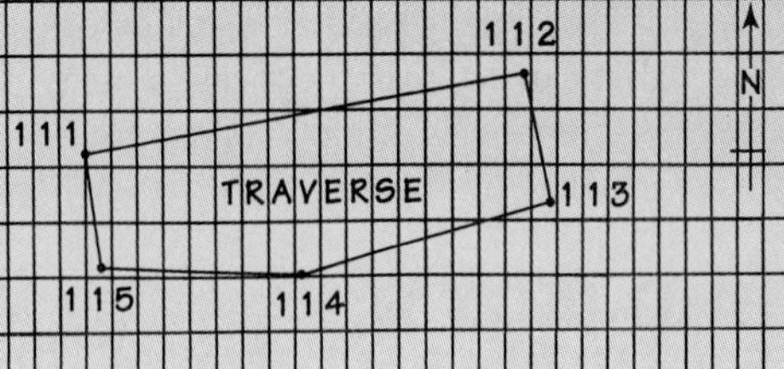

TRAVERSE

FIGURE G-3 Sample field notes for Project 3 (traverse distance)

*Projects 3 and 5 can be combined, and you can use EDM-equipped theodolites or total stations. The traverse courses can be measured using EDM and a prism (pole-mounted or tribrach-mounted). Each station will be occupied with a theodolite-equipped EDM instrument, or total station, and each pair of traverse courses can be measured at each setup. Traverse computations will use the mean distances thus determined and the mean angles obtained from each setup. Refer to Chapters 3, 7, and 9.

Procedure for Steel Taping

- Each course of the traverse will be measured twice, forward (direct) and then immediately back (reverse), with the two measurements agreeing to within 0.03 ft or 0.008 m. If the two measurements do not agree, they will be repeated until they do and before the next course of the traverse is measured.

- When measuring on a slope, the high-end surveyor normally holds the tape directly on the mark, and the low-end surveyor has to use a plumb bob to keep the tape horizontal.

- The low-end surveyor uses the hand level to keep the tape approximately horizontal by sighting the high-end surveyor and noting how much lower she or he is in comparison. The plumb bob is then set to that height differential. Use chaining pins or other markers to mark intermediate measuring points on the ground temporarily. Use scratch marks or concrete nails on paved surfaces.

- If a range pole is first set behind the far station, the rear surveyor can keep the tape aligned properly by sighting at the range pole and directing the forward surveyor on line.

- Record the results as shown in Figure G-3, and then repeat the process until all five sides have been measured and booked. When booked erroneous measurements are to be discarded, cross them out with a straightedge; don't erase.

- If the temperature is something other than standard, correct the mean distance for temperature (C_T); that is, $C_T = .00000645(T - 68°F)$ L_{ft} or $C_T = 0.0000116(T - 20°C)L_m$.

Reference   Chapter 3.

G.5 PROJECT 4: DIFFERENTIAL LEVELING

Description   You will use the techniques of differential leveling to determine the elevations of a temporary benchmark (TBM), and of the intermediate stations (if any) identified by the instructor. See Figure G-4 for sample field notes.

Equipment   Survey level, rod, and rod level (if available).

Purpose   To give you experience using levels and rods and recording all measurements correctly in the field book.

Procedure

- Refer to Figure G-4.

- Start at the closest municipal or college benchmark (BM) (description given by the instructor), and take a backsight (BS) reading to establish a height of instrument (HI).

- Insert the description of the BM (and all subsequent TPs), in detail, under Description in the field notes.

- Establish a turning point (TP 1) generally in the direction of the defined terminal point (TBM 33) by taking a foresight (FS) on TP 1.

- When you have calculated the elevation of TP 1, move the level to a convenient location, and set it up again. Take a BS reading on TP 1, and calculate the new HI.

- The rod readings taken on any required intermediate points (on the way to or from the terminal point) will be booked in the Intermediate Sight (IS) column, unless some of those intermediate points are also being used as turning points (TPs). See, for example, TP 4.

- If you can't "see" the terminal point (TBM 33) from the new instrument location, establish additional turning points (TP 2, TP 3, etc.), and repeat the above steps until you can take a reading on the terminal point.

- After you have taken a reading (FS) on the terminal point (TBM 33) and calculated its elevation, move the level slightly and set it up again. Now, take a BS on the terminal point (TBM 33), and prepare to close the level loop back to the starting benchmark.

- Repeat the leveling procedure until you have surveyed back to the original BM. If you use the original TPs on the way back, book them

No. PROJECT #4
DIFFERENTIAL LEVELING

No.
Date MARCH 1, 2013 Page 11

STA	BS	HI	IS	FS	ELEV	DESCRIPTION
						BROWN–INST.
						SMITH–ROD
						TEMP.=65°F
BM 21	0.54	182.31			181.77	BM. BRONZE PLATE ON E. WALL OF S.E.
						STAIRWELL OF PHASE 1 BLDG., ABOUT
						1 m ABOVE THE GROUND.
TP 1	0.95	175.04		8.22	174.09	N. LUG ON TOP FLANGE OF HYD.@
						E/SIDE OF BUS SHELTER
TP 2	0.80	168.76		7.08	167.96	SPIKE IN S/SIDE OF HP @ 237 FINCH
						AVE.
TP 3	0.55	160.20		9.11	159.65	SPIKE IN S/SIDE OF HP @ 245 FINCH
						AVE.
111			4.22			TRAVERSE STATION I.B.
112			4.71			TRAVERSE STATION I.B.
113			2.03			TRAVERSE STATION I.B.
114			1.22			TRAVERSE STATION I.B.
TP 4	3.77	163.45		0.52	159.68	TOP OF I.B. @ STA. 115
TBM 33				1.18	162.27	BRASS CAP ON CONC. MON.-CONTROL
						STATION 1102
TBM 33	1.23	163.50			162.27	
TP 4	2.71	162.39		3.82	159.68	
TP 3	8.88	168.53		2.74	159.65	
TP 5	11.86	177.38		3.01	165.52	TOP OF N.E. CORNER OF CONCRETE
						STEP @ 233 FINCH AVE.
TP 1	10.61	184.72		3.27	174.11	
BM 21				2.94	181.78	(e=+0.01)
ΣBS	= 41.90		ΣFS	= 41.89		
	ΣBS,	41.90	– ΣFS,	41.89 =	0.01	181.77 +.01 = 181.78, CHECK

After the arithmetic check has been successfully applied, compute the elevations for the intermediate sights—IS.

FIGURE G-4 Sample field notes for Project 4 (differential leveling)

by their original numbers; you do not have to describe them again. If you use new TPs on the way back, describe each TP in detail under Description, and assign each one a new number.

- If the final elevation of the starting BM differs by more than 0.04 ft or 0.013 m from the starting elevation (after the calculations have been checked for mistakes by performing an "arithmetic check," also known as a "page check"), repeat the project. Your instructor may give you a different closure allowance, depending on the distance leveled and the type of terrain surveyed.

Notes

- Keep BS and FS distances from the instrument roughly equal.
- "Wave" the rod (or use a rod level) to ensure a vertical reading.

- Eliminate parallax for each reading.
- Use only solid (steel, concrete, or wood) and well-defined features for TPs. If you cannot describe a TP precisely, do not use it!
- Perform an arithmetic check on the notes before assessing closure accuracy.

Reference   Chapter 2.

G.6 PROJECT 5: TRAVERSE ANGLE MEASUREMENTS AND CLOSURE COMPUTATIONS

Description   You will measure the angles of a five-sided field traverse (see also Project 3), using techniques consistent with the desired precision ratio of 1/5,000.

Equipment   Theodolite or total station and a target device (range pole, plumb bob string target, or prism).

Purpose   To introduce you to the techniques of setting up a theodolite or total station over a closed traverse point, turning and "doubling" interior angles, and checking your work by calculating the geometric closure. Then compute (using latitudes and departures) a traverse closure to determine the precision ratio of your fieldwork. (If you are using a total station with a traverse closure program, use that program to check your calculator computations.) For traverse computation purposes, assume a direction for one of the traverse courses, or use one supplied by your instructor. See Figure G-5 for sample field notes.

Procedure

- Using the same traverse stations that were used for Project 3, measure each of the five angles (direct and double).

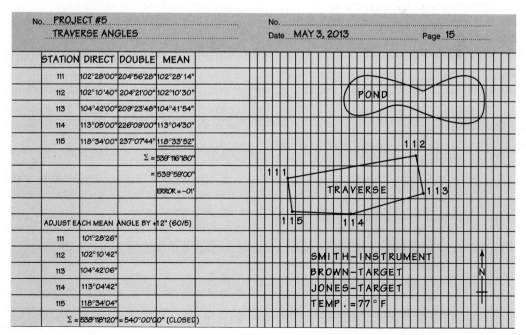

FIGURE G-5 Sample field notes for Project 5 (traverse angles)

- Read all angles from left to right. Begin the first (of two) angles at 0°00′00″ ("direct"), and begin the second angle ("double") with the value of the first angle reading.

- Transit the telescope between the direct and double readings.

- Divide the double angle by 2 to obtain the "mean" angle. If the mean angle differs by more than 30″ (or a value given by your instructor) from the direct angle, repeat the procedure.

- When all the mean angles have been booked, add the angles to determine the geometric closure.

- If the geometric closure exceeds 01′(30″$\sqrt{N}$) find the error.

- Combine the results from Projects 3 and 5 to determine the precision closure of the field traverse (1/5,000 or better is acceptable for taping surveys). Use an assumed direction for one of the sides (discussed earlier).

G.7 PROJECT 6: TOPOGRAPHIC FIELD SURVEYS

Topographic field surveys can be accomplished in several ways, for example:

- Cross sections and tie-ins, with a manual plot of the tie-ins, cross sections, and contours.

- Theodolite/EDM, with a manual plot of the tie-ins and contours.

- Total station, with a computer-generated plot on a digital plotter.

Purpose   Each type of topographic survey shown in this section is designed to give you experience in collecting field data (location details and elevations) using different specified surveying equipment and surveying procedures. These different approaches to topographic surveying have the same objective: the production of a scaled map or plan showing all relevant details and height information (contours and spot elevations) of the area surveyed. Time and schedule constraints will normally limit most programs to include just one or two of these approaches.

G.7.1 Cross Sections and Tie-Ins Topographic Survey

Description   Using the techniques of right-angled tie-ins and cross sections (both referenced to a baseline), you will locate the positions and elevations of selected features on the designated area of the campus. (See Figures G-6–G-8.)

Equipment   Cloth or fiberglass tape, steel tape, and two plumb bobs. A right-angle prism is optional.

Procedure

- Establish your own baseline using wood stakes or pavement nails. You can also use a curb line, as shown in Figure G-6, as the survey baseline (use keel or other nonpermanent markers to mark the baseline).

- The point of intersection of your baseline with some other line (or point, as defined by your instructor) will be 0 + 00.

- Measure the baseline stations (e.g., 50 ft or 20 m) precisely with a steel tape, and mark them clearly on the ground. Use keel marks, nails, or wood stakes.

- Determine the baseline stations of all features left and right of the baseline by estimating 90° (swung-arm technique) or by using a right-angle prism.

- When you have recorded the baseline stations of all the features in a 50- to 100-ft, or 20- to 30-m interval (the steel tape can be left lying on the ground on the baseline), determine and record the offset (o/s) distances left and right of the baseline to each feature. Tie in all detail to the closest 0.10 ft or 0.03 m. (A fiberglass tape can be used for these measurements.)

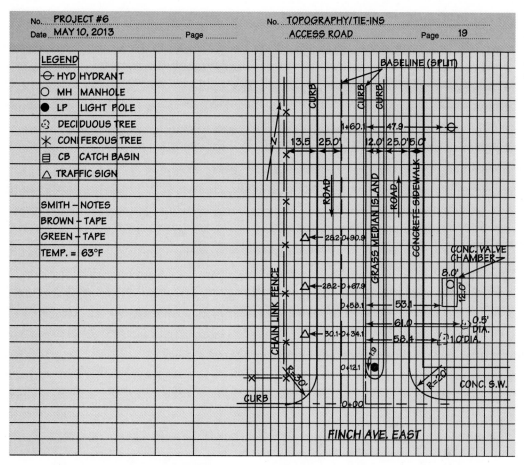

FIGURE G-6 Sample field notes for Project 6 (topography tie-ins)

- Do not begin the measurements until the sketches have been made for the survey area.

- Elevations should be determined using a level and a rod. The level should be set up in a convenient location where a benchmark (BM) and a considerable number of intermediate sights (ISs) can be "seen." See Figure G-7.

- Hold the rod on the baseline at each station and at all points where the ground slope changes (e.g., top and bottom of curb, edge of walk, top of slope, bottom of slope, limit of survey). See the typical cross section illustrated in Figure G-8.

- When all the data (that can be "seen") have been taken at a station, the rod holder then moves to the next (50-ft or 20-m) station and repeats the process.

- When the rod can no longer be seen, the instrument operator calls for the establishment of a turning point (TP). After taking a foresight to the new TP, the instrument operator moves the instrument closer to the next stage of work and a backsight is taken to the new TP before continuing with the cross sections. In addition to cross sections at the even-station intervals (e.g., 50 ft, 20 m), take full or partial sections between the even stations if the lay of the land changes significantly.

STA.	B.S.	H.I.	I.S.	F.S.	ELEV.	DESCRIPTION
BM 3	8.21	318.34			310.13	S. W. CORNER OF CONC. VALVE CHAMBER
						@ 0+53.1
0+00			3.34		315.00	₵, ON ASPH.
			0.03		318.31	75.0' LT, ON ASPH.
			7.35		310.99	50.0' RT, ON ASPH.
0+50			6.95		311.39	₵, ON ASPH.
			7.00		311.34	25' LT, BOT. CURB
			0.3		318.0	38.5' LT, @ FENCE, ON GRASS
			7.5		310.8	6' RT, ₵ OF ISLAND, ON GRASS
			8.32		310.02	12' RT, BOT. CURB
			8.41		309.93	37.0' RT, BOT. CURB
			8.91		309.43	37.0' RT, TOP OF CONC. WALK
			9.01		309.33	42.0' RT, TOP OF CONC. WALK
			9.3		309.0	46.9' RT, TOP OF HILL, ON GRASS
			11.7		306.6	56.6' RT, @ BUILDING WALL, ON GRASS

No. PROJECT #6
Date MAY 17, 2013 Page

No. TOPOGRAPHY/CROSS SECTIONS
FINCH AVE. E. TO 3+00N. Page 23

FIGURE G-7 Sample field notes for Project 6 (topography cross sections)

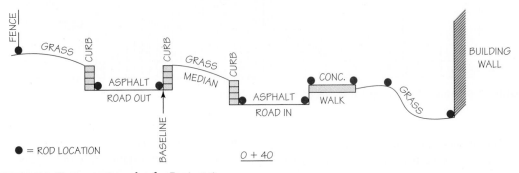

FIGURE G-8 Cross-section plot for Project 6

G.7.2 Theodolite/EDM Topographic Survey

Description Using electronic distance measurement (EDM) instruments and optical or electronic theodolites, you will locate the positions and elevations of all topographic detail and a sufficient number of additional elevations to create a representative contour drawing of the selected areas. See the sample field notes in Figure G-9.

Equipment Theodolite, EDM, and one or more pole-mounted reflecting prisms.

Procedure

- Set the theodolite at a control station (northing, easting, and elevation known), and backsight on another known control station.

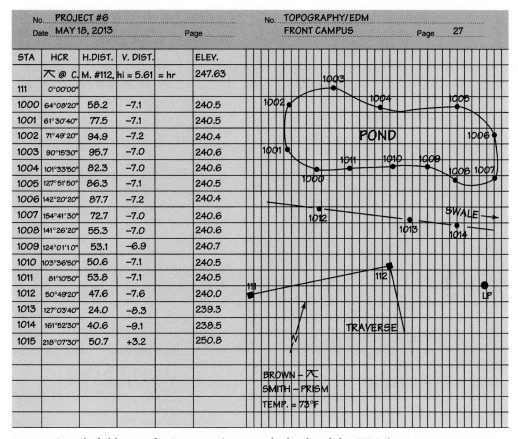

STA	HCR	H.DIST.	V. DIST.		ELEV.
	⌇ @ C.M. #112, hi = 5.61		= hr		247.63
111	0°00'00"				
1000	64°08'20"	58.2	−7.1		240.5
1001	61°30'40"	77.5	−7.1		240.5
1002	71°49'20"	94.9	−7.2		240.4
1003	90°15'30"	95.7	−7.0		240.6
1004	101°33'50"	82.3	−7.0		240.6
1005	127°51'50"	86.3	−7.1		240.5
1006	142°20'20"	87.7	−7.2		240.4
1007	154°41'30"	72.7	−7.0		240.6
1008	141°26'20"	55.3	−7.0		240.6
1009	124°01'10"	53.1	−6.9		240.7
1010	103°36'50"	50.6	−7.1		240.5
1011	81°10'50"	53.8	−7.1		240.5
1012	50°49'20"	47.6	−7.6		240.0
1013	127°03'40"	24.0	−8.3		239.3
1014	161°52'30"	40.6	−9.1		238.5
1015	218°07'30"	50.7	+3.2		250.8

FIGURE G-9 Sample field notes for Project 6 (topography by theodolite/EDM)

- Set an appropriate reference angle (e.g., 0°00'00" or some assigned azimuth) on the horizontal circle.

- Set the height of the reflecting prisms (HR) on the pole equal to the height of the optical center of the theodolite/EDM (hi). If the EDM is not coaxial with the theodolite, set the height of the target (target/prism assembly) equal to the optical center of the instrument. Take all vertical angles to the prism target, or to the center of the prism if the EDM is coaxial with the theodolite.

- Prepare a sketch of the area to be surveyed.

- Begin taking readings on the appropriate points and enter the data in the field notes (see Figure G-9) and enter the "shot" number in the appropriate spot on the accompanying field-note sketch. Keep shot numbers sequential, perhaps beginning with 1,000. Work is expedited if two prisms are employed. While one prism holder is walking to the next "shot" location, the instrument operator can take a reading at the other prism holder.

- When all field shots (horizontal and vertical angles and horizontal distances) have been taken, sight the reference backsight control station again to verify the original angle setting; also verify that the height of the prism is unchanged.

- Reduce the field notes to determine station elevations and course distances, if required.

- Plot the topographic features and elevations at $1' = 40''$ or 1:500 metric (or at another scale, as given by your instructor).
- Draw contours over the surveyed areas. See Chapter 8.

G.7.3 Total Station Topographic Survey

Description   Using a total station and one or more pole-mounted reflecting prisms, you will tie in all topographic features and any additional ground shots (including break lines) that are

required to define the terrain accurately. See Sections 5.15, 5.14.5, and 5.15.5, as well as Figure G-10.

Equipment   Total station and one or more pole-mounted reflecting prisms.

Procedure

- Set the total station over a known control point (northing, easting, and elevation known).* Turn on the instrument, and index the circles if necessary (by transiting the telescope and revolving the instrument 360°). Some newer total stations do not require this operation.

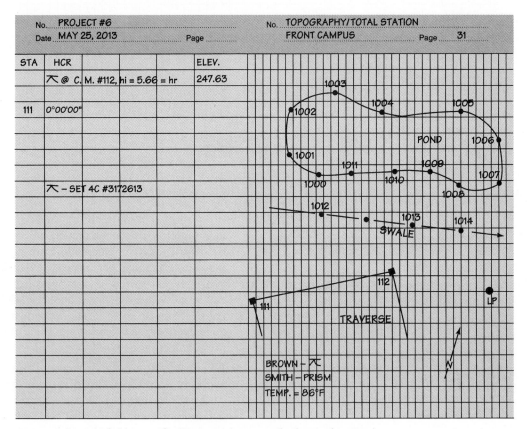

FIGURE G-10 Sample field notes for Project 6 (topography by total station)

---

*As an alternative, the station can be set in any convenient location, and its position can be determined using the onboard resection program after the required number of visible control stations are sighted.

- Set the program menu to the type of survey (topography) being performed and to the required instrument settings. Select the type of field data to be stored (e.g., N, E, and Z; E, N, and Z; etc.). Set temperature and pressure settings, if required.

- Check the configuration settings [tilt correction, coordinate format, zenith vertical angle, angle resolution (e.g., $5''$), $c + r$ correction (no), units (ft/m, degree, mm Hg), auto power off (say, 20 minutes)].

- Identify the instrument station from the menu. Insert the date, station number coordinates, elevation, and hi. It may have been possible to upload all control station data prior to going out to the field. In that case, scan through the data, and select the appropriate instrument station and backsight station(s). Enter the height of the instrument (hi), and store or record all the data.

- Backsight to one or more known control point(s) (point number, north and east coordinates, and known elevation). Set the horizontal circle to $0°00'00''$ or to some assigned reference azimuth for the backsight reference direction. Store or record the data.

- Set the initial topography point number in the instrument (e.g., 1,000), and set for automatic point number incrementation—if desired. Adjust the height of the reflecting prism (HR) equal to the instrument hi—if desired.

- Begin taking intermediate sights. Provide an attribute code (consistent with the software code library; see Table G-2 for an example) for each reading. Some software programs enable attribute codes to provide automatically for feature "stringing" (e.g., curb1, edge of water1, fence3). See Chapter 5. Most total stations have an automatic mode for topographic surveys, where one button-push will measure and store all the point data as well as the code and attribute data. The code and attribute data of the previous point are usually presented to the surveyor as a default setting. If the code and attribute data are the same for a series of readings, the surveyor only has to press "enter" and not enter all that identical data.

- Put all or some selected point numbers on the field sketch. These field notes will be of assistance later in the editing process if mistakes have occurred in the numbering or the coding.

- When all required points have been surveyed, check into the control station originally backsighted to ensure that the instrument orientation is still valid.

- Transfer the field data into a properly labeled file in a computer.

- After opening the data-processing program, import the field data file and begin the editing process and the graphics generation process (this is automatic for many programs).

- Create the TIN and contours.

- Either finish the drawing with the working program or create a dxf file for transfer to a CAD program. Then finish the drawing.

- Prepare a plot file, and then plot the data (to a scale assigned by your instructor) on the lab digital plotter.

G.8 PROJECT 7: BUILDING LAYOUT

Description   You will lay out the corners of a building and reference the corners with batter boards. See Figure G-11.

Equipment   Theodolite, steel tape, plumb bobs, wood stakes, light lumber for batter boards, C clamps, keel or felt pen, level, and rod. (A theodolite/EDM, or total station and prism can replace the theodolite and steel tape.)

Purpose   To give you experience in laying out the corners of a building according to dimensions taken from a building site plan and in constructing batter boards, referencing both the line and grade of the building walls and floor.

Table G-2   Typical Code Library

Control		Utilities	
TCM	Temporary control monument	HP	Hydro pole
CM	Concrete monument	LP	Lamp pole
SIB	Standard iron bar	BP	Telephone pole
IB	Iron bar	GS	Gas valve
RIB	Round iron bar	WV	Water valve
NL	Nail	CABLE	Cable
STA	Station		
TBM	Temporary benchmark		
Municipal		**Topographic**	
℄ (CL)	Centerline	GND	Ground
RD	Road	TB	Top of bank
EA	Edge of asphalt	BB	Bottom of bank
BC	Beginning of curve	DIT	Ditch
EC	End of curve	FL	Fence line
PC	Point on curve	POST	Post
CURB	Curb	GATE	Gate
CB	Catch basin	BUSH	Bush
DCB	Double catch basin	HEDGE	Hedge
MH	Manhole	BLD	Building
STM	Storm sewer manhole	RWALL	Retaining wall
SAN	Sanitary sewer manhole	POND	Pond
INV	Invert	STEP	Steps
SW	Sidewalk	CTREE	Coniferous tree
HYD	Hydrant	DTREE	Deciduous tree
RR	Railroad		

Procedure

- After the front and side lines have been defined by your instructor and after the building dimensions have been given, set stakes X and Y on the front property line, as shown in Figure G-11.

- Set up the theodolite at X, sight on Y, turn 90° (double), and place stakes at A and B.

- Set up the theodolite at Y, sight on X, turn 90° (double), and place stakes at C and D.

- Measure the building diagonals to check the accuracy of the layout. Adjust and remeasure if necessary.

- After the building corners have been set, offset the batter boards a safe distance (e.g., 6 ft or 2 m), and set the batter boards at the first-floor elevation, or as given by your instructor.

- After the batter boards have been set, place line nails on the top of the batter boards as follows:

  1. Set up on A, sight B, place nail 2, transit the telescope, and set nail 1.

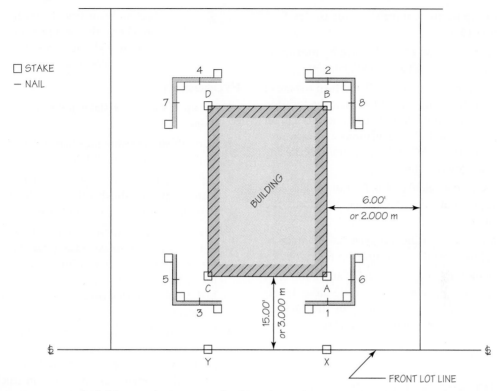

FIGURE G-11 Sample field notes for Project 7 (building layout)

2. From the setup on *A*, sight *C*, place nail 5, transit the telescope, and place nail 6.

3. Set up on *D* and place nails 3, 4, 7, and 8 in a similar fashion.

G.9 PROJECT 8: HORIZONTAL CURVE

Description Given the centerline alignment of two intersecting tangents (including a stationing reference stake), you will calculate and lay out a horizontal curve.

Equipment Theodolite, steel tape, plumb bobs, wood stakes, and range pole or string target. (A theodolite/EDM, or total station and prism can replace the theodolite and steel tape.)

Purpose To give you experience in laying out a circular curve at specified station intervals after first calculating all the necessary layout measurements from the given radius and the measured location of the PI and the Δ field angle.

Procedure

- Intersect the two tangents to create the PI.
- Measure the station of the PI.
- Measure (and double) the tangent deflection angle (Δ).
- After receiving the radius value from your instructor, compute *T* and *L*. Then compute the station of the BC and EC. See Section 13.3.
- Compute the deflections for even stations at 50-ft or 20-m intervals. See Section 13.4.

- Compute the equivalent chords. See Section 13.5.

- Set the BC and EC in the field by measuring out T from the PI along each tangent.

- From the BC, sight the PI and turn the curve deflection angle ($\Delta/2$) to check the location of the EC. If the line of sight does not fall on the EC, check the calculations and measurements for the BC and EC locations. Make any necessary adjustments.

- Using the calculated deflection angles and appropriate chord lengths, stake out the curve.

- Measure from the last even station stake to the EC to verify the accuracy of the layout.

- Walk the curve, looking for any anomalies (e.g., two stations staked at the same deflection angle). The symmetry of the curve is such that even minor mistakes are obvious in a visual check.

G.10 PROJECT 9: PIPELINE LAYOUT

Description   You will establish an offset line and construct batter boards for line-and-grade control of a proposed storm sewer from MH 1 to MH 2. Stakes marking those points will be given for each crew in the field.

Equipment   Theodolite, steel tape, wood stakes, and light lumber and C clamps for the batter boards. (A theodolite/EDM or total station and prism can replace the theodolite and steel tape.)

Purpose   To give you experience in laying out offset line-and-grade stakes for a proposed pipeline. You will learn how to compute a grade

sheet and construct batter boards, and check the accuracy of your work by sighting across the constructed batter boards.

Procedure

- Set up at the MH 1 stake and sight the MH 2 stake.

- Turn off 90°, measure the offset distance (e.g., 10 ft or 3 m), and establish MH 1 on the offset.

- Set up at the MH 2 stake, sight the MH 1 stake, and establish MH 2 on the offset. Refer to Figure 13-37 for guidance.

- Give the MH 1 offset stake a station of 0 + 00, measure to establish grade stakes at the even stations (50 ft or 20 m), and check the distance from the last even station to the MH 2 stake to check that the overall distance is accurate.

- Using the closest benchmark (BM), determine the elevations of the tops of the offset grade stakes. Close back to the benchmark within the tolerance given by your instructor.

- Assume that the invert of MH 1 is 7.97 ft or 2.430 m (or another assumed value given by your instructor) below the top of the MH 1 grade stake.

- Compute the invert elevations at each even station, and then complete a grade sheet similar to those shown in Table 13-10 and 13-11. Select a convenient height for the grade rod.

- Using the "stake to batter board" distances in the grade sheet, use the supplied light lumber and C clamps to construct batter boards similar to those shown in Figure 13-39. Use a small carpenter's level to keep the crosspieces horizontal. Check to see that all crosspieces line up in one visual line. A perfect visual check (all batter boards line up behind one another) is a check on all the measurements and calculations.

EARLY SURVEYING

H.1 EVOLUTION OF SURVEYING

Surveying is a profession with a very long history. Since the beginning of property ownership, boundary markers have been required to differentiate one property from another. Historical records dating to almost 3000 B.C. show evidence of surveyors in China, India, Babylon, and Egypt. The Egyptian surveyor, called the *harpedonapata* ("rope stretcher"), was in constant demand because the Nile River flooded more or less continuously, destroying boundary markers in those fertile farming lands. Egyptian surveyors used ropes with knots tied at set graduations to measure distances.

Ropes were also used to lay out right angles. The early surveyors discovered that the 3:4:5 ratio provided right-angle triangles. A 12-unit rope had knots tied at unit positions 3 and 7, as shown in Figure H-1. To lay out XZ at 90° to XY, one surveyor held the three-unit knot at X, the second surveyor held the seven-unit knot at Y, and the third surveyor, holding both loose ends of the rope, stretched the rope tightly to find the location of point Z. These early surveyors knew that multiples of 3:4:5 (e.g., 30:40:50) produced more accurate positioning.

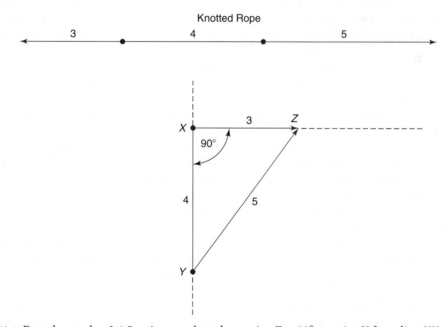

FIGURE H-1 Rope knotted at 3:4:5 ratio—used to place point Z at 90° to point X from line XY

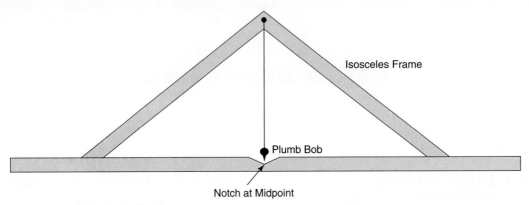

FIGURE H-2 Early Egyptian level

Another ancient surveying instrument consisted of three pieces of wood in the form of an isosceles triangle, with the base extended in both directions (see Figure H-2). A plumb bob suspended from the apex of the frame lined up with a notch cut in the midpoint of the base only when the base was level. These levels came in various sizes, depending on the work to be done.

It is presumed that the great pyramids were laid out with knotted-rope levels as described here, and various forms of water-trough levels for the foundation layout. These Egyptian surveying techniques were empirical solutions that were field proven. It remained for the Greeks to provide the mathematical reasoning and proofs to explain why the field techniques worked. Pythagoras is one of many famous Greek mathematicians; he and his school (about 550 B.C.) developed theories regarding geometry and numbers. They were also among the first to deduce that the earth was spherical by noting the shape of the shadow of the earth that was cast on the moon. The term *geometry* is Greek for "earth measurement," clearly showing the relationship between mathematics and surveying. In fact, the history of surveying is closely related to the history of mathematics and astronomy.

By 250 B.C., Archimedes recorded in a book known as the *Sand Reckoner* that the circumference of the earth was 30 myriads of stadia (i.e., 300,000 stadia). He had received some support for this value from Eratosthenes, who was a mathematician and a librarian at the famous library of Alexandria in Egypt. According to some reports, Eratosthenes' technique was as follows. Eratosthenes knew that a town called Syene was 5,000 stadia south of Alexandria. He also knew that at summer solstice (around June 21), the sun was directly over Syene at noon because there were no shadows. That phenomenon was demonstrated by noting that the sun's reflection was exactly centered in the well water.

Eratosthenes assumed that at the summer solstice, the sun, the towns of Syene and Alexandria, and the center of the earth all lay in the same plane (see Figure H-3). At noon on the day of the summer solstice, the elevation of the sun was measured at Alexandria as being $82\frac{4}{5}^\circ$. The angle from the top of the rod to the sun was then calculated as being $7\frac{1}{5}^\circ$. Because the sun is such a long distance from the earth, it can be assumed that the sun's rays are parallel as they reach the earth. With that assumption, it can be deduced that the angle from the top of the rod to the sun is the same as the angle at the earth's center, $7\frac{1}{5}^\circ$. Because $7\frac{1}{5}^\circ$ is one-fiftieth of 360°, it follows that the circular arc subtending $7\frac{1}{5}^\circ$. (the distance from Syene to Alexandria) is one-fiftieth of the circumference of the earth. The circumference of the earth is thus determined to be 250,000 stadia. If the stadia being used were one-tenth of a mile (different values for the stadium existed, but one value was roughly one-tenth of our mile), then it is possible that Eratosthenes had calculated the

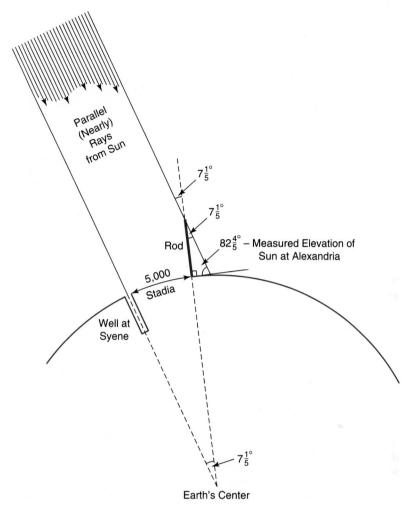

FIGURE H-3 Illustration of Eratosthenes' technique for computing the earth's circumference

earth's circumference to be 25,000 miles. Using the Clarke ellipsoid with a mean radius of 3,960 miles, the circumference of the earth would actually be $C = 2 \times 3.1416 \times 3,960 = 24,881$ miles.

After the Greeks, the Romans used practical surveying techniques for many centuries to construct roadways, aqueducts, and military camps. Some Roman roads and aqueducts exist to this day. For leveling, the Romans used a *chorobate*, a 20 (approximate) wooden structure with plumbed end braces and a 5-ft (approximate) groove for a water trough (see Figure H-4). Linear measurements were

often made with wooden poles 10–17 ft long. With the fall of the Roman Empire, surveying and most other intellectual endeavors became lost arts in the Western world.

Renewed interest in intellectual pursuits may have been fostered by the explorers' need for navigational skills. The lodestone, a naturally magnetized rock (magnetite), was first used to locate magnetic north; later, the compass would be used for navigation on both land and water. In the mid-1500s, the surveyors' chain was first used in the Netherlands, and an Englishman, Thomas Digges,

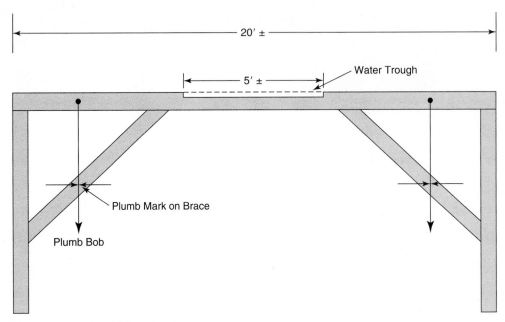

FIGURE H-4 Roman level (chorobate)

first used the term *theodolite* to describe an instrument that was graduated in 360° and used to measure angles. By 1590, the plane table (a combined positioning and plotting device) was created by Jean Praetorius; the plane tables used in the early 1900s were little different. The telescope, which was invented in 1609 by Galileo (among others), could be attached to a quadrant (angle-measuring device), thus permitting the technique of triangulation—a simpler method of determining long distances. Jean Picard (1620–1682) was apparently the first to use a spiderweb crosshair in a telescope. He also used vernier scales to improve the precision of angular measurement. James Watt, inventor of the steam engine, is also credited with being the first to install stadia hairs in the survey telescope.

The first dumpy levels were created in the first half of the 1700s by combining a telescope with a bubble level. The repeating style of theodolite (see Chapter 5 and Section H.3) was seen in Europe in the mid-1800s, but it soon lost favor because scale imperfections caused large cumulative errors. Direction theodolites (see Chapter 5)

were favored because high accuracy could be achieved by reading angles at different positions on the scales, thus eliminating the effect of scale imperfections. Refinements to theodolites continued over the years, with better optics, micrometers, coincidence reading, lighter-weight materials, and so on. Heinrich Wild is credited with many significant improvements in the early 1900s that greatly affected the design of most European survey instruments produced by the Wild, Kern, and Zeiss companies.

In the United States, William J. Young of Philadelphia is credited with being one of the first to create the transit (see Figure H-8 on page 553) in 1831. The transit differed from the early theodolites because the telescope was shortened so that it could be revolved (transited) on its axis. This simple but brilliant adaptation permitted the surveyor to produce straight lines accurately by simply sighting the backsight and transiting the telescope forward (see Section 5.12). When this technique was repeated—once with the telescope normal and once with the telescope inverted—most of the potential errors (e.g., scale graduation imperfections, crosshair misalignment, standards misalignment) in the instrument could be removed by averaging.

When using a repeating instrument, angles could also be quickly and accurately accumulated (see Chapter 5). The transit proved to be superior for North American surveying needs. If the emphasis in European surveying was on precise control surveys, the emphasis in North America was on enormous projects in railroad and canal construction and vast projects in public land surveys, which were occasioned by the influx of large numbers of immigrants. The American repeating transit was fast to use, practical, and accurate, and thus was a significant factor in the development of the North American continent. European and Japanese optical theodolites, which replaced the traditional American vernier transit beginning in the 1950s, have now themselves been largely replaced by electronic theodolites and total stations.

Electronic distance measurement (EDM) was first introduced by Geodimeter (Trimble), Inc., in the 1950s. It replaced triangulation for control survey distance measurements and the steel tape for all but short distances in boundary and engineering surveys. Global positioning system (GPS) surveys are now used for most control survey point positioning. Aerial surveys became very popular after World War II. This technique is a very efficient method of performing large-scale topographic surveys and accounts for the majority of such surveys, although total station surveys are now competitive at lower levels of detail density.

H.2 DUMPY LEVEL

The dumpy level (see Figure H-5) was at one time used extensively on all leveling surveys. Although this simple instrument has been replaced, to a large degree, by more sophisticated instruments, it is shown here in some detail to aid in the introduction of the topic. For purposes of description, the level can be analyzed with respect to its three major components: telescope, level tube, and leveling head.

The telescope assembly is illustrated in Figure H-6(a), (b), and (c). These schematics also describe the telescopes used in theodolites/transits. Rays of light pass through the objective (1) and form an inverted image in the focal plane (4). The image thus formed is magnified by the eyepiece lenses (3) so that the image can be seen clearly. The eyepiece lenses also focus the crosshairs,

FIGURE H-5 Dumpy level
(Courtesy of Keuffel & Esser Company, Morristown, NJ)

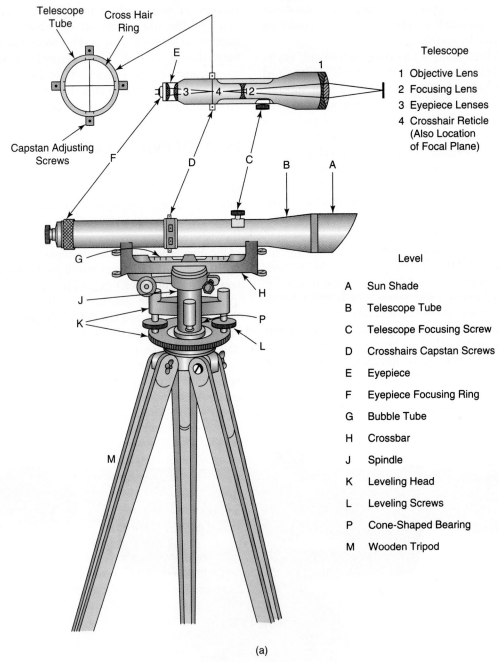

Telescope Tube

Cross Hair Ring

Capstan Adjusting Screws

Telescope

1 Objective Lens
2 Focusing Lens
3 Eyepiece Lenses
4 Crosshair Reticle
(Also Location of Focal Plane)

Level

A   Sun Shade
B   Telescope Tube
C   Telescope Focusing Screw
D   Crosshairs Capstan Screws
E   Eyepiece
F   Eyepiece Focusing Ring
G   Bubble Tube
H   Crossbar
J   Spindle
K   Leveling Head
L   Leveling Screws
P   Cone-Shaped Bearing
M   Wooden Tripod

(a)

FIGURE H-6 (a) Dumpy level

(Adapted from *Construction Manual*, Ministry of Transportation, Ontario)

Diagram of Optical System

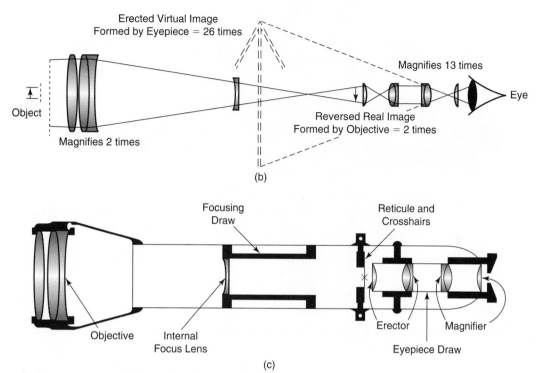

(b)

(c)

FIGURE H-6 (*Continued*) (b) Diagram of an optical system for a level or theodolite telescope (Courtesy of Sokkia Co. Ltd.). (c) Telescope (Courtesy of Sokkia Co. Ltd.)

which are located in the telescope in the principal focus plane. The focusing lens (negative lens, 2) can be adjusted so that images at varying distances can be brought into focus in the plane of the reticle (4). In most telescopes designed for use in North America, additional lenses are included in the eyepiece assembly so that the inverted image can be viewed as an erect image. The minimum focusing distance for the object ranges from 1 to 2 m, depending on the instrument.

The line of collimation (line of sight) joins the center of the objective lens to the intersection of the crosshairs. The optical axis is the line taken through the center of the objective lens and perpendicular to the vertical lens axis. The focusing lens (negative lens), which is moved by focusing screw C [Figure H-6(a)], has its optical axis the same as the objective lens.

The crosshairs [Figure H-6(a), 4] can be thin wires attached to a crosshair ring or, as is more usually the case, the crosshairs are lines etched on a circular glass plate enclosed by a crosshair ring. The crosshair ring, which has a slightly smaller diameter than the telescope tube, is held in place by four adjustable capstan screws. The crosshair ring (and the crosshairs) can be adjusted left and right or up and down simply by loosening and then tightening the two appropriate opposing capstan screws.

Four leveling foot screws are utilized to set the telescope level. The four foot screws surround the center bearing of the instrument [Figure H-6(a)] and are used to tilt the level telescope using the center bearing as a pivot.

Figure H-7 illustrates how the telescope is positioned during the leveling process. The telescope is

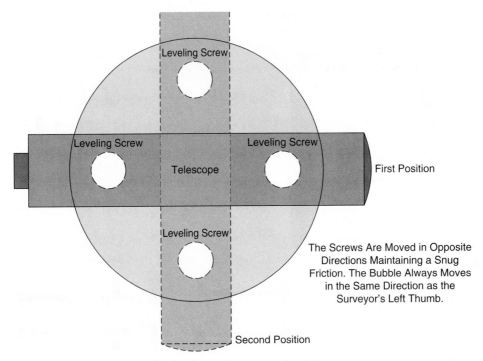

FIGURE H-7 Telescope positions when leveling a four-screw level instrument

first positioned directly over two opposite foot screws. (See Figure 2-6 for telescope positioning of three-screw instruments.) The two screws are kept only snugly tightened (overtightening makes rotation difficult and could damage the threads) and are rotated in opposite directions until the bubble is centered in the level tube. Loose foot screws indicate that the rotations have not proceeded uniformly. At worst, the foot screw pad can rise above the plate, making the telescope wobble. The solution for this condition is to turn the loose screw until it again contacts the base plate and provides snug friction when turned in opposition to its opposite screw.

The telescope is first aligned over two opposing screws, and the screws are turned in opposite directions until the bubble is centered in the level tube. The telescope is then turned 90° to the second position, which is over the other pair of opposite foot screws, and the leveling procedure is repeated. This process is repeated until the bubble remains centered in those two positions. When the bubble remains centered in these two positions, the telescope is then turned 180° to check the adjustment of the level tube.

If the bubble does not remain centered when the telescope is turned 180°, the level tube is out of adjustment. The instrument can still be used, however, by simply noting the number of divisions that the bubble is off center and by moving the bubble half the number of those divisions. For example, if you turn the leveled telescope 180° and note that the bubble is four divisions off center, the instrument can be leveled by moving the bubble to a position of two divisions off center. The bubble will remain two divisions off center, no matter which direction the telescope is pointed. It should be emphasized that the instrument is, in fact, level if the bubble remains in the same position when the telescope is revolved, regardless of whether or not that position is in the center of the level vial. See Section 5.9.1 for the adjustments used to correct this condition.

H.3 ENGINEER'S VERNIER TRANSIT

H.3.1 The Engineer's Transit: General Background

Prior to the mid-1950s, and before the development and widespread use of electronic and optical theodolites, most engineering surveys for topography and layout were accomplished using the engineer's transit (see Figure H-8). This instrument had open circles for horizontal and vertical angles; angles were read with the aid of vernier scales. This four-screw instrument was positioned over the survey point by using a slip-knotted plumb bob string attached to the chain hook hanging down from the instrument.

Figure H-9 shows the three main assemblies of the transit. The upper assembly, called the alidade, includes the standards, telescope, vertical circle and vernier, two opposite verniers for reading the horizontal circle, plate bubbles, compass, and upper-tangent (slow-motion) screw. The spindle of the alidade fits down into the hollow spindle of the circle assembly. The circle assembly includes the horizontal circle that is covered by the alidade plate except at the vernier windows, the upper clamp screw, and the hollow spindle previously mentioned.

The hollow spindle of the circle assembly fits down into the leveling head. The leveling head includes: the four leveling screws; the half-ball joint, about which opposing screws are

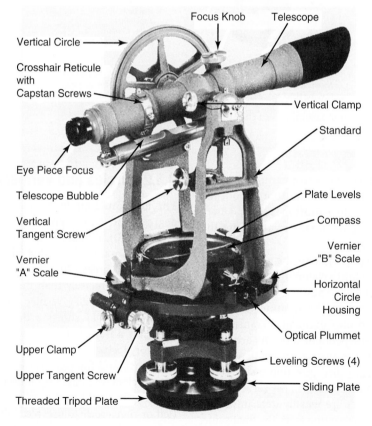

FIGURE H-8 Engineer's transit

(Courtesy of Keuffel & Esser Company, Morristown, NJ)

The ALIDADE ASSEMBLY, which includes

    Telescope
    Vertical Circle
    Standards
    Verniers
    Vernier Cover
    Plate Levels
    Inner Center
    Upper Tangent

The CIRCLE ASSEMBLY, which includes

    Horizontal Limb
    Outer Center
    Upper Clamp

The LEVELING HEAD ASSEMBLY, which includes

    Leveling Head
    Leveling Screws
    Shifting Plate
    Friction Plate
    Half Ball
    Tripod Plate
    Lower Clamp
    Lower Tangent

The Inner Center of the Alidade Assembly fits into the Outer Center of the Circle Assembly and can be rotated in the Outer Center. The Outer Center fits into the Leveling Head and can be rotated in the Leveling head.

FIGURE H-9 Three major assemblies of the transit
(Courtesy of Sokkia Co. Ltd.)

manipulated to level the instrument; a threaded collar that permits attachment to a tripod; the lower clamp and slow-motion screw; and a chain with an attached hook for attaching the plumb bob.

The upper clamp tightens the alidade to the circle, whereas the lower clamp tightens the circle to the leveling head. These two independent motions permit angles to be accumulated on the circle for repeated measurements. Transits that have these two independent motions are called repeating instruments. Instruments with only one motion (upper) are called direction instruments. Since the circle cannot be previously zeroed (older instruments), angles are usually determined by subtracting the initial setting from the final value. It is not possible to accumulate or repeat angles with a direction theodolite.

H.3.2 Circles and Verniers

The horizontal circle is usually graduated into degrees and half-degree s or 30 minutes (see Figure H-10), although it is not uncommon to find the horizontal circle graduated into degrees and one-third degrees (20'). To determine the angle value more precisely than the least count of the circle (i.e., 30' or 20'), vernier scales are employed.

Figure H-11 shows a double vernier scale alongside a transit circle. The left vernier scale is used for clockwise circle readings (angles turned to the right), and the right vernier scale is used for counterclockwise circle readings (angles turned to the left). To avoid confusion about which vernier (left or right) scale to use, recall that the vernier to be used is the one whose graduations are increasing in the same direction as are the circle graduations.

FIGURE H-10 Part of a transit circle showing a least count of 30 minutes. The circle is graduated in both clockwise and counterclockwise directions, permitting the reading of angles turned to both the left and the right

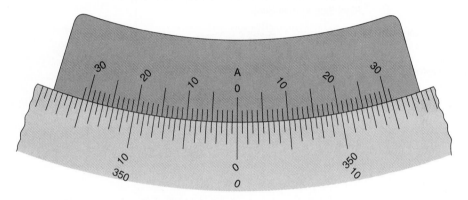

FIGURE H-11 Double vernier scale set to zero on the horizontal circle

The vernier scale is constructed so that 30 vernier divisions cover the same length of arc as do 29 divisions (half-degrees) on the circle. The width of one vernier division is $(29/30) \times 30' = 29'$ on the circle. Therefore, the space difference between one division on the circle and one division on the vernier represents 01'. In Figure H-11, the first division on the vernier (left or right of the index mark) fails to line up exactly with the first division on the circle (left or right) by 01'. The second division on the vernier fails to line up with the corresponding circle division by 02', and so on. If the vernier were moved so that its first division lined up exactly with the first circle division (30' mark), the reading would be 01'. If the vernier were moved again the same distance of arc (1'), the second vernier mark would now line up with the appropriate circle division line, indicating a vernier reading of 02'. Generally, the vernier is read by finding which vernier division line coincides exactly with any circle line, and then by adding the value of that vernier

line to the value of the angle obtained from reading the circle to the closest 30' (in this example).

In Figure H-12(a), the circle is divided into degrees and half-degrees (30'). Before even looking at the vernier, we know that its range will be 30' (left or right) to cover the least count of the circle. Inspection of the vernier shows that 30 marks cover the range of 30', indicating that the value of each mark is 01'. (Had each of the minute marks been further subdivided into two or three intervals, the angle could then have been read to the closest 30'' or 20''.) If we consider the clockwise circle readings (field angle turned left to right), we see that the zero mark is between 184° and 184°30'; the circle reading is therefore 184°. Now to find the value to the closest minute, we use the left-side vernier. Moving from the zero mark, we look for the vernier line that lines up exactly with a circle line. In this case, the 08' mark lines up; this is confirmed by noting that both the 07' and 09' marks do not line up with their corresponding circle mark, both by the same amount. The angle for this illustration is 184° + 08' = 184°08'.

If we consider the counterclockwise circle reading in Figure H-12(a), we see that the zero mark is between 175°30' and 176°; the circle reading is therefore 175°30'. To that value, we add the right-side vernier reading of 22', to give an angle of 175°52'. As a check, the sum of the clockwise and counterclockwise readings should be 360°00'.

All transits are equipped with two double verniers (A and B) located 180° apart. Although increased precision can theoretically be obtained by reading both verniers for each angle, usually

Note: Clockwise Angles (i.e., Angles Turned to the Right) Utilize Only the Left Side Vernier Scale, Counterclockwise Angles (i.e., Angles Turned to the Left) Utilize Only the Right Side Vernier Scale.

Vernier Readings

Clockwise	184°08'
Counterclockwise	175°52'
	359°60' = 360°00'

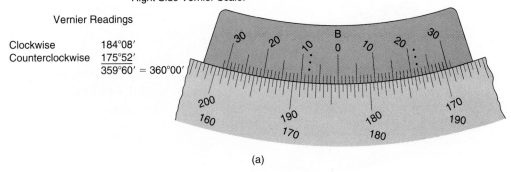

(a)

Graduated 30 Minutes Reading to One Minute
Double Direct Vernier

Vernier Readings

Clockwise	342°35'
Counterclockwise	17°25'
Check:	360°00'

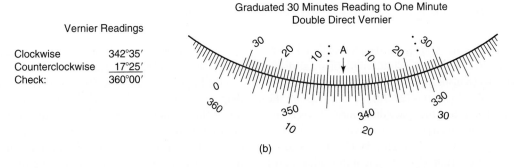

(b)

Graduated 20 Minutes Reading to 30 Seconds
Double Direct Vernier

Clockwise	229°50'30"
Counterclockwise	130°09'30"
Check:	360°00'00"

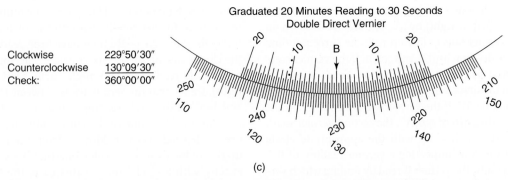

(c)

FIGURE H-12 Sample vernier readings. Triple dots identify aligned vernier graduations

Note: Appropriate vernier scale graduation numerals are angled in the same direction as the referenced circle graduation numerals.

only one vernier is employed. Furthermore, to avoid costly mistakes, most surveying agencies favored the use of the same vernier, the A vernier, at all times.

As noted earlier, the double vernier permits angles to be turned to the right (left vernier) or to the left (right vernier). By convention, however, field angles are normally turned only to the right. The exceptions to this occur when deflection angles are being employed, as in route surveys, or when construction layouts necessitate angles to the left, as in some curve deflections. There are a few more specialized cases (e.g., star observations) when it is advantageous to turn angles to the left, but as stated earlier, the bulk of surveying experience favors angles turned to the right. This type of consistency provides the routine required to foster a climate in which fewer mistakes occur, and in which mistakes that do occur can be recognized readily and eliminated.

The graduations of the circles and verniers as illustrated were in wide use in the survey field. However, there are several variations to both circle graduations and vernier graduations. Typically, the circle is graduated to the closest 30′ (as illustrated), 20′, or 10′ (rarely). The vernier has a range in minutes covering the smallest division on the circle (30′, 20′, or 10′), and can be further graduated to half-minute (30″) or one-third minute (20″) divisions. A few minutes spent observing the circle and vernier graduations of an unfamiliar transit will easily disclose the proper technique required for reading [see also Figure H-12(b) and (c)].

The use of a magnifying glass (5×) is recommended for reading the scales, particularly for the 30′ and 20′ verniers. Vernier transits were largely replaced by optical theodolites in the 1970s and 1980s; optical theodolites were largely replaced by electronic transits/theodolites in the 1990s.

H.3.3 Telescope

The telescope [see Figure H-6(b) and (c)] in the transit is somewhat shorter than that in a level, with a reduced magnifying power (26× versus the 30× often used in the level). The telescope axis is supported by the standards, which are of sufficient height to permit the telescope to be revolved (*transited*) 360° about the axis. A level vial tube is attached to the telescope so that, if desired, it may be used as a level.

The telescope level has a sensitivity of 30″ to 40″ per 2-mm graduation, compared to a level sensitivity of about 20″ for a dumpy level. When the telescope is positioned so that the level tube is under the telescope, it is said to be in the direct (normal) position; when the level tube is on top of the telescope, the telescope is said to be in a reversed (inverted) position. The eyepiece focusing ring is always located at the eyepiece end of the telescope, whereas the object focus knob can be located on the telescope barrel just ahead of the eyepiece focus, midway along the telescope, or on the horizontal telescope axis at the standard.

H.3.4 Leveling Head

The leveling head supports the instrument. Proper manipulation of the leveling screws allows the horizontal circle and telescope axis to be placed in a horizontal plane, which forces the alidade and circle assembly spindles to be placed in a vertical direction. When the leveling screws are loosened, the pressure on the tripod plate is removed, thus permitting the instrument to be shifted laterally a short distance $\left(\frac{3}{8}\text{ in.}\right)$ This shifting capability permits the surveyor to position the transit center precisely over the desired point.

H.4 OPTICAL THEODOLITES

H.4.1 Repeating Optical Theodolites

Optical theodolites are characterized by three-screw leveling heads, optical plummets, light weight, and glass circles being read either directly or with the aid of a micrometer; angles (0° to 360°) are normally read in the clockwise direction (see Figure H-13). In contrast to the American engineer's transit, most theodolites do not come equipped with compasses or telescope levels. Most

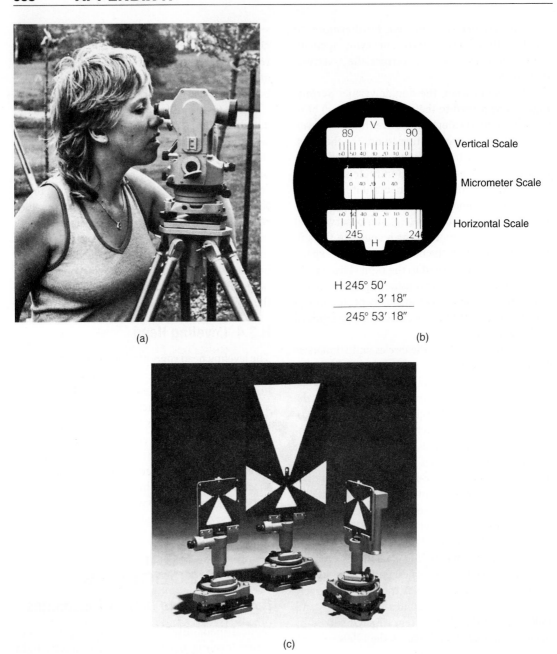

H 245° 50'
          3' 18"
  245° 53' 18"

(a)   (b)

(c)

FIGURE H-13 (a) Twenty-second micrometer theodolite, the Sokkia TM 20. (b) Horizontal and vertical scales with micrometer scale. (c) A variety of tribrach-mounted traverse targets. Targets and theodolites can be interchanged easily to save setup time (forced-centering system)

(Courtesy of Sokkia Co. Ltd.)

theodolites are now equipped with a compensating device that automatically indexes the horizontal direction when the vertical circle has been set to the horizontal setting of 90° (or 270°).

The horizontal angular setting (in the vertical plane) for theodolites is 90° (270°), whereas for the transit it is 0°. A word of caution: Although all theodolites have a horizontal setting of 90° direct or 270° inverted, some theodolites have their zero set at the nadir, while others have the zero set at the zenith. The method of graduation can be quickly ascertained in the field by simply setting the telescope in an upward (positive) direction and noting the scale reading. If the reading is less then 90°, the zero has been referenced to the zenith direction; if the reading is more than 90°, the zero has been referenced to the nadir direction.

The graduations of both the vertical and horizontal circles are, by means of prisms and lenses, projected to one location just adjacent to the telescope eyepiece, where they are read by means of a microscope. Light, which is necessary in the circle-reading procedure, is controlled by an adjustable mirror located on one of the standards. Light required for underground or night work is directed through the mirror window by attached lamps powered by battery packs.

Optical theodolites have two horizontal motions. The alidade (see Figure H-9) can be revolved on the circle assembly with the circle assembly locked to the leveling head. This permits the user to turn and read horizontal angles. Alternatively, the alidade and circle assembly can also be clamped together freely revolving on the leveling head. This permits the user to keep a set angle value on the circle while turning the instrument to sight another point.

The theodolite tripod has a flat base through which a bolt is threaded up into the three-screw leveling base called a tribrach, thus securing the instrument to the tripod. Most theodolites have a tribrach release feature that permits the alidade and circle assemblies to be lifted from the tribrach and interchanged with a target or prism [see Figure H-13(c)]. When the theodolite (minus its tribrach) is placed on a tribrach vacated by the

target or prism, it will be instantly over the point and nearly level. This system, called **forced centering**, speeds up the work and reduces centering errors associated with multiple setups.

Optical plummets can be mounted in the alidade or in the tribrach. Alidade-mounted optical plummets can be checked for accuracy simply by revolving the alidade around its vertical axis and noting the location of the optical plummet crosshairs (bull's eye) with respect to the station mark. Tribrach-mounted optical plummets can be checked by means of a plumb bob. Adjustments can be made by manipulating the appropriate adjusting screws, or the instrument can be sent out for shop analysis and adjustment.

Typical specifications for repeating micrometer optical theodolites are listed below:

Magnification: 30×

Clear objective aperture: 1.6 in. (42 mm)

Field of view at 100 ft (100 m): 2.7 ft (2.7 m)

Shortest focusing distance: 5.6 ft (1.7 m)

Stadia multiplication constant: 100

Bubble sensitivity

    Circular bubble: 8′ per 2 mm

    Plate level: 30″ per 2 mm

Direct circle reading: 01″ to 06″ (20″ in older models) from 0° to 360°

Like the engineer's transit, the repeating theodolite has two independent motions (upper and lower), which necessitates upper and lower clamps with their attendant tangent screws. However, some theodolites come equipped with only one clamp and one slow-motion or tangent screw; these instruments have a lever or switch that transfers clamp operation from upper to lower motion and thus probably reduces the opportunity for mistakes due to wrong-screw manipulation.

When some tribrach-equipped laser plummets are used in mining surveys, the alidade (upper portion of instrument) can be released from the tribrach with the laser plummet projecting upward to ceiling stations. After the tribrach is properly centered under the ceiling mark, the alidade is then replaced into the tribrach, and the instrument is ready for final settings. The visual plumb line helps the surveyor to position the instrument over (under) the station mark more quickly than when using an optical plummet. Some manufacturers place the laser plummet in the alidade portion of the instrument, permitting an easy check on the beam's accuracy by simply rotating the instrument and observing whether the beam stays on the point (upward plumbing is not possible with alidade-mounted laser plummets).

H.4.2 Angle Measurement with an Optical Theodolite

The technique for turning and doubling (repeating) an angle is the same as for a transit. The only difference in procedure is that of zeroing and reading the scales. In the case of the direct reading optical scale, zeroing the circle is simply a matter of turning the circle until the 0° mark lines up approximately with the 0′ mark on the scale. Once the upper clamp has been tightened, the setting can be precisely accomplished by manipulation of the upper tangent screw.

In the case of the optical micrometer instruments (see Figure H-13), it is important to first set the micrometer to zero, and then to set the horizontal circle to zero. When the angle has been turned, it will be noted that the horizontal (or vertical) circle index mark is not directly over a degree mark. The micrometer knob is turned until the circle index mark is set to a degree mark. Movement of the micrometer knob also moves the micrometer scale; the reading on the micrometer scale is then added to the even degree reading taken from the circle. The micrometer scale does not have to be reset to zero for subsequent angle readings unless a new reference backsight is taken. The vertical circle is read in the same way, using the same micrometer scale. If the vertical index is not automatically compensated (as it is for most repeating theodolites), the vertical index coincidence bubble must be centered by rotating the appropriate screw when vertical angles are being read.

H.4.3 Direction Optical Theodolites

The essential difference between a direction optical theodolite and a repeating optical theodolite is that the direction theodolite has only one motion (upper), whereas the optical repeating theodolite has two motions (upper and lower). Since it is difficult to precisely set angle values on this type of instrument, angles are usually determined by reading the initial direction and the final direction, and by then determining the difference between the two.

Direction optical theodolites are generally more precise; for example, the Wild T-2 shown in Figure H-14 reads directly to 01″ and by estimation to 0.5″, whereas the Wild T-3 shown in Figure H-15 reads directly to 0.2″ and by estimation to 0.1″. In the case of the Wild T-2 (Figure H-14) and the other 1″ theodolites, the micrometer is turned to force the index to read an even 10″ [the grid lines shown above (beside) the scale are brought to coincidence], and then the micrometer scale reading (02′44″) is added to the circle reading (94°10′) to give a result of 94°12′44″. In the case of the T-3 (Figure H-15), both sides of the circle are viewed simultaneously; one reading is shown erect, the other inverted. The micrometer knob is used to align the erect and inverted circle markings precisely. Each division on the circle is 04′; but if the lower scale is moved half a division, the upper also moves half a division, once again causing the markings to align, that is, a *movement of only* 02′. The circle index line is between the 73° and 74° mark, indicating that the value being read is 73°. Minutes can be read on the circle by counting the number of divisions from the erect 73° to the inverted value that is 180° different than 73° (i.e., 253°). In this case, the number of divisions between these two numbers is 13, each having a value of 02′ (i.e., 26′). These optical instruments have been superceded by precise electronic theodolites.

H.4.4 Angles Measured with a Direction Theodolite

As noted earlier, since it is not always possible to set angles on a direction theodolite scale precisely, directions are observed and then subtracted one

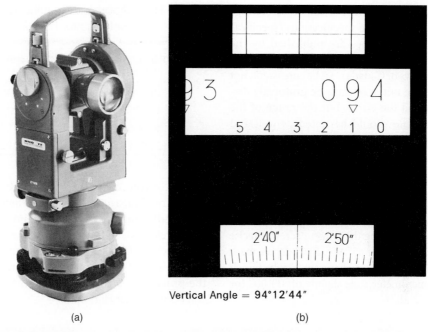

Vertical Angle = 94°12′44″

(a)   (b)

FIGURE H-14 (a) Wild T-2, a 1″ optical direction theodolite. (b) Vertical circle reading (Courtesy of Leica Geosystems, Inc.)

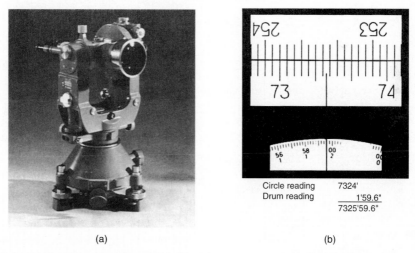

Circle reading	7324′
Drum reading	1′59.6″
	7325′59.6″

(a)   (b)

FIGURE H-15 (a) Wild T-3 precise theodolite for first-order surveying. (b) Circle reading (least graduation is 4 minutes) and micrometer reading (least graduation is 0.2 seconds) (Courtesy of Leica Geosystems, Inc.). On the micrometer, a value of 01′59.6″ can be read. The reading is, therefore, 73°27′ 59.6″

from the other in order to determine angles. Furthermore, if several sightings are required for precision purposes, it is customary to distribute the initial settings around the circle to minimize the effect of circle graduation distortions. For example, using a directional theodolite where both sides of the circle are viewed simultaneously, the initial settings (positions) would be distributed per $180°/n$, where n is the number of settings required by the precision specifications (specifications for precise surveys are published by the National Geodetic Survey—United States, and the Geodetic Surveys of Canada). To be consistent, not only should the initial settings be uniformly distributed around the circle, but the range of the micrometer scale should be noted as well and appropriately apportioned.

For the instruments shown in Figures H-14 and H-15, the settings would be as given in Table H-1. These initial settings are accomplished by setting the micrometer to $00'00''$ (or $02'30''$, $05'00''$, $07'30''$), and then setting the circle as close to zero as possible using the tangent screw. Precise coincidence of the 0° (or 45°, 90°, 135°) degree mark is achieved using the micrometer knob, which moves the micrometer scale slightly off $00'00''$ (or $2'30''$, $5'00''$, $7'30''$).

Table H-1 Approximate Initial Scale Settings for Four Positions

10' Micrometer, Wild T-2	2' Micrometer, Wild T-3
0°00'00"	0°00'00"
45°02'30"	45°00'30"
90°05'00"	90°01'00"
135°07'30"	135°01'30"

The direct readings are taken first in a clockwise direction; the telescope is then transited (plunged), and the reverse readings are taken counterclockwise. In Figure H-16 the last entry at position #1 is 180°00'12" (R). If the angles (shown in the abstract) do not meet the required accuracy, the procedure is repeated while the instrument still occupies that station.

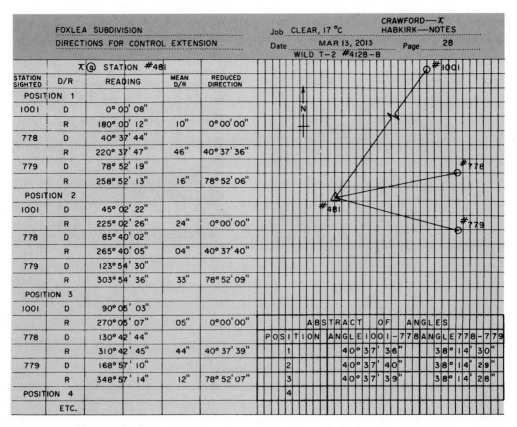

FIGURE H-16 Field notes for directions

Index

A

AASHTO, 134
Abney hand level (clinometer), 76
Absorbed energy, 358
Accuracy, 17
  construction surveys, 349
  EDM, 84, 85
  effect of angle magnitude on computed
    distances, 341
  GPS, 201
  positional (ISO 4463), 346
  positioning standards, 308
  standards for, 167
Accuracy ratio, 18, 167
Adjacency, 280
Adjustment
  of level, 48
  of level loops, 52
  of total station, 118
  of transit and theodolite, 118
  of traverse, 158, 169, 170
  of weighted measurements, 497
Adverse possession, 481
Aerial photographs
  altitude, 379
  camera, 375
  focal length (f), 377
  fiducial marks, 380
  flight lines, 381
  flying height, 379, 381
  ground control, 383
    existing photographs, 383
    new photography, 384
  mosaics, 380, 386
  orthophotos, 383
  overlap of, 381
  photogrammetry, 392
  parallax, 387, 390
  principal point, 380
  relief displacement of, 380
  scale of, 377

stereoscopic plotting, 387, 392
  analytical stereoplotter, 392
  soft-copy photogrammetry, 392
stereoscopic viewing, 387
surveys, 5, 375
Airborne digital imagery, 375, 395, 398
Airborne laser bathymetry, 276
Air photo interpretation, applications for the
  surveyor and engineer, 399
Alidade, 553
Alignment error, 76
Alluvium, 478
Altitude, 379
Aluminum marker, 349
Ambiguity resolution, 204
American Congress on Surveying and Mapping
  (ACSM) angle, distance, and closure
  requirements, 312, 313
American Land Title Association (ALTA), 312
Angle, 107
  adjustment (balancing) of, 159
  closure, 92
  deflection, 94
  doubling an angle, 114
  electronic measurement, 32
  exterior, 92
  horizontal, 91
  interior, 92
  laying off, 120
  measurement by direction theodolite, 560
  measurement by electronic theodolite,
    7, 110, 113
  measurement by transit or theodolite,
    553, 560
  nadir, 91
  by repetition, 114
  vertical, 91, 92, 110
  zenith, 91
Antenna reference height (ARH), 202
Arc of circular curve, 482
Archimedes, 546

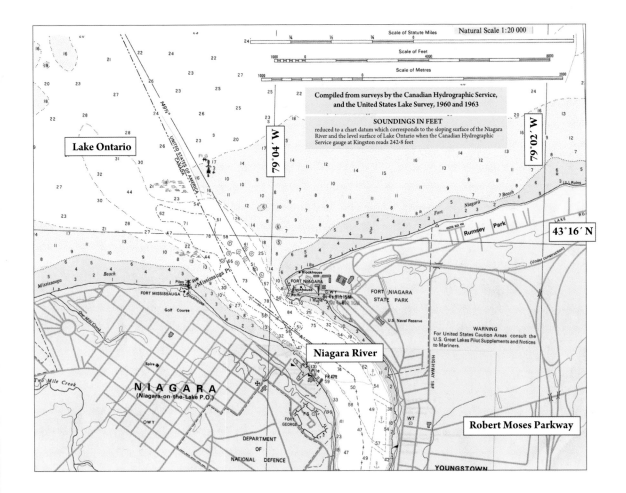

Hydrographic map of the lower Niagara River.
This map is adapted from one produced by the
Canadian Hydrographic Service and the United States
Lake Survey, 1960 and 1963.

FIGURE F.1

Aerial photograph, at 20,000 ft, showing the mouth of the Niagara River.
(Courtesy of U.S. Geological Survey, Sioux Falls, S.Dak.)

FIGURE F.2

Aerial photograph, at 20,000 ft, showing the Niagara Falls area.
(Courtesy of U.S. Geological Survey, Sioux Falls, S. Dak.)

FIGURE F.3

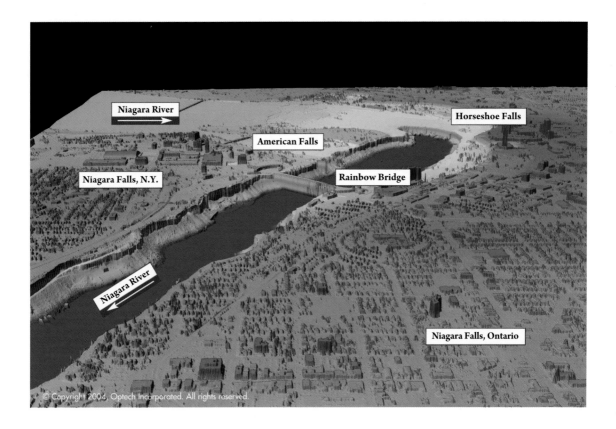

Color-coded elevation data of Niagara Falls, surveyed using lidar techniques by ALTM 3100. (Courtesy of Optech Incorporated, Toronto)

FIGURE F.4

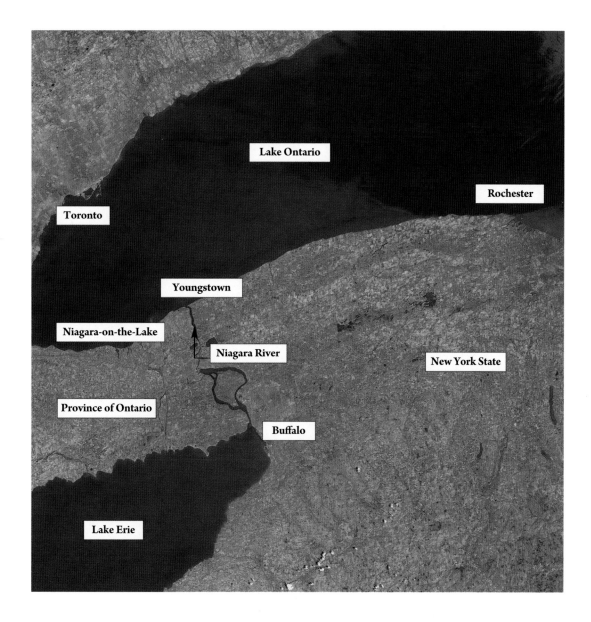

Landsat image showing western New York and the Niagara region of Ontario.
(Courtesy of U.S. Geological Survey, Sioux Falls, S.Dak.)

FIGURE F.5

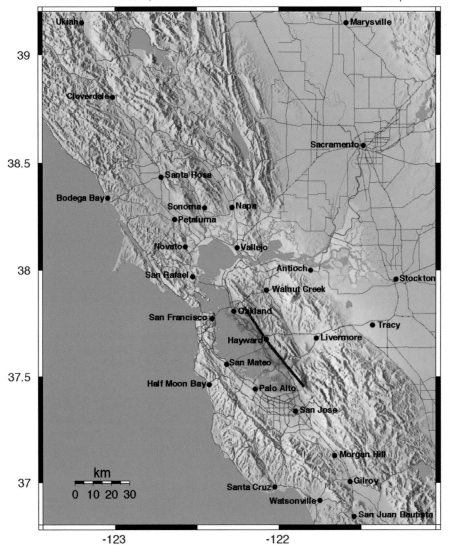

-- Earthquake Planning Scenario --
Rapid Instrumental Intensity Map for HRC_HS Scenario
Scenario Date: Tue Dec 3, 2002 04:00:00 AM PST   M 6.7   N37.57 W121.97   Depth: 0.0km

PLANNING SCENARIO ONLY -- PROCESSED: Tue Dec 3, 2002 12:48:05 PM PST

PERCEIVED SHAKING	Not felt	Weak	Light	Moderate	Strong	Very strong	Severe	Violent	Extreme
POTENTIAL DAMAGE	none	none	none	Very light	Light	Moderate	Moderate/Heavy	Heavy	Very Heavy
PEAK ACC.(%g)	<.17	.17-1.4	1.4-3.9	3.9-9.2	9.2-18	18-34	34-65	65-124	>124
PEAK VEL.(cm/s)	<0.1	0.1-1.1	1.1-3.4	3.4-8.1	8.1-16	16-31	31-60	60-116	>116
INSTRUMENTAL INTENSITY	I	II-III	IV	V	VI	VII	VIII	IX	X+

Scenario ShakeMap illustrating the strength and regional extent of shaking that can be expected from a future M 6.7 earthquake on the southern Hayward fault. (Courtesy of USGS.)

FIGURE F.6

Crater Lake Bathymetry

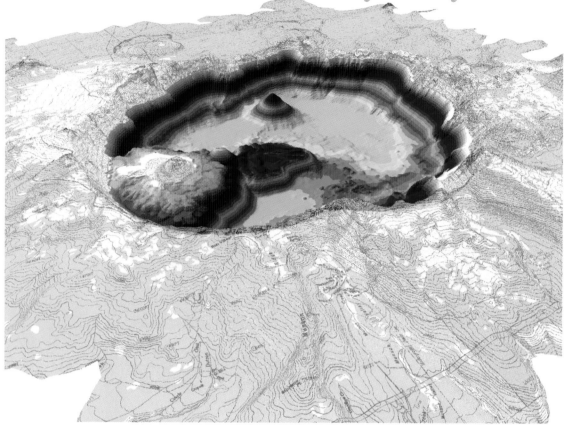

Crater Lake

Bathymetry
Elevation

- 1850 - 1900
- 1800 - 1850
- 1750 - 1800
- 1700 - 1750
- 1650 - 1700
- 1600 - 1650
- 1550 - 1600
- 1500 - 1550
- 1450 - 1500
- 1400 - 1450
- 1350 - 1400
- 1300 - 1350
- 1289.105 - 1300

This map shows the information from a Bathymetry survey of Crater lake and the adjacent topogrpahic informatin draped over a surface model of the area.

The Bathymetry information is based on a survey of Crater Lake in July 2000 done by USGS and the University of New Hampshire's Center for Coastal and Ocean Mapping in cooperation with the National Park Service and can be accessed at http://oe.oregonexplorer.info/craterlake/bathymetry.html

The topographic information is from Digital Raster Graphic (DRG) of the adjacent USGS quadrangle maps. These may be found at http://www.usgs.gov/pubprod/

The surface model was generated from National Elevation Dataset (NED) which can be downloaded through http://eros.usgs.gov/

FIGURE F.7

USE of LiDAR for Flood Study

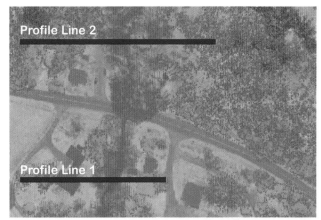

Figure A

Profile Line 1

Profile Line 2

Profile Line 2 - Bare Earth Only

Figure A shows elevation colored LiDAR points with intensity value shading. The intensity values better define buildings and roadways
Profiles are also colored based on elevation. Note Roofs of buildings in Profile 1

FIGURE F.8

SYMBOLS

B̶L̶	baseline
C̶L̶	centerline
S̶L̶	street line
Δ N	change in northing
Δ E	change in easting
Δ λ″	change in longitude (seconds)
φ, λ	latitude, longitude
$\overline{\Lambda}$	instrument
P̄	occupied station (instrument)
P̊	reference sighting station
※	point of intersection
=	is equal to
≠	is not equal to
>	is greater than
<	is less than
≈	is approximately equal to
Σ	the sum of

THE GREEK ALPHABET

Name	Uppercase	Lowercase
alpha	A	α
beta	B	β
gamma	Γ	γ
delta	Δ	δ
epsilon	E	ε
zeta	Z	ζ
eta	H	η
theta	Θ	θ
iota	I	ι
kappa	K	κ
lambda	Λ	λ
mu	M	μ
nu	N	ν
xi	Ξ	ξ
omnicron	O	o
pi	Π	π
rho	P	ρ
sigma	Σ	σ
tau	T	τ
upsilon	Υ	υ
phi	Φ	φ
chi	X	χ
psi	Ψ	ψ
omega	Ω	ω